HANDBOOK OF
COMMERCIAL CATALYSTS
HETEROGENEOUS CATALYSTS

HOWARD F. RASE

The W.A. Cunningham Professor in Chemical Engineering, Emeritus
Department of Chemical Engineering
The University of Texas
Austin, Texas

CRC Press
Boca Raton   London   New York   Washington, D.C.

Library of Congress Cataloging-in-Publication Data

Catalog record is available from the Library of Congress.

Preface

This handbook is written for chemical engineers and chemists who are involved in selecting or improving an existing process. It has been my goal to supply information based on the open literature that you can use to quickly gain background on a particular process or catalyst. For each reaction that is presented, the following topics are discussed.

- Product uses
- Chemistry
- Mechanism
- Catalyst type
- Catalyst suppliers and licensors
- Catalyst deactivation
- Catalyst regeneration
- Process units (description)
- Process kinetics

With such information, you can be prepared to begin rational analysis of an existing or planned reaction system and logically discuss catalyst characteristics and operations with technical representatives of catalyst manufacturers as well as your own colleagues.

The original definition of a handbook, *a conveniently carried and concise reference book,* has only partially survived as more important and valuable information becomes available. I have, however, been determined to make this handbook concise and easy to use as a means for quickly finding information on 150 major industrial processes using heterogeneous catalysts. This goal has been accomplished by using the same outline for each process, by including hundreds of tables and figures that serve to consolidate important concepts, by listing catalyst suppliers and licensors for each process, by using a single writing style and organization made possible by being the sole author, and by providing copious references for further study.

Despite the major developments in understanding, many of the phenomena that occur on a catalytic surface, and much of successful catalyst development and use, continues to be part science and part art. In fact, detailed understanding of a new catalyst often follows its initial invention. Hence, another goal for this handbook has been to facilitate the effective combination of significant studies by the research and development community with practical knowledge or art. Much of the latter is proprietary for each process, but valuable general insights have been published and included in this book when considered appropriate and useful. Excellent literature, including use recommendations, is also available from catalyst suppliers.

In contrast to the study of semiconductors used in the electronics industry, where the model system for laboratory study is the same as the real system, most catalysts are polycrystalline and not amenable to many advanced surface-science techniques which require single crystals (catalytic converter catalyst used in auto exhaust systems is an exception). Studies on single crystals, however, when combined with other insights, can yield valuable postulates about the mechanism of commercial catalysts. As always, there are various possible interpretations. But a postulated mechanism or reaction scheme based on rational concepts, even not totally verified, can provide a framework for developing useful explanations for observed catalyst behavior in industrial reactors and lead to improved operation and/or even improved catalyst formulations.

Finally, in addition to practical experience within one's own organization, technical representatives of catalyst suppliers and licensors are valuable fountains of knowledge on catalysts produced by their companies. Their help in catalyst selection as well as catalyst development should be sought in the early stages of a project as well as in ongoing efforts directed at process improvement.

Howard F. Rase
Austin, Texas

About the Author

Howard F. Rase, after receiving his B.S. in chemical engineering at the University of Texas in 1942, served as a Chemical Engineer for the Dow Chemical Company, a Process Engineer for Eastern States Petroleum Company, and a Process and Project Engineer for Foster Wheeler Corporation. In 1949, he entered the graduate program in Chemical Engineering at the University of Wisconsin and received his PhD in 1952. He then joined the faculty of Chemical Engineering at the University of Texas, where he served for 45 years, specializing in plant design, catalysis, and process kinetics, and with research emphasis on catalyst characteristics, development, and deactivation. He has written six technical books for professional engineers [*Project Engineering of Process Plants* (with M. H. Barrow), *Piping Design for Process Plants, Philosophy and Logic of Chemical Engineering, Chemical Reactor Design for Process Plants* (Vols. 1 and 2), and *Fixed-Bed Reactor Design and Diagnostics: Gas-Phase Reactions*] and served as a consultant to industry for 47 years on reactor design, catalysis, and process improvement.

Dedication

To my wife Beverly and our granddaughters,
Carolyn Elizabeth and Kathryn Victoria

How to Use This Handbook

Index

To find information on heterogeneous catalytic processes, use either the index of products or index of reactants. These indices provide a quick access to page numbers at which sections related to a given product or reactant can be found.

Table of Contents

Each chapter covers a specific reaction type along with separate coverage of a number of commercially significant examples. Each of these examples is subdivided in the chapters and in the table of contents in most cases as follows: product uses, reaction chemistry and thermodynamics, reaction mechanism, catalyst type (including suppliers and licensors), process-unit descriptions, and process kinetics. The table of contents provides, therefore, a convenient means for quickly locating specific information in each of the above categories for 150 commercial heterogeneous catalytic processes organized into 19 major reaction types designated as chapters.

A Word about Kinetics, Mechanisms, and Thermodynamics

Kinetic equations have been included for each reaction considered when sufficient experimental data warrants the usefulness of the proposed form. Simplicity is preferred, which is often possible over the narrow range of optimum operating conditions used in practice.

A logical mechanism or reaction scheme is also included when available, since it is a valuable tool in reasoning about catalyst performance and improvement. Unfortunately, the development of a reasonable mechanism is difficult and requires sophisticated and tedious observations. In earlier decades, the development of a kinetic expression that fit experimental data was often referred to as *confirming a mechanism.* Such so-called mechanisms were based on a general conceptual framework such as Langmuir–Hinshelwood or Rideal–Eley kinetics. In many cases, these kinetic forms have proved useful in process modeling, but they do not constitute proof of a mechanism. In fact, Power–Law kinetics has, in many cases, been useful in process modeling as well, and generally no mechanistic meaning is assigned.

Unless otherwise noted, thermodynamic data for reactions reported herein as heats of reaction and equilibrium constants were calculated from the tabulations by Stull, Westrum, and Sinke* of enthalpies and equilibrium constants of formation of the reactants and products in the ideal gaseous state from 298 to 1000 Kelvins. Accurate design calculations must correct for non-idealities for systems deviating from the ideal gaseous state. Modern design programs provide rigorous and readily usable routines for obtaining accurate thermodynamic data over a wide range of conditions.

---

* Stull, D. R.; Westrum, E. F., jr. and Sinke, G. C., *The Thermodynamic Properties of Organic Compounds.*

Contents

1 Acetoxylation

Acetoxylation involves the replacement of hydrogen by an acetate group in an oxygen containing atmosphere and the formation of water. The reaction occurs with vinyl, allyl, benzyl, and aryl hydrogens.[1]

1.1 ETHYLENE + ACETIC ACID → VINYL ACETATE

Vinyl acetate consumes the largest fraction of acetic acid manufactured. The major uses of vinyl acetate include homopolymerization to polyvinyl acetate (PVA) used in the production of adhesives, paints, and binders, and it is an important ingredient of water-based paints. The glass transition of PVA is below room temperature, and the painted coating forms a film after the water base evaporates. Copolymers of vinyl acetate and vinyl chloride are used in flooring, and a small amount in PVC pipe.[2] A significant portion of polyvinyl acetate is converted by saponification to polyvinyl alcohol, which is used to produce fibers (Japan) as well as textile sizing, adhesives, emulsifiers, and paper coatings.[2] Polyvinylbutyral is made by reacting butyral aldehyde with the hydroxyl groups in PVA and is used for the inner shatter prevention layer of safety glass.[2]

CHEMISTRY (VAPOR PHASE)

$$CH_2 = CH_2 + CH_3\,COOH + \tfrac{1}{2}\,O_2 \longrightarrow CH_2 = CHO\overset{\overset{\textstyle O}{\|}}{C}CH_3 + H_2O$$

ethylene   acetic acid   vinyl acetate

130–180°C @ 5–12 bar
Catalyst: Pd on activated carbon or on silica, or Al_2O_3 plus promoters (see "Catalyst Types").

MECHANISM

Although some studies suggest palladium (II) acetate, $Pd_2(CH_3CO_2)_2$, as the catalytically active species,[5] it appears that the reaction occurs between dissociatively adsorbed acidic acid and dissociatively adsorbed ethylene on palladium.[6] See Figure 1.1.

Detailed proofs of the various steps have not been forthcoming, but the investigators did definitely confirm that $Pd_2(CH_3CO_2)_2$ was not an active species. In fact, catalytic activity is maintained only in those conditions where Pd(II) acetate does not exist.[7] Such conditions are low partial pressures of acetic acid and oxygen, higher temperatures, and addition of potassium acetate solution to the feed.[7] Apparently, Pd(II) acetate is associated with aggregation of palladium and ultimate deactivation.

Main reaction   *Side reaction*

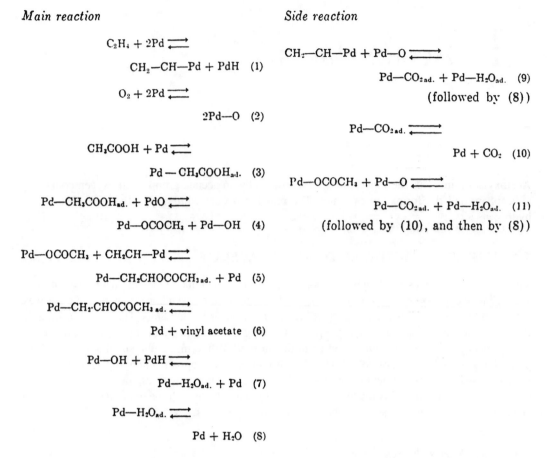

$$C_2H_4 + 2Pd \rightleftharpoons$$

$$CH_2-CH-Pd + PdH \quad (1)$$

$$O_2 + 2Pd \rightleftharpoons$$

$$2Pd-O \quad (2)$$

$$CH_3COOH + Pd \rightleftharpoons$$

$$Pd - CH_3COOH_{ad.} \quad (3)$$

$$Pd-CH_3COOH_{ad.} + PdO \rightleftharpoons$$

$$Pd-OCOCH_3 + Pd-OH \quad (4)$$

$$Pd-OCOCH_3 + CH_2CH-Pd \rightleftharpoons$$

$$Pd-CH_2CHOCOCH_{3\,ad.} + Pd \quad (5)$$

$$Pd-CH_2\cdot CHOCOCH_{3\,ad.} \rightleftharpoons$$

$$Pd + \text{vinyl acetate} \quad (6)$$

$$Pd-OH + PdH \rightleftharpoons$$

$$Pd-H_2O_{ad.} + Pd \quad (7)$$

$$Pd-H_2O_{ad.} \rightleftharpoons$$

$$Pd + H_2O \quad (8)$$

$$CH_2-CH-Pd + Pd-O \rightleftharpoons$$

$$Pd-CO_{2\,ad.} + Pd-H_2O_{ad.} \quad (9)$$

(followed by (8))

$$Pd-CO_{2\,ad.} \rightleftharpoons$$

$$Pd + CO_2 \quad (10)$$

$$Pd-OCOCH_3 + Pd-O \rightleftharpoons$$

$$Pd-CO_{2\,ad.} + Pd-H_2O_{ad.} \quad (11)$$

(followed by (10), and then by (8))

FIGURE 1.1 Proposed Mechanism for Vapor-Phase Catalyzed Synthesis of Vinyl Acetate from Ethylene. Reprinted by permission: Nakamura, S., and Yasui, T., *Journal of Catalysis* 17, 366 (1970), Academic Press, Inc.

CATALYST TYPES AND LICENSORS

The two major processes in use (Bayer/Hoechst and Quantum) differ primarily in the preparation of the proprietary catalysts. Both employ palladium along with alkali metal acetates on silica alumina or on activated charcoal. The Bayer/Hoechst catalyst is prepared by depositing a Pd salt on the carrier followed by reduction to the metal, whereas the Quantum catalyst is prepared by depositing palladium acetate on the carrier.[3,4] In addition to alkali acetates, both catalysts contain promoters variously reported as cadmium, platinum, rhodium, and gold.[3,4] The catalyst is produced in the form of tablets or extrudates depending on the nature of the carrier. A fluidized catalyst was introduced by BP Chemicals (London) in late 1998. See "Process Units."

CATALYST DEACTIVATION

Traces of acetylene in the ethylene feed is a strongly adsorbed poison, but the catalyst can be reactivated by oxygen treatment.[6]

The alkali acetates incorporated in the catalyst, which promote activity and selectivity, migrate in the direction of flow and must be renewed over an operational cycle to reach run times of 2-1/2 to 3+ years.[3,4] Renewal is accomplished by addition of these salts in solution via the gaseous feed to the reactor.

PROCESS UNITS

These processes were introduced in the late sixties and many improvements have been made by various operators. Such actions are not in the public domain, but the general operating procedure has been described[3,4,6].

The exothermic reaction, like so many partial oxidations, is carried out in a fixed bed multitubular reactor. Boiler feed water is used as the shell-side fluid. Operating conditions are reported to be in the range 140–180°C, and pressure is in the range of 5 to 12 bar. Although the reaction rate increases with pressure, higher pressures force the flammability limit to lower oxygen content and thus lower conversion to vinyl acetate. Hence, an optimum pressure is constrained by the catalyst characteristics and the flammability of the feed mixture.

Feed to the reactor is composed of fresh feed plus recycle. A typical reported total feed in mole percent is 10–20% acetic acid, 10–30% CO_2, and 50% ethylene.[4] The remainder is oxygen, which is fed at 1.5% below the flammability limit, which varies with operating conditions for a particular unit. Operating temperature also affects efficient energy recovery. A low temperature may only provide preheat for boiler feed water, while a higher temperature can produce a a more useful saturated steam.

Ethylene that is fed in excess reaches conversions of only 8–10%, while oxygen conversion is in the range of 90%.[4] Acetic acid conversion is up to 30%. The total yield, which accounts for recycled acetic acid, is reported as high as 99%.

Product gas is cooled and the condensate consisting of vinyl acetate and water product is readily separated in a two-phase separator drum. The crude vinyl acetate is then piped to a distillation section for purification. The remaining vapor phase is washed, and then CO_2 is removed in a potash solution, followed by recovery and recycling of a portion of the CO_2, which is valuable for temperature and reaction control.

In late 1998, BP Chemicals (London) introduced a fluid-catalyst process (LEAP) using a similar chemistry but having the usual fluidized-bed advantage of precise temperature control and ease of catalyst regeneration. Lower investment costs and longer sustained catalyst activity are claimed. Operating conditions are 150–200°C at 8–10 bar, and yield is 99% on acetic acid and 92–94% on ethylene.[15]

PROCESS KINETICS

Based on the mechanism shown in Figure 1.1, kinetic data were fit using a rate determining step of the combination of dissociately adsorbed ethylene and acetic acid.[6]

The rate of vinyl acetate formation is

$$R_{VAC} = \frac{k_5 K_1 K_2^{0.5} K K_4 p_{C_2H_4} p_{O_2}^{0.5} p_{CH_3COOH}}{[1 + K_1 p_{C_2H_2} + (K_2 p_{O_2})^{0.5} + K_3 p_{CH_3COOH}]^2}$$

where p = atm

R_{VAC} = g/1 hr

This complex multiconstant equation should be amenable to simplification for the commercial operating range.

1.2   BUTADIENE + ACETIC ACID → 1,4-DIACETOXY-2-BUTENE → 1,4-BUTANEDIOL → TETRAHYDROFURAN

The production of 1,4-butanediol is possible by several different routes. In the U.S.A. and Europe, the Reppe acetylene process continues to dominate because of the use of C_4 hydrocarbons for other profitable products. In Japan, however, acetoxylation of butadiene is the preferred process step to 1,4-butanediol and tetrahydrofuran. The major uses of 1,4-butanediol include the production of tetrahydrofuran and polyesters. Tetrahydrofuran is an excellent solvent for a wide variety of polymers and is also used in the manufacture of polytetramethylene glycol, which is an important reactant in forming certain polyurethanes and stretch fibers.

The polyester, polybutene-terephthalate, is formed by the polymerization of 1,4-butanediol with terephthalic acid. It is more flexible than polyethylene terephthalate and is used for injection molding applications.

CHEMISTRY (LIQUID PHASE)

Unlike the Reppe process, which primarily produces 1,4-butanediol, a portion of which can be converted to tetrahydrofuran, the acetoxylation process unit can be operated to produce both 1,4-butanediol and tetrahydrofuran (THF) from the same intermediate (1,4-acetoxyhydroxybutane). The relative amounts of each can be adjusted by changes in operating temperature and residence time.[3]

The acetoxylation intermediate is hydrogenated and then hydrolyzed to produce the desired products[9,10,11,12] as depicted in Figure 1.2.

CATALYST TYPE

The proprietary acetoxylation catalyst is palladium on activated charcoal promoted by tellurium. It is reported to be in the form of granules, probably 4 × 8 mesh. The hydrogenation step employs a standard hydrogenation catalyst, either supported nickel or palladium. In the case of nickel, zinc or some other additive may be added to moderate activity. The hydrolysis reactions are catalyzed by specially prepared cation and anion-exchange resins.[12] Diacetocyclization is catalyzed by a cation-exchange resin. Other acetoxylation catalysts have been described, including Pd-Sb-V-CsCl-KOAC.[14]

Licensor

The licensor is Mitsubishi-Kasei Corporation.

CATALYST DEACTIVATION

Catalyst life for the acetoxylation catalyst is one year, but the catalyst can be regenerated after removal by treatment with oxidation and reducing agents.[12] Although some polymerization of butadiene could be expected, it is minimized by the liquid feed acting as a useful wash of polymer as it is formed. The high activity of the catalyst was made possible the use of moderate temperatures that avoided excessive coking and assured the best selectivity. The catalyst promoted with CsCl and KOAC exhibits movement of these components after long use, which reduces the resistance to coke formation that these components provide.[14]

Activated charcoal proved to be the best catalyst carrier. It was not adversely affected by acetic acid and did not produce significant amounts of high boilers.[10]

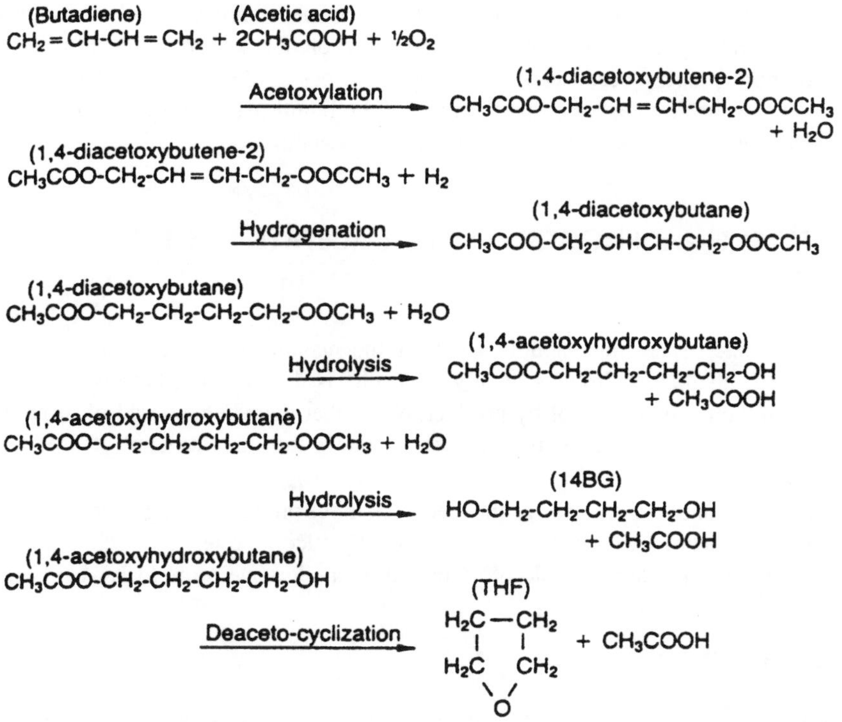

(Butadiene)   (Acetic acid)
$CH_2 = CH\text{-}CH = CH_2 + 2CH_3COOH + \frac{1}{2}O_2$

(1,4-diacetoxybutene-2)

$\xrightarrow{\text{Acetoxylation}}$ $CH_3COO\text{-}CH_2\text{-}CH = CH\text{-}CH_2\text{-}OOCCH_3$
                        $+ H_2O$

(1,4-diacetoxybutene-2)
$CH_3COO\text{-}CH_2\text{-}CH = CH\text{-}CH_2\text{-}OOCCH_3 + H_2$

(1,4-diacetoxybutane)

$\xrightarrow{\text{Hydrogenation}}$ $CH_3COO\text{-}CH_2\text{-}CH\text{-}CH\text{-}CH_2\text{-}OOCCH_3$

(1,4-diacetoxybutane)
$CH_3COO\text{-}CH_2\text{-}CH_2\text{-}CH_2\text{-}CH_2\text{-}OOCH_3 + H_2O$

(1,4-acetoxyhydroxybutane)

$\xrightarrow{\text{Hydrolysis}}$ $CH_3COO\text{-}CH_2\text{-}CH_2\text{-}CH_2\text{-}CH_2\text{-}OH$
                        $+ CH_3COOH$

(1,4-acetoxyhydroxybutane)
$CH_3COO\text{-}CH_2\text{-}CH_2\text{-}CH_2\text{-}CH_2\text{-}OOCH_3 + H_2O$

(14BG)

$\xrightarrow{\text{Hydrolysis}}$ $HO\text{-}CH_2\text{-}CH_2\text{-}CH_2\text{-}CH_2\text{-}OH$
                        $+ CH_3COOH$

(1,4-acetoxyhydroxybutane)
$CH_3COO\text{-}CH_2\text{-}CH_2\text{-}CH_2\text{-}CH_2\text{-}OH$

(THF)

$\xrightarrow{\text{Deaceto-cyclization}}$ $\begin{matrix} H_2C\!-\!CH_2 \\ |\quad\ | \\ H_2C\quad CH_2 \\ \diagdown\ \diagup \\ O \end{matrix}$ $+ CH_3COOH$

FIGURE 1.2 Reaction Scheme for the Production of 1,4-Butanediol and Tetrahydrofuran via Acetoxylation. Basis: Mitsubishi Chemical Ind. 14 BG/THF process. Reprinted by permission: Tansabe, Y., *Hydrocarbon Processing,* p. 189, Sept. 1981.

PROCESS UNITS[9–12]

The acetoxylation reactor is a fixed-bed reactor that operates at 70°C and 70 bar. Butadiene, acetic acid, and air are mixed and fed together. The air feed rate is adjusted to maintain a safe non-flammable mixture. Recycled acetic acid removed from the product stream provides an additional heat sink for temperature control. Acetic acid is removed from the 1,4-diacetoxybutene-2 product by distillation, and the 1,4-diacetoxybutene-2 is hydrogenated in a trickled-bed. The 1,4-diacetoxybutane is then hydrolyzed over the ion-exchange resins to 1,4-acetoxy-hydroxybutane (a monoacetate) and 1,4-butanediol at 60°C and 50 bar. The ratio of these two products can be changed by adjusting operating conditions.

After removing the excess water and acetic acid by distillation, a second stage of hydrolysis is practiced to improve yields by reacting the remaining diacetate. Excess water is again removed by distillation, followed by a distillation that separates the monoacetate from the 1,4-butanediol. The butanediol is purified and the monoacetate sent to the diaceto-cyclization step to produce tetrahydrofuran. The combined yields of 1,4-butanediol and THF based on diacetoxybutane fed is 99%. The yield of 1,4-diacetoxybutene-2 is 90% based on butadiene fed and the yield of diacetoxybutane is 98% based on diacetoxybutene fed.[12]

PROCESS KINETICS

Extensive studies in the vapor phase on Pd catalyst with various promoters have revealed that minimal adsorption of butadiene favored the desired reaction because coke formation was thereby inhibited.[14] In such cases, the rate of reaction in the vapor-phase process tends to be proportional to the acetic acid and oxygen partial pressures. The reduction in coke formation is further advanced in the liquid-phase process because of the washing effect of the liquid reactants, particularly butadiene. It is reasonable to assume that the rate for the liquid-phase process will also be proportional to acetic acid concentration and oxygen partial pressure.

1.3 PROPYLENE + ACETIC ACID → ALLYL ACETATE → ALLYL ALCOHOL

Acetoxylation is one of several reaction paths for the manufacture of allyl alcohol. The once-dominant process, alkaline hydrolysis of allyl chloride, is disappearing because of its corrosive environment and the large amount of by-product, NaCl, that must be disposed of or recovered for chlorine. The acetoxylation process, by contrast, produces the same amount of acetic acid as is used in the reaction.

Allyl alcohol major uses have been as the feedstock in the production of glycerol, and diethylene glycol bis(allyl carbonate) for optical lenses.[3,13] Also, allyl alcohol esters are used in polymers, and allyl alcohol is used to produce epichlorohydrin and in an alternate route to 1,4-butanediol.[3,13]

CHEMISTRY (VAPOR PHASE)

The acetoxylation reaction to the acetates is analogous to the vinyl-acetate process, except propylene is used rather than ethylene.

1.
$$CH_2 = CH_2 + CH_3 COOH + \tfrac{1}{2} O_2 \longrightarrow CH_2 = CHOCCH_3 + H_2O$$
ethylene   acetic acid   vinyl acetate

150–250°C @ 5–10 bar
Catalyst: Pd on activated charcoal
$\Delta H = -45$ kcal/mole
Liquid phase hydration
60–80°C
Catalyst: acidic ion-exchange resin

2.
$$CH_2 = CHCH_2 + OCCH_3 + H_2O \rightleftharpoons CH_2 = CHCH_2 OH + CH_3COOH$$
allyl acetate   allyl alcohol   acetic acid

Liquid phase hydration
60–80°C
Catalyst: acidic ion-exchange resin

MECHANISM

See section on "Vinyl Acetate" process.

CATALYST TYPE

The catalyst is similar to that used for vinyl acetate production. It is palladium along with an alkali metal acetate deposited on activated charcoal. Promoters are reported to be iron or bismuth compounds.[3]

Licensors

Licensors are Showa Denka, Daicel Chemical Industries, Hoechst, and Bayer.

CATALYST DEACTIVATION

Propylene feed stocks often come from steam cracking units and can contain small amounts of acetylene due to upsets in the selective hydrogenation of acetylene impurities at the cracking unit. Acetylene is a strongly adsorbed poison, but the catalyst can be reactivated by oxygen treatment.

PROCESS UNIT

The acetoxylation is accomplished in the vapor phase in a fixed-bed reactor, probably a multitubular reactor with cooling by boiler feed water, since the operating temperature is such that valuable steam can be produced. Operating conditions are variously reported as 150–250°C, depending on the process.[3,13] See the description of the vinyl acetate process for more analogous detail.

The hydrolysis to allyl alcohol is done using the cooled liquid phase separated from the allyl acetate reactor effluent. The reactor is an adiabatic fixed-bed unit packed with the acidic ion exchange resin in the form of granules.

PROCESS KINETICS

It is reasonable to suggest that a similar kinetic expression as used for vinyl acetate might apply.

REFERENCES (ACETOXYLATION)

1. Rylander, R. W., in *Catalysis*, Vol. 4, J. R. Anderson and M. Boudart, eds., p. 2, Springer-Verlag, New York, 1988.
2. Chenier, P. J. *Survey of Industrial Chemistry*, 2nd ed., New York, 1992.
3. Weissermel, K., and Arpe, H. I., *Industrial Organic Chemistry*, 3rd ed., VCH, New York, 1997.
4. Roscher, G., in *Ullmann's Encyclopedia of Industrial Chemistry*, 5th ed., Vol. A27, p. 413, VCH, New York, 1996.
5. Samanos, B.; Boutry, P., and Montarsal, R., 66th *Intr. Symp. of Catalytic Oxidation*, London, July, 1970.
6. Nakamura S. and Yasui, T., *J. Catal.* 17, 366 (1970).
7. Nakamura, S. and Yasui, T., *J. Catal.* 23 315 (1971).
8. Bedell, K. R. and Rainbird, H. A., *Hydrocarbon Proc.*, p. 141, Nov. 1972.
9. Brownstein, A. M., *Chemtech,* August, 1991, p. 506.
10. Mitsubishi Kasel Corp., *Chemtech,* Dec. 1988, p. 759.
11. Brownstein, A. M., and List, H. L., *Hydrocarbon Proc.*, Sept. 1977, p. 159.
12. Tanabe, Y., *Hydrocarbon Proc.,* Sept. 1981, p. 187.
13. Nagato, N., in *Kirk-Othmer Encyclopedia of Chemical Technology*, 4th ed., Vol. 2, p. 144, Wiley, New York, 1992.
14. Shinohara, H., *Applied Catal.*, 50 (2), 199 (1989); 24 (1–2), 17 (1986); 14 (1–3), 145 (1985); 10 (1), 27 (1984).
15. Chementator Section: *Chemical Engineering*, p. 17, December, 1998.

2 Alkylation

2.1 INTRODUCTION

The term *alkylation* refers to the replacement of a hydrogen atom bonded to a carbon atom of a paraffin or aromatic ring by an alkyl group.[1] Most alkylations are acid catalyzed either by a homogeneous liquid strong-acid catalyst (H_2SO_4, H_3PO_4, HF, $AlCl_3$) or a solid strong-acid, which has been used where feasible and economical (zeolites, supported acids, acidic ion-exchange resins). Developments with heterogeneous catalysts were driven, in part, by efforts to minimize waste disposal problems associated with spent acids. Not all alkylation processes, however, have been amenable to the use of a heterogeneous acid catalyst which, in some reaction systems, produces coke at a rapid rate. Thus, the production of alkylate gasoline from isobutane and olefins continues to employ either liquid sulfuric or hydrofluoric acid. Major improvements in handling and reducing acid consumption, and thus waste acid quantities, have proved successful in such processes.[1]

2.1.1 SOLID ACID-CATALYST MECHANISM

The mechanistic concepts for solid-acid catalysts derive largely by analogies from the many studies on homogeneous acid catalysis which present a more readily verifiable system for detailed study. In recent years various studies on solid catalysts using, for example, deuterium exchanged reactants have confirmed many of the earlier analogies. The steps shown in Figure 2.1 present a reasonable mechanism for benzene alkylation by ethylene that is applicable to other similar alkylations.[2] The first step shows the formation of the carbenium ion by chemisorption of ethylene on an active Brønsted site. This step rapidly reaches an equilibrium concentration of surface carbenium ions. The second step is the reaction of the benzene ring with the surface carbenium ion, and this is the rate controlling step. The third step involves the desorption of the product alkylated benzene and the regeneration of the Brønsted acid site. This third step is also rapid and at equilibrium.[2]

Figure 18.26 (Chapter 18) illustrates qualitatively the relative strengths of acid sites for the several carbenium ion reactions.[2] Alkylation and cracking require the same level of high acid strength, which is not surprising, since alkylation is the reverse reaction of cracking. Cracking is endothermic and requires high temperatures, whereas alkylation is exothermic and requires low temperatures (below 400°C). Thus, alkylation can be made exclusive by low-temperature operation. However, low temperature favors polymerization and catalytic coke formation, which can be suppressed by catalysts with low coke forming tendencies.

Alkylation of benzene by higher olefins such as propylene are more easily accomplished because the secondary carbenium ion is more reactive. Hence, lower-temperature operation is possible.

2.1.2 CATALYST TYPES AND USES

The various solid catalysts that are used or that have been used are summarized in Table 2.1. Additional details will be found under the several process descriptions following this section.

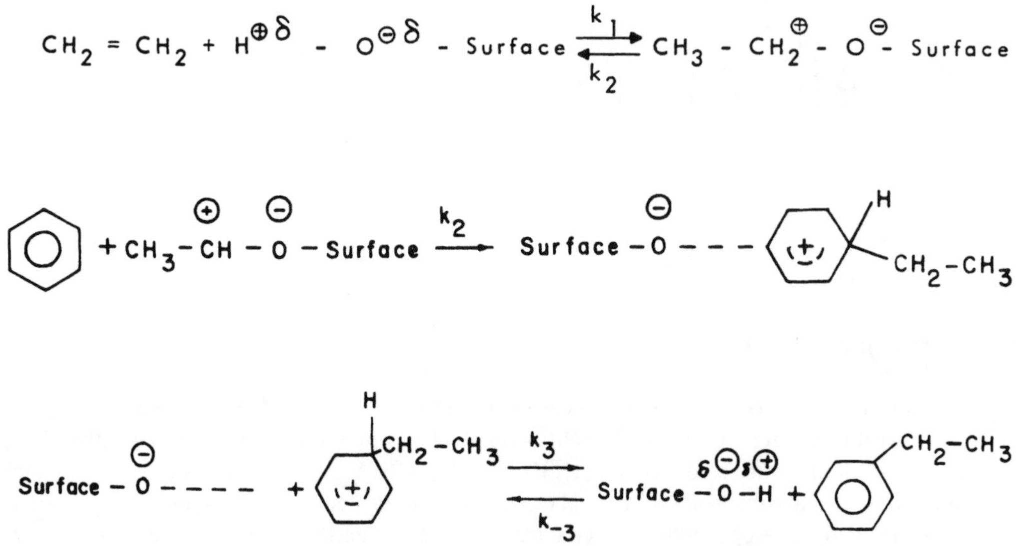

FIGURE 2.1 Mechanism for Ethylbenzene from Ethylene and Benzene by Solid Alkylation Catalysts. Equations reproduced from Wojciechowski, B. W. and Corma, A., *Catalytic Cracking: Catalysts, Chemistry, and Kinetics,* Marcel Dekker, New York, 1986, by courtesy of Marcel Dekker, Inc.

2.1.3 THE ROLE OF ZEOLITES IN ALKYLATION

Traditional acid-catalyzed alkylations involve liquid acids or solid supports impregnated with acids such as phosphoric acid. These processes, although efficient, can cause corrosion and create major disposal problems. By contrast zeolites create no environmental problems and, in fact, 34 naturally occurring zeolites make up a major portion of the Earth's crust. This environmentally neutral character of zeolites certainly made them a potentially valuable catalyst in the acid form. But the discovery of shape selectivity of zeolites gave catalyst development chemists another tool for optimizing catalyst characteristics.

The literature on zeolites is massive, and no attempt will be made here to summarize the fascinating detail of the science and engineering of zeolite catalysis. The following major issues, however, deserve the reader's attention.

Shape Selectivity

Shape selectivity can be attained by a variety of procedures. See Figure 2.2.[7,8]

- *Molecular exclusion (reactant selectivity).* Select a zeolite with pore size and structure such that some of the molecules in the reactor feed can enter the pores and react, and others cannot.
- *Molecular exclusion (product selectivity).* Select a zeolite with opening pore size and inner pore size such that one or more of the products is too bulky to diffuse out of the cavity. The other products readily diffuse out. Those remaining behind may crack to smaller molecules or form coke that deactivates the catalytic surface.
- *Transition-state selectivity.* Select a zeolite with an inner space of a size that is inadequate for a particular transition state but adequate for another, thereby favoring one product over another.

TABLE 2.1
Solid Alkylation Catalyst Types and Uses*

Catalyst	Description	Suppliers	Licensor	Uses
Boron trifluoride on γ–Al$_2$O$_3$	Small amount of BF$_3$ on alumina	UOP LLC	UOP LLC (Alkar Process)	Alkylation of benzene with ethylene (vapor-phase process) to produce **ethylbenzene**
Zeolite, Y-type (faujasite)	USY-type (ultra stable) pore size: 7.4A SiO$_2$/Al$_2$O$_3$>3	UOP LLC	(original catalyst developed by Unocal)	Alkylation of benzene with ethylene (liquid-phase process) to produce **ethylbenzene**
Zeolite, ZSM-5 (pentasil)	Specially modified medium pore zeolite high silica/alumina ratio (>20) pore size: 5.5A extrudates	Mobil	Mobil-Badger, (now Mobil-Raytheon Engineers & Constructors) (Third Generation EP Process)	Alkylation of benzene with ethylene (vapor-phase process) to produce **ethylbenzene**
EBMAX or EBEMAX catalyst	MCM-22 pentasil zeolite extrudates	Mobil United Catalysts	Mobil-Raytheon Engineers & Constructors (EBMAX)	Alkylation of benzene with ethylene to produce **ethylbenzene** (liquid-phase)
Packaged zeolite	Pentasil zeolite packaged	CDTECH	ABB Lummus Global	Catalytic distillation simultaneous alkylation of benzene with (ethylene or propylene) to produce **ethylbenzene** or **cumene**
Phosphoric Acid on Kieselguhr (SPA Process, i.e. solid phosphoric acid process)	65-70% H$_3$PO$_4$-n-silica. Extrudates: 1/4 and 3/16-in. spheres: 1/4 × 5/16 in., 3/16 × 1/4 in.	United Catalysts	UOP LLC (SPA Process i.e. solid phosphoric acid process)	Alkylation of benzene with propylene (vapor-phase process) to produce **cumene**
Q-Max catalyst	beta zeolite extrudates	UOP LLC	UOP LLC	Alkylation of benzene with propylene to produce **cumene** (liquid-phase)
MCM-22 catalyst	Pentasil zeolite extrudates	Mobil	Mobil-Raytheon Engineers & Constructors	Alkylation of benzene with propylene to produce **cumene** (liquid-phase)
3-DDM	Dealuminated mordenite to create controlled 3-dimensional structure	Dow	Dow/Kellogg	Alkylation of benzene with propylene to produce **cumene** (liquid-phase)
Solid-acid catalyst (not revealed)	Heterogeneous acid catalyst	UOP LLC	UOP (Detal Process)	Alkylation of higher molecular-weight olefins with benzene to produce **linear alkyl-benzenes** for detergent manufacture

*Based on information from Refs. 1 and 3–6. All of these catalysts are proprietary. Only limited information on catalyst details exists in the open literature, and some such information is speculative.

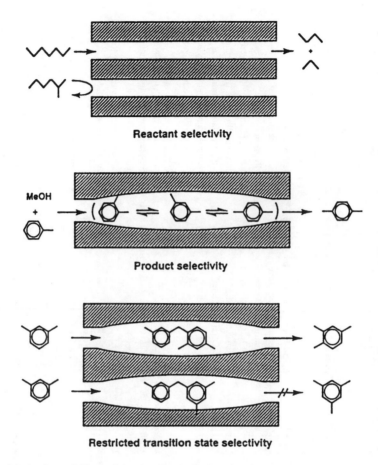

Reactant selectivity

Product selectivity

Restricted transition state selectivity

FIGURE 2.2 Mechanism of Shape-Selective Catalysis. Reprinted by permission: Sugi, B. Y. and Kubota, Y., *Catalysis: Specialized Periodical Reports,* Vol. 13, p. 56, The Royal Society of Chemistry, Cambridge, England, 1997.

- *Configurational diffusion controlled selectivity.* Select a zeolite of such structure that a large difference exits between rapidly diffusing reactant and product molecules and other molecules present that diffuse much more slowly. The slower diffusing molecules can thus become only minor players in the reaction process.

Many other shape-selective attributes have been postulated and have been reviewed.[8]

Modification of Zeolites[8,9]

In the production of synthetic zeolites, the catalyst developer has several tools for modifying characteristics of a zeolite to fit the needs of a desired catalyst system. More detailed discussions are available, but the following provides a brief review.

- SiO_2/Al_2O_3 ratio
  The ratio can be modified by chemical dealumination or by steaming, which removes framework aluminum. The steaming process when properly applied can develop meso-pores that facilitate the diffusion of larger molecules.

- Crystallite Size
  Crystals can be made smaller by various techniques, including lower temperature in the presence of saccharides or by thermal shock. Large crystallites are favored by low-speed agitation and longer crystallization time. Lower crystallite size increases rate of diffusion.
- Substitution of other Atoms in the Framework
  Other atoms (B, Cr, Ga, Ge, Fe, P, and Ti) have been substituted for either Al or Si atoms during synthesis or post synthesis modification. A variety of different properties can be attained.
- Cation-Ion Exchange
  Synthetic zeolites are mostly produced with Na or K cations occupying the cation positions. Acid catalysis requires that these cations be replaced by hydrogen protons. In the manufacturing process, this ion exchange is accomplished following spray drying using a 5–10% ammonium chloride or nitrate solution. This step is followed by drying and calcining in an inert gas during which NH_3 is released, leaving behind acid sites (H+) referred to as the hydrogen form (e.g., ZSM-5 $\rightarrow$ HZSM-5). If organic compounds are used in preparation of the initial zeolite, they must be removed by calcining prior to ammonium exchanges.
- Binders
  Both zeolite catalysts and adsorbents are combined with a binder, usually a clay such as halloisite, montmorillonite, or attapulgite. The clay in gel form is mixed with 10–20% of zeolite as a paste and then spray dried. The porosity of the binder is greater than the zeolite and does not, in most cases, limit the net rate of diffusion to the active sites on the zeolite.
- Pore Size
  Pore size can be manipulated by post-synthesis procedures such as depositing organo-silicones in the pore structure to alter the diffusivity differences between isomers such as the xylenes so that one isomer (p-xylene) will have a much greater diffusivity than the larger o-xylene or m-xylene. Treatment with phosphorus, which attaches to the zeolite active acid site via the framework oxygen, has been used to reduce the effective pore size of the channels and pore openings of ZSM-5 and thus produces a p-xylene selective catalyst.[8]

2.2 ALKYLATION OF BENZENE AND TOLUENE

2.2.1 BENZENE + ETHYLENE → ETHYLBENZENE

All but about 1% of ethylbenzene is used for dehydrogenation to styrene. The remaining small amount is primarily used as a solvent for some types of paint. Essentially, all ethylbenzene is produced by alkylation of benzene with ethylene. A small amount is also produced from mixed xylenes by superfractionation, which is highly energy intensive.

Chemistry

Main Reaction

benzene   ethylene   ethylbenzene

Temp., K	ΔH, kcal	K_p
500	−24.97	24.668×10^3
600	−24.83	33.729×10^2
700	−24.66	17.378
800	−24.48	1.914

Operating Conditions[1,5,11,12]

Process	Catalyst	Temperature	Pressure
Alkar (vapor phase), UOP LLC, (process no longer offered)	BF_3/alumina	100–150°C	25–35 bar
Lummus-UOP (liquid phase)	USY zeolite extrudate	≈270°C	38 bar
Mobil-Badger* (vapor phase)	ZSM-5 zeolite extrudates	400–450°C	15–30 bar
EB-MAX (liquid phase), Mobil/Raytheon	MCM-22 extrudates	not reported	
Catalytic Distillation, ABB Lummus, Global	packaged zeolite	not reported	

*Depending on plant economics, temperatures as low as 350°C and pressure as low as 8 bar may be used.

Side Reactions

The following side reactions are possible:

- *Polyethylbenzenes.* A portion of the ethylbenzene formed is further alkylated to di-, tri-, tetra-, etc. benzenes.

$$C_6H_5CH_2CH_3 + C_2H_4 \leftrightarrows CH_3CH_2C_6H_4CH_2CH_3$$

ethylenbenze   ethylene   diethylbenzene

$$\Delta H = -23.46 \text{ kcal @ } 600°K$$

$$K_p = 19.543 \text{ @ } 600°K$$

- *Transalkylation.* Polyethylbenzenes are converted to ethylbenzene and benzene.

$$CH_3CH_2C_6H_4CH_2CH_3 + C_6H_5 \leftrightarrows 2C_6H_5CH_2CH_3$$

diethylbenzene   benzene   ethylbenzene

$$\Delta H = 1.38 \text{ kcal @ } 600°K$$

$$K_p = 17.022 \text{ @ } 600°K$$

By using an excess of benzene, transalkylation is favored.
- *Oligomerization.* This polymerization-type reaction is the most unwanted reaction, since the oligomers are precursors to other by-products, the most deleterious of which is coke formation.

$$2C_2H_4 \rightarrow C_4H_8$$

$$\Delta H = -17.49 \text{ kcal @ } 600°K$$

$$K_p = 144.5 \text{ @ } 600°K$$

The oligomers can also alkylate benzene, producing higher alkylbenzenes. Coke formation by oligomer reactions constitutes a reaction pathway that can lead to catalyst deactivation. Other possible side reactions include cracking of oligomers, dehydrogenation, and isomerization.[5] Process improvements have focused strongly on minimizing such undesired reactions.

Other side reactions that occur to some degree are isomerization (EB $\rightarrow$ xylenes), dehydrogenation, and other alkylations. Many of the products of these reactions occur only in very small amounts. Benzene feed often contains small amounts of toluene (seldom over 1000 ppm). It, of course, is alkylated and becomes dealkylated in the transalkylation reactor. Toluene then becomes part of the ethylbenzene product, but it causes no problem in the subsequent styrene-producing process. Xylene impurities in the product ethylbenzene are much lower than produced in the $AlCl_3$ process.

Mechanism

See Figure 2.1.

Catalyst Suppliers and Licensors

See Table 2.1. The catalysts are proprietary and the processes are licensed. Limited information on catalysts details exists in the open literature.

Yield and Purity

The overall yield for most of these processes is in the range of 98.5–99.5, and product purity is reported as high as 99.9%.

Catalyst Deactivation

The major deactivating agent is coke. Coke formation is inhibited by the relatively narrow pores of zeolites used in alkylation reactions. The limited space in the channels provides less space for the formation of large coke precursor molecules. Of course, coke formation is catalyzed by acid sites, and some attempt to optimize acid strength is a possible procedure for reducing coke-forming reactions.[13]

Organic nitrogen compounds, being strong bases, will poison zeolite and other acid catalysts.

Catalyst Regeneration

When necessary, coke may be removed by carefully burning it in an air-nitrogen mixture. Great care must be exercised, especially if water vapor as steam is present, since there is danger in it destroying some of the zeolite crystallinity.[12] Licensors' recommendations must be carefully followed.

Process Units[1,6,11,12]

Until about 1980, most ethylbenzene plants were based on the homogeneous catalysis of aluminum chloride. The process was successful but had the disadvantage of a highly corrosive liquid system requiring expensive lined or Hasteloy reactors. Spent catalyst had to be washed and neutralized and then disposed, which action is becoming increasingly costly and has fueled the efforts to develop noncorrosive catalyst systems involving solid catalysts. Few new $AlCl_3$-based plants have been built since 1980, and none since 1990, but 40% of EB plants continued to use their $AlCl_3$ catalyzed alkylation.

Vapor-phase processes with solid catalysts can use dilute ethylene streams and have adequate ethylene concentration at the catalyst surfaces. Essentially pure ethylene, of course, can be used, and most vapor-phase plants use polymer-grade ethylene.

Alkar Process (Vapor Phase)

The first such process using a solid catalyst was developed as early as 1958. It employed a catalyst composed of alumina with boron trifluoride adsorbed on its surfaces in relatively small amounts. This system is not corrosive, but even small amounts of water in the feed will not only tend to remove BF_3 from the catalyst but will also create a corrosive mixture over time. The catalyst is susceptible to poisoning by CO, sulfur compounds, oxygenates, and water, all of which can be present in dilute refinery ethylene streams and must be removed prior to use. The process proved advantageous for the use of dilute refinery streams (8–10% ethylene) attaining 100% ethylene conversion and 99%+ yields.

A multitubular reactor was used with shell-side cooling. Benzene feed was dehydrated and combined with the ethylene containing steam along with make-up BF_3 to replace that lost continuously from the catalyst. The recovery section removed carry-over BF_3 from the reactor product, separated benzene for recycle to the reactor and as feed to the transalkylation reactor for conversion of polyethylbenzenes, (separated from the reactor product), to ethylbenzene. Yields were in the 99%+ range and product purity was typically 99.95 ethylbenzene.

The last plant was built in 1979, and some have been revamped using other processes. The process was unique and highly successful, but waste disposal became an increasingly expensive problem. As the grandfather of solid-catalyst alkylation, it started a major new trend in the industry.

Mobil/Badger ZSM-5 Process (Vapor Phase)

The original ZSM-5 process was the first zeolite process for alkylation of aromatics (first plant in 1981). It, in many ways, revealed the value of shape selectivity made possible by synthetic zeolites. The medium pore-size zeolite (ZSM-5) proved ideal for avoiding or minimizing undesired reactions such as oligomerization and also exhibited only a minor effect on product yield and quality when nonaromatics are present in the feed to the reactor.

Initially, two reactors were provided so that coke could be removed by burning in a used bed while flow is switched to the second bed. Over time, changes in operating procedures, and probably zeolite synthesis, along with zeolite post-synthesis modifications, have culminated in the third-generation process that uses only one alkylation reactor, since time between regenerations is now two years rather than only several months.

Briefly, the vapor-phase process consists of a main multibed adiabatic reactor with cold-shot cooling between beds. Fresh benzene and recycle benzene are combined and preheated in a furnace. This heated stream then flows to the reactor inlet where it is joined by a portion of the ethylene that has not been heated. The remainder of the ethylene is introduced between the beds to control the bed temperature. At higher temperatures, the reverse reaction of dealkylation becomes favorable (e.g., above 600°C), as do other undesired reactions.

Excess benzene is used in a ratio of 5:1 benzene-to-ethylene (values up to 20 have been reported) to reduce the tendency to form additional amounts of polyethylbenzenes. Excess benzene also favors transalkylation.

Reactor effluent flows to the purification system where excess benzene is separated overhead in the first column and recycled to the main reactor. The bottoms flows to an ethylbenzene column where ethylbenzene is separated overhead, and the bottoms flows to a polyethylbenzene (PEB) column where PEB is separated overhead and sent to a smaller reactor for transalkylation to ethylbenzene from which water and light materials are stripped prior to recycling to the first column.

The original design provided for direct recycle of polyethylbenzene to the main reactor where transalkylation can also be accomplished. The third-generation design provided a separate transalkylation reactor. Although the same catalyst is effective for both alkylation and transalkylation, separate operating conditions for transalkylation improve the reaction efficiency. Generally, a higher operating temperature for transalkylation is indicated for operating conditions for the main reactor. Higher alkylbenzenes such as C_8 and C_9, if formed, reach equilibrium and are simply recycled with the PEB stream. The catalyst is active in decomposing nonaromatics to light gases, which are easily purged. Also, branched-chain alkylbenzenes such as cumene are easily dealkylated, thereby recovering the benzene. Bottoms from the PEB column is small in quantity and is generally used as fuel.

Since the vapor-phase process must operate at a higher temperature, benzene reactor feed must be preheated in a furnace. Vapor-phase operation, however, allows for the use of less costly refinery stream with 10–15% ethylene, which would require high pressures for liquid-phase operation. The only purification required of such streams is the removal of higher olefins that will alkylate and reduce EB yield. Most plants use high-purity ethylene with ethane no higher than 2000 ppm as the impurity. Fortunately, ethane causes no problem in the process and remains inert and is removed with the off gases.

Mobil/Raytheon E&C, EB Process, EBMAX (Liquid Phase)

Dilute ethylene streams cannot be used for liquid-phase processes, because complete dissolution of ethylene is not possible at economical operating conditions with low gas-phase ethylene partial pressures.

This is a new process, with the first plant commissioned in 1995 by Raytheon Engineers and Constructors (the former Badger component of Raytheon) using a new Mobil catalyst named by Mobil MCM-22. Because the process is operated in the liquid phase, lower temperatures are used, and a preheating furnace is not required. This unique catalyst has a high alkylation activity but does not catalyze oligomerization or cracking. These characteristics make it possible to operate at a lower benzene-to-ethylene ratio with obvious savings in energy costs and lower equipment sizes for both reactors and distillation towers. The layout of the plant is similar to that described above for the Mobil/Badger third-generation process with a multibed adiabatic reactor and a transalkylation reactor for converting recycled polyethylene benzenes to ethylbenzene.

Since oligomerization does not occur, aliphatic impurities are negligible in the product. Amounts of C_8 and other heavier impurities are very low.

Lummus/UOP EB Process (Liquid Phase)

This process was developed by Unocal in the 1980s and was the first zeolite-based liquid-phase process. An ultraselective USY-type zeolite is used. It is now licensed by ABB Lummus Crest and UOP, and the first plant was built in 1990.

The recovery section is similar to that described for previously described EB processes based on zeolites. Two dual-bed reactors are operated in series with ethylene cold-shot cooling between beds and an intercooler between reactors. The operating temperature (270°C) is close to the critical temperature to take advantage of the higher catalyst activity. Water, however, is added to the feed stream to reduce the tendency to form oligomers.

The product EB is low in xylenes but, as is the case in all these processes, higher aromatics must be removed as residue, which lowers yield, depending on the quantity formed. Careful operation, however, in accordance with licensor recommendations can reduce the net residue production.

CD Tech/ABB Lummus Global and Chemical Research & Licensing (Two-Phase Process)

This process involves catalytic distillation, which has proved to be the preferred process for the manufacture of MTBE (methyl tert-butyl ether, the gasoline additive). Catalyst is loaded into fiberglass cylindrical containers or bales and placed in a pattern in a column so that liquid flows down through the bales and ethylene vapor moves upward. Since oligomerization is a higher-order reaction for ethylene than alkylation, the distillation action causes a lower concentration of ethylene in the liquid. In so doing, oligomerization is retarded, and large excesses of benzene, which accomplishes the same result in other processes, can be avoided. The lower ethylene concentration, however, also retards the main reaction to ethylbenzene and requires more catalyst loading and a standard fixed-bed finishing reactor to complete the alkylation. The first commercial plant began operation in 1994.

Fina/United Catalysts

This is a new process with an improved catalyst.

Process Kinetics

As is the case with other zeolite catalyst processes, product selectivity is significantly influenced by the macroporous structure of the zeolite used. As an example, ZSM-5, (medium pore size) favors the primary alkylation reactions, whereas beta-zeolite (large pore size) favors secondary alkylation reactions.[37] For the medium-pore zeolite, the rate expression of choice is based on a Langmuir–Hinshellwood mechanism.[37]

$$r_A = \frac{k_s C_A K_A C_B K_B}{(1 + C_A K_A + C_B K_B)^2}$$

where k_A = surface reaction rate constant, mol g cat^{-1} s^{-1}
C_A = concentration of ethylene, moles, L^{-1}
C_B = concentration of benzene, moles, L^{-1}
K_A = adsorption constant of ethylene, mol^{-1}L
K_B = adsorption constant of benzene, mol^{-1}L
r_A = rate of alkylation, moles g cat^{-1}s^{-1}

At a high ratio of ethylene-to-benzene in the feed, the equation was found to simplify to the following:[37]

$$r_A = \frac{A_1 C_B}{(A_2 + C_B K_B)^2}$$

where A_1 and A_2 are constants.

Actual practice employs a significant excess of benzene, for which case the following equation might apply:

$$r_A = \frac{A_1' C_A}{(A_2' + K_A C_A)^2}$$

A further simplification using stoichiometric rate equations has been proposed and may be helpful in any effort to model both the main reaction and side reactions.[38]

$$r_1 = k_1(P_B P_E - P_{EB}/K_1)$$

$$r_2 = k_2(P_{EB} P_E - P_{DEB}/K_2)$$

$$r_3 = k_3 P_E$$

where   P = partial pressure
                B, E, EB, DEB = benzene, ethylene, ethylbenzene, and diethylbenzene
                   r_1, r_3, r_3 = rate of formation of EB, DEB, and side reaction of ethylene,
                                respectively

It is important to consider mass transfer in the commercial scale reaction system, especially for liquid phase operation. This issue is most easily accounted for by an effectiveness factor and a liquid-solid mass-transfer coefficient.[39]

2.2.2 BENZENE + PROPYLENE → CUMENE (ISOPROPYLBENZENE)

Cumene is used almost exclusively for production of phenol by the Hock process, which was discovered in 1944 and commercialized in 1953.[4] Cumene is oxidized to cumene hydroperoxide which is then converted under acidic conditions to one mole each of cumene and acetone. As with other alkylations, acid-catalyzed processes are used to produce the cumene.

Chemistry

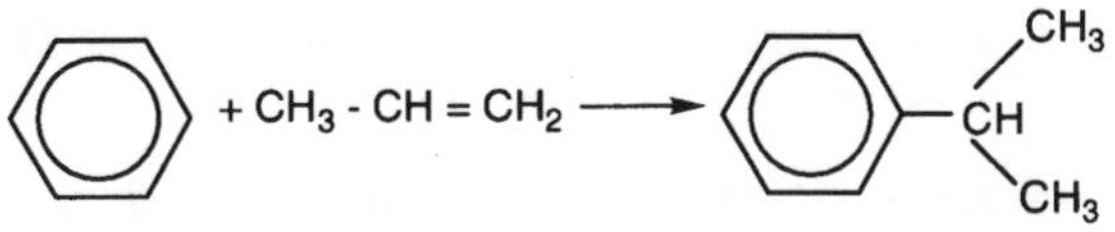

| | benzene | propylene | isopropylbenzene (cumene) |

Temp., K	ΔH, kcal	K_p
400	23.58	19.364×10^4
500	23.14	5.297×10^2

Operating Conditions (see also Table 2.1)

Process	Catalyst	Temperature	Pressure
SPA, UOP LLC (vapor phase)	65–70% H_3PO_4 on silica	200–260°C	30–40 bar
Q-MAX, UOP LLC	beta zeolite	not reported	
MCM-22, Raytheon E&C/Mobil (liquid phase)	mesoporous zeolite	not reported	

Side Reactions

Depending on the process, side reactions occur in varying degrees. Side products such as dipropyl-benzene, oligomers, and heavier alkylbenzenes can be separated and used for high-octane stock, although they reduce yield.

Mechanism

Refer to Figure 2.1, the mechanism for ethylbenzene. In the case of cumene from propylene and benzene, a secondary cation (CH_3CHCH_3) is formed by attack at the double bond, which is chemisorbed at the active site. Reaction with the π cloud of benzene occurs, and deprotonation follows, yielding the desorbed cumene product and the restored active surface.[15]

Catalyst Suppliers and Licensors

See Table 2.1.

Process Units

The catalysts are proprietary, and the processes are licensed. Limited information exists on catalyst details in the open literature.

UOP SPA Process (Solid Phosphoric Acid)

This process actually was originated in the early 1930s by UOP as a means for oligomerizing olefins to product in the gasoline range (polymer gasoline). The process was called the *catalytic condensation process*. It was discovered that, when phosphoric acid and silica were mixed and heated, a solid catalytic material was produced[3] and proved to be effective. During World War II, the need for high-octane components for aviation fuel was critical, and it was known that cumene was a valuable high-octane constituent for aviation gasoline units. Many existing polymer gasoline catalytic-condensation plants were easily converted to the manufacture of cumene by alkylation of benzene with propylene using the same catalyst. Later, when a new demand for cumene developed because of the successful Hock phenol process, many of these plants continued to be used to produce cumene, and others were constructed. The process continues to be the dominant route to cumene. In the early 1990s, 99% of cumene producers used this process. This dominance is slowly declining, as new zeolite catalyst processes have been introduced.

Catalyst Deactivation (SPA Process)   The phosphoric acid is no doubt in the form of a crystalline hydrate. With use of the usual operating temperature, water is removed, and the catalyst begins to disintegrate. Thus, an important aspect of the operation is a continuous controlled addition of water. If this operation is done carefully, catalyst life of 6–12 months is realized.[16] Ultimately, the catalyst disintegrates as noted by an increase in pressure drop.

Coke also can form on the catalyst, but catalyst replacement often is necessary before coking becomes a problem.

Propylene feed must be free of ethylene and other olefins so as to avoid troublesome product impurities when alkylated. Nitrogen compounds poison the catalyst and must be removed by washing.[14] Thiophene can be present in benzene and is alkylated to isopropylthiophene, which has a boiling point close to cumene. Its presence in cumene feed to the phenol process is detrimental, and it must be removed by mineral-acid washing if present in quantities greater than 0.15%.[14]

Reactors and Separation Section (SPA Process)   The reactor consists of four adiabatic beds with means for introducing water between beds for both cold-shot cooling and for rehydrating the catalyst.[14] A supplementary inert-liquid quenching stream can also be used.

Early in the operation of the Hock process for phenol from cumene, it was discovered that oligomer impurities in the cumene suppressed the rate of cumene oxidation. As in other benzene alkylation processes, excess benzene was used to deter the side reactions. However, other means were needed to reduce oligomer formation. In the early 1950s, the reactors were designed for upflow with the addition of an inert dispersant in the propylene-benzene feed.[14] Since the exothermic reaction results in temperature increase in the direction of flow, upflow reduces backmixing and reduces the rate of oligomerization.

The feed propylene may contain propane and some other lighter saturated hydrocarbons, but other olefins must not be present, since they will alkylate benzene and produce unwanted alkyl-benzenes. The separation section following the reactor consists of a flash and rectifying system that removes propane and other light ends. The propane is of good quality for commercial fuel use. A significant amount of excess benzene for recycle is separated via the flash system. The enriched cumene is fed to the benzene column, where the remainder of the benzene is recovered. The cumene from the bottoms is clay treated to remove unwanted heavy hydrocarbons. The clay-treated effluent is then fractionated in a rerun column to separate additional impurities which are close boilers to cumene, and an energy intensive separation is required.[16] The heavies removed can be used as fuel.

The SAP catalyst does not catalyze transalkylation, but a yield of 94–96 wt% is achieved. Many of these plants have a separate transalkylator, which enables a yield of 97–98%.[1] The transalkylator uses a mixed metallic oxide catalyst on silica[14] to convert the ~3% of diisopropyl-benzene to cumene.

UOP Q-MAX Process[6,3]

The Q-MAX process is a relatively new liquid-phase process (1992), with the first plant operational in 1996. A new proprietary zeolite catalyst is used, which is thought to be a beta zeolite. The same catalyst is used for the transalkylation step. Two down-flow reactors in series are used each with two adiabatic beds in series. Propylene is introduced at the inlet of each reactor and between beds. Benzene is fed in modest excess at the inlet of the first reactor.

The separation section is similar to that described for the SAP process, except clay treatment is not required. It consists of a depropanizer, a benzene separation column, and a cumene column, which produces cumene overhead and polysiopropylbenzene (PIB) and heavier aromatics in the bottoms. A final column removes the PIBs (mostly diisopropylbenzenes) overhead and a heavy aromatic fraction, about 1% or less of the overall yield. The overhead is sent to the transalkylator to convert back to cumene. Yields of cumene as high as 99.6 wt% are reported, and cumene purities of 99.97+ wt%.

The catalyst can be regenerated and then only every 18 months. Catalyst life is greater than five years.

Raytheon MCM-22 Process[6,3]

This liquid-phase process uses a modified Mobil catalyst called MCM-22, similar to that used in the ethylbenzene process. It, as in the companion ethylbenzene process, does not catalyze many of the unwanted side reactions such as oligomerization; excess benzene requirements are modest. The separation section is similar to that described for the Q-MAX process. A product of 99.96% purity is claimed.

Dow/Kellogg 3-DDM Process[6,3,17]

This liquid-phase process, like the above previous two, permits low excess benzene feed and exhibits low coking tendencies and high yields with small amounts of heavies. The separation section is similar to that described above for the Q-MAX process. Cumene purity of 99.94 wt% minimum is reported.

Interestingly, some information on the catalyst has been reported.[12] The 3-DDM catalyst is said to be a dealuminated mordenite accomplished in a manner to convert the two-dimensional tubular pores to a controlled three-dimensional structure. The resulting shape selectivity favors cumene. Side reactions mainly form p-diisopropylbenzene and a small amount of m-diisopropylbenzene, both of which are transalkylated to form additional cumene.[12] Only a small heavies purge is produced, as is the case with the other zeolite-type processes.

CD Tech Process

This two-phase process is similar to that described for ethylbenzene.

Process Kinetics

Liquid-phase processes involving proprietary zeolite catalysts have become the most common route to cumene. The alkylation of benzene with isopropylbenzene is a much faster reaction than ethylene alkylation. Mass transfer effects are significant, and the following rate equation combines both mass transfer and reaction, using a simple first-order form for the latter.[39]

$$r_0 = (1/k_L a_p + 1/\eta k_1)^{-1} C_0$$

where a_p = external catalyst surface area/mass

C_o = concentration of olefin

k_1 = intrinsic first-order rate constant

k_L = liquid-solid mass transfer coefficient

r_o = rate of olefin consumption

η = catalyst particle effectiveness factor

This equation is also been applied to the liquid-phase reaction involving ethylene.

Smaller particle size increases a_p and η, which improves the net rates and can be particularly valuable for the cumene case.

2.2.3   BENZENE + LINEAR OLEFIN (C_{10}–C_{14}) → LINEAR ALKYLBENZENE

The long-chain linear alkylbenzenes are used in producing linear alkylbenzenesulfonates, the major uses of which are in household detergents (74%) and industrial cleansers (15%). They are biodegradable and took the place of the highly effective nonlinear alkylbenzenesulfonate which, because of its highly branched chain, was not biodegradable.[15]

Most linear alkylenzenes (LABs) are produced by homogeneous catalysis using either HF or AlCl$_3$, with HF catalyzed alkylation the most frequently used process. But, as in other benzene alkylations, a move toward a heterogeneous catalyst began with the introduction of the UOP Detal process.[1,3,6]

Chemistry (LAB)

$$R - CH = CH - R^1 + \bigcirc \longrightarrow R - CH - CH_2 - R^1$$
$$(C_{10}\text{-}C_{18})$$

The linear olefins are produced by dehydrogenation of n-alkanes over platinum-on-nonacidic alkaline alumina.[17] The monoalkenes thus produced have the internal double bond randomly distributed in the resulting mixture. Some diolefins are also produced and must be partially hydrogenated to mono-olefins. The original linear alkanes are obtained by molecular-sieve separation from hydrocarbon streams also containing isoparaffins and cycloalkanes.

The alkylation process occurs at mild conditions over a heterogeneous catalyst that is not described in the open literature. It can be safely said that the catalyst has acidic characteristics that favor the alkylation reaction as well as simultaneous partial isomerization of olefins. The resulting alkylate will exhibit a statistical distribution of the benzene ring along the paraffinic chain.[4] Hence, the product composition will be the same whether the mixed linear olefin feed is used or a feed composed of only alpha olefins. Since the former olefin mixture is cheaper than alpha olefins, it is used exclusively.

Catalyst Deactivation (LAB)

The alkylation catalyst is said to lose stability in the presence of diolefins and also of aromatics other than benzene that are produced as by-products of the dehydrogenation reaction of the alkane feed. The heavy aromatics thus formed end up as heavies that must be separated from the product linear alkylbenzene and reduce the yield of desired detergent alkylate.[3] The problem is overcome by removing the aromatics formed in the dehydrogenation process prior to passing the alkenes to the alkylation reactor.

Catalyst Types, Suppliers, and Licensors

See Table 2.1.

Process Unit (LAB)

The UOP Detal alkylation process has been described in a general way.[6] The feed to the alkylation reactor consists of a linear paraffin-olefin mixture, since dehydrogenation of paraffin to olefin is not complete, so that undesired secondary reactions are kept to a minimum. This feed is joined by recycle benzene plus fresh benzene and passed downflow in a liquid-phase through a fixed-bed reactor where alkylation occurs. The reactor effluent flows to a benzene column that removes the excess benzene overhead for recycle. The paraffins are removed in the next column and recycled to the dehydrogenation unit. The final column distills the alkylate overhead and heavy alkylate is removed as bottoms.

2.2.4 TOLUENE + ETHYLENE → p-ETHYLTOLUENE → p-METHYLSTYRENE

Poly (p-methylstyrene) has certain characteristics that make it more useful for some applications than polystyrene. These include higher flash point and higher glass-transition temperature. The highly selective alkylation of toluene to p-ethyltoluene (vinyl toluene) is made possible by modified Mobil ZSM-5 zeolite catalyst, which results in a catalyst of lower pore-channel size. These smaller channels permit the para-isomer to diffuse at a rate three times that of the other ethyltoluene isomers, which results in a selectivity of 97%.[17] The process is licensed by Mobil/Raytheon.

Chemistry

toluene   p-ethyltoluene

Temp., K	ΔH, kcal	K_p
400	−25.13	26.06×10^5
500	−25.05	4728
600	−24.49	70.96

Process Units

This process is an interesting example of the power of shape-selective catalysts to produce astounding results. Because poly (p-methylstyrene) had characteristics that could propel it into major large-scale use, great commercial interest was aroused in the early 1980s. It was fueled further by the lower cost of toluene than benzene. Unfortunately, the dehydrogenation process turned out to be more expensive to operate, including higher energy costs than for styrene. Yields were also lower.[5] The motivation to make p-methylstyrene on a large scale subsided. It is now produced on a small scale for specialty polymers.[5]

2.3 ALKYLATION OF PHENOLS

Phenol is more easily alkylated than benzene, because the OH group on phenol is a strong donor electron that makes alkylation of phenol possible at relatively mild conditions. Alkylated phenols are used as reactants in the production of various resins, insecticides, herbicides, and many other synthetic products.[1]

2.3.1 PHENOL + METHANOL → CRESOL + XYLENOL

Cresols and xylenols have been recovered from oil-refinery spent caustic streams produced in sweetening processes used to remove mercaptans contained in thermal and catalytic cracked distillates and from coal tar.[21] The crude cresylic acid recovered from oil refinery streams consists of 20% phenol, 18% o-cresol, 30% m and p-cresol, and 22% xylenols. That from coal tar includes 45% phenol, 35% cresols, and 15% xylenols. In recent years, the supply from coal tar exceeds that from refinery cracked distillate. Since the late 1960s, however, the demand has exceeded the supply from refinery waste streams and coal tar, both because of the progressive decline in the quantity of these streams and the increased demand for cresols and xylenols.

The various cresols serve as important precursors to valuable products.[22] Ortho-cresol is used in producing herbicides and insecticides, epoxy-o-cresol novolak resins as sealing for integrated circuits, dye intermediates, antiseptic, and fragrance by alkylation with propylene, antioxidants, and directly as a valuable solvent. Meta-cresol is used in producing insecticides, fragrances and flavors, perfume fixative, antioxidants for polyethylene and polypropylene, and disinfectants and preservatives. Para-cresol is used in producing BHT (a nonstaining light-resistant antioxidant) as well as other antioxidants and fragrances. Mixtures of cresols are used as solvents for synthetic resin coatings and as disinfectants and wood preservatives. Finally, methylcyclohexanol and hexanone produced by hydrogenation of cresols are used in the paint and textile industries.[22]

Xylenols also have many uses.[22] The 2,6 xylenol is used in producing polyphenylene oxide resins, which are noted for high impact resistance, thermal stability, and fire resistance. Other significant uses are in the production of polycarbonates, insecticides, and antioxidants. The 3,4 and 3,5 xylenols are intermediates in the production of a number of different insecticides, disinfectants, and preservatives. Xylenol mixtures rich in 3,5 isomer are used in the production of xylenol-formaldehyde resins.

Although there are other synthetic routes to cresol and xylenols, the most prevalent procedure is by methylation of phenol with methanol, for which there are several processes.

2.3.2 BASE-CATALYST PROCESS (VAPOR PHASE)

The catalyst is MgO promoted with other oxides such as Mn, Cu, Ti, U, and Cr. The MgO vapor-phase process is the process of choice for producing essentially pure o-cresol or 2,6 xylenol in the liquid product exiting the reactor.

Chemistry

phenol methanol   o-cresol

ΔH @ 700°K = −16.4 kcal

K_p @ 700°K = 1.46×10^5

phenol methanol   2,6 xylenol

Side Reactions	ΔH @ 700 K, kcal
$CH_3OH \rightarrow CO + 2H_2$	24.47
$CO + H_2O \rightarrow CO_2 + H_2$	−9.05
$CO + 2H_2 \rightarrow CH_4 + H_2O$	−52.70

Mechanism

Base catalyzed (MgO) alkylations are said to occur in the case of alkylation of phenol with methanol by dissociation of phenol into a phenoxide and a proton on the catalyst surface. The proton attaches to the adsorbed methanol to form a methyl cation or formate, which then reacts with phenol to produce o-cresol and 2,6 xylenol. It has been suggested that 2,6 xylenol is favored, because the spread between the 2 and 6 positions is similar to that between the sites of the methyl cations.

Experimental evidence strongly suggests that the reactions to o-cresol and 2,6 xylenol occur in parallel rather than consecutively.[23] Water as steam is added in the ratio of 1−2 moles per mole of phenol to reduce coke formation. A higher ratio (1:5−6) serves to favor 2,6 xylenol over o-cresol.[21] This situation can possibly be explained by the fact that both water in the hydroxyl form (OH), o-cresol, and methanol adsorb on metal ions.[23] It is possible that a high amount of adsorbed water hinders the formation of o-cresol but not that of 2,6 xylenol. In fact, this increased preference for the xylenol is exactly what occurs and thus makes the assumption of parallel reactions plausible. If the reactions were consecutive, both products would be hindered to some degree. By competing with methanol, water hinders the excessive decomposition of methanol into CO and hydrogen.

Catalyst Type

The process is proprietary. In the open literature, the catalyst is described as MgO combined with other metal oxides as promoters in small amounts of other oxides such as Cu, Ti, U, Zr, Cr, and silicon.[23] For example, 17% of Cu increases the yield of 2,6 xylenol.[22] The catalyst is probably produced in the form of pellets.

The licensors and owners are General Electric, Croda Synthetic Chemicals, and Nippon Cresol.

Deactivation

Coke formation is the major source of catalyst deactivation, but activity can be restored by careful burning of the coke.

Process Unit

Phenol, methanol, and water are heated and passed as a vapor mixture through a multitubular heat-transfer reactor. Hot oil or salt melts are required as the heat-transfer fluid on the shell side because of the high operating temperature (420−460°C). Operating pressure is slightly above atmospheric.

A typical product at 440°C and a feed of 1:4:2.4 (phenol, methanol, water) and a phenol conversion of 82% is reported as 69.5%, 2,6 xylenol, 23% o-cresol, 2,4 xylenol, and 6.6% 2,4,6 trimethylphenol. If a higher ratio of methanol-to-phenol (1:5−6) is used and recycle of o-cresol, a yield of 95% of 2,6 xylenol based on phenol can be obtained.[22]

Since 2,6 xylenol and o-cresol have different boiling points, separation of these products is readily accomplished by vacuum distillation. Unreacted methanol is readily removed by a simple distillation and recycled. Unreacted phenol must be separated from water and then recycled. About 1−2 moles of methanol feed per mole of xylenol formed is lost to decomposition via side reactions.

Process Kinetics

Because of the excess methanol in the feed, first-order rate constants have been used with some success and convenience.

2.3.3 ACID-CATALYST PROCESS (γ-AL$_2$O$_3$)

A solid acidic oxide such as γ-Al$_2$O$_3$ catalyzes the methylation of phenol in a vapor-phase operation. Operating conditions of 300–320°C and a feed ratio of 1 part phenol to 2 parts methanol produce a product of 82% o-cresol, 17% 2,6 xylenol, and 1% m/p-cresol.[22] A higher 2,6 xylenol-to-o-cresol can be produced by recycling o-cresol and raising the methanol/phenol ratio. If the temperature is increased significantly, phenol conversion is increased, but selectivity declines. Other products begin to dominate (m, p-cresol, other xylenols and unwanted side-reaction products).

There are several advantages and disadvantages to this process.[22]

Advantages	Disadvantages
Catalyst does not need as frequent regenerations.	Operations above 60% conversion produces di-, tri-, and tetramethyl phenols.
Longer catalyst life.	Low conversion (60% max.) requires more recycle, which means larger equipment and higher energy costs.
Methanol does not decompose into CO and H$_2$.	Presence of p-cresol in product requires a special complex distillation to separate the p-xylene impurity from the 2, 6 xylenol.

Apparently, the disadvantages outweigh the advantages, for the process is no longer used.

Other solid acid catalysts with strong acid sites have been suggested such as silica-alumina, zeolites, and phosphoric-acid-Kieselguhr. The strong acidity adds isomerization and disproportion activity which favors high m and para-cresol mixtures at high temperatures (450°C). At lower temperature (250–300°C), an m/p cresol mix is favored with high para content, 85% when a shape-selective zeolite (HKY-type) is used.[22]

2.3.4 META/PARA MIXTURES

Most meta and para xylenes are obtained as mixtures from natural sources, coal tar, and catalytic-cracked distillates. Ortho cresol is readily separated from the m-p isomers by distillation. But the separation of meta and para-cresol requires heroic efforts because of their very close boiling points. These include selective adsorption of p-cresol, fractional crystallization, or chemical manipulation via additional compounds. The results are very good. For example, the adsorption process can produce streams of p and m-cresol, each with purities of 99% or more.

2.4 HIGHER ALKYLPHENOLS

Alkylphenols with alkyl chains of three or more carbon atoms are produced by alkylation of phenol with the appropriate olefin. Many of the products are used in producing smaller volume special chemicals, but some are made in large-tonnage continuous processes with a heterogeneous catalyst.

2.4.1 PHENOL + ISOBUTYLENE → P-TERT-BUTYLPHENOL

p-tert-butylphenol (TBP) is used as an antioxidant. Phosphoric acid esters of TBP are used as UV stabilizers, as a chain regulator for polycarbonate resin production, and in the production of phenolic resins (large-scale use).

Chemistry

$$\text{phenol} \quad \text{isobutylene} \quad \text{p–tert–butylphenol}$$

$\Delta H = 19.1 \text{ kcal}$

Catalyst Type and Suppliers

Acid catalysts are used.[1] For the vapor-phase solid catalyst process, zeolites, activated clays, or strongly acidic ion-exchange resins can serve as catalysts. An ion-exchange catalyst (dehydrated sulfonated polystyrene-polyvinyl benzene) operation has been described.[22] The catalyst is produced in the form of beads (0.3–1.5 mm). Suppliers include Bayer, Dow Chemical, Grace Davison, Rohm & Haas, and United Catalysts.

Process Unit

Two adiabatic reactors in series are used. The first reactor operates at 90–100°C. A recycle of phenol is fed along with fresh feed to produce a ratio of 1.4–1 phenol to isobutylene. Part of the effluent from Reactor 1 is recycled via a cooler to maintain the temperature. The net effluent from Reaction 1, which includes unreacted phenol p-tertbutylphenol and side reaction products (2-tert-butylphenol and 2,4 ditertbutylphenol, is fed preheated (120°C) to the second reactor held at 120°C, where isomerization and transalkylation occur essentially isothermally. Temperatures above 140°C will destroy the catalyst efficiency. Most of the side-reaction products are reacted to additional p-tert-butylphenol (PTBP). Distillation columns follow to produce >98% of the desired product (PTBP). The yield based on net phenol fed is 95%.

Other processes have been described using phosphoric-acid impregnated Kieselguhr.[1] Only one reactor is used, but a higher ratio of phenol-to-olefin aids in maintaining the required operating temperature. Both a heater and a cooler are used on the inlet feed line to fine tune the reaction temperature.

Process Kinetics

The ion-exchange resin catalyzed production of 4-tert-octylphenol via phenol and diisobutylene has been extensively studied.[40] The rate of reaction was found to be first order in both phenol and diisobutylene. It is reasonable to suggest that the same might be true for the reaction of isobutylene with phenol.

2.4.2 Other Higher Alkylphenols

Two other large-volume alkylphenols synthesized by heterogeneous catalysts are summarized in Table 2.2.

2.5 ALKYLATION OF POLYNUCLEAR AROMATICS

Polynuclear aromatics, when alkylated, will yield a larger variety of isomers than a single aromatic molecule, because polynuclear aromatics have more positions on which to attach the alkyl group. This situation is a challenge for zeolite catalyst research.

TABLE 2.2
Other Higher Alkylphenols

Product	Reactants	Process	Major Use
p-nonylphenol	Phenol, nonane	• Catalyst: macroporous polystyrene-polydivinylbenzene ion-exchange resin. • Suppliers: Dow, Rohm & Haas, Bayer • Two reactors in series (70°C in, 120°C out for Reactor 1, 100 to 115°C for Reactor 2). • Product purity after distillation, 98%+ • Yield 45%	Production of nonionic surfactants
p-tert-octylphenol	Phenol, diisobutene	• Catalyst: macroporous polystyrene-polydivinylbenzene ion exchange resin. • Suppliers: Dow, Rohm & Haas, Bayer • Reactor — 100–105°C by use of cooling coils • Catalyst kept in hydrated form by adding water (1–2%) to feed to decrease catalyst activity, thereby controlling temperature. • Diisobutene conversion: 95%	Production of nonionic surfactants used in emulsion polymorization of acrylic and vinyl monomers

Source: Based on information from Refs. 1 and 22.

An excellent review on the alkylation of biphenyl and naphthalene using various zeolites such as HM (mordenite), HY (faujasite), and HZSM-5 (pentasil) provides some useful general rules for developing an alkylation process based on zeolite catalysis.[10]

- Minimization of steric restrictions of transition states composed of polynuclear aromatics, alkylating agent, and acid sites in the zeolite pore. For example, acid strength can be reduced by dealuminization.
- Prevention of reaction at the external surface.
- Control of number of acid sites inside the pore.
- Control of acid strength.

Zeolite catalyst are showing promise and should be followed for possible commercial applications which now, in many cases, involve homogeneous acid catalysts.

2.6   ALKYLATION OF AROMATIC AMINES

Many alkylated aromatic amines are produced via homogeneous catalysts (aluminum anilides or Friedal-Crafts alkylation). But it has been demonstrated[24] that a number of aryl amines can be synthesized using solid-acid catalysts such as large-pore zeolites (HY zeolite). One important example is the production of 5-tert-butyl, 2,4-toluenediamine for which a selectivity of 85% at a conversion of 84% 2,4-toluenediamine is realized. This product is already used in a commercial mixture of 2,4 and 2,6 isomers in performance-polymer applications. It could also be valuable as a replacement for the more reactive diethyl m-toluenediamine in reaction injection molding as a chain extender in the molding of complex auto body parts.[24] The essentially pure and less active 5-tert-butyl, 2,4, toluendiamine (TBTD) would allow more time for filling complex molds. Other applications for a relatively pure TBTD could develop and commercial production become attractive.

2.7 TRANSALKYLATION AND DISPROPORTIONATION

2.7.1 TOLUENE → XYLENE + BENZENE OR TOLUENE + TRIMETHYLBENZENE → XYLENE

The primary source of xylenes is catalytic reformate from which benzene, toluene, xylene, and smaller amounts of ethylbenzene are recovered by extractive distillation or an extraction process augmented by distillation. A C_8 cut composed of ortho, meta, and para xylene and ethylbenzene is subjected to super fractionation, which separates the o-xylene (normal bp 14°C) from the meta (normal bp 139.1°C) and p-xylene (normal bp 138.5). Other means such as fractional crystallization or selective adsorption must be used to separate p-xylene from m-xylene. The remaining raffinate now composed mainly of meta xylene is isomerized to produce additional ortho and para-xylene for separation.

Para-xylene is used mainly to produce terephthalic acid and dimethyl terephthalate, both of which are used in making polyethylene terephthalate an important polymer for fibers, films, and resins. Ortho-xylene oxidation is a major route to phthalic anhydride, a widely used plasticizer. A significant portion of the meta-xylene is recycled to the isomerization unit, since its use in producing isophthalic acid and isophthalonitrile is small in comparison to the m-xylene available. Isophthalic acid is a raw material for unsaturated strong polyester resins. Also, a fungicide is made beginning with isophthalonitrile[15,18].

Direct routes to xylenes or essentially pure p-xylene via disproportionation or transalkylation of toluene are offered by a number of licensors. Initially, noble-metal catalysts supported on alumina or rare earth catalysts were used to produce equilibrium xylene mixtures from toluene. More recently, several processes that selectively form p-xylene[17,18] have been introduced. Although toluene is often cheaper than benzene, there remains the fact that the major source of xylenes is a part of the catalytic reformer product. Much innovation in separation and recovery of pure isomers continues to be focused in that area where selective zeolite adsorbents are playing an increasing role.

Chemistry

toluene   benzene   xylenes   toluene   trimethylbenzene   xylenes

Disproportionation   Transalkylation

Thermodynamics (p-xylene as an example)

Temp., K	ΔH, kcal	K_p
Disproportionation:		
600	0	7.551×10^{-2}
700	−0.11	7.481×10^{-2}
800	−0.78	7.278×10^{-2}
Transalkylation:		
600	−0.35	0.2318
700	−0.49	0.2193
800	−0.82	0.2100

Since the thermodynamics are not very favorable, toluene must be recycled to have a high ratio of toluene to products. Temperature has essentially no effect on the equilibrium constant, and, because no change in moles occurs, pressure has no effect on equilibrium composition. Both of these variables do affect the kinetics.

Of course, a mixture of the three xylene isomers is formed by such a process, and separation of the isomers must be practiced. This situation was an opportunity for a zeolite catalyst. Zeolites ultimately were used as a vehicle to separate the isomers, since a properly selected pore size could preferentially admit p-xylene over the other isomers. Could a zeolite catalyst do both transalkylation and isomerization and preferentially discharge p-xylene? This question was ultimately answered in the affirmative.

Process Units

Several processes were developed over a period of years to take advantage of the lower cost of toluene. All this development took place during a period of major contributions to the separation of each isomeric xylene from mixed xylenes in reformate, the dominant source.

Toray/UOP Tatoray Process[3,17,18,19]

This process was implemented in 1969 by the inventor, Toray. UOP assisted in the development and became the licensor. Toluene or a mixture of toluene and trimethylbenzene is fed together with H_2 at a ratio of 5–12:1 H_2/toluene to a preheat furnace and then to a single fixed-bed reactor where the vapor-phase reaction takes place. No intermediate cooling is required, because the main reaction as shown above has a very low enthalpy of reaction. The high hydrogen content in the feed is effective in preventing rapid coke formation.

The catalyst in the form of extrudates is said to be a noble metal-on-alumina or zeolite, or a rare-earth catalyst. Various improved catalyst formulations have been developed. Suppliers are UOP LLC or Johnson Matthey (noble metal). The current one is designated as TA-4, which is not informative concerning the catalyst characteristics. Operating conditions are reported to be 350–530°C at 20–50 bar. The xylene wt% product distribution is 23–25% p-xylene, 50–55% meta-xylene, and 23–25% ortho-xylene.

Effluent from the reactor is cooled and separated into light gases (mainly H_2), which are recycled and liquid product. The liquid is then processed in a stripper, which removes C_5 and lower hydrocarbons in the overhead and produces a benzene, toluene, xylene, trimethylbenzene product that is further separated into the products and recycle toluene, usually in existing towers in the aromatics complex.

When only methyl substituted aromatics are involved, the equilibrium conditions can be calculated readily as shown in Figure 2.3 at 700 K. The optimum conversion is 46 to 48 wt% per pass. Although a higher level up to 59% is possible, the catalyst suffers deactivation above 50%, and excessive production of heavy-aromatic by-products occurs.

The newer catalyst is reported to require regeneration only after a year of operation, and, with careful regeneration, catalyst life can extend to more than two years.

Catalyst Deactivation   In addition to coke formation, saturated light ends can deactivate the catalyst, since they cause cracking reactions and deactivating side products. In addition, the following should be avoided:

- water (depresses transalkylation)
- olefins (coke precursors)
- chloride (causes aromatics cracking)
- nitrogen compounds (poison active sites)

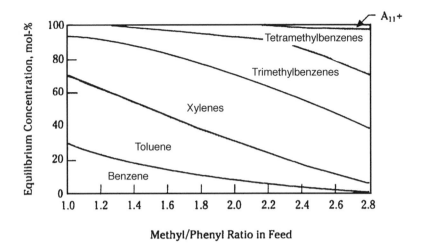

FIGURE 2.3   Equilibrium Distribution of Methyl Substituted Aromatics at 700 K. Reprinted by permission: Jeannert, J. J., *Handbook of Petroleum Refining Processes*, 2nd edition, R. A. Meyers, editor, p. 2.57, © McGraw-Hill Companies, Inc., New York, 1996.

Regeneration   The new catalyst is reported to require regeneration no more than once a year by careful burning. With care in regeneration and avoidance of catalyst poisons, catalyst life will extend beyond the original two years reported in 1970.

Lyondell/HRI/IFP Xylene-Plus Process[3,18]

This vapor-phase process was developed by ARCO and licensed by Lyondell/HRI/IFP. It also operates with toluene or toluene/trimethylenzene. The catalyst is said to be a non-noble metal and is supplied by IFP. The reactor consists of a moving bed with continuous regeneration, and thus no H_2 is required. Toluene conversion is reported as 30% per pass. Product xylene wt% distribution is, 26 p-xylene, 50 meta-xylene, and 24 ortho-xylene. Recovery of product and unreacted feed recycle is similar to the previous description except without hydrogen.

Mobil Processes

The Mobil Technology Company, in the 1970s, developed the first process that used a zeolite catalyst and produced a high yield of p-xylene by virtue of the shape selectivity of the pore structure. The licensor is Mobil/Raytheon. As shown in Figure 2.2, the pore admits the toluene, the isomers are formed, but only the para isomer can escape. Isomerization of these molecules left behind creates additional p-xylene. The first process was a liquid-phase process. This process was followed by several improved processes that ultimately increased the selectivity to p-xylene. To accomplish this feat, the ZSM-5 catalyst was precoked with toluene to create effectively smaller pore sizes. The result is a relative diffusivity of p-xylene much higher than the other isomers, creating thereby a selectivity of 80–90% at 25–30% toluene conversion. The xylene product is reported to have a purity of 95% p-xylene. Operating conditions are 400–470°C and 20–35 bar. Hydrogen to inhibit additional rapid coking is fed at a 1–3:1 ratio of H_2 to hydrocarbon. A number of units are in operation with the designation MSTDP.

A further improved process named MTPX was introduced in 1995. In this case, the optimum catalyst selective pore structure was created in the catalyst manufacturing. One reference[18] speculates that organosilicates were used to develop the desired pore structure. Mobil built a unit at its Beaumont refinery but was not licensing the process as of this writing.

A Disproportionation Process

ABB Lummus Global licenses a disproportionation process for alkenes using a non-noble metal on silica as pellets. The catalyst is supplied by Engelhard.

Process Kinetics

The reaction system is complex and also involves isomerization. Some simplification of the kinetics is useful, and researchers have used rate expressions with reactants and products each expressed to the power of the stoichiometry.[41] Thus, for disproportionation,

$$r_1 = k_1 \left(P_T^2 - \frac{P_B P_X}{K_1} \right)$$

and transalkylation,

$$r_2 = k_2 \left(P_T P_{TR} - \frac{P_X}{K_2} \right)$$

where subscripts T, B, X, and TR refer to toluene, benzene, xylenes, and trimethylbenzene, respectively.

2.8   LOWER ALIPHATIC AMINES BY ALKYLATION OF AMMONIA (AMINATION)

The lower aliphatic amines, methyl through butyl, are very reactive compounds and are involved in the synthesis of a wide variety of commercially valuable compounds. Methylamines are produced in the largest amount, with dimethylamine dominating because of its use in the production of the important and effective solvents, N, N, dimethylacetamide and N, N, dimethylformamide.[4] Trimethylamine the third isomer has only a few uses. Ethyl, propyl and butylamines are produced in lesser quantities than the aggregate of methylamines.

2.8.1   METHANOL + AMMONIA → METHYLAMINES

The major route to methylamines is the heterogeneously catalyzed alkylation of ammonia by methanol.

Chemistry

Alkylation is conducted in the vapor phase within the range of 300–425°C and 6–30 bar, depending somewhat on plant pressure economics, since reaction equilibrium is not affected by total pressure.

	Temp., K	ΔH, kcal	K_p
1. Monomethylamine			
$NH_3 + CH_3OH \rightarrow CH_3NH_2 + H_2O$	600	−3.52	36.308
	700	−3.465	25.23
2. Dimethylamine			
$CH_3NH_2 + CH_3OH \rightarrow (CH_3)_2NH + H_2O$	600	−8.03	158.49
	700	−7.91	64.269

	Temp., K	ΔH, kcal	K_p
3. Trimethylamine			
$(CH_3)_2NH + CH_3OH \rightarrow (CH_3)_3N + H_2O$	600	−18.03	100.69
	700	−9.59	35.975
Disproportionation reactions also occur[25]			
4. $CH_3NH_2 + CH_3NH_2 \rightarrow (CH_3)_3NH + NH_3$	600	−4.01	4.365
	700	−3.42	3.729
5. $CH_3NH_2 + (CH_3)_2NH \rightarrow (CH_3)_3N + NH_3$	600	−6.0	5.164
	700	−5.84	2.535
6. $(CH_3)_2NH + (CH_3)_2NH \rightarrow (CH_3)_3N + CH_3NH_2$	600	−1.99	1.183
	700	−1.98	0.929

Thermodynamics

The six reactions essentially reach equilibrium under commercial operating conditions and suggest a tendency to produce trimethylamine because of Reactions 4, 5, and 6, which convert the other amines to trimethylamine. Since, as noted previously, trimethylamine has only minimal commercial value, the use of excess ammonia to reverse Reactions 4 and 5 is a useful strategy. Reaction 6 will also reverse as high trimethylamine content occurs.

Simultaneous solution of these six equations yields a useful quantitative picture as shown in Figure 2.4, which plots the weight percent of the reacting components versus nitrogen to carbon ratio in the feed at several temperatures.[25] The other common procedure for favoring desired products is to recycle unwanted excess products such as trimethylamine to the reactor. Both of these procedures are practiced.

Mechanism

Because silica-alumina is well known as a dehydration catalyst, there is some suggestion that dimethylether is an intermediate in the reaction system.[25–27]

$$2CH_3OH \leftrightarrows CH_3\text{-}O\text{-}CH_3 + H_2O$$

Dimethylamine can be formed from the dimethylether as well as from methanol.

$$CH_3\text{-}O\text{-}CH_3 + NH_3 \leftrightarrows (CH_3)_2NH + H_2O$$

The Brønsted sites would certainly be the sites for both ammonia and amine chemisorption, and the distribution and strength of these sites is important to the ultimate activity of the catalyst. Promoters that are proprietary probably are important in adjusting acid strength.

Catalyst Type and Suppliers

The catalyst is an amorphous silica-alumina reported to be promoted by molybdenum sulfide and silver phosphate.[28]

Suppliers: Engelhard, Grace Davison, Akso Nobel
Licensors: Acid-Amine Technologies, Inc.

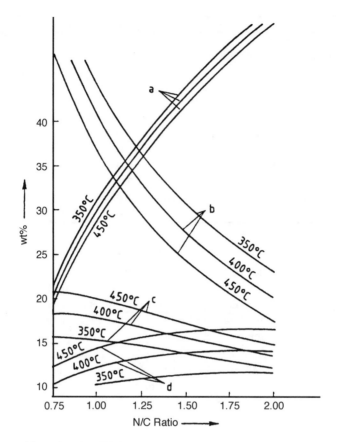

FIGURE 2.4 Composition of an Equilibrium Mixture of Methylamines. N/C = nitrogen/carbon ratio. a) ammonia, b) trimethylamine, c) dimethylamine, d) monomethylamine. Reprinted by permission: Van Gysel, A. B. and Musin, W., in *Ullmann's Encyclopedia of Industrial Chemistry,* 5th ed., Vol. A16, p. 535, Wiley/VCH, Weinheim, Germany, 1984.

Catalyst Deactivation

Catalyst life is reported to be 2–3 years, which suggests slow loss of surface area may be the only significant deactivating process. Silica-alumina with its active Brønsted acid sites will catalyze coke formation at high temperatures (above 450°C) due to poorly monitored operations. Coke will, of course, cause significant activity loss and must be removed by careful burning.

Process Units

A mixture of ammonia and methanol in a ratio of two-six moles of ammonia to methanol is fed to preheater exchangers along with excess recycle products (e.g., mono-and trimethylamine). Reactants as vapor then pass to a fixed-bed adiabatic reactor. Excess ammonia serves to moderate temperature rise, and most of the exothermic heat generated is recovered by exchanging reactor effluent with reactor feed.

The reactor operates in a temperature range of 300–425°C and a pressure of 6–30 bar, depending on plant economics. Although large recycle rates are required, the availability of ammonia at elevated pressure makes it possible to reduce equipment size in the design stage.

Separation of light gas H_2 and CO is followed by a series of distillation columns that would impress a separations expert. Azeotropic and extractive distillation combine with regular distillation columns to produce very pure products in the range of 99.6%+.[6] All three isomers are separated, and unwanted excess of any one of them can be recycled so that more of desired products can be obtained.

Process Kinetics

Simultaneous equilibrium calculations provide the best means for predicting product composition. If rate equations are desired for calculating temperature and conversion along the bed length, assumption of second-order rate equations have proved useful with each component partial pressure to the power of one.[27] In a study using Na-mordenite as the catalyst, the rate was first-order in NH_3 and zero order in methanol.[26]

2.8.2   OTHER METHYLAMINE PROCESSES

The value in developing a selective catalyst has been recognized for some time. The obvious choice is a shape-selective zeolite. A number of studies have been reported, most with the goal of limiting trimethylamine (TMA) production by preventing access of TMA through the selection of an appropriate pore size.[28–30] A La-exchanged mordenite limited the production of trimethylamine production to 10% of the total amines formed. The pore size, 0.39 nm, was larger than all but the trimethylbenzene. A modified ZSM-5 catalyst with pore openings smaller than the trimethylamine gave a selectivity of 98% operating at 300–400°C. At this writing, no shape-selective successful commercial catalytic process has been developed for methylamines. Although the zeolite is more active than the amorphous silica-alumina catalyst and can operate at lower temperature, coke formation in the pores has been a problem in some tests.

2.8.3   OTHER LOWER-ALKYLAMINES PROCESSES

Although there are many possible lower alkylamines other than the methylamines, most of the important ones of these are the $C_2–C_4$ alkylamines. These have a wide variety of uses examples of which are most impressive for mono- and di-methylamine (insecticides, fungicides, pesticides, photo-developer, solvents, nonionic detergents, ion-exchange resin, automotive coating, catalysts, epoxy-resin accelerator).

In addition to the alkylation of ammonia with an alcohol as described for methylamines using an acid catalyst (amorphous silica-alumina), alkylation can also be accomplished by hydrogenation catalysts under reducing conditions. Again, the reaction is conducted in a fixed-bed reactor containing a catalyst such as supported cobalt, nickel, copper, or copper chromites. Operating conditions for methanol are in the range 130–250°C, which is somewhat lower than for the acid-catalyst process, but hydrogen must be continuously passed over the catalyst at a ratio 2,5:1 of H_2-to-methanol feed to prevent catalyst deactivation due to the formation of nitride and carbon deposits on the catalyst.[32,33] In addition, hydrogen has been shown to prevent disproportionation reactions at sufficient partial pressure, and thereby assures high selectivity to the desired amines.[34] The reactant alcohol also suppresses disproportionation.[35]

Separation operations are similar to the acid-catalyzed process and unwanted isomers are recycled as required for both processes.

Mechanism

A mechanism has been proposed involving (I) a carbonyl intermediate formed by dehydrogenation of the alcohol, (II) reactions between the adsorbed carbonyl and the amine, (III) release of water, and (IV) hydrogenation of the resulting enamine species.[35] See Figure 2.5. Various alcohols can be used (e.g., C_1–C_5), as well as various aliphatic amines in the place of ammonia.

Catalyst Suppliers and Licensors

Suppliers: Engelhard, United Catalysts, Celanese (was Hoechst)
Licensor: Acid-Amine Technologies

A Zeolite Process

A successful process is now in operation using a modified pentasil zeolite, a boron-silicate, to produce tert-butylamine.[31] This amine is difficult to produce via the alcohol and has usually been manufactured by the addition of HCN to 2-methylpropene in an acid environment. The process is effective but is corrosive and requires a great deal of effort and expense in providing corrosion-resistant equipment and safe facilities for handling HCN.[32] By contrast, the zeolite process does not present severe environmental or safety problems.

Process Kinetics

Based on isotope labeling experiments, an assumption was made of a rate determining step involving the abstraction of an α-hydrogen atom and transfer to an active site.[35] The resulting rate equation is based on a single-site Langmuir adsorption.

$$r = \frac{kK_{ROH}P_{ROH}}{(1 + \Sigma K_i P_i)^2}$$

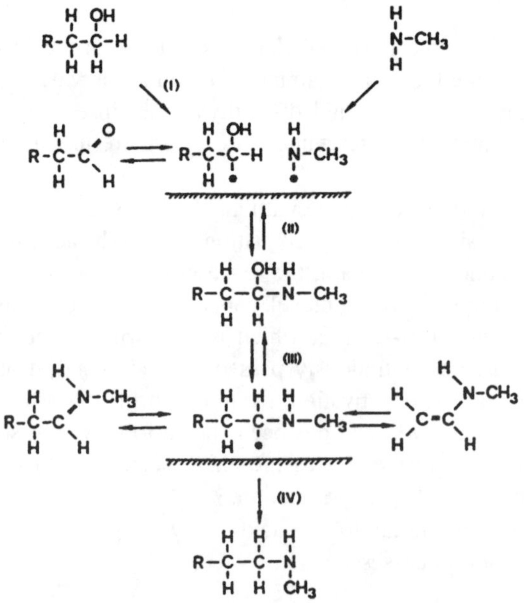

FIGURE 2.5 Reaction Pathway for Formation of a Secondary Aliphatic Amine from an Aliphatic Alcohol and a Primary Amine. Reprinted from Baiker, A., in *Heterogeneous Catalysis and Fine Chemicals: Studies in Surface Science and Catalysis,* Vol. 41, p. 283, copyright 1988, with permission from Elsevier Science.

where K_i = adsorption equilibrium constant for species I

P_i = partial pressure for species I

K_{ROH} = adsorption equilibrium constant of alcohol

P_{ROH} = partial pressure for alcohol

REFERENCES (ALKYLATION)

1. Hammershaimr, H. U.; Imal, T.; Thompson, G. J.; and Vera, B. V. in *Kirk–Othmer Encyclopedia of Chemical Technology,* 4th edition, Vol. 2, p. 85, Wiley, New York, 1992.
2. Wojciechowski, B. W. and Corma, A., *Catalytic Cracking: Catalysts, Chemistry and Kinetics*, Marcel Dekker, New York, 1986.
3. Meyers, R. A., *Handbook of Petroleum Refining Processes*, 2nd edition, McGraw-Hill, New York, 1996.
4. Weissermel, K. and Arpe, H. I., *Industrial Organic Chemistry,* 3rd edition, VCH, New York, 1997.
5. Chen, S. S., in *Kirk–Othmer Encyclopedia of Chemical Technology*, 4th edition, Vol. 22, p. 956, Wiley, New York, 1997.
6. Yearly *Petrochemical Handbook*, Hydrocarbon Processing, March, 1991–1997.
7. Sugi, Y. and Kubota, Y., in *Catalysis: Specialists Periodical Reports,* Vol. 13, p. 55, Royal Society of Chemistry, London, 1997.
8. Chen, N. Y.; Degnan, T. F., Jr. and Smith, C. M., *Molecular Transport and Reaction in Zeolites*, VCH, New York, 1994.
9. Stiles, A. B. and Koch, T. A., *Catalyst Manufacture*, 2nd edition, Marcel Dekker, New York, 1995.
10. Chen, N. Y.; Garwood, W. E. and Dwyer, F. G., *Shape Selective Catalysis in Industrial Applications*, Marcel Dekker, New York, 1989.
11. Coty, R.R.; Welch, V.A.; Ram, S. and Singer, J. in *Ullmann's Encyclopedia of Industrial Chemistry* 5th edition, Vol. A10 p. 35, VCH, New York, 1987.
12. Dwyer, F. G. and Lewis, P. L., in *Encyclopedia of Chemical Processing and Design*, J. J. McKetta, editor, Vol. 20, p, 82, Marcel Dekker, New York,1984.
13. Naccache, C. in *Deactivation and Poisoning of Catalysts*, J. Ondar and H. Wise, editors, Marcel Dekker, New York, 1985.
14. Brayford, D. in *Encyclopedia of Chemical Processing and Design*, J. J. McKetta, editor, Vol. 14, p. 39, Marcel Dekker, New York, 1982.
15. Chenier, P. J., *Survey of Industrial Chemistry*, 2nd edition, VCH, New York, 1992.
16. Canfield, R. C. and Unruh, T. L., *Chem. Engineering*, p. 32, March 21, 1983.
17. Olah, G. A. and Molnar, A., *Hydrocarbon Chemistry*, Wiley, New York, 1995.
18. Cannella, W. J., in *Kirk–Othmer Encyclopedia of Chemical Technology: Supplement*, 4th edition, p. 831, Wiley-Interscience, New York, 1998.
19. Otani, S., *Chem. Eng.,* p. 118, July 27, 1970.
20. Fiege, H. and Heinz-Werner, V. in *Ullmann's Encyclopedia of Industrial Chemistry*, Vol. A19, p. 313, VCH, New York, 1991.
21. Clonts, K. E. and McKetta, R. A. in *Encyclopedia of Chemical Processing and Design*, J. J. McKetta, editor, Vol. 13, p. 212, Marcel Dekker, New York, 1981.
22. Fiege, H., in *Ullmann's Encyclopedia of Industrial Chemistry*, 5th edition, Vol. A8, p. 26, VCH, New York, 1987.
23. Knyazeva, E. M., Khasanova, N. S., Koval, L. M. and Sudakova, N. N., *Russian J. of Phys. Chem.*, 58 (4), 592 (l984).
24. Burgoyne, W. F.; Dixon, D. D. and Case, J. P., *CHEMTECH*, p. 690, November, 1980.
25. Van Gysel, A. B. and Musin, W. in *Ullmann's Encyclopedia of Industrial Chemistry*, 5th edition, Vol. A16, p. 535, VCH, New York 1984.
26. Weigert, F. J., *J. Catal*, 103, 20 (1987).
27. Keane, Jr., M.; Sonnichsen, G. C.; Abrams, L.; Corbin, D. R.; Gier, T. E. and Shannon, R. D., *Appl. Catal.,* 32, 361 (1987).

28. Kung, H. H. and Smith, K. J. in *Methanol Production and Use*, W. H. Cheng and H. H. Kung, editors, p. 195, Marcel Dekker, New York, 1994.
29. Mochida, K.; Yasutake, A.; Fujitsu, H. and Takeshita, K., *J. Catal.*, 82, 313 (1983).
30. Segawa, K. and Tachibana, H., *J. Catal.*, 131, (2), 482 (1991).
31. Sato, H., *Catal. Rev. Sci. Eng.*, 39, (4), 385 (1997).
32. Helen, G., in *Ullmann's Encyclopedia of Industrial Chemistry*, 5th edition, Vol. A2, p. 1, VCH, New York, 1985.
33. Baiker, A., *Ind. Eng. Chem. Prod. Res. Dev.*, 20, 615 (1981).
34. Baiker, A., and Kigenski, *Catal. Rev. Sci. Eng., 27,* 653 (1985).
35. Baiker, A. in *Studies in Surface Science and Catalysis,* Vol. 41, p. 283, Elsevier, Amsterdam, 1988.
36. Turcotte, M. G. and Johnson, T. A., in *Kirk–Othmer Encyclopedia of Chemical Technology*, 4th edition, Vol. 2, p. 369, Wiley, New York, 1992.
37. Smirniotis, P.G. and Ruckenstein, E., *Ind. Eng. Chem. Res.*, 34, 1517 (1995).
38. Maria, G.; Pop, G.; Musca, G. and Boeru, P., *Studies in Surface Science and Catalysis,* 75 part B, 1665 Elsevier, Amsterdam,1993.
39. Ercan, C.; Dautzenberg, F.M.; Yeh, C.Y. and Barner, H.E., *Ind. Eng. Chem. Res., 37*, 1724 (1998).
40. Patwsrdhsm, A. and Sharma, M., *Ind. Eng. Chem. Res.*, 29 (1), 29 (1990).

3 Ammonolysis

Ammonolysis is a rather general term analogous to hydrolysis. Any reaction involving the splitting of a compound by means of ammonia can be called ammonolysis. Since it is such a broadly defined term, it provides a convenient place to group some miscellaneous reactions involving ammonia.

3.1 PHENOL + AMMONIA → ANILINE

This is an alternate process for producing aniline introduced by Halcon/Scientific Design. The dominant process, hydrogenation of nitrobenzene to aniline is presented in Chapter 12 along with a brief list of the uses of aniline.

Chemistry

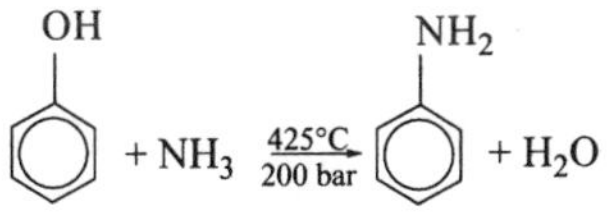

OH   NH$_2$

$+ NH_3 \xrightarrow[200\ bar]{425°C} + H_2O$

ΔH @ $700°K = -2.87$ kcal
K_p @ $700°K = 9.727$

Catalyst Type

The proprietary catalyst is said to be a Lewis acid catalyst consisting of Al_2O_3-SiO_3, possibly a zeolite, along with mixed oxides of Mg, B, Al, Ti, and cocatalyst of Ce, V, or Ti.[1] Moderate-acidity catalyst produces the best selectivity.[7]

Licensor: Scientific Design

Process Unit

Vaporized and fresh phenol plus recycle gaseous ammonia are mixed and fed to an adiabatic fixed-bed reactor in the ratio of 20:1 of NH_3-to-phenol. This high ratio assures high conversion for this reversible reaction, reduces by-product formation (diphenylamine, triphenylamine, and carbazole), and aids in controlling reactor temperature to a modest value to prevent ammonia decomposition.[2–4]

Effluent from the reactor, after heat exchange, passes to a separation section where ammonia is removed and recycled followed by water removal. A final distillation separates product aniline from impurities.

The catalyst exhibits long life and does not normally require regeneration.[1] Yield based on ammonia is said to be 96% and on phenol 80 percent.[2]

A large plant in the U.S.A. and another in Japan have been operating for some years. The traditional nitrobenzene route produces higher yields and lower energy costs, but its catalyst life is short in comparison to the phenol process.[2]

Zeolite Catalyst

As is often the case, much fundamental and applied research has been conducted with the goal of developing a zeolite catalyst that would eliminate side-reaction products. At this date, no such process has been announced. However, a copper zeolite has shown promise,[5] and its developments bear watching.

Process Kinetics

Several studies based on moderate-acidity alumina or silica-alumina indicated a pseudo-first-order rate with respect to phenol and zero-order in ammonia; since ammonia is fed at a large excess, the observations seem logical[7].

3.2   META-CRESOL + AMMONIA → META-TOLUIDINE

The toluidines are produced mainly by hydrogenation of nitrotoluenes. An alternate route by Mitsui Petrochemical (2000 tons/year) for producing m-toluidine has been reported[1] using ammonolysis in a process similar to phenol ammonolysis described above. Meta-toluidine is used in the production of various dyes and specialty chemicals.

m-cresol   m-toluidine
                3-ammotoluene

Catalyst Type

Patent literature suggests that the catalysts is TiO_2-SiO_2 in the form of pellets[8]. The licensor is Mitsui Petrochemical.

Process Unit

One can reasonably speculate that the reactor for this gas-phase reaction is a single fixed-bed reactor and that ammonia is fed in excess to prevent side reactions.

3.3   DIETHYLENE GLYCOL + AMMONIA → MORPHOLINE

Morpholine has many uses as an intermediate in the production of pharmaceuticals, insecticides, vulcanization accelerators, and emulsifiers. Morpholine itself is used for removing acid gases from hydrocarbon gas streams and is an effective extraction agent for aromatics in mixed hydrocarbon streams.

Chemistry

diethylene glycol   morpholine
                           tetrahydro-1,4-oxazine

Operating conditions are reported to be 150–400°C and 30–400 bar.[1]

Catalyst Type and Suppliers

Based on patent literature, the catalysts are variously described as supported Ni or Co and also Cu or Cr based.[1,6]

Suppliers: Engelhard, United Catalysts, Celanese (formerly Hoechst)

Probable Plant Operators (Based on Patent Holdings)
Wyandote, Goodyear, BASF

Process Unit

No further details are known other than the fact that hydrogen is fed with the reactants to prevent catalyst deactivation. Using excess ammonia should favor the main reaction and make possible the use of a single fixed-bed adiabatic reactor for this exothermic reaction.

Process Kinetics

Kinetics of the reaction over a copper catalyst has been reported as follows:[9]

$$r = kK_D P_D / (1 + K_D P_D + K_W P_W)^2$$

Range: 200–275°C and 20 atm

where subscripts D and W refer to diethylene glycol and water, respectively.

REFERENCES (AMMONOLYSIS)

1. Weissermel, H., Arpe, J., *Industrial Organic Chemistry*, 3rd edition, VCH, New York, 1997.
2. Lawrence, F. R. and Marshall, W. J., in *Ullmann's Encyclopedia of Industrial Chemistry*, 5th edition, Vol. A2, p. 303, VCH, New York, 1985.
3. Toseland, B. A. and Simpson, M. S., in *Kirk-Othmer Encyclopedia of Chemical Technology*, 4th edition, Vol. 2, p. 433, Wiley, New York, 1992.
4. *Hydrocarbon Proc., Petrochemical Handbook, '91*, p. 136, March, 1991.
5. Burger, M. H. W. and van Bekkum, H., *J. Catal.* 148, 65 (1984).
6. Mercker, H. J., in *Ullmann's Encyclopedia of Industrial Chemistry*, 5th edition, Vol. A2, p. 14, VCH, New York, 1985.
7. Ho, L.W. and Chang, K.R., *Proc. Int. Conf. Pet. Refin., Petrochem. Process.*, Vol. 3, 11159, 1991.
8. Mitsui Tostsu Chemicals, JP 48067229-19730913.
9. Kliger, G.A. and Glebov, L.S., *Kinetics and Catalysis (translation of Kinet. Katal.)*, 37 (6), 786 (1996).

4 Ammoxidation

Ammoxidation involves the combined action of ammonia and oxygen on a methyl group or methane to produce a nitrile. The most important of such nitriles are acrylonitrile and hydrogen cyanide. A number of aromatic nitriles also find use as intermediates in the synthesis of valuable products.

4.1 PROPYLENE → ACRYLONITRILE

Acrylonitrile is a major commodity chemical that is used in a variety of applications, the largest uses of which are given as follows in declining order:[1-4]

1. *Polyacrylonitrile* fibers are used as fabrics blended with polyester for sweaters and sportswear where the warmth and feel mimic that of wool without the care problems associated with wool. It is also used for carpets and draperies.
2. *Plastics* (ABS [acrylonitrile-butadiene-styrene] resins) are used for automotive parts, pipe fittings, and appliances; and SAN (styrene-acrylonitrile) resins which, because of high clarity provide a substitute for glass in automobile instrument panels, instrument lenses, and housewares.
3. *Adiponitrile* (produced by hydrodimerization of acrylonitrile) is used to produce hexa-methylene-diamine for nylon 66.
4. *Acrylamide* (produced by partial hydrolysis of acrylonitrile) is used to make water soluble polymers for paper making, ore flotation, flocculating agents for water treatment, and special polymers for paints and resins.
5. *Nitrile rubber* (acrylonitrile-butadiene copolymer) is used for special industrial applications requiring resistance to oil, flexibility at low temperatures, and heat resistance up to 120°C.

The development of the propylene ammonoxidation process initially by Sohio (now BP) provided major economic advantages, particularly for areas of the world with significant oil and gas reserves and large capacities for producing propylene. The propylene route is now the dominant process. Efforts to produce acrylonitrile from propane, a lower-cost feedstock, are in progress.[4] In fact, a large-scale demonstration unit is being constructed at BP's Great Lake, Texas facility. Since propane can be used directly in this new process without first converting it to propylene, considerable reduction in production costs are realized.[36]

Chemistry (Vapor Phase)

$$CH_2 = CH - CH_3 + NH_3 + 1.5O_2 \rightarrow CH_2 = CH - C \equiv N + 3H_2O$$

Operating conditions: 0.5 - 2 bar and 400 - 500°C

Operating conditions: 0.5–2 bar and 400–500°C

Temp., K	ΔH, kcal	K_p
600	–125.22	6.966×10^{48}
700	–122.17	3.062×10^{42}
800	–120.69	1.365×10^{38}

Side reactions produce HCN, acetonitrile, and some combustion products (CO and CO_2) of propylene. Catalysts are Bi_2O_3/MoO_3 with iron and other additives that reduce acetonitrile production. Different catalysts are also used such as Te, Ce, and Mo oxides on silica and Fe-Sb or U-Sb oxides. [3,4]

Mechanism

The major catalyst components Bi and Mo are the same as used for the oxidation of propylene to acrolein. The reasons for the high selectivity to acrylonitrile in the presence of ammonia have intrigued a number of investigators for over 30 years. Much work has been done under conditions where both acrolein and acrylonitrile are formed. In this way, it has been possible to define the conditions for maximizing acrylonitrile production and guide the development of promoters that will enhance selectivity. The use of iron compounds, for example, has led to major improvements in selectivity.

Extensive investigations have been reported involving kinetic studies, various spectroscopic techniques and probe-molecule reaction studies.[5–10] Out of this work have come postulates of detailed mechanisms for both ammoxidation and oxidation of propylene as depicted in Figure 4.1. The following steps apply for ammoxidation with species identifying numbers corresponding to those in the figure.[6]

- Mo = O moieties of the active site converted to Mo = NH in a fast reaction (Number 28).

$$NH_3O^{-2} \rightarrow NH^{-2} + H_2O$$

- α-hydrogen abstraction by oxygen atoms associated with bismuth to form the π-allylic Mo complex (Number 30). This is the slow and controlling step.
- The π-allylic complex (Number 30) undergoes further hydrogen abstraction to form the O-N allylic Mo species (Number 31).
- Number 31 undergoes a 1,4 hydrogen shift to produce 3-iminopropylene (Number 32).
- The 3-iminopropylene remains adsorbed while undergoing an additional 1,4 hydrogen shift after reoxidation of the Mo occurs.
- The resulting species (Number 33) then splits off acrylonitrile and the remaining entity (Number 27) is replenished with oxygen by dissociative chemisorption of gaseous oxygen on reoxidation sites, thus completing the cycle. The two lone pairs of the Bi-O-Bi surface complex are thought to be involved in the reduction of oxygen and the reoxidation of the catalyst.[8] Reoxidation occurs by oxygen chemisorbed from the vapor phase.[11]

These insights have led to some practical results in providing guidance for improved selective catalysts and development of the most favorable operating conditions. Reaction kinetic studies have shown that the ratio of acrylonitrile-to-acrolein in the product is dependent on the ammonia-to-propylene ratio in the feed.[5] This observation suggests a relation to the number of ammonia molecules activated in the acrylonitrile cycle.[5]

The energy for ammonia activation is high, and higher-temperature operation (above 350°C) assures high selectivity to acryonitrile. More recent studies have shown that ammonia not only

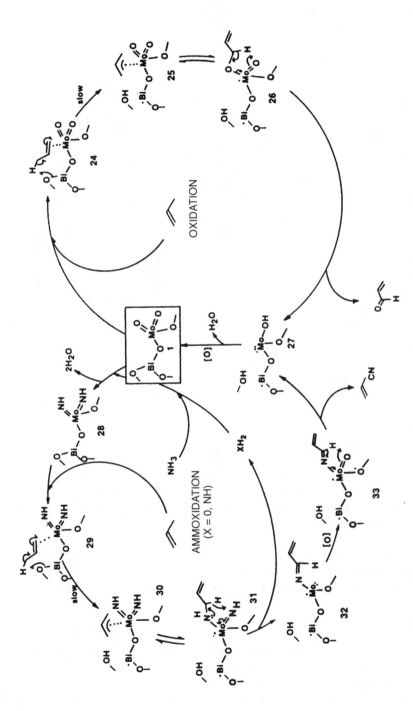

FIGURE 4.1 Mechanism of Selective Propylene Ammoxidation and Oxidation over Bismuth Molybdates. Reprinted from *Applied Catalysis* 15, 127 (1985), Grasselli, R. K., "Selectivity and Activity Factors in Bismuth-Molybdate Oxidation Catalysis," p. 127–139, copyright 1985, with permission from Elsevier Science.

takes part in the ammoxidation reaction, it also affects the catalyst characteristics. It blocks catalyst sites for wieldy bound oxygen that is active in total oxidation reactions and forms new active sites as well.[14] The net result with modern catalysts such as Bi-Mo-Fe oxides with promoters is a very low acrolein production. More significant by-products are HCN and acetonitrile. These are postulated to result from cracking of acrylonitrile on acid sites.[15]

Catalyst Types

The partial oxidation catalysts (in particular, those used for ammoxidation) have gone through many years of study and improvements. The actual catalysts in use at any time are proprietary, but development themes available from the patent literature show the changes that have created remarkably successful catalysts. The early catalysts were Bi_2O_3-MoO_3 on silica produced in slurry form and spray dried to yield microspheres in the range of 60–200 mesh for use in a fluidized bed.[13] The usual calcining step followed, yielding tough particles with adequate attrition resistance.

The mode of production probably remains roughly similar, but newer catalysts have various promoters and additives that enhance the selective activity and produce a catalyst of remarkably long life. An early improvement involved the addition of ferric nitrate and potassium carbonate, which end up as oxides upon calcination. The more recent catalysts are also Bi_2O_3-MoO_3 based and contain less Bi but additional components such as bivalent cations (Co, Ni, Mn, and Mg) and trivalent iron (Fe^{3+}) or Ce^{4+} which enhance lattice oxygen transfer.[3,11,12] Other catalysts include Te^{4+} or Sb^{3+} as a replacement for bismuth and Sb^{5+} as a replacement for a large portion of the Mo^{5+}.[3] Potassium and phosphorous are included in many acrylomotrile catalysts as promoters. Catalyst composed of UO_2-Sb_2O_3 have also been used.

Licensors

BP Chemicals (formerly Sohio) (also supplies catalyst)
Ugine Kuhlmann (fixed-bed process)
Snamprogetti (fixed-bed process)

Deactivation

The catalyst has an unusually long life if the Mo that gradually sublimes from the catalyst is restored periodically or continuously in a side-stream draw-off of fluidized catalyst. Replacement is accomplished by introducing a spray or vapor containing MoO_3 as the ammonium salt or vaporized MoO_3.[13] Similarly, tellurium catalysts can be restored by replacing the tellurium (TeO_4) that has volatilized.[13]

As in any fluidized bed system, attrition can create fines that will be lost to the fines removal system. The catalyst preparation techniques that have been patented address this issue by employing careful procedures that have been found to yield catalysts with low attrition rates.

Process Units

Most acrylonitrile plants use fluidized-bed reactors with internal heat-transfer coils for removing the reaction heat. With proper fluidization, optimum temperature can be maintained at a constant value in the neighborhood of 450°C. Operating ranges are variously reported as 400–500°C and 0.5–2 bar.[2,3,4] The combined exothermic heat of reaction must include not only the partial oxidation to acrylonitrile but also that generated by the side reaction of complete oxidation of a portion of propylene to CO and CO_2.[3] The overall selectivity to acrylonitrile based on propylene is reported to be 70 percent.[3]

The cooling coils in the fluidized bed are fed with boiler feedwater that is converted to valuable medium- or high-pressure saturated steam that can be superheated in the convection section of a nearby furnace.

The product recovery process involves removal and neutralization of excess ammonia followed by a series of distillations for recovery of acrylonitrile and removal of side-products, mainly HCN and acetonitrile, both of which have commercial value.

Several units are operating with tubular fixed-bed reactors and use molten salt as the heat-transfer medium. Steam is then generated from the pump-around fluid in an external exchanger. To attain the kind of temperature control possible in a fluidized bed and prevent over oxidation, the inert component of the fixed-bed catalyst can include material of high conductivity. Alternatively, it is possible that water addition or recycle of off gases could be practiced. Carbon dioxide has been reported to also improve partial-oxidation selectivity by strong adsorption on sites active in total combustion of a hydrocarbon.[16]

Process Kinetics

Early in the development of the acrylonitrile process, kinetics for propylene ammoxidation were developed. Under synthesis conditions, it was found that the several rates (main and side reactions) exhibit first-order dependence on propylene and zero-order dependence on ammonia and oxygen if they are fed in at least stoichiometric amounts.[37, 38] More recently, a valuable application of these data has been used in developing a process model for studying by-product reduction procedures to improve yield and reduce unwanted side products that become part of a waste stream.[38]

Table 4.1 summarizes the reactions and first-order rate forms. The study described is an excellent example of analyzing and applying published kinetic data in a simple and straightforward manner so that a reasonable process model can be implemented.

Langmuir–Hinshelwood equations have also been presented for the main reaction.[39]

TABLE 4.1
Model Reactions and Rate Equations for Propylene Ammoxidation

Propylene Ammonia Oxygen Acrylonitrile Water

$$CH_2=CH-CH_3 + NH_3 + 3/2\ O_2 \rightarrow CH_2=CH-CN + 3H_2O$$ [1] $(-r_1) = k_1 C_{C_3H_6}$

Acrolein

$$CH_2=CH-CH_3 + O_2 \rightarrow CH_2=CH-CHO + H_2O$$ [2] $(-r_2) = k_2 C_{C_3H_6}$

Acetonitrile

$$CH_2=CH-CH_3 + NH_3 + 9/4\ O_2 \rightarrow CH_3-CN + 1/2\ CO_2 + 1/2\ CO + 3H_2O$$ [3] $(-r_3) = k_3 C_{C_3H_6}$

$$CH_2=CH-CHO + NH_3 + 1/2\ O_2 \rightarrow CH_2=CH-CN + 2H_2O$$ [4] $(-r_4) = k_3 C_{CH_2CHCHO}$

Hydrocyanic acid

$$CH_2=CH-CN + 2O_2 \rightarrow CO_2 + CO + HCN + H_2O$$ [5] $(-r_5) = k_5 C_{CH_2CHCN}$

$$CH_3-CN + 3/2\ O_2 \rightarrow CO_2 + HCN + H_2O$$ [6] $(-r_6) = k_6 C_{CH_3CN}$

Reprinted from *Waste Management (N.Y.)* 13 (1), 1993, Hopper, J., Yaws, C.L., Ho, T.C., and Vichailak, M., "Waste Minimization by Process Modification," p. 5, copyright 1993 with permission from Elsevier Science.

4.2   METHANE → HCN

Although a major source of hydrogen cyanide *(hydrocyanic acid)* is a by-product of acrylonitrile production, significant additional amounts are required to satisfy its demand as an important intermediate reactant. Important uses include the formation of acetone cyanohydrin used in the major process for producing methyl methacrylate, the hydrocyanation of butadiene to form adiponitrile, the production of cyanogen chloride from HCN and Cl_2 (which can be trimerized to cyanuric chloride and used in the manufacture of a family of triazine herbicides), and the manufacture of chelating agents and sodium cyanides.

Chemistry (Vapor Phase)

$$CH_4 + NH_3 + 1.5\ O_2 \rightarrow HCN + 3H_2O$$

    Catalyst: platinum-rhodium wire screens
    1100–1200°C @ atmospheric pressure
    ΔH @ 1000 K = −1112.38 kcal
    K_p @ 1000 K = 2.443 x 10^{29}

The above equation is an overall stoichiometric equation for what is a complex series of reactions.

Mechanism

Although no detailed mechanism including the catalyst components and their characteristics has been proposed, an adroit decade-long study involving the identification of 13 separate reactions has been summarized.[10] Its validity has been tested based on separate measurements on polycrystalline Pt wires.

    The reactions studied are summarized in Table 4.2 along with rate equations, which are discussed on page 57. The major reaction producing HCN is the second reaction listed in Table 4.2.

$$NH_3 + CH_4 \rightarrow HCN + 3H_2$$

The bimolecular reactions 6 and 11 in the table,

$$CH_4 + 1.5O_2 \rightarrow CO + 2H_2O$$

$$CH_4 + NO \rightarrow HCN + 1/2H_2 + H_2O$$

are very fast and exhibit similar rates. The oxidation of methane to CO and H_2O supplies a major portion of the heat to offset the endothermicity of the main HCN producing reaction. Since Reaction 2 in Table 4.2 can be accomplished without excess methane and oxygen present, clearly, the practical reason for including excess CH_4 and air in the reaction mixture was to supply the necessary heat to maintain the required elevated temperatures by direct-contact heat transfer.

    Reactions of significance, such as the hydrolysis and polymerization of HCN, are not included in Table 4.2 since they are marginalized by short contact time and rapid quenching. Just like all the reactions in Table 4.2, these reactions are exothermic but are not equilibrium limited at the usual operating temperatures. Hence, they can only be prevented from occurring by rapid cooling, which renders the reaction rates negligible even though the equilibrium constant increases.

$$HCN + H_2O \rightarrow NH_3 + CO \qquad K_p\ @ 1000°K\ =\ 164.06$$

TABLE 4.2
Reaction Summary

Reaction	Rate Expression	Reference
$NH_3 \rightarrow \frac{3}{2}H_2 + \frac{1}{2}N_2$	$\dfrac{4.9 \times 10^{18} \exp(-2130/T)P_{NH_3}}{[1 + 0.044\exp(2390/T)P_{CH_4}/P_{NH_3}^{1/2}]^3}$ if $P_{CH_4} > 0$	Hasenberg & Schmidt, 1985–1987
	$\dfrac{4.9 \times 10^{18} \exp(-2130/T)P_{NH_3}}{1 + 4.35 \times 10^5 \exp(8400/T)P_{NH_3} + 9.85 \times 10^{-6}\exp(13850/T)P_{H_2}^{3/2}}$ if $P_{CH_4} = 0$	Loffler & Schmidt, 1985
$NH_3 + CH_4 \rightarrow HCN + 3H_2$	$\dfrac{7.80 \times 10^{18} \exp(-1950/T)P_{CH_4}P_{NH_3}^{1/2}}{[1 + 0.044\exp(2390/T)P_{CH_4}/P_{NH_3}^{1/2}]^4}$	Hasenberg & Schmidt, 1985–1987
$NH_3 + \frac{5}{4}O_2 \rightarrow NO + \frac{3}{2}H_2O$	$\dfrac{2.1 \times 10^{16} \exp(10850/T)P_{NH_3}P_{O_2}^{1/2}}{1 + 4.0 \times 10^{-5}\exp(12750/T)P_{NH_3}}$	Pignet & Schmidt, 1974, 1975 Hasenberg & Schmidt, 1985–1987
$NH_3 + \frac{3}{2}NO \rightarrow \frac{5}{4}N_2 + \frac{3}{2}H_2O$	$\dfrac{1.48 \times 10^{17} \exp(3875/T)P_{NO}P_{NH_3}^{1/2}}{[1 + 5 \times 10^{-5}\exp(7950/T)P_{NO} + 0.0145\exp(2880/T)P_{NH_3}^{1/2}]^2}$	Takoudis & Schmidt, 1983
$\frac{1}{2}O_2 + H_2 \rightarrow H_2O$	$1.5 \times 10^{19} P_{O_2}P_{H_2}$	Blieszner, 1979
$CH_4 + \frac{3}{2}O_2 \rightarrow CO + 2H_2O$	$\dfrac{4 \times 10^{19} \exp(-5000/T)P_{CH_4}P_{O_2}^{1/2}}{1 + 5 \times 10^{-10}\exp(15000/T)P_{CH_4}}$	Hasenberg & Schmidt, 1985–1987
$NO + H_2 \rightarrow \frac{1}{2}N_2 + H_2O$	$\dfrac{3.5 \times 10^{18} \exp(7300/T)P_{H_2}P_{NO}}{[1 + 2.7 \times 10^{-4}\exp(9750/T)P_{NO} + 15\exp(1100/T)P_{H_2}^{0.7}]^2}$	Papapolymerou & Schmidt, 1985

TABLE 4.2
Reaction Summary (continued)

Reaction	Rate Expression	Reference
$NO \rightarrow \frac{1}{2}N_2 + \frac{1}{2}O_2$	$\dfrac{5.53 \times 10^{16} \exp(-2625/T)P_{NO}}{1 + 6.95 \times 10^{-4} \exp(4125/T)P_{NO} + 1.56 \exp(4775/T)P_{O_2}}$	Mummey & Schmidt, 1981
$NO + CO \rightarrow \frac{1}{2}N_2 + CO_2$	$\dfrac{3.5 \times 10^{17} \exp(2900/T)P_{NO}}{1 + 4 \times 10^{-9} \exp(15000/T)P_{CO}}$	Klein et al., 1985
$CO + \frac{1}{2}O_2 \rightarrow CO_2$	$\dfrac{2.5 \times 10^{15} \exp(16000/T)P_{CO}P_{O_2}}{[3 \times 10^{-7}\exp(15000/T)P_{CO} + 300\exp(6000/T)P_{O_2}]^2}$	Bliezner, 1979 Schwartz et al., 1986
$CH_4 + NO \rightarrow HCN + \frac{1}{2}H_2 + H_2O$	$\dfrac{1.8 \times 10^{20} \exp(5000/T)P_{CH_4}P_{NO}}{1 + 5 \times 10^{-10}\exp(15000/T)P_{CH_4}}$	Hasenberg & Schmidt, 1985–1987
$CO + H_2O \rightarrow CO_2 + H_2$	$\dfrac{3.65 \times 10^{17} \exp(-1595/T)P_{CO}P_{H_2O}^{1/2}}{[1 + 0.048\exp(3037/T)P_{CO}]^2}$	Blieszner, 1979
$CH_4 + 3NO \rightarrow \frac{3}{2}N_2 + CO + H_2O$	$\dfrac{1.25 \times 10^{15} \exp(5000/T)P_{CH_4}P_{NO} + 3 \times 10^{20} \exp(-750/T)P_{NO}P_{CH_4}^{1/2}}{1 + 1 \times 10^{-11} \exp(20000/T)P_{CH_4}}$	Hasenberg & Schmidt, 1985–1987

Source: Waletzko, N. and Schmidt, L.D., *AIChE Journal* 34 (7) 1146, 1988. Reproduced with permission of the American Institute of Chemical Engineers.

Catalyst Type and Suppliers

The catalyst consists of a series of screens or gauzes woven from 0.076 mm (0.003 in.) Pt (90%)/Rh (10%) wire in an 80 mesh pattern. The screen or gauze pack consists of 10–50 gauze/layers,[18] with the higher number of gauzes required at higher pressures. The lower number appears to be the most common. Gauze pack diameters are reported to be as large as 2 m or 6.4 ft.[21]

Catalyst Suppliers

Engelhard
Johnson Matthey

Catalyst Activity

A period of 2 to 3+ days under operating conditions is required to activate the catalyst to its full potential.[20] During this induction period, the surface area and the activity increase steadily due to the development of whiskers of Pt-Rh on the surface.[24] The resultant high activity persists over 2000+ hours. At this point, further surface area increases no longer contribute to activity, since they then are caused by such phenomena as carbon deposition and breaking of whiskers.

Once the catalyst deactivates appreciably, it is removed and returned for recovery of the precious metals by the supplier and credited to the user's account.

Deactivation

The gauze catalyst is subject to deactivation by carbon formation, which is catalyzed by iron deposits that may have been caused by erosion of upstream rust deposits. Iron concentrations as small as 35 ppm can reduce catalyst activity and also catalyze NH_3 and HCN decomposition.[21] Carbon deposits can cover active sites and thereby reduce activity. Additional causes of catalyst deactivation are sulfur and phosphorous compounds,[23] which are well known poisons for platinum. Higher hydrocarbon impurities, even in small amounts, will cause significant deactivation due to carbon formation. Carbon deposition occurs more readily on lower-temperature regions of the gauze, namely the outer perimeter of the gauze pack.[21]

Process Units

The original process was developed by L. Andrussow and patented by I.G. Farhenindustrie in 1933, in the United States. Over the years, each producer of HCN has made various improvements, particularly in the recovery section, but all the resulting processes continue to be designated as the Andrussow process which, in the aggregate, constitutes 79% of the total production of hydrogen cyanide.

The reactor consists of the gauze pack described in the previous section followed by a waste-heat boiler installed at the outlet of the reactor, so that product gas can be quenched rapidly. Premixed filtered air, methane, and ammonia are charged to the reactor operating in the range of 2 atm and 1100–1150°C.[21,22] Endothermic heat required by the NH_3-CH_4 reaction is provided by the combustion of a portion of the methane.

The reactant mixture is close to the stoichiometry of the overall reaction. It is not clear from the open literature what the various producers find optimal for their own reactor units. The original patent reported 12% NH_3, 13% methane, and 75% air[21] or $CH_4/NH_3 = 1.08$ and air/($CH_4 + NH_3$) = 3.0, which is in the range of some reported reactor conditions. Optimum conditions have been suggested to be above the flammability limit[22], which varies with operating conditions but is around an air-to-($CH_4 + NH_3$) ratio of 3.25. If operating in this region, great care must be given to avoiding the explosive limit.

Reaction temperature, when operating at a steady state, can be controlled by changes in reactant ratios or by changes in flow rate.[20] Another convenient variable for systems with preheaters is, obviously, control of the preheat temperature.

Table 4.3[23] provides estimates of reactor off-gases as well as residual gases after HCN recovery. The Andrussow process produces side reaction products due to the oxidation reactions; and, of course, large amounts of nitrogen are present because of the use of air as the oxygen source. The BMA process is discussed below.

TABLE 4.3
Estimates of Reactor Off-Gases and Residual Gases after HCN Recovery

	BMA		Andrussow*	
Compound	After reaction	Residual	After reaction	Residual
HCN	22.9	$<10^{-2}$	8.0	$<10^{-2}$
NH_3	2.5	$<10^{-2}$	2.5	>0
H_2	71.8	96.2	22.0	24.6
N_2	1.1	1.5	46.5	51.9
CH_4	1.7	2.3	0.5	0.6
CO			5.0	5.6
H_2O			15.0	16.8
CO_2			0.5	0.6

*Calculated

Compositions are in volume percent.

Reprinted by permission: Klenk, H.; Griffiths, A.; Huthmacher, K.; Itzel, H.; Knorre, H.; Voight, C. and Weiberg, O., *Ullmann's Encyclopedia of Industrial Chemistry,* 5th edition, Vol. A8, p. 161, Wiley/VCH, Weinheim, Germany, 1992. Compositions are in volume percent.

The contact time in the reactor is only a few milliseconds, and the effluent passes instantly into the quench system, which must reduce the product mixture to below 400°C to prevent decomposition of HCN by hydrolysis. The quench unit attached directly below the gauze pack also serves as a waste-heat boiler with multiple tubes through which the product passes. The steam generated is a valuable source of plant energy. A mass ratio of steam to HCN produced is reported to be in the range of 10 to 20.[22] The yield of HCN based on NH_3 is reported to be typically 70 percent,[20] but some patents claim yields as high as 80 percent.

The cooled product is then freed of excess ammonia to prevent polymerization of the hydrogen cyanide.[23] This operation can be accomplished by introducing H_2SO_4 to form ammonium sulfate that is of marginal value and costly to dispose. More recently, the off-gas has been treated with a monoammonium phosphate solution in an absorber that reacts with the NIH_3 to produce diammonium phosphate. This reaction can be reversed, and the ammonia released for recycle by stripping the monoammonium phosphate solution with steam.[23] The HCN in the gas phase leaving the NH_3 absorber is passed to a countercurrent water-wash absorber to capture the HCN. Finally, the HCN is stripped from the water and condensed.[23]

Alternative Processes

Several processes were developed on the logical basis that a number of side reaction products could be essentially eliminated if the reaction were conducted without the addition of air or oxygen. In such a situation, methane or natural gas could be combusted separately and the heat produced transferred through the reactor wall to supply the heat for the major endothermic reaction of methane and ammonia to hydrogen cyanide.

$$NH_3 + CH_4 \rightarrow HCN + H_2 \qquad \Delta H = 65.34 kcal \; @ \; 1000°K$$

$$K_p = 0.159$$

Another advantage would be the production of valuable hydrogen instead of water.

Two processes were introduced that accomplished these advantageous goals. The Degussa BMA process *(Blausaure-Methan-Ammoniak)* essentially involves a number of direct-fired furnaces each containing up to 10 reactor bundles composed of 10 to 30 sintered alumina tubes coated on the inside with platinum.[22] The tubes are described as 20 mm in diameter and 200 cm in length, and the reaction is conducted at 1200–1300°C and slightly above atmospheric pressure. The first 25% of the tube serves to heat the reactants to reaction temperature. This zone is followed by a reaction zone and a third zone close to and cooled by the tube support.[21] Yields of 80–85% based on NH_3 and 90% based on methane are reported.[23]

Comparison of product compositions for the Andrussow and Degussa BMA processes as shown in Table 4.3[23] demonstrates the advantages of a higher yield and a valuable residual gas product exhibited by the Degussa process. The residual gas is a rather pure stream of hydrogen that can be used directly in other processes. By contrast, the Andrussow process produces a residual stream contaminated with large amounts of H_2 and N_2 as well as CO and water. However, the Degussa process requires much higher investment and must invoke clever efforts to recover waste heat to maintain economic viability.

At last count, there were several Degussa plants in Europe and the U.S.A. By contrast, there are at least ten companies with multiple plants worldwide that use the Andrussow process with individual variations and improvements.

A second alternative process uses a fluidized bed of coke particles and is noncatalytic. One unit is currently in operation.[4]

Process Kinetics

Detailed kinetics of the Andrussow process using surface reaction equations shown in Table 4.2[18,19] have been published. In addition to the obvious value of a study on an important industrial reaction, the experimental effort and subsequent predictive value demonstrates that surface reaction equations can be valuable on well defined surfaces, a condition that cannot often be reached under practical operating conditions. The HCN synthesis has certain favorable simplifying conditions.[18]

1. The catalyst is self-supported metal wires rather than supported particles.
2. The reactions are relatively simple ones between small polyatomic molecules whose kinetics have been studied under well defined conditions.
3. The temperature is sufficiently high that the surface has low coverages of most adsorbed species and may be self-cleaning.

Kinetic equations for Reactions 1 and 2 of Table 4.2 can be used to model the Degussa BMA process successfully.[18] A model was carefully crafted based on first principles. Both the flux limits and mass-transfer limits were deduced by the use of an effective sticking coefficient. Model predictions led to the conclusion that the highest practical temperature be used and that higher pressures only increase selectivity slightly.[18] Modest pressures sufficient to reduce equipment size are therefore chosen.

For those preferring power-law equations, the rate of HCN formation was found to be first order in both NH_3 and CH_4 when it is restricted to the important period up to the maximum conversion to HCN, after which the secondary reaction of HCN hydrolysis becomes significant.[27] Since such a period corresponds to the optimum contact time, the simplified single kinetic expression can be useful. Early in the study of this reaction, it was observed that mass transfer could become controlling at high temperatures[25,26] and, as previously, mentioned flux and mass-transfer issues need to be considered (see Ref. 18).

4.3   ISOBUTYLENE → METHACRYLONITRILE

The production of methacrylonitrile via isobutylene ammoxidation, using similar procedures as described for propylene ammoxidation to acrylonitrile, is practiced in regions of the world such as Japan, where isobutylene is not in high demand as feedstock for MTBE production. Methacrylonitrile can be used as an alternative route to methacrylic acid and methylmethacrylate via hydrolysis to methacrylanide, which can then be esterified to methylmethacrylate or hydrolyzed to methacrylic acid.[17]

4.4   AROMATIC METHYL COMPOUNDS → NITRILES

Some aromatic nitriles offer convenient routes to valuable compounds. The nitriles are readily prepared by ammoxidation of the methyl groups, which yields valuable products or precursors to valuable products. See Table 4.4.

Chemistry and Mechanism

The overall reactions shown in Table 4.4 are all similar. They are exothermic and are typical oxidation reactions that require a catalyst that can supply oxygen. The reaction system, although not studied in its mechanistic details, is clearly complex. Just as in the much-studied case of propylene ammoxidation, the catalyst oxidation characteristics must be maintained by addition of air or oxygen in the feed or by rejuvenation of the catalyst with oxygen in a separate vessel.

A reaction scheme for m-xylene ammoxidation has been proposed[28] and is shown in Figure 4.2. The isophthalonitrile was shown to be stable and formed both by a reaction parallel to the formation of m-tolunitrile and by a consecutive reaction from m-tolunitrile.[28] Unreacted tolunitrile can be recycled. Since tolunitrile is less stable, it is the source of the side-reaction products-carbon oxides and hydrogen cyanide. The xylene reactant also produces the same side-reaction products.

A study of beta-picoline[29] includes a more detailed mechanism involving both the overall reaction and the reduction and oxidation of the catalyst.

$$RCH_3(g) + NH_3(g) + S_{OX} \xrightarrow{k'_1} RCN(g) + 3H_2O + S_{red}$$

$$RCH_3(g) + S_{OX} \xrightarrow{k'_2} CO_2(g) + N_2(g) + H_2O + S_{red}$$

$$O_2 + S_{red} \xrightarrow{k'_3} S_{OX}$$

where S_{ox} and S_{red} represent the oxidized and reduced state of the catalyst.

It is reasonable to assume that all the alkyl aromatic ammoxidations involve similar reaction steps as suggested in these two references. Because of the adjacent methyl groups on o-xylene, the

TABLE 4.4

Aromatic Nitriles from Aromatic Methyl Compounds by Ammoxidation[4,34,35]

Aromatic Methyl Compounds	Aromatic Nitrile	Reaction Examples	Uses
toluene	benzonitrile	CH_3–C$_6$H$_5$ + NH_3 + $1.5O_2$ → C$_6$H$_5$–CN + $3H_2O$	• Solvent. • Can be hydrolyzed to benzoic acid, a process used when naturally occurring benzonitrile was used. Direct oxidation of toluene is the preferred route to benzoic acid.
o-xylene	phthalonitrile	o-(CH$_3$)$_2$C$_6$H$_4$ + $2NH_3$ + $3O_2$ → o-(CN)$_2$C$_6$H$_4$ + $6H_2O$	• An intermediate in the manufacture of phthalocyanine dyes.
m-xylene	isophthalonitrile (product: m-dicyanobenzene)		• Used in a process for producing a herbicide-fungicide (tetrachloro-1,3 dicyanobenzene). • Hydrogenated to m-xylene diamine, which is converted to a diisocynate used in producing polyurethanes. • The diamine is also used as a monomer in production of specialty polyamides.
p-xylene	terephthalonitrile (product: p-dicyanobenzene)		• Hydrogenated to p-xylene diamine. • Can be converted to terephthalic acid, which is used to produce polyesters with glycols. Requires a series of steps involving hydrolysis to ammonium terephthalate and subsequent cleavage to the acid and NH_3. The process is not in commercial use.
β-picoline	3-cyanopyridine	(3-methylpyridine) + NH_3 + $1.5O_2$ → (3-cyanopyridine) + $3H_2O$	• Hydrolyzed to nicotinic acid (niacin), the anti-pellagra vitamin and enzyme cofactor.
2-chlorotoluene	2-chlorobenzonitrile	(2-chlorotoluene → product: 2-chlorobenzonitrile)	• Intermediate in synthesis of azodyes.
4-chlorotoluene 2,6 dichlorotoluene	4-chlorobenzonitrole 2,6 dichlorobenzonitrile		• Intermediate in synthesis of pigments for plastics. • Herbicide.

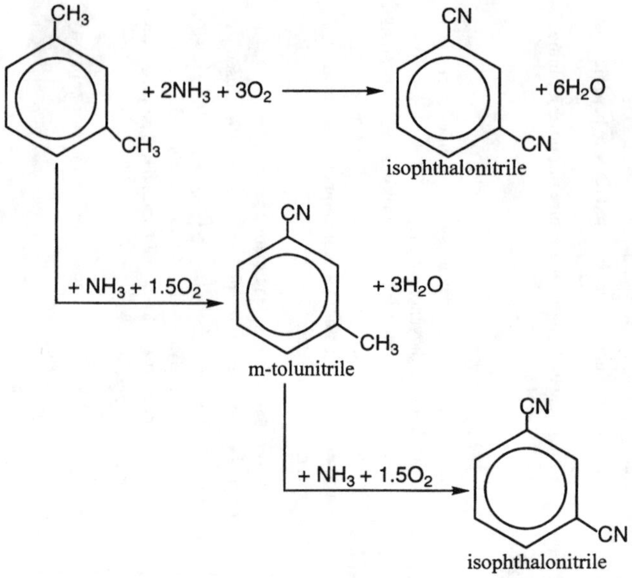

FIGURE 4.2 A Proposed Reaction Scheme for Ammoxidation of m-Xylene.

reaction scheme is complicated by additional reactions of the product, phthalonitrile, with water and ammonia.[30]

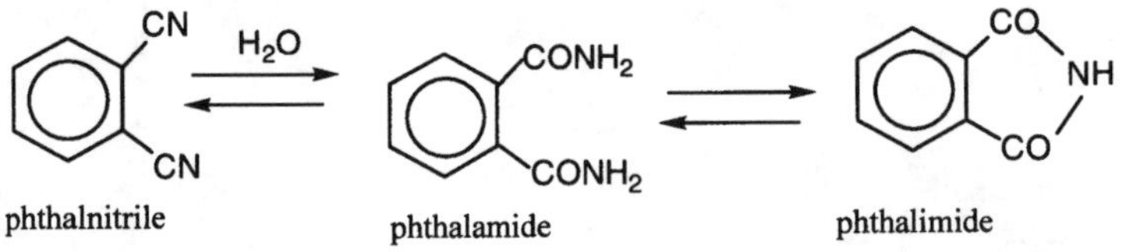

Catalysts and Catalyst Suppliers

The processes are proprietary as are the catalysts. The catalysts used in published reaction studies include $V_2OAl_2O_3$, $V_2O_5/Cr_2O_3/Al_2O_3$.

Licensors

　ABB-Lummus Global (all methyl aromatics listed)

　Showa Denko (m- and p-xylenes)

　Japan Catalytic Chemical Industry (o-xylene)

　BASF (o-xylene)

　Mitsubishi-Badger now Raytheon E&C (m-xylene)

　Degussa (beta-picoline)

Suppliers

　Engelhard

Process Units

Fixed-bed multitubular vapor-phase reactors or fluidized beds have been used. Operating temperatures range from 300–450°C, depending on the reactant being ammoxidized. The ABB-Lummus process separates the two functions of ammonolysis and oxidation, which is referred to as *oxidative ammonolysis*. Fluidized beds are used with the catalyst continuously circulated between the reactor where the oxygen from the catalyst supplies the necessary oxygen to produce the overall reactions shown in Table 4.4. Instead of supplying free oxygen in the reactor to reoxidize the catalyst, that oxygen is supplied to the regenerator. In this manner, nonselective radical oxidations are avoided, improved operational safety is served by eliminating the explosion hazards associated with free oxygen in the reactor, and a less costly recovery system results because of the lower amount of noncondensables in the product stream.[31-33] The catalyst used has been developed to produce many of the listed nitriles. Operation of an ABB-Lummus commercial plant using m-xylene to produce isophalonitrile has been reported.[30]

Process Kinetics

Kinetic studies on vapor-phase ammoxidation of m-xylene and 4-picoline have been described.[28,29] Because of the similarity of the various aromatic-nitrile forming reactors, the observations and proposed rate forms may be useful in modelling other similar reactions.

For m-xylene, the following experimental observations were reported.[28]

- The rate of reaction of m-xylene is nearly independent of the concentration of m-xylene under the conditions studied. Therefore, it is zero-order in m-xylene.
- The rate of reaction of m-xylene and the rate of formation of individual products are not influenced by either the ammonia or oxygen concentration.
- The rate of reaction of m-tolunitrile is inhibited by m-xylene.
- The reaction of m-tolunitrile is first-order at low m-tolunitrile concentrations.

Rate equations based on these observations and the reaction scheme in Figure 4.2 were presented.

$$r_X = k_X$$

$$r_M = k_1 - k_X P_M / P_X$$

$$r_D = k_5 + k_2 P_M / P_M$$

$$r_B = k_3 + k_4 P_M / P_X$$

where P = the partial pressure of the subscript-indicated component

and subscripts are as follows:

      X = xylene
      D = isophthalonitrile
      M = m-tolunitrile
      B = carbon oxides and HCN

Langmuir–Hinshelwood equations derived with surface reaction controlling reduced to the above equations if the adsorption of m-xylene is assumed to be strong compared to other components. The simplified equations should be tried first.

The study on picoline presents only an equation for its disappearance based on the overall reaction and the oxidation-reduction of the catalyst.[29]

REFERENCES (AMMOXIDATION)

1. Chenier, P. J., *Survey of Industrial Chemistry*, 2nd edition, VCH, New York, 1992.
2. Langvardt, P. W. in *Ullmann's Encyclopedia of Industrial Chemistry*, 5th edition, Vol. Al, p. 177, VCH, New York, 1985.
3. Brazdil, J. F. in *Kirk-Othmer Encyclopedia of Chemical Technology*, 4th edition, Vol. 1, p. 352, Wiley, New York, 1991.
4. Weissermel, K. and Arpe, H. J., *Industrial Organic Chemistry*, 3rd edition, p. 304, VCH, New York, 1997.
5. Barrington, J. D.; Kartisek, C. T. and Grasselli, R. K., *J. Catal.*, 87, 363 (1984).
6. Grasselli,R.K., *Applied Cataly.*, 15 127 (1985).
7. Brazdil, J. F.; Glaeser, L. C. and Grasselli, R. K., *J. Catal.*, 81, 142 (1983).
8. Glaeser, L. C.; Brazdil, J. F.; Hazle, M. A.; Mehicic, M. and Grasselli, R. K., *J. Chem. Soc., Faraday Trans.I*, 81, 2903 (1985).
9. Brazdil, J. L. and Grasselli, R. K., *J. Catal.*, 79, 104 (1983).
10. Grasselli, R. K. and Barrington, J. D., *Adv. Catal.*, 30, 133 (1981).
11. Schuit, G. C. A. and Gates, B. C., *Chemtech.*, 13, 693 (1981).
12. Gates, B. C., *Catalytic Chemistry*, Wiley, New York, 1992.
13. Stiles, A. B. and Koch, T. A., *Catalyst Manufacture*, 2nd edition, Marcel Dekker; New York, 1995.
14. Rizayer, R. C.; Marnedov, I. A.; Vislovsk, V. P. and Sheinin, V. E., *Appl. Catal.* 83, 103 (1992).
15. Grasselli, R. K., personal communication, Nov. 13, 1997.
16. Skolovskii, V. D.; Davydov, A. A. and Orsitser, O. *Catal Rev. Sci. Eng.*, 37, 425 (1996).
17. Gross, A.W. and Dobson, J. C., *Kirk-Othmer Encyclopedia of Chemical Technology,* 4th edition, Vol. 16, p. 495, Wiley, New York, 1995
18. Waletzko, N. and Schmidt, L. D., *AIChE Journal*, 34(7), 1146 (1988).
19. Hasenberg, D. and Schmidt, L. D., *J. Catal.* 104, 441 (1987).
20. Satterfield, C. N. *Heterogeous Catalysis in Industrial Practice*, 2nd edition, McGraw-Hill, New York, 1991.
21. Dowell, A. M. III, Tucker, D. H.; Merritt, R. E. and Teich, C. I. in *Encyclopedia of Chemical Processing and Design,* J. J. McKetta, editor, Vol. 27, p. 1, Marcel Dekker, New York, 1988.
22. Pesce, L. D. in *Kirk-Othmer Encyclopedia of Chemical Technology,* 4th edition, Vol. 7, p. 753, Wiley, New York, 1993.
23. Klenk, H., Griffiths, A.; Huthmacher, H. I.; Knowe, H.; Koigt, C. and Weiberg, O. in *Ullmann's Encyclopedia of Industrial Chemistry*, 5th edition, Vol. A-8, p. 161, VCH, New York, 1987.
24. Rylander, P. N. in *Catalysis: Science and Technology*, Vol. 4, p. 30, Springer Verlag, New York, 1983.
25. Koberstein, E., *Ind. Eng. Chem. Process Des. Develop.*, 12 (4), 444 (1973).
26. Pignet, T. P. and Schmidt, L. D., *Chem. Eng. Sci.* 29, 1123 (1974),
27. Pen, P. Y. K., *J. Catal.* 21, 27 (1971).
28. Ito, M. and Sano, K., *Bulletin Chem. Soc. Japan,* 40, 1307 (1967).
29. Das, A. and Kar, A., *Ind. Eng. Chem. Process Des. Dev. 19,* 689 (1980).
30. Sze, M. C. in *Encyclopedia of Chemical Processing and Design*, J. J. McKetta, editor, Vol. 31, p. 155, Marcel Dekker, New York, 1990.
31. Paristian, J. E.; Pnzio, J. F.; Stavropoulos, N. and Sze, M. C., *Chemtech,* p. 174, March, 1981.
32. Sze, M. C. and Gelbein, A. P., *Hydrocarbon Process.,* p. 103, Feb., 1976.
33. Gelbein, A. P.; Sze, M. C. and Whitehead, R. T., *Chemtech,* p. 479, August, 1973.
34. Pollak, P.; Romeder, G.; Hagedorn, F. and Gellke, H. P. in *Ullmann's Encyclopedia of Industrial Chemistry,* 5th edition, Vol A17, p. 363, VCH New York, 1991.
35. Franck, H. G. and Stadelhofer, J. W., *Industrial Aromatic Chemistry*, Springer Verlag, New York, 1988.
36. *Chemical Engineering,* p. 19 October, 1998.
37. Callahan, J. L.; Grassell, R. K.; Milbergr, E. C. and Strecker, H.A., *Ind. Eng. Chem. Prod. Res. Develop.,* 9, 134 (1970).
38. Hopper, J.; Yaws, C. L.; Ho, T. C. and Vichailak, M., *Waste Manage. (NY)* 13 (1), 3 (1993).
39. Lankhuyzen, S. P.; Florack, R. M. and van der Baan, H.S., *J. Catal.,* 42, 20 (1976).

5 Carbonylation

Carbonylation is a broad term that simply refers to reactions involving the introduction of a carbonyl group (C=O) into a molecule. Industrially, two major processes make up this general reaction type, hydroformylation or the oxo reaction (the production of aldehydes from CO, H_2, and olefins) and carboxylation (reaction of olefins with CO in the presence of H_2O and a proton supplying catalyst to produce carboxylic acids). Essentially, all of these reactions are carried out with homogeneous catalysts. One development using a zeolite heterogeneous catalyst is in operation.

5.1 CARBOXYLATION OF OLEFINS TO CARBOXYLIC ACIDS

Carboxylation of olefins is accomplished primarily by the homogeneous Koch process using H_3PO_4/BF_3 catalyst. A heterogeneous process using a pentasil zeolite is reported to be operated by BASF at 250–300°C and 300 bar.[1,2]

The branched carboxylic acids have high thermal stability. The C_4 acid and the higher branched acids are used in synthetic oils and transmission fluids and in the formation of resins and paints. Peroxyesters of pivalic acid serve as free-radical initiators for vinyl chloride polymerization. Its largest use is in production of herbicides in which case its esters provide resistance to hydrolysis.[2] Finally, a significant portion is used in synthesis of a number of pharmaceuticals.

Chemistry

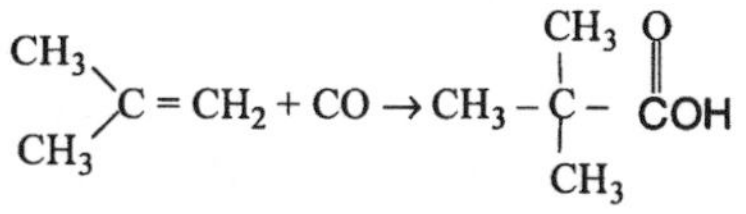

iso-butene

pivalic acid
trimethylacetic acid
neopentanoic acid

Mechanism

It is reasonable to suggest that the much studied homogeneous Koch-reaction mechanism applies to the heterogeneous reaction involving Brønsted sites on zeolites. The Koch reaction is suggested to occur via a carbenium ion from the isobutylene with the proton supplied by the acid catalyst. The carbenium ion then reacts with CO to produce an acyl cation which reacts with water to form the pivalic acid and releases the proton.[2,3] The presence of water is necessary.[4]

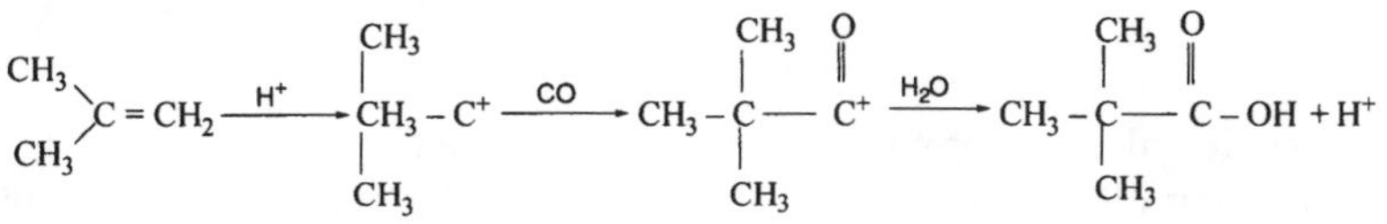

Recent NMR studies confirm the necessity of the presence of water and demonstrate that the reaction can be accomplished with good yield near atmospheric pressure. This situation could permit a lower-pressure process and deserves additional study.[4,5]

Process Unit

One would assume that the reaction is carried out in an adiabatic fixed-bed reactor. Based on the operating pressure reported, the reaction must be conducted as a trickle bed with CO in the gas phase.

Process Kinetics

Since the interaction of CO and chemisorbed iso-butylene is postulated to be a rapid step, adsorption of iso-butylene may be the controlling step. If such is the case, the rate of production of trimethy-acetric acid may be first-order in iso-butylene.

The application of zeolites such as ZSM-5 offer intriguing options. Although Co in zeolite pores reacts selectively to form tertiary carboxylic acids from smaller olefins such as iso-butylene, this is not the case for large olefins such as octene-1, which mainly produces linear C_9 and C_{17} carboxylic acids.[4] Apparently, the octene-1 molecule is already rather large relative to ZSM-5 pore size and cannot form the even more bulky tertiary C_9 carboxylic acid. Obviously, zeolite pore size can have a major effect in influencing selectivity.

5.2   CARBON MONOXIDE + CHLORINE → PHOSGENE

The major use of phosgene (45% in the U.S.A.) is in the production of toluene diisocyanate via nitration of dinitrotoluene, followed by hydrogenation of the dinitrotoluene to toluenediamine. The diamine is phosgenated to toluenedisocyanate.[6] Combination with polypropylene glycol or other polyether alcohols yields polyurethanes. Another 35% is used in the production of other isocyantes, while 10% is used in polycarbonate manufacture. The remaining 10% is accounted for in the manufacture of other isocyanates, pharmaceuticals, agricultural chemicals, and dye stuffs. Much R & D effort has been directed to finding routes to isocyanates other than involving the highly toxic phosgene, and some have been successfully commercialized.[6]

Chemistry

Phosgene is usually produced on site by the activated-carbon catalyzed reaction of carbon monoxide and chlorine.

$$CO + Cl_2 \rightarrow COCl_2 \quad \Delta H = -26.22 \text{ kcal } @ 500°K$$

$$K_p = 22.542 \times 10^3$$

- Non-condensable impurities (make recovery of phosgene difficult)
- Water (will form HCl in the reactor)
- Hydrocarbons and hydrogen (can ignite a reaction of chlorine with steel that is destructive)
- Sulfides (react to produce sulfur chlorides that will surely poison the activated carbon)

Process Unit

The reactor is a multitubular reactor with catalyst in the shell and coolant in the tubes. The reaction is very rapid and is controlled by mass transfer and heat transfer rather than reaction kinetics. The primary process control is the temperature of the cooling medium and its flow rate.

Effluent from the reactor is condensed producing liquid phosgene and uncondensed gases, which are scrubbed for removal of traces of phosgene. The resulting water gas is further treated for waste phosgene decomposition by caustic scrubbing, water washing over activated carbon, or combustion.[7]

Numerous safety precautions must be followed that generate costs that could be avoided by processes that can provide other routes to the desired isocyanates.

The reaction is exothermic and is conducted at around 500 K and slightly above atmospheric pressure. A slight excess of carbon monoxide is fed to ensure that all the chlorine is consumed.

Catalyst Type and Suppliers

Activated carbon, granules

Suppliers

Engelhard, United Catalysts

Licensors

Caloric GmbH, Haldor Topsoe

Catalyst Poisons

Since activated carbon is such a high-area catalyst, it strongly adsorbs many impurities that could reduce its effectiveness. Apparently, protection of equipment and safety considerations preclude allowing a variety of impurities from entering the reactor. These include the following:[7]

- Water (decomposes phosgene; produces corrosive HCl)
- Hydrocarbons and hydrogen (react with chlorine to produce corrosion of steel)
- Sulfides (form sulfur chlorides)

REFERENCES (CARBONYLATION)

1. Weissermel, H., Arpe, J., *Industrial Organic Chemistry,* 3rd edition, VCH, New York, 1997.
2. Keenan, M. J. and Krevalis, M. A. in *Kirk-Othmer Encyclopedia of Chemical Technology,* 4th edition, Vol. 5, p. 194, Wiley, New York, 1993.
3. Olah, G. A. and Molnar, A., *Hydrocarbon Chemistry,* Wiley, New York, 1995.
4. Stepano, A. G.; Luagin, M. V.; Romannikov, V. N. and Zamaraev, K. I. *J. Am. Chem. Soc.* 117, 3613 (1995).
5. Lurgin, M. V.; Romannikov, V. N.; Sidelnikov, V. N. and Zamareav, K. I., *J. Catalysis,* 164, 411 (1996).
6. Weissermel, L. and Arpe, H.J., *Industrial Organic Chemistry,* VCH, New York, 1997.
7. Dunlap, K.L. in *Kirk-Othmer Encyclopedia of Chemical Technology,* Vol. 18, p. 645, Wiley, New York, 1996.

6 Dehydration of Alcohols

Although dehydration of alcohols is used in certain fine chemical productions as a convenient procedure for introducing a double bond in a compound, its large-scale commercial use is confined to only several processes. One involves the production of tetrahydrofuran from 1,4-butanediol. This acid catalyzed reaction usually employs liquid acids in a homogeneous reaction. The other large-scale process dehydrates 1-phenylethanol to produce styrene using a heterogeneous catalyst.

6.1 1-PHENYLETHANOL $\rightarrow$ STYRENE

The Oxirane process, for producing propylene oxide via peroxidation of isobutane or ethylbenzene, forms useful by-products. In the case of ethylbenzene, a substantial amount of 1-phenylethanol is formed. Catalytic dehydration readily converts this alcohol into the more valuable product styrene.

Chemistry

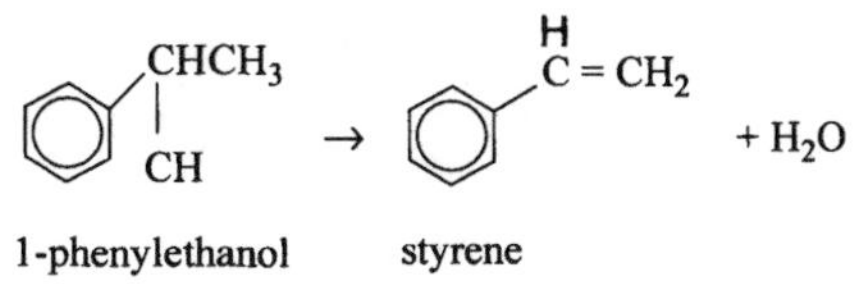

1-phenylethanol   styrene

180–300°C @ modest pressure, 1-2 bar
supported TiO_2 catalyst

Catalyst Type and Suppliers

TiO_2/Al_2O and $TiO_2/silica$

Licensors
ARCO, Shell, Nizhnekamsk (CIS)

Process Unit

The dehydration is carried out in the vapor phase in a fixed-bed adiabatic reactor in the range of 1-2 bar and 180-300°C.[1,2] Conversions of as high as 97% and selectivities of 89–95% have been reported.

REFERENCES (DEHYDRATION OF ALCOHOLS)

1. Hubbell, D. S. in *Encyclopedia of Chemical Processing and Design*, J. J. McKetta, editor, Vol. 55, p. 217, Marcel Dekker, New York, 1996.
2. Weissermel, J., Arpe, J., *Industrial Organic Chemistry*, 3rd edition, VCH, New York, 1997.

7 Dehydrochlorination

Dehydrochlorination simply describes a reaction that removes HCl from a chlorine-containing organic compound. The most prominent such reactions occur in an alternate procedure for producing linear olefins used in manufacturing linear alkyl benzenes for production of detergents.

7.1 C₁₀–C₁₃ CHLORIDE → LINEAR OLEFINS

$$R'CH - CH_2R'' \rightarrow R'CH = CHR'' + HCl$$
$$\quad\,\, |$$
$$\quad\,\, Cl$$

200–350°C, liquid phase
catalyst: silica-alumina

Catalyst Type and Suppliers

Silica-alumina and Pd-on-alumina
Grace-Davison, PQ Corp., Johnson Matthey

8 Dehydrogenation

Although dehydrogenation is the reverse of hydrogenation, and theoretically the same catalysts should be operable in either direction, there are substantial practical differences between the two processes. These differences become clear when the following facts are considered:

- Since dehydrogenation is endothermic, high temperatures and lower pressures favor dehydrogenation.
- Low partial pressures of reactants favor dehydrogenation in the gas phase.
- Removal of hydrogen by purging from a liquid reactant system or reaction of hydrogen with an acceptor reactant as it is formed favors dehydrogenation.
- Dehydrogenation catalysts are less sensitive to catalyst poisons, possibly because of the lower chemisorption of potential poisons at the high operating temperatures.[1]

8.1 DEHYDROGENATION OF ETHYLBENZENE

8.1.1 ETHYLBENZENE → STYRENE

The dehydrogenation of ethylbenzene is, on a tonnage basis, the major catalytic dehydrogenation process. Styrene is the fourth in the series of most used monomers.

ethylene > vinyl chloride > propylene > styrene

Polystyrene is the major product (63%) produced from styrene. Its uses are well known to most consumers: protective packaging, electronic housings, toys, insulation, and disposable cups and utensils. Copolymers have unusually valuable properties. Styrene-butadiene rubber (SBR) is the most widely used synthetic rubber. It is a major component of tire treads for automobiles, where it is blended with polybutadiene, yielding a tread surface of excellent grip. Natural rubber, because of its better flexing properties, is blended with SBR for use in forming the carcass.[3] The liner is made from butyl rubber because it is impermeable to air. Radial tires use a higher percentage of natural rubber in the blends with SBR.[3] SBR rubber is also used for electrical wire insulation and footwear. Styrene-butadiene latex is used for paper coatings and carpets. Styrene-acrylonitrile is molded for the exterior cover of various home appliances and housewares, as is acrylonitrile-butadiene-styrene, which is a also cast for automobile parts, pipe, electronic devices, and canoes.[2]

Copolymers of styrene such as styrene-acrylonitrile (SAN) and acrylonitrile-butadiene-styrene (ABS) can be produced with excellent crystal-like clarity and are easily dyed.

Chemistry

Dehydrogenation of ethylbenzene to styrene is an equilibrium-limited reaction. It is highly endothermic ($\Delta H = 29.83$ kcal/g mole) and involves an increase in moles. Several side reactions occur that can be minimized by careful control of operating conditions and catalyst selectivity. The suggested reaction scheme is as follows.[4,12]

			900 K	
			ΔH, kcal	K_p
$C_6H_5CH_2CH_3 \rightarrow C_6H_5CHCH_2 + H_2$	(1)		29.83	0.376
$C_6H_5CH_2CH_3 \rightarrow C_6H_6 + C_2H_4$	(2)		25.22	0.238
$C_6H_5CH_2CH_3 + H_2 \rightarrow C_6H_5CH_3 + CH_4$	(3)		−15.18	8.874×10^{-39}
$C_2H_4 + 2H_2O \rightarrow 2CO + 4H_2$	(4)		55.43	4.103×10^{-5}
$CH_4 + H_2O \rightarrow CO + 3H_2$	(5)		59.59	1.291
$CO + H_2O \rightarrow CO_2 + H_2$	(6)		− 8.53	2.193

Reaction 1 is the main reaction that accomplishes the selective dehydrogenation of the ethyl side chain without hydrogenating the benzene ring with hydrogen that accumulates as the reaction progresses. Figure 8.1 illustrates the effect of temperature, operating pressure, and steam to ethylbenzene ratio on the equilibrium conversion. Of course, the relevant pressures influencing reaction rate are the partial pressures of the reactants. In addition to high-temperature operation, equilibrium conversion is improved for a reaction with an increase in moles by lower partial pressures, by reducing operating pressure, and by increasing the steam-to-ethylbenzene ratio in the feed. Steam, however, represents a significant operating cost factor, and improvements in the process have invariably been accomplished by lower steam rates and a commensurate lowering of total pressure to counteract the higher reactant and product mole fractions. In fact, most adiabatic styrene reactors now operate at sub-atmospheric pressures that allow steam rates as low as 4 or 6:1 moles steam per mole of ethylbenzene. Earlier reactors were operated in the range of a 14–18 steam-to-ethylbenzene mole ratio.

There is a limit to how low the steam rate can be lowered, because steam serves several other valuable purposes as follows:

- It provides the required heat to the reaction mix to replace the endothermic heat consumed by the reaction.

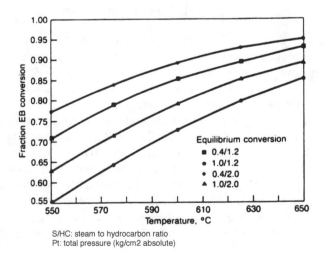

S/HC: steam to hydrocarbon ratio
Pt: total pressure (kg/cm2 absolute)

FIGURE 8.1 Equilibrium Conversion of Ethylbenzene to Styrene [at Various Mass Ratios of Steam-to-Ethylbenzene and Total Pressures (P_t at 0.4 and 1.0 kg/cm^2 and S/EB at 1.2 and 2.0)]. Reprinted by permission: Sundaram, K. M., Sardina, H., Fernandez-Boujin, J. M. and Hildreth, J. M., *Hydrocarbon Processing,* p. 83, Jan. 1981. P_t = total pressure, S = steam, and EB = ethylbenzene.

- It maintains the iron oxide catalyst in the optimum oxidation state.
- It removes coke buildup on the catalyst via the water gas reaction,

$$H_2O + C \rightarrow CO + H_2$$

Since preheating ethylbenzene to the desired operating temperature will cause extensive thermal cracking, steam also serves conveniently as a means for rapidly heating ethylbenzene to the desired inlet temperature for the requisite mixture. Steam is heated in a furnace to a higher temperature (~750°C or higher) and then combined with preheated ethylbenzene at a lower temperature (~525°C).

Reactions 2 and 3, which are respectively cracking and hydrocracking reactions, can occur both catalytically and thermally. Reactions 4, 5, and 6 are typical water-gas reactions. All of these side reactions can be minimized by selective catalysts and by avoiding excessively high reaction temperatures (above 650°C). Other by-products that are produced in much smaller quantities include phenylacetylene, cumene, n-propylbenzene, α-methylstyrene, biphenyl, stilbene, and phenanthrane.[2] Thermal cracking is a significant contributor to the formation of these side-reaction aromatics, and, again, avoidance of excessively high temperatures will reduce their production.

Catalyst Types

The commercial catalyst of choice for most styrene plants is a mixture of iron oxide and potassium carbonate with small amounts of one or more of the following promoters: Cr_2O_3, Ce_2O_3, MoO_3, CaO, MgO, and V_2O_5.[5,6] Pigment-grade iron oxide is kneaded with potassium carbonate and water to produce a paste that can be extruded or further prepared for pilling.[7] Heat treatment and calcining at the catalyst plant causes decomposition of the potassium carbonate.

$$K_2CO_3 \Leftrightarrow K_2O + CO_2$$

It has been suggested[6] that potassium ferrite sites are formed on the surface of the iron oxide[6] by reacting with K_2O to produce the active sites.

$$K_2O + Fe_2O_3 \rightarrow K_2Fe_2O_4$$

Others have suggested that the K_2O increased catalytic activity by promoting electron transfer at the solid-gas interface.[8] One issue, however, is clear. The K_2O promoted catalyst exhibits an order of magnitude greater activity than unpromoted iron oxide. This high activity, however, is reached after three to four days of operation at atmospheric pressure, 600°C, and 2.0 weight ratio of steam-to-hydrocarbon. The actual planned design operating conditions can then be imposed.

The Cr_2O_3 promoter in the range of 1–3 wt% acts as a structural stabilizer by preventing sintering and the resulting loss of surface area.[8] Alternatively, other binders, such as cement, can be used to assure structural strength.[2] Other promoters are used including, respectively, MgO with Cr_2O_3 and Ce_2O_3, MoO_3, and MgO without Cr_2O_3.

The K_2O component acts as an effective water-gas shift catalyst and continuously removes carbonaceous deposits as CO and CO_2. The long catalyst life of one to two years is attributed to this property.

Catalyst Surface Area

Calcination of the extrudates or pills at the catalyst plant is done at a high enough temperature (900–950°C) to reduce the surface area to 1.5 to 3 m²/g. In addition, combustible fibrous materials can be added that, under calcining conditions, vaporize and leave macropores. The combination of low surface area due to calcining, which tends to eliminate micropores, and the overt creation of macropores serves to enhance selectivity. At the high temperatures required to assure favorable equilibrium conditions for styrene formation, slower thermal and catalytic reactions including cracking reactions are significant enough to cause increased unwanted side reaction products if given enough time in the catalyst pores. By confining pore structure to macropores, diffusion in and out of the pores becomes more rapid and thereby minimizes side reaction products and thus improves selectivity.

The intrinsic reaction rate of the major reaction is high and, even with large pores, the actual rate is affected by intraparticle diffusion. For any given catalyst composition and operating conditions, higher conversions are obtained by using smaller diameter extrusions. This option is, of course, limited by an acceptable maximum pressure drop.

Catalyst Suppliers

The major suppliers of ethylbenzene dehydrogenation catalysts provide a variety of promoted iron-oxide catalysts to serve the specific needs of each operating plant. The offerings can be divided into two general categories, high-activity, high-selectively and high-stability catalysts.[5,20] The high activity catalysts permit high styrene yields at lower operating temperature and lower steam ratios. The high-selectivity catalysts require some temperature increase in exchange for much improved selectivity. Other catalysts optimize high-activity and high selectivity characteristics. It should be remembered that in all cases selectivity tends to decline at higher conversions approaching equilibrium since the rate of dehydrogenation of ethylbenzene will decline while the rate of the side reactions which are not equilibrium limited will increase. Potential users should seek the advice of supplier's technical representatives to optimize the economics of their particular unit.

Suppliers

Criterion, United Catalysts, Engelhard

Licensors

ABB Lummus Global, UOP LLC, Fina/Raytheon, Lungi, BASF, Dow/Engelhard

Typical Catalyst Characteristics

		Components Wt% (Ranges)		
Form	Size, in	Fe_2O_3	K_2O	Promoters
Extrudates (plain or ribbed for lower ΔP)	1/8, 3/16, 1/4, 3/8	45–75	27–10	Various (see text)

Catalyst Deactivation

Poisons

Ethylbenzene dehydrogenation catalysts are subject to poisoning by halides (usually chlorides and fluorides) and compounds of sulfur, phosphorous, and silica. Solids contained in steam caused by low-quality boiler feed water can also deactivate the catalyst by coating the active sites.[5] Ethylbenzene is produced either by Friedel–Crafts alkylation of benzene with ethylene using $AlCl_3$ in the liquid phase along with co-catalysts such as ethyl chloride, BF_3, $ZnCl_4$, $SnCl_4$, H_3PO_4, or by zeolite catalyzed alkylation. Although organic sulfur compounds at times can easily contaminate the petrochemical feeds of benzene and ethylene, they are temporary poisons, and the catalyst will return to previous activity once the sulfur compounds are no longer in the feed. The halides, by contrast, act as permanent poisons. The feed to the reactor should have no more than 1 ppm of chloride to perform satisfactorily over a long period of operation. Halides react with K_2O to produce KCl, which apparently lowers the promoting effect of K_2O.[8] In addition, significant amounts of KCl tend to be transferred as a vapor to the cooler part of the bed which is toward the outlet. Such large quantities can totally destroy the activity of that portion of the bed as well as seriously affect selectivity in the remainder of the bed because of K_2O loss.

Potassium Migration

During prolonged use the potassium promoter tends to migrate from the periphery toward the center of the catalyst pellet. In so doing, the highly concentrated potassium content in the center region, and its absence in the peripheral region, ultimately leaves both regions inactive and confines the active region to a narrow band between the center and the periphery.[8] At this condition, the catalyst presents little or no diffusion resistance and can continue operation for some time until further loss of activity occurs.

Steam treatment of used catalysts has been shown to cause chemical-vapor transport from regions of high K_2O concentration to those of low concentration.[9]

$$K_2O(s) + H_2O(g) \rightarrow 2KOH(g)$$

The KOH will migrate to the low concentration region containing smaller agglomerates of K_2O. Because of the lower surface area-to-mass ratio of the larger agglomerates, they will have a lower temperature and drive the exothermic equilibrium to the right. Conversely, as the KOH moves to the region of smaller agglomerates, the reverse reaction will occur, and K_2O is deposited. Backflow of steam through a bed removed from service can also cause redistribution of K_2O from the lower-temperature outlet region throughout the remainder of the bed.[10]

Process Units

Reactors used in dehydrogenation of ethylbenzene include both adiabatic and isothermal designs. Adiabatic reactors are the most widely used, especially in the United States. Isothermal reactors have found favor in Europe with several producers. With the newer catalysts conversions of 60 to 75% and selectivities of 85 to 95% are possible with either reactor type.[2]

Adiabatic Reactor System

Adiabatic reactors for styrene production, as is the case for many other adiabatic units, are staged so that reheating of the reaction mix can be accomplished either by direct contact of additional steam or by means of a superheated-steam exchanger. Two separate reactors or single reactors with separate beds and the necessary internals for adding reheating steam may be used.

Superheated steam at around 830°C is mixed with preheated ethylbenzene (~530°C). The ethylbenzene can be heated to the lower temperature in a heat exchanger along with a portion of the steam to be charged. The higher temperature superheated steam is heated in a direct-fired furnace. It is important to avoid heating the ethylbenzene alone to higher temperatures and allowing it to have excessive contact type in the transfer line to the reactor, for ethylbenzene will undergo cracking reactions and reduce the yield of styrene.

Most adiabatic reactors are radial reactors, which are essential for low pressure-drop operation. Since the volumetric flow increases as reaction proceeds due to the increase in moles, the flow is directed from the inside of the annular bed radially outward.

Isothermal Reactor

Two major types of isothermal reactors for styrene production have been described.[2,9] The Lurgi reactor uses 20,000 to 30,000 tubes, 1 to 2-1/2 in. (2.5 to 6.4 cm) diameter and 8 to 10 ft (2.4 to 3 m) lengths packed with catalyst and surrounded by flowing molten salt solution of carbonates of sodium, lithium, and potassium. The molten salt is circulated through an external heater to maintain its temperature at about 630°C, 20° above the reaction temperature, which is held constant.

Steam is used for the same reasons given for the adiabatic system except it is not needed to provide the heat energy to replace the endothermic reaction now being supplied by the heating medium. Thus, less steam is consumed than for the adiabatic reactor.

The other major isothermal process is used by BASF in Europe. It differs from the Lurgi in that it uses flue gas from a fired heater at 760°C (1400°F) as the heating medium. This higher temperature is necessary to compensate for the low heat capacity of the flue gas. The packed tubes are fewer in number and larger; 4–8 in. (10–20 cm) diameter and 8.2–13.1 ft (2.5–4 m) length.[9]

Both of these isothermal processes can claim some improvement in yield and, because of better temperature control and optimization, savings in steam cost. However, the capital costs are higher than the adiabatic units, and the maximum practical size of a single isothermal reactor limits the total capacity to substantially less than a single modern adiabatic design.

Adiabatic Reactor with Oxygen Addition (Oxidative Dehydrogenation)

A process has been described in which oxygen is added between stages in a separate bed containing an oxidation catalyst reported to be composed of Pt, Sn, and Li on alumina. Sufficient oxygen is added to combine with the hydrogen and form steam. Thus, the equilibrium is shifted further toward styrene by removing the hydrogen. In addition, heat is supplied by the oxidation for the second dehydrogenization bed. Concerns have been expressed about safety, formation of phenylacetylene (may effect styrene polymer quality), and extra cost for the Pt catalyst. One plant has been built in Japan. The dehydrogenation catalyst is thought to be similar to that for standard existing dehydrogenation-only units.

Operating Conditions

All of the more recent reactors just described are operated at subatmospheric pressure (0.5 atm outlet) to take advantage of the improved equilibrium conditions. Also, lower pressures permit lower steam rates that can yield substantial energy savings. Since temperatures above 610°C mark the beginning of undesirable cracking reactions,[2] isothermal reactors are operated conveniently at 610°C. Adiabatic reactors, however, experience temperature decline along the bed, and it is necessary to enter the beds with a steam-ethylbenzene mixture at a temperature as high as 630–640°C.[9] Thus, some cracking does occur, but it is minimized by the relatively short contact time at the entering temperature.

As the catalyst slowly deactivates over time, it becomes necessary to gradually raise the operating temperature in all the reactor types to keep up production. This procedure ultimately approaches the design temperature for the reactor, and continued operation can then occur only by lowering the feed rate (lowering the space velocity). In this way, lower production may be acceptable until the planned not-too-distant turn-around date.

Process Kinetics

Various rate expressions have been proposed for the main reaction and the side reactions.[11–16] In general, they take the Langmuir–Hinshelwood, Hougen–Watson (LHHW) form or use simple molecular (power law) forms as follows:

LHHW Form

$$C_6H_5CH_2CH_3 \Leftrightarrow C_6H_5CHCH_2 + H_2 \qquad\qquad r_1 = \frac{k_{p1}K_E(P_E - P_HP_S/K_1)}{1 + K_EP_E + K_SP_S}$$

$$C_6H_5CH_2CH_3 \rightarrow C_6H_6 + C_2H_4 \qquad\qquad r_2 = \frac{k_{p2}K_EP_E}{1 + K_EP_E + K_SP_S}$$

$$C_6H_5CH_2CH_3 + H_2 \rightarrow C_6H_5CH_3 + CH_4 \qquad\qquad r_3 = \frac{k_{p1}K_EP_E}{1 + K_EP_E + K_SP_S}$$

where E = ethylbenzene
　　　 H = hydrogen
　　　 S = styrene

Side reactions 4, 5, and 6 have often been written as simple power-law rate expressions as shown below; or if one assumes these reactions to be very rapid, they can be combined with reactions 2 and 3 to yield the following, assuming that half of the ethylene is converted to methane ($1/2\ C_2H_4 + H_2 \rightarrow CH_4$):

$$(2') \quad C_6H_5CH_2CH_3 + 2H_2O \rightarrow CO_2 + 3H_2 + C_6H_5CH_3$$

$$(3') \quad C_6H_5CH_2CH_3 + 2H_2O \rightarrow CO_2 + C_6H_6 + 2H_2 + CH_4$$

An industrial model for simulation and optimization of a styrene unit has been described with brief references to the kinetics.[18] Since most of the reactions, at the high temperatures required for dehydrogenation, occur not only catalytically but to some degree thermally as free-radical reactions, the model formulated the thermal reactions as simple molecular power-law equations. Catalytic rates for the same apparent overall reactions were written in forms similar to LHHW equations.

Power-Law

Typical power-law forms have also been used.[15]

$$r_1 = k_2(P_E - P_SP_H/K_1)$$
$$r_2 = k_2P_E$$
$$r_3 = k_3P_EP_H$$
$$r_4 = k_4P_C^{1/2}P_W$$
$$r_5 = k_5P_MP_W$$
$$r_6 = k_6P_{CO}P_W$$

Complexity

As is true with so many commercial reactions, the reaction system is complex. Initially, the reactions are strongly controlled by intraparticle diffusion. But because of changes that occur in the catalyst, including K_2O migration, the process gradually becomes reaction controlled while catalyst activity is declining. To base any practical model on intrinsic kinetics and effectiveness factors seems futile. Time would be better spent on obtaining a useful empirical expression for activity as a function of time-on-stream and operating temperature history. The actual activity decline is related directly to accumulated coke and the loss of active components such as K_2O by agglomeration.[17]

8.2   STYRENE DERIVATIVES FROM OTHER ALKYL AROMATICS

Several styrene derivatives are commercially produced by dehydrogenation over iron oxide catalysts, but in much smaller quantities. These products include the vinyltoluene isomers and divinylbenzene. These compounds are more reactive than styrene and thus cause more side reaction products than styrene in the dehydrogenation reaction.[19]

8.2.1 DIETHYLBENZENE → DIVINYLBENZENES

diethylbenzene   divinylbenzene

The process is conducted above 600°C using superheated steam in a fixed-bed reactor.[9] Diethylbenzene is a side reaction product in the production of ethylbenzene and occurs in all three isomeric forms.

Dehydrogenation produces a number of side reaction products including benzene, toluene, xylene, styrene, and toluene derivatives. All three isomers (ortho, meta, and para diethylbenzene) are produced, but the ortho isomer converts to naphthalene, which must be separated from the other isomers because it has no useful polymerizable functions and also causes an objectionable odor.[9,19] The refined product consists of meta and para divinylbenzene and ethylvinylbenzene with various amounts of divinylbenzene, depending on the commercial grade.

Because of the two polymerizable groups, divinylbenzene is used in small amounts as a copolymer with styrene to produce crosslinking and a product of low solubility, heat resistance, surface hardness, and high impact and tensile strengths. These characteristics are ideally suited for ion exchange resins for water conditioning, which constitutes its major use.

Process Kinetics

Detailed Langmuir-Hinshelwood rate equations have been developed for the two-step mechanism of the formation of ethylvinylbenzene from diethylbenzene and then divinylbenzene from ethylvinylbenzene.[21] The initial rates can be expressed rather simply.

$$r_1 = \frac{k_1 K_D P_D}{\left(1 + K_E P_E\right)^2}$$

$$r_2 = \frac{k_2 K_D P_D}{\left(1 + K_E P_E\right)^2}$$

where r_1 = rate of reaction of diethylbenzene to ethylvinylbenzene

r_2 = rate of reaction of ethylvinylbenzene to divinylbenzene

subscripts D and E = diethylbenzene and ethylvinylbenzene

Equations involving higher conversions in which the reverse reaction and other adsorbed products must be considered are provided, along with a suggested model for simulation.[21]

8.2.2 ETHYLTOLUENE → VINYLTOLUENES

ethyltoluene   vinyltoluene

Ethyltoluene is produced by the Friedel–Crafts or zeolite-catalyzed alkylation of toluene with ethylene to form the isomers of ethyltoluene. Again, the ortho isomer must be removed prior to dehydrogenation, since it forms indene, which causes undesired characteristics in the final copolymers.[9]

The meta and para ethyltoluenes are then dehydrogenated using iron oxide catalysts and steam. The purified commercial product contains 60% meta and 40% p-vinylbenzene and is used in producing a copolymer with styrene. These copolymers exhibit high heat resistances, good flow properties, and increased solubility in aliphatic solvents.[9,19]

8.2.3 ISOPROPYLBENZENE (CUMENE) → α-METHYLSTYRENE

isopropylbenzene
cumene

α-methylstyrene

The styrene derivative α-methylstyrene is used as a monomer in conjunction with styrene and other monomers to produce specialized products. When copolymerized with acrylonitrile/butadiene/styrene (ABS) polymers, it increases the temperature at which heat distortion occurs. In copolymers for coatings and resins, it lowers the reaction rates to a practical value and improves product clarity.[9,19]

Dehydrogenation of cumene is a viable process, similar to all other styrene dehydrogenations to styrene or styrene derivatives, but it is conducted on a small scale because the quantities required are modest. However, most of the alpha-methylstyrene used today is obtained as a by-product from the production of phenol from cumene.

Catalyst Suppliers

The catalyst used in dehydrogenation in the production of various styrene derivatives are essentially the same as used in producing styrene. See page 70.

8.3 DEHYDROGENATION OF LOWER ALKANES

8.3.1 GENERAL BACKGROUND

Dehydrogenation of alkanes to olefins, like other dehydrogenations in the vapor phase, requires high temperatures, low pressures, and a low hydrogen-to-hydrocarbon ratio. This latter requirement impinges on the need for maintaining catalyst activity by hydrogen in sufficient amounts to slow the formation of coke.

Thermodynamics

Figures 8.2 and 8.3 present equilibrium plots for several lower alkanes.[1] These highly endothermic reactions, with an increase in moles and low values of K_p, require relatively high temperatures and either subatmospheric pressures or atmospheric pressures or above with high steam content in the feed. Ethane requires very high temperatures, which enables thermal steam cracking to be the preferred and cheaper route. Higher olefins, C_3, C_4, and C_5, can be conveniently recovered from both thermal and catalytic cracking units. High demand, however, has driven the development of separate catalyzed dehydrogenation processes for converting alkanes to olefins or diolefins.

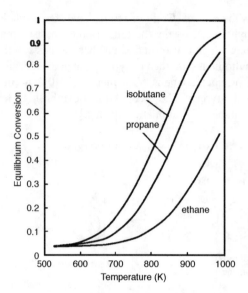

FIGURE 8.2 Equilibrium Conversions for Dehydrogenation of Ethane, Propane, and Isobutane as a Function of Temperature at 1 atm. Reprinted by permission: Resasco, D. E. and Haller, G. L., *Catalysis: Specialist Periodical Report,* Vol. 11, p. 381, The Royal Society of Chemistry, Cambridge, England, 1994.

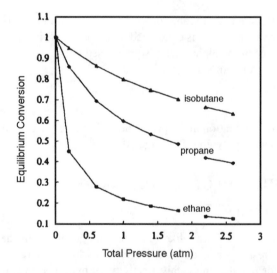

FIGURE 8.3 Equilibrium Conversion for Dehydrogenation of Ethane, Propane, and Isobutane as a Function of Pressure at 873 K. Reprinted by permission: Resasco, D. E. and Haller, G. L., in *Catalysis: Specialist Periodical Report,* Vol. 11, p. 382, The Royal Society of Chemistry, Cambridge, England, 1994.

The reaction pathways are much more complex for dehydrogenation of C_4 and C_5 than for C_2 and C_3 paraffins, because skeletal isomers and dienes enter the picture.[2] Figure 8.4 shows the complex equilibrium conditions for C_4 straight-chain hydrocarbons. Analogous equilibrium product distributions apply for C_5 alkanes and alkenes. Clearly, catalysts of low selectivity for skeletal isomerization and coke formation are essential. The ultimate thermodynamic product of all alkanes and alkenes is coke, and alkenes are the most susceptible to such fate. Hydrogen inhibits coke formation to some degree, but large amounts reduce the net rate of dehydrogenation. The catalyst

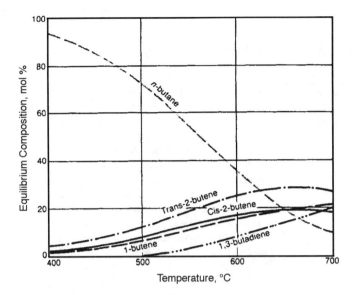

FIGURE 8.4 Equilibria for n-Butane Dehydrogenation (1 atm). Reprinted by permission: Vora, B. V. and Imai., T., *Hydrocarbon Processing,* p. 171, April 1982.

support should be nonacidic, or made so by incorporation of an alkali additive, since the rate of coke formation is enhanced by acid sites.[1] Even so, coke formation remains a major source of catalyst deactivation, which necessitates frequent catalyst regeneration. Once olefins form, they constitute, by virtue of the unsaturation, active species for secondary reactions. The temperature required for practical dehydrogenation rates facilitates cracking, polymerization, and coke formation. The successful catalyst recipes involve additives that dramatically reduce such undesired reactions.

Catalyst Types and Processes

Most alkane dehydrogenation processes are proprietary as are the catalysts used. A summary from information in the open literature (see Refs. 1,2, and 4) is presented in Table 8.1. This table is referred to in the following sections on the various olefins and diolefins produced by dehydrogenation.

Catalyst Suppliers

The specific catalyst suppliers are usually ones directly involved with the proprietary technology licensor. Since these suppliers could change, various suppliers who could offer similar catalysts upon request by a licensor are included.

   Chromia-alumina catalysts: United Catalyst, Engelhard
   Pt-on-alumina catalysts: United Catalyst, Engelhard, Johnson Matthey, Precious Metals
   Other catalyst not related to a licensed process:
      Fe_2O_3-K_2CO_3-Cr_2O_3: United Catalyst and Engelhard (for isoprene production)

8.3.2   PROPANE → PROPYLENE

$$CH_3CH_2CH_3 \rightarrow CH_2 = CHCH_3 + H_2$$

$$\Delta H = 30.93 \text{ kcal}, \; K_p = 0.556 \text{ @ } 900 \text{ K}$$

TABLE 8.1
Proprietary Lower Alkane Dehydrogenation Processes*

Process Name	Licensor	Description	Catalyst Type
Catadiene and Catofin	ABB Lummus Global	These two processes are similar in principle. Catadiene refers to the process for producing butadiene or isoprene from butane and isopentane, respectively. Catofin refers to a similar process for producing isobutylene from isobutane and propylene from propane. These processes operate in the gas phase at subatmospheric pressure (0.1–0.25) and 558–650°C. Neither steam nor hydrogen are used to prevent coke formation. Instead, five fixed-bed reactors (horizontal drums) are operated in parallel with two on stream, two being regenerated, and one being purged. Each reactor, when receiving feed operates for approximately four minutes per cycle. A total cycle of reaction, regeneration, and purging requires 22 minutes. Although a small fraction of feed goes to coke, the heat evolved during regeneration produces and stores in the bed the endothermic energy requirement for the next reaction cycle. The number of reactors and cycle length can be varied.	Chromia-alumina cylindrical pellets supplied by United Catalysts (cylinders or shaped).
Oleflex	UOP LLC	The process employes four radial fixed-bed reactors in series with interstage heaters between reactors. Catalyst is circulated continuously from the first through each successive reactor, and from the last to a regenerator and then back to the first reactor. Coke is removed in the regenerator along with vapor thus produced along with carbon oxides. Regeneration conditions also cause the Pt to be redistributed throughout the catalyst particles, thus preserving activity. Process is used for dehydrogenation of C_2-C_4 olefins or ispentane feedstock. Reaction conditions: 600–630°C at slightly above atmospheric pressure.	Thought to be 0.3 wt% Pt-on-alumina with alkali to eliminate the acidity of the alumina. Zn and Cu are used as promoters. The particles are spheroidal in shape to facilitate circulation.
STAR	Phillips Petroleum	Steam active reforming (STAR) operates at high steam rates and at essentially isothermal conditions possible by multiple catalyst packed tubes installed in direct-fired furnace modules. Depending on the design there can be more than one module. Process is used for C_2-C_4 olefins. Reaction conditions: 600°C at pressures up to 3.5 atm.	0.2–0.6 w% Pt on Zn-aluminate, a support that is nonacidic, has high stability, and does not catalyze isomerization. Sn is used as a promoter, which reduces coke formation.

*All these processes involve recycle of unreacted reactant. Based on descriptions in Refs. 1, 2, 4, 15, and 16.

Although a major portion of propylene production is obtained from steam cracking units, the demand for polypropylene continues to rise. Depending on the product mix and feedstock characteristics in refineries or petrochemical plants, it may be advantageous to produce some propylene by catalytic dehydrogenation of propane. Also, areas of the world not near adequate sources of propylene from steam cracking will find propane catalytic dehydrogenation a valuable alternative.

The Catofin and Oleflex processes have been used for exclusive production of polymer-grade propylene. See Table 8.1.

8.3.3 ISOBUTANE → ISOBUTYLENE (ISOBUTENE)

Beginning with C_4 and higher alkanes, the dehydrogenation can become complex, involving skeletal isomerization, diene formations, and some cracking.[2]

CH₃CH(CH₃)₂ —-H₂→ CH₂ = C(CH₃)₂ ΔH = 29.32 kcal, K_p = 1.762
isobutane   isobutylene

ΔH - 1.86 kcal   ΔH - 3.85 kcal
K_p = 2.027 K_p = 0.41

CH₃CH₂CH₂CH₃ —-H₂→ CH₃CH₂CH = CH₂ —-H₂→ CH₂ = CHCH = CH₂
n-butane   n-butene   butadiene

ΔH - 31.31 k=l   ΔH = 28.50
K_p = 0.356 K_p = 0.290

Values of ΔH and K_p at 900°K.

If isobutylene is the desired product, the catalyst must have high selectivity for isobutylene, which means low activity for skeletal isomerization and coke formation.

Isobutylene is used to produce the polymer polyisobutylene known commonly as *butyl rubber*, which, because of its air impermeability, is used to form the liner of tires or as inner tubes.[3] It is also used as a liner for large reservoirs, as engine mounts, and in the production of the antioxidant butylated hydroxytoluene (BHT).

Most isobutylene has been obtained by separation from C_4 mixtures produced by steam cracking units. Isobutylene constitutes one-quarter of the C_4 fraction by weight of naphtha steam cracking product. Light hydrocarbon feeds produce about the same fraction of isobutylene in the C_4 product. Ethane and propane are converted, respectively, to 2 and 4 wt% C_4-hydrocarbons while naphtha produces up to 12 percent.[5,7] But the large number and size of steam cracking units creates a major amount of isobutylene in either case. The total C_4 hydrocarbon cut from fluid catalytic cracking is also larger, and 15% of it is isobutylene by weight. In recent years, the demand for the antioxidant gasoline additive, methyl t-butyl ether (MTBE) has required an additional major source of isobutylene, namely, dehydrogenation of isobutane.

Isobutylene is very reactive and can conveniently be reacted using acid addition of methanol to form MTBE. Any unreacted isobutane and other C_4 hydrocarbons that will not react can be easily separated from the MTBE. It is for this reason that MTBE units are located adjacent to isobutane dehydrogenation units.

The most frequently offered processes for producing isobutylene by dehydrogenation of isobutane are Catofin, Oleflex, and STAR. See Table 8.1 for a description of each.

8.3.4 N-BUTANE → 1-BUTENE → 1-3 BUTADIENE

$$CH_3CH_2CH_2CH_3 \Leftrightarrow CH_2 = CHCH_2CH_3 + H_2 \rightarrow CH_2 = CHCH = CH_2 + H_2$$

At 900 K: ΔH = 31.31 kcal   ΔH = 28.50 kcal
                 K_p = 0.356 K_p = 0.290

The major use of butadiene is in the manufacture of styrene-butadiene rubber latex, followed by polybutadiene rubber. Lesser uses include production of nitrile-butadiene rubber, polychloroprene rubber, acrylonitrile-butadiene-styrene polymer, and hexamethylenediamine.[5]

The C_4 fraction produced by steam cracking of ethane, propane, and naphtha constitutes, respectively, 2, 4, and 12 wt% of the product mix. The weight percent of butadiene in these several products is, respectively, approximately 2, 3, 5 percent. Because of the large production of ethylene and propylene, most requirements for butadiene can now be met by separating the butadiene from steam cracker C_4 fractions by extractive distillation.[6] Although Western Europe produces most ethylene and propylene by steam cracking of naphtha, about more than one-half of U.S. production uses ethane and propane. It is possible that European plants will increase the use of ethane and propane as North Sea exploration yields more light gas production. A corresponding decline in butadiene from steam cracking will then occur. As these trends develop, more isobutane dehydrogenation capacity will be needed.

The most used dehydrogenation process is the Catadiene process. See Table 8.1 for a description. Since butadiene has been produced in large quantities beginning in the U.S.A. during World War II, (1943), many catalysts and processes have been developed including Fe_2O_3-on-bauxite, Fe_2O_3-Cr_2O_3 with K_2O (Shell, similar to styrene catalyst), and calcium-nickel phosphate—Cr_2O_3 (Dow). Phillips developed a two-stage process with chromia-alumina to produce butene, which is then dehydrogenated to 1,3 butadiene in the second stage over an Fe/Mg/K/Cr mixed-oxide catalyst in the presence of excess (10–20 per mole of butene).

Oxidative Dehydrogenation of Butene to Butadiene

Oxidative dehydrogenation of butene has been practiced by Phillips and Petro-Tex using added oxygen to supply heat and shift the equilibrium by effectively removing hydrogen by reacting it exothermally to water vapor. Also, the oxygen burns off coke deposits on the catalyst. Operation at lower temperatures than in the traditional process is possible. Selectivities of up to 93% at conversions of 65–75%. A fixed-bed adiabatic reactor was used.[14] These plants were shut down in the U.S.A. as adequate supplies of butadiene became available from steam cracking units.

8.3.5   ISOPENTANE → ISOPRENE

$$CH_3CH(CH_3)CH_2CH_3 \rightarrow CH_2 = C(CH_3)CH = CH_2 + 2H_2 \qquad \Delta H = 70.85 \text{ kcal}$$

isopentane	isoprene	K_p = 3.917 x 10-4
2-methylbutane	2-methyl-l,3-butadiene	@ 900°C

Isoprene is polymerized to cis-1,4-polyisoprene, often called *synthetic polyisoprene,* or *synthetic natural rubber,* because natural rubber contains 93–95 wt% of this same polyisoprene. It is used in applications similar to those filled by natural rubber. These applications include tires, belts, gaskets, rubber sheeting, rubber bands, bottle nipples, footwear, sealants, and caulking.[8] Isoprene is also used in producing a number of copolymers. See Table 8.2.

Most isoprene is obtained by liquid-liquid extraction and extractive distillation of the C_5 fraction of the products of steam cracking of mainly naphtha or heavier feedstocks such as gas oil to produce ethylene and propylene. Such recovery proves more energy efficient than dehydrogenation of isopentene.

The dehydrogenation process when used is similar to that for producing butadiene (see Table 8.1). Catalysts other than Cr_2O_3-on-alumina have been used to produce isoprene from isopentane These include a Shell process employing, at 600°C, Fe_2O_3-K_2CO_3-Cr_2O_3 and Sr-Ni-phosphate.[9,15]

TABLE 8.2
Isoprene Copolymers

Copolymer	Description	Uses
Isobutylene-isoprene (butyl rubber)	Isoprene used in small amounts to provide necessary unsaturation for vulcanization and creation of a cross-linked molecular network	Tire liners, inner tubes and other applications requiring low gas permeability
Styrene-isoprene-styrene	Block copolymer	Adhesives and footwear such as shoe soles (can be injection molded at high temperatures without a curing step)

Alkane Dehydrogenation Catalyst Deactivation

Coking

The major source of catalyst deactivation is rapid coke formation, which requires frequent catalyst regeneration. Although steam provides endothermic heat for the reaction, removes coke via the water-gas reaction, and lowers the partial pressure of the reacting system, it cannot be used with chromia-alumina catalyst. Water vapor inhibits the activity of chromia-alumina; and not only must steam be excluded, but also the feed to the reactor must be carefully dried.[10] It is rather clear that coking is facilitated by acid supports. Even weak acid supports such as ZrO_2 cause rapid coke formation and, clearly, alumina (a stronger acid) must produce even more coke in a shorter time.[11] Alkali ions (promoters) are used to reduce the alumina acidity, but coke formation still occurs, although at a reduced rate.

Platinum-alumina catalysts also produce coke fairly rapidly, but steam can be used as a diluent that also removes some of the coke. It is thought that coke forms initially on the platinum surface and then migrates over time to the alumina surface, where it does not seriously affect the active metal surface.[1] Of course, there is a limit to the amount of coke that will migrate, and at some point the Pt surfaces become covered, and the catalyst must be regenerated. The addition of tin (Sn) as a promoter seems to increase the mobility of the carbon deposit so that more Pt surface remains available for a longer period. The promoter also increases the H_2 content of the coke, which tends to increase the reactivity of the coke toward H_2 and can result in the formation of methane and removal of corresponding portions of coke.[1,17] On the negative side, the presence of H_2 lowers equilibrium conversion and also reduces the rate of the forward reaction. Some optimal balance of H_2-to-hydrocarbon probably exists that is created by careful tuning of promoter addition in catalyst preparation. Although rates of both dehydrogenation and coking decline with H_2/hydrocarbon ratio, the coking rate declines much more than the rate of dehydrogenation.[1]

Regeneration

Regeneration is frequent in alkane dehydrogenation, and great care must be exercised in avoiding excessive temperatures (>650°C). Higher temperatures can, in the case of the chromia-alumina catalyst, convert the Cr_2O_3 to an inactive form and the high-area δ-alumina to low-area α-alumina.[11] The Pt-alumina catalysts are also subject to alumina similarly losing surface area and the Pt sintering and agglomerating. Reversal of agglomeration is accomplished by several different methods, depending on the licensed technology. Even normal and controlled cycles will ultimately produce some agglomeration that accumulates over time.

Process Kinetics

Any process model for alkane dehydrogenation must include a useful activity expression, since rapid coke formation has a profound effect on catalyst activity. Obtaining valid data on such a rapidly changing system is difficult. The use of a gradientless Berty-type reactor can be helpful,[12] but testing the model on an operating unit is necessary to confirm and adjust the constants. In this matter, intrinsic rate forms can be combined with deactivation rates to yield a model that can prove useful.

As an example, the following rate forms have been proposed for dehydrogenation over Cr_2O_3-alumina catalysts.

Butene → Butadiene[13]

$$r_D = \frac{k_B(P_B - P_D P_H/K)}{(1 + K_B P_B + K_H P_H + K_D P_D)^2} \exp(-F_1 C_C)$$

$$r_C = \frac{k_C P_B^{0.743} + k_D P_D^{0.853}}{1 + 1.6957\sqrt{P_H}} \exp(-F_2 C_C)$$

where k_B, k_C, k_D = rate constants
K_B, K_H, K_D = chemisorption constants
B, H, D = butene, hydrogen, and butadiene, respectively
K = equilibrium constant
F_1 and F_2 = deactivation constants
C_C = coke content, mass/mass of catalyst (requires an empirical equation for coke content as a function of operating conditions and time on stream)

Isobutane → Isobutene[12]

$$r_E = k(P_I - P_E P_H/K)\phi/(1 + K_{EH}P_E P_H + K_E P_E) - (d\phi)/(dt)$$

$$= (k_1 P_I + k_2 P_E)\phi$$

where
ϕ = catalyst activity factor
subscripts I, E, H = isobutane, isobutene, and hydrogen, respectively

Both of these forms can fit both reaction systems. It should be possible to develop simple Power-Law rate forms and include side reactions such as cracking and other thermal reactions.

Recent activity in kinetic studies on dehydrogenation of isobutane has focused on Pt/Al_2O_3 catalysts because of their use in newer plants for providing isobutene for MTBE production. The reader is referred to these studies.[17–19]

Isopentane → Isoprene

An interesting study employed a function of the diffusivity, which could be correlated as a linear function of the coke content of the catalyst.[20] Thus, a rate expression based on negligible intraparticle diffusion could be combined with the actual function of diffusivity to calculate the actual catalyst performance under plant conditions.

8.3.6   HIGHER ALKANES (C_6–C_{20}) → OLEFINS

In producing linear alkylbenzenes for detergents, it is necessary to obtain the required C_{10}-C_{14} olefins for alkylation of benzene either directly by oligomerization of ethylene or by dehydrogenation of C_{10}-C_{14} paraffins. The major process for the dehydrogenation route is a proprietary UOP technology called "Pacol-Olex." The process can also be used to yield olefins as low as C_6. Lower olefins C_3-C_5 can be formed from the corresponding paraffins using a similar design under milder conditions.[16]

Chemistry[15,16]

$$n(C_{10}\text{-}C_{14} \text{ paraffins}) \rightarrow (C_{10}\text{-}C_{14} \text{ olefins}) + nH_2$$

$\Delta H = 30$ kcal
400–600°C @ 3 bar
Catalyst: Pt-on-basic alumina with promoter

Thermodynamics

As is true for all paraffin dehydrogenations, the endothermic reaction is favored by high temperature, low pressure, and low hydrogen partial pressure. Such demands are always a problem, for the product olefins readily form coke at such conditions. Hence, it is essential to maintain a higher hydrogen partial pressure by recycling hydrogen to the reactor. In addition, the nonacidic nature of the support prevents excessive acid-catalyzed coke formation.

Short contact times help reduce not only coke formation but also other side-reaction products. Short residence times, however, reduce the conversion, which is variously reported to be 10–15%.[15]

Mechanism

There is reason to believe that the usual catalytic dehydrogenation mechanism with noble metals applies. The overall reaction scheme with side reactions, however, is quite complex, as shown in Figure 8.5. The challenge of imparting high selectivity is clearly revealed in this figure.

Catalyst Type and Supplier

The proprietary catalyst is known to be platinum on a basic alumina support along with promoter.[5] It is spherical in shape.

Licensor

UOP LLC (Pacol-Olex™)

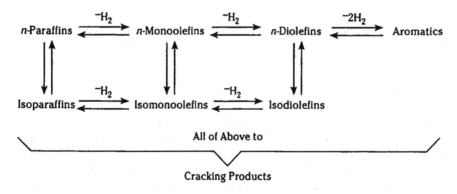

FIGURE 8.5   Reaction Scheme for Dehydration of Higher Alkanes (UOP Pacol Process). Reprinted by permission: Pujado, P. R., in *Handbook of Petroleum Refining Processes,* R.A. Meyers, ed., 2nd edition, p. 5.12, copyright McGraw-Hill Companies, Inc., New York, 1996.

Catalyst Deactivation

A major deactivating process is that of coking which, of course, would be even a greater problem if an acid alumina support were used. The catalyst life is about 30–60 days, which suggests that decline in activity occurs related to irreversible changes such as Pt agglomeration.

Process Unit

Initially, the reactor system consisted of two parallel adiabatic fixed-bed reactors of the radial type. A radial design was used to minimize pressure drop at the high space velocities used. One reactor was always in operation. The second reactor was switched via block valves to be in service when the other reactor required catalyst replacement. More recently, a single reactor has been used with a fresh catalyst feeding hopper above the reactor and a deactivated catalyst receiving hopper below, both with H_2 and N_2 purge connections. In this design, catalyst can be rapidly changed, with the catalyst protected from exposure to the atmosphere.[16]

In addition, improved selective catalysts are being offered that enable operation at higher flow rates and higher conversion by means of lower pressures, which improves equilibrium conversion. The result is lower utility costs, higher capacity in existing plants, and lower capital investment in new plants because of the smaller equipment sizes.[16]

8.4 ALCOHOLS TO ALDEHYDES OR KETONES

Primary and secondary alcohols can be dehydrogenated to give corresponding aldehydes and ketones. Dehydrogenation was at one time a major process for producing aldehydes and ketones; but it has in many instances, given way to new more energy efficient processes. Like many first members in a series, formaldehyde is different than the other aldehydes. Two processes are in use, one is described as oxidative-dehydrogenation and one as oxidation. Both are discussed under in Chapter 15, "Oxidation."

Catalyst Types

A variety of catalysts have been used for dehydrogenation of alcohols. Table 8.3 summarizes some of the more frequently used catalysts and the suppliers.

Catalyst Deactivation

The catalysts build up coke deposits slowly and must be regenerated every three to six months.[1] Catalyst life is reported to be two to three years. Slow poisoning may occur due to side reaction products and water.

Process Kinetics

Earlier studies indicated that C_2, C_3, and C_4 alcohol dehydrogenation could be expressed by an identical rate form and that simple power-law equations were useful in some cases.[7,8] More recently, a detailed study has been published using a ZnO-Cr_2O_3 catalyst and involving not only primary and secondary alcohols through C_4 but also a variety of substituent side chains.[10] These include 3,3-dimethyl-1-propanol, 2-(N,N-dimethylamino) ethanol, 2-phenylethanol, 2-methoxyethanol, 2-phenoxyethanol, 2,2,2-trifluoroethanol, and tetrahydrofurfuryl alcohol. All of these alcohols gave excellent fits to a Langmuir–Hinshelwood form, with only the alcohol and the product aldehyde having a retarding effect on the rate, since H_2 exhibited no retardation.[10]

TABLE 8.3
Alcohol Dehydrogenation Catalyst

Catalyst	Typical Compositions, Wt%		Form	Aldehyde or Ketone Product	Suppliers
Copper chromite	Cu	Cr	Tablet: 1/8 × 1/8 in.	Acetaldehyde	United Catalysts
	40	23	(3 × 3 mm)*	Acetone	Engelhard
	35	31			Celanese
	34	28			
	60	10			
Copper/alumina	Cu: 82		Tablet: 1/8 × 1/8 in.	Acetaldehyde	Engelhard
	Al₂O₃: 8		(3 × 3 mm)*	Acetone (recommended for situations where Cr toxicity presents a waste problem)	
ZnO	100% ZnO		Extrudate: 3/16 in.	Methy ethyl ketone	United Catalyst
ZnO/CuO	Various		Extrudate: 3/16 in.	Methyl ethyl ketone	United Catalysts
Sponge nickel or Raney nickel	Residual aluminum: 4–10%		Powder 30–35 microns	Acetone	Activated Metals Grace Davison
Pt/alumina	0.05% Pt		Tablets, extrudates 1/8 in.	Methy ethyl ketone	Engelhard Johnson Matthey United Catalysts

*Other sizes are available.

$$r = \frac{kK_A P_A}{1 + K_A P_A + K_R P_R}$$

where subscripts A = alcohol and R = aldehyde or ketone.

Interest in the complex alcohols is centered on a source of substituted aldehydes and ketones for producing various specialty chemicals.

8.4.1 ETHANOL → ACETALDEHYDE

$$CH_3CH_2\,OH \rightarrow CH_3CHO + H_2 \qquad \Delta H = 17.04 \text{ kcal and } K_p = 7.82 \times 10^{-2} \text{ @ 500 K}$$

ethanol   acetaldehyde $\Delta H = 17.25$ kcal and $K_p = 1.127$ @ 600 K

Acetaldehyde is primarily consumed in the manufacture of acetic acid by oxidation. It is also used to produce vinyl acetate (via the diacetate), cellulose acetate, tetraphthalic acid (monomer for a polyester), crotonaldehyde, crotonyldenediane (fertilizer), acetate resins and esters, and pyridine derivatives.[1,9]

Dehydrogenation of ethanol to produce acetaldehyde has mostly been replaced by the homogeneously catalyzed directed two-phase oxidation of ethylene. Ethanol dehydrogenation can continue to prove useful in countries with no petrochemical industry but ample ethanol production from grains.

Process Description

Copper chromite catalysts have been generally selected. Vapor-phase operation is practiced at a temperature range of 260–290°C and atmospheric pressure.[3] Conversion can be as high at 50%, and recycle of unreacted ethanol is necessary. The overall yield is 85 to 95 mole percent. Either adiabatic fixed beds or multitubular reactors may be used. The heated tubular reactor seems to be preferred, because it approximates isothermal conditions.

Most acetaldehyde is now produced by the oxidation of ethylene in a homogeneous liquid catalyst system.

8.4.2 ISOPROPYL ALCOHOL → ACETONE

$$CH_3CH(OH)CH_3 \rightarrow CH_3COCH_3 + H_2$$ $\Delta H = 13.6$ kcal and $K_p = 2.286$ @ 500 K

isopropyl alcohol   acetone $\Delta H = 13.67$ kcal and $K_p = 22.49$ @ 600 K

2-propanol   2-propanone

                              dimethyl ketone

Acetone is an important solvent and a major chemical intermediate in the production of methacrylate, bisphenol A, and many aldol condensation products used as solvents.[2, 4, 9] It was for many years produced by dehydrogenation of isopropyl alcohol, but the cumene hydropenoxide process for phenol production also yields one mole of acetone for each mole of phenol and is the major source of acetone. Dehydrogenation of isopropyl alcohol in the late 1980s constituted only 4% of U.S. production and 19% of Western Europe production.[4] Dependence on a by-product as a source can, of course, become a problem if the demand for the major product (phenol, in this case) declines.

Process Description

Isopropanol dehydrogenation is accomplished using the water azeotrope (87.8% of isopropanol), which is vaporized and fed to a multitubular reactor with heating medium circulating outside the tubes. Reaction temperature is maintained at 300°C and pressure at 1.5 to 3.0 atmospheres when using copper chromite catalysts.[5] Conversion is 85% per pass, and overall yield is 98 percent after recycle of unreacted isopropanol and recovery in the separation section.

Sponge (Raney) nickel has also been used in powder form as a catalyst. The process is conducted with the catalyst suspended in a high-boiling solvent for isopropanol. The alcohol is introduced slowly in this batch process, and the acetone product vaporizes off and is condensed.[3]

8.4.3 SEC-BUTYL ALCOHOL → METHYL ETHYL KETONE (MEK)

$$CH_3CH_2CH(OH)CH_3 \rightarrow CH_3CH_2COCH_3 + H_2$$ $\Delta H = 17.53$ kcal and $K_p = 1.53$ @ 500 K

sec-butyl alcohol   methyl ethyl ketone $\Delta H = 13.7$ kcal and $K_p = 14.96$ @ 600 K

2-butanol   2-butanone

The major use of methyl ethyl ketone is as a coating solvent for all synthetic and natural resins used in lacquer formulations.[6] Adhesives, magnetic tapes, printing inks, lube oil dewaxing make up a list in descending order for this versatile solvent. Because of the growing use of water-based coatings, growth in the MEK market is expected to remain stationary or decline to some degree.

Catalyst

Dehydrogenation of sec-butyl alcohol is the major process employed worldwide, although one producer uses direct oxidation of butane. A wide assortment of catalysts have been used over many years, including zinc oxide, zinc-copper, copper chromite, and platinum-on-alumina. Catalyst life is reported to be three to five years, with three- to six-month cycles between regeneration. Catalyst deactivation can occur by exposure to water, butene oligomers, and di-sec-butyl ether.[6]

Process Description

Multitubular reactors are used with heating medium on the shell side and reactants in the gas phase flowing through the tubes. Fixed-bed reactors in series with reheating between stages have also been used. Hydrogen is scrubbed from the reactor-effluent separator vapor by 2-butanol feed and the product MEK is recovered from the liquid product.

Temperatures in the range of 250–450°C and pressures of 1 to 3 atmospheres are used, depending on the catalyst selected. Yields of 95% have been reported.

8.5   OTHER DEHYDROGENATIONS

Dehydrogenation like hydrogenation is an important step in the production of intermediates required in the synthesis of many useful products. Several examples are described in Table 8.4. The following are two other important classes of dehydrogenation reactions.

TABLE 8.4
Other Dehydrogenations

Reactant	Product	Uses	Conditions	Catalyst
1,4 butanediol $HOCH_2(CH_2)_2CH_2OH \longrightarrow$ (γ-butyrolactone) $+ H_2$	γ-butyrolactone	• Solvent for polymers of acrylonitrile, cellulose acetate, methy methacrylate, and styrene • Used in paint removers and drilling oils • Precursor to polyvinyl-pyrolidane, piperidine, phenylbutyric and thiobutyric acids	Vapor phase 200–250°C atm pressure Yield: 93%	Cu/Al_2O_3
cyclohexanol (cyclohexane)—OH $\longrightarrow$ (cyclohexane)=O $+ H_2$	cyclohexanone	• Precursor to c-caprolactam via cyclohexanone oxime. Used in the production of nylon 6. • Solvent for various resins, crude rubber, waxes, fats, and shellac.	Vapor phase 400–450°C atm pressure Yield: 87%	Zn/Al_2O_3 or Cu/Al_2O_3
cyclodecanol (cyclodecane)OH $\longrightarrow$ (cyclodecane)O $+ H_2$	cyclodecanone	• Precursor to lauryl lactam used in production of nylon 12	Liquid phase 230–245°C atm pressure	Cu/Al_2O_3 or Cu/Cr

Source: Information based on Refs. 2–5.

8.5.1   AROMATIZATION

Cyclohexane and substituted cyclohexanes undergo dehydrogenation to produce aromatics. Most reactions of this type occur in catalytic reforming, see Chapter 18 for details.

8.5.2   Fine Chemicals

Dehydrogenation is a valuable tool in the synthesis of fine chemicals. The reactants and products, except hydrogen are in the liquid phase. Continuous stirring and refluxing assures rapid removal of hydrogen[1], and prevents reverse reaction so that the reaction is not equilibrium limited. Operating temperature can then be set based on a desired reaction rate and known region of high selectivity.

REFERENCES, SECTIONS 8.1 AND 8.2 (ALKYL AROMATICS)

1. *Engelhard Catalysts*, Engelhard Minerals and Chemicals Corp., Newark, NJ, 1977.
2. Li, C. H. and Hubbell, D. in *Encyclopedia of Chemical Processing and Design,* ed. J.J. McKetta, Vol. 55, p. 197, Marcel Dekker, New York, 1996.
3. Chenler, P. J., *Survey of Industrial Chemistry*, 2nd edition, VCH Publishers, New York, 1982.
4. Sheel, J. C. P. and Crowe, C. M., *Can. J. Chem. Eng.* 42, 183 (1969).
5. *United Catalyst Product Bulletin: G-64 & G-84 Catalysts*, United Catalysts, Inc., Louisville, KY.
6. Hirano, T., *Applied Catalysis*, 26, 65 (1986).
7. Stiles, A. B. and Koch, T. A., *Catalyst Manufacture*, 2nd edition, Marcel Dekker, New York, 1995.
8. Lee, E. H., *Catal. Rev.*, 8, 285 (1973).
9. Franck, H. G. and Stadelhofer, J. W., *Industrial Aromatic Chemistry*, Springer-Verlag, Berlin, 1988.
10. Herzog, B.D.and Rase, H F, *Ind. Eng. Chem. Prod. Res. Dev.,* 23, 187 (1984).
11. Davidson, B. and Shah, M J, *IBM J Res. Develop.,* 2, 289 (1965).
12. Carra, S. and Froni, L., *Ind. Eng. Chem. Process Res. Dev.,* 1, 281 (1965).
13. Sheel, J. G. P. and Crowe, C.M., *Can. J. Chem. Eng.,* 47, 183 (1969).
14. Model, D. J., *Chem. Eng. Compute.,* 1, 100 (1972).
15. Claugh, D. E. and Ramirez, W. F., *MChE Journal*, 22, 1097 (1976).
16. Heggs, P. J. and Abdullah, N., *Chem. Eng. Res. Des.* 64 258 (1986).
17. Aikhwartr, S.; Elshishini, S. G.; Elnashaie, S. S. E. H. and Abdalla, B. K., *Studies in Surface Science and Catalysis,* Vol. 73, *Progress in Catalysis*, p. 351, Elsevier, New York, 1992.
18. Sundaram, K. M.; Sardina, H.; Fernandez-Baujin, J. M. and Hildreth, J. M., *Hydrocarbon Process,* Jan., 1991, p. 93.
19. Lewis, P. J.; Hampton, C. and Koch, P. in *Kirk-Othmer Encyclopedia of Chemical Technology,* 3rd edition, Vol. 21, p. 770, Wiley, 1983.
20. *Dehydrogenation Catalysts,* Criterion Catalysts Bulletin, Houston, Texas.
21. Forni, L. and Valerio, A., *Ind. Eng. Chem. Process Des. Develop.,* 10, (4) 552 (1971).

REFERENCES, SECTION 8.3 (LOWER ALKANES)

1. Resasco, D.E. and Haller, G. L., in *Catalysis* Vol. 11, p. 379, J. J. Spivey and S. K. Agarwal, senior reporters, Royal Society of Chemistry, Cambridge, England, 1994.
2. Vera, B. V. and Imal, T., *Hydrocarbon Process,* April, 1982, p. 171.
3. Chenier, P. D., *Survey of Industrial Chemistry*, 2nd edition, VCH, New York, 1992.
4. "Petrochemial Process '97," *Hydrocarbon Process,* March, 1997.
5. Obenaus, F.; Droste, W. and Neumeister, J., in *Ullmann's Encyclopedia of Industrial Chemistry,* 5th edition, Vol. A4, p. 483, VCH, New York.
6. Muller, H. J. and Loser, E., in *Ullmann's Encyclopedia of Industrial Chemistry,* 5th edition, Vol. A4, p. 431, VCH, New York, 1985.
7. Hatch, C. F. and Matar, S., *Hydrocarbon Process,* p. 135, Jan., 1978.
8. Senyak, M. J. in *Concise Encyclopedia of Polymer Science and Engineering*, p. 504, J. I. Kroschwitz, editor, Wiley, New York, 1990.
9. Weitz, H. M. and Loser, E. in *Ullmann's Encyclopedia of Industrial Chemistry,* 5th edition, Vol. A14, p. 632, VCH, New York, 1989.

10. Rylander, P. N., in Ullmann's *Encyclopedia of Chemical Technology,* 5th edition, Vol. A13, p. 487, VCH, New York, 1991.

11. Thomas, C. L.; *Catalytic Processes and Proven Catalysts,* Academic Press, New York, 1970.

12. Zwahlen, A. G. and Agnew, J. B., *Ind. Eng. Chem. Res.,* 31, 2088 (1992).

13. Durez, F. J. and Froment, G. F., *Ind. Eng. Chem. Process Des. Dev.,* 15, 291 (1976).

14. Wissermel, K. and Arpe, H.-I., *Industrial Organic Chemistry,* 3rd edition, VCH, New York, 1997.

15. Olah, G. A. and Molnar, *Hydrocarbon Chemistry,* Wiley, New York, 1995.

16. Pujado, P. R. in *Handbook of Petroleum Refining Processes,* R. A. Meyers, editor, 2nd edition, McGraw-Hill, New York, 1997.

17. Cortright, R. D. and Dumesic, J. A., *J. Catal.,* 148 (2), 771 (1994).

18. Boeh, H. and Zimmerman, H., *Erdoel, Erdgas, Kohle,* 111(2), 90 (1995).

19. Loc, L. C.; Gaidaia, N. A.; Kiperman, S. L.; Thoang, H. S.; Podklenera, N. and Georgievski, V. *Kinet Catal.* 36 (4) 517 (1995).

20. Noda, H.; Ozaki, M.; Tone, S. and Otake, T., *Bull. Japan. Pet. Inst.,* 17 (1) 88 (1975).

REFERENCES, SECTION 8.4 (ALCOHOLS)

1. Reintard, J., Laib, R. J. and Bolt, N. M. in *Ullmann's Encyclopedia of Industrial Chemistry,* 5th edition, Vol. A1, p. 31, VCH, New York, 1985.

2. Faith, W. L., Keyes, D. B. and Clark, D. L., *Industrial Chemicals,* Wiley, New York, 1963.

3. Miller, S. A., *Chem. Proc. Eng.* 49 (3), 75 (1968).

4. Howard, W.L. in *Kirk-Othmer Encyclopedia of Chemical Technology,* 4th edition, Vol. 1, p. vis, Wiley, New York, 1991.

5. Fair, J. R. in *Encyclopedia of Chemical Processing and Design,* J. J. McKetta, editor, Vol. 27, p. 482, Marcel Dekker, New York, 1988.

6. Braithwaite, J., in *Kirk-Othmer Encyclopedia of Chemical Technology,* 4th edition, Vol. 14, p. 985, Wiley, New York, 1995.

7. Thaller, L. H. and Thodos, G., *AIChEJ,* 6, 369 (1960).

8. Franckaets, J. and Froment, G. F., *Chem. Eng. Sci.,* 19, 807 (1964).

9. Chenier, P.J., *Survey of Industrial Chemistry,* 2nd edition, VCH, New York, 1992.

10. Gulkova, D. and Kruas, M., Collect. *Czech. Chem. Commun.,* 57 (11), 2215 (1992).

REFERENCES, SECTION 8.5 (OTHER DEHYDROGENATIONS)

1. Rylander, P. N., in *Ullmann's Encyclopedia of Industrial Chemistry,* edition, Vol A13, p. 487, VCH, New York, 1989.

2. Meyers, R. A., *Handbook of Petroleum Refining Processes,* 2nd edition, McGraw-Hill, New York, 1997.

3. Weissermel, K. and Arpe, H.-J., *Industrial Organic Chemistry,* 3rd edition, VCH, New York, 1997.

4. Olah, G. A. and Molnar, A., *Hydrocarbon Chemistry,* Wiley, New York, 1995.

5. *The Merck Index,* Merck & Company, Rahway, N.J., 1989.

9 Epoxidation

Epoxidates are cyclic ethers with three-membered rings. They are also called *oxiranes*. *Epoxidation* describes the reaction that is used to produce the epoxide from an alkane or alkane side chain via a hydroperoxide or a peroxycarboxylic acid (peracid).

9.1 ETHYLBENZENE → PROPYLENE OXIDE + STYRENE[2,3]

Most propylene oxide is made by homogeneous reactions (the non-catalytic chlorohydrin process and various homogeneous catalyzed epoxidations of isobutane or ethylbenzene). Epoxidation processes (i.e., oxirane) yield two major products: propylene oxide plus tert-butyl alcohol with isobutane as the feed and propylene oxide plus 1-phenylethanol. The tert-butyl alcohol can be dehydrated to valuable isobutylene, and 1-phenylethanol is dehydrated to styrene. The heterogeneous-catalyzed expoxidation is called the *Styrene Monomer-Propylene Oxide Process,* because the carbinol thus produced is readily dehydrated to styrene.

The major use of propylene oxide is in the manufacture of propylene glycol, which is used extensively in pharmaceuticals, foods, and cosmetics as a humectant, solvent, and emulsifier. It is also used in dairies and breweries as a nontoxic antifreeze and in production of synthetic resins.[4]

Chemistry[1–3]

Oxidation of ethylbenzene in a homogeneous reaction system produces ethylbenzene hydroperoxide, which is then reacted with propene to yield the expoxide, propylene oxide.

ethylbenzene hydroperoide   propylene oxide   methylphenol carbinel

In a separate reaction, the carbinol can be dehydrated to styrene.

Catalyst and Licensors

Compounds of V, W, Mo, or Ti on silica

Licensors

Shell, Arco, and Nizhnekamsk (all SMPO processes)

REFERENCES (EPOXIDATION)

1. Valbert, J. R., Zajecek, J. G. and Orenbuch, D. J., in *Encyclopedia of Chemical Processing and Design*, J. J. McKetta, editor, Vol. 45, p. 88, Marcel Dekker, New York, 1993.
2. Ohlah, G. A. and Molnar, A., *Hydrocarbon Chemistry*, Wiley, New York, 1995.
3. Weissermel, H. and Arpe, J., *Industrial Organic Chemistry*, 3rd edition, VCH, New York, 1997.
4. Merk Index, 11th edition, Merk and Company, Rahway, N.J., 1989.

10 Hydration

Addition of water to unsaturated hydrocarbons is called *hydration*. Direct hydration of olefins (ethylene, isopropylene, sec-butylene, and tert-butylene) is used to produce the corresponding alcohols. Direct hydration is often accomplished via a solid acid catalyst.

10.1 ETHYLENE → ETHANOL

In earlier days, ethanol was produced by fermentation of carbohydrates such as starch and molasses. As the petrochemical industry developed, the major source of ethanol became a synthetic process, initially based on a homogeneous process in which ethylene is esterified with H_2SO_4 and then hydrated to ethanol *(indirect hydration)*. This process is still in use, but direct hydration with a heterogeneous catalyst is more common worldwide. In recent decades, fermentation processes have made a comeback for use as an additive (10%) to gasoline in the U.S.A. in areas of high corn production. Actually, the product does not compete on an economic basis, but it has been subsidized by U.S. federal and state programs as an aid to farmers and corn and grain processors.

Ethanol is widely used as a solvent (59%) for toiletries, perfumes, cosmetics, inks, detergents, and household cleaners. The second category of uses is as an intermediate in producing various valuable chemicals (ethylacrylate, ethylamines, and glycol ethers[1,2]).

Chemistry[2,3]

$$C_2H_4 + H_2O \Leftrightarrow CH_3CH_2OH \text{ @ 250–300°C and 60–70 bar}$$

Temp., K	ΔH, kcal	K_p
500	−11.17	1.383×10^{-2}
600	−11.12	1.991×10^{-3}

The high pressure used requires equilibrium constants based on fugacities, for which see "Thermodynamics," below.

Catalyst: phosphoric acid impregnated on Kieselguhr
Side Reaction: $2CH_3CH_2OH \rightarrow (CH_3CH_2)_2O + H_2O$

Thermodynamics

The equilibrium is favored by high pressure and low temperature (see Figure 10.1), but practical rates require higher temperatures in the range indicated above. With the low equilibrium constants, elevated pressure is necessary to force the reaction to the right. Higher pressures ultimately increase polymerization. Hence, a practical upper limit must be found.

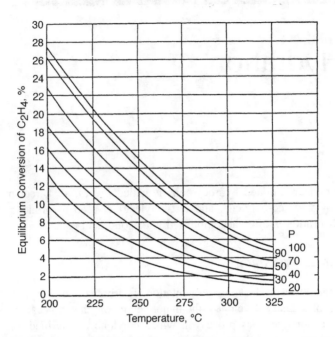

FIGURE 10.1   Equilibrium Conversion of Ethylene as a Function of Temperature at Various Lines of Constant Pressure in Atmospheres for $C_2H_4/H_2O = 2$ in feed. Reprinted from *Genie Chimique* Vol. 78(6), Muller J. and Waterman, H.I., "Equilibre D'Hydration de L'Ethylene," p. 183, copyright 1957, with permission from Elsevier Science.

Mechanism

As is often assumed for acid-impregnated supports, it is reasonable to suggest that the homogeneous acid-catalyst carbenium ion mechanism applies.[2,4]

$$CH_2 = CH_2 + H_3O^+ \Leftrightarrow \left[H_2C \overset{H}{=} CH_2 \right]^+ + H_2O$$

fast and at equilibrium

$$\left[H_2C \overset{H}{=} CH_2 \right]^+ \Leftrightarrow \left[H - \overset{H}{\underset{H}{C}} - \overset{H}{C} - H \right]$$

slow and rate determining step

$$H - \overset{H}{\underset{H}{C}} - \overset{H}{C} - H + 2H_2O \Leftrightarrow CH_3CH_2OH + H_3O^+$$

fast and at equilibrium

Catalyst Type and Suppliers

Liquid phosphoric acid absorbed in the pores of a Kieselguhr or silica support

Suppliers

   United Catalysts, UOP LLC, Grace Davison

Possible Licensors

   Union Carbide, Hülls, Shell, BP

Catalyst Deactivation

As with acid impregnated catalyst, loss of phosphoric acid over time causes some deactivation, but the acid can be replaced by adding acid to the bed via spray nozzles. Above an operating pressure of 70 to 80 bar, polymer formation can occur and coat the catalyst, thereby destroying active surface.

Process Unit[2,3,5]

An adiabatic fixed-bed reactor is used. The feed is preheated and fed in a ratio of ethylene/water varying from 1.3–3.0, depending on reaction conditions. The temperature rise is about 10–20°C, and the equilibrium conversion is about 12% while the actual conversion is approximately 7%. The product is cooled, and the uncondensed vapor is washed to remove residual ethanol and then recycled as unreacted ethylene to the reactor. The liquid stream containing most of the ethanol is then sent to distillation where a 95% by volume ethanol/water azeotrop is produced. Selectivity to alcohol is said to be 97 percent. Diethyl ether that might have formed is also recycled to the reactor where the reverse reaction occurs to form ethanol.

   Because of the low conversion, recycle of unreacted ethylene is high along with small amounts of impurities, which must be kept to a reasonable level. This situation requires continuous purging, the amount of which can be reduced by using a relatively pure ethylene feed.

Process Kinetics

A simple power-law kinetic equation has been proposed.[4]

$$r = k\left(P_E - \frac{P_A}{L_p P_W}\right)$$

where subscripts

$$E = \text{ethylene}$$
$$A = \text{alcohol}$$
$$W = \text{water}$$

Later studies confirmed this form but showed that k varies with water partial pressure.[6]

10.2 PROPYLENE → ISOPROPYL ALCOHOL*

Prior to the introduction of the phenol-from-cumene process, which simultaneously produces acetone, most acetone was made from isopropyl alcohol. Hence, the major uses of isopropyl alcohol are now as an all-purpose, inexpensive solvent for both water and hydrocarbon affinity. It is used for this purpose in industrial products, cosmetics, aerosol sprays, paints, varnishes, and printing inks.[1,7] Its use as a disinfectant is known to most consumers, but few realize that it is more effective than the more costly ethyl alcohol.

---

* Normal propyl alcohol is produced via hydroformylation of ethylene and then hydrogenation of the n-propylaldehyde formed.

The homogeneous catalyzed indirect-hydration process via isopropyl hydrogen sulfate continues to be used in some plants. In fact, it was the basis of the world's first petrochemical process built by Exxon in 1920. The direct-hydration process was introduced in 1951 by ICI, using a tungsten oxide-zinc oxide promoted catalyst on silica.[3,7,8] The direct hydration processes now in use employ heterogeneous catalysts based on H_3PO_4 impregnated within Kieselguhr or acidic ion-exchange resins. The ion-exchange process, in particular, avoids much of the corrosion and waste disposal problems associated with free acids.

Chemistry

$$CH_3CH = CH_2 + H_2O \Leftrightarrow (CH_3)_2CHOH$$

Temp., K	ΔH, kcal	K_p
400	−12.37	0.1132
500	−12.34	5.047×10^{-3}

Catalyst Types and Suppliers

	Operating Conditions	
Catalyst	Temp. Range	Pressure Range
H_3PO_4–on-Kieselguhr	230–290°C	200–250 bar
Cation-exchange (sulfonic) resin	130–155°C	60–100 bar

Catalyst Suppliers

Phosphoric Acid-on-Kieselguhr: UOP LLC, Grace Davison, United Catalysts
Cation Exchange Resin: Dow, Rohm & Haas

Licensors

Phosphoric Acid Process: Union Carbide, Hülls, Shell, BP
Cation-Exchange Resin: Condea

Mechanism

The mechanism presented for ethanol should apply for both of these processes. The secondary propyl carbenium ion is more stable than the ethyl carbenium ion and thus forms more rapidly. Since carbenium ion formation is the controlling step, isopropyl alcohol formation is more rapid.

Catalyst Deactivation (Ion-Exchange Resin)

The most significant loss of resin activity is caused by attrition of the resin beads. Ion-exchange resin life is reported to be eight months. In the case of H_3PO_4 on Kieselguhr, excessive temperature can cause polymer formation. Loss of H_3PO_4 over time is counteracted by addition of H_3PO_4 during operation.

Process Unit[3,7,8]

See section on Ethanol for analogous H_3PO_4/Kieselguhr process.

Cation-Exchange Catalyst Process

Preheated water and liquid propylene are premixed in a molar ratio of 12:1 to 15:1 (water to propylene). Heat supplied by the water vaporizes the propylene, and the two components are introduced at the top of an adiabatic fixed trickle-bed reactor. The reactor outlet is subjected to a series of separators (high and low pressure) along with distillations that provide a recycle propylene stream and an isopropyl alcohol product stream of 91% alcohol. Anhydrous alcohol can also be obtained by azeotropic distillation.

Per-pass conversion of propylene is 75%, compared to 5 or 6% for the supported-acid process. Thus, less recycle is required, and less pure propylene may be used for the trickle-bed process. Selectivity to isopropyl alcohol is 93%, and diisopropyl ether (3.5 wt%) is the main impurity. Total conversion based on feed and recycle is 94%.

The higher conversion per pass is attributable to the higher activity of the ion-exchange resin and thus lower permissible operating temperature and a resulting higher equilibrium constant. The large excess of water also serves to increase the equilibrium conversion.

10.3 N-BUTENES → SEC-BUTANOL

The various butyl alcohols are produced as follows:[3,7,8]

1-butanol	• Hydrogenation of n-butyraldehyde produced by the homogeneously catalyzed oxo process (Chap. 12)
Sec-butanol	• Indirect hydration by homogeneous sulfuric acid process • Direct hydration of n-butenes using an acidic ion-exchange resin (this section)
Iso-butanol	• Hydrogenation of iso-butyraldehyde produced as a coproduct in the oxo process along with n-butyraldehyde (Chap. 12) • Indirect hydration of 1 and 2-butene by homogeneous sulfuric acid process
Tert-butanol	• By-product of ARCO propylene oxide process • Indirect hydration of iso-butene by homogeneous sulfuric acid process

This summary leaves for discussion in this section one heterogeneous catalyzed hydration, that of n-butenes to sec-butanol.

Sec-butanol (sec-butyl alcohol) major use is in the production of methyl ethyl ketone. It is also used as a solvent alcohol or ester. The ion-exchange catalyst process is now preferred to earlier sulfuric acid processes.

Chemistry

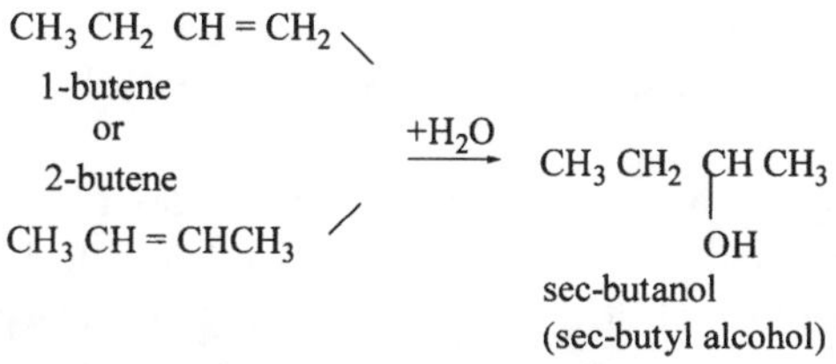

$$CH_3\ CH_2\ CH = CH_2$$
1-butene
or
2-butene
$$CH_3\ CH = CHCH_3$$

$\xrightarrow{+H_2O}$

$$CH_3\ CH_2\ \underset{\underset{OH}{|}}{CH}\ CH_3$$
sec-butanol
(sec-butyl alcohol)

150–170°C inlet

50–70 bar

Temp., K	ΔH, kcal	K_p
400	−12.13	0.264
500	−10.14	1.472×10^{-2}

Catalyst Type and Suppliers[10]

The n-butenes are not as reactive as propylene, and a higher-activity cation exchange resin had to be developed with very strong acidity, stability, and resistance to acid loss from the catalyst. The proprietary catalyst has been described as sulfonated polystyrene crosslinked with divinylbenzene, which stabilizes the catalyst. An optimum degree of crosslinking (8 to 12%) was determined.[10] Increasing crosslinking improves stability but also increases susceptibility to hydrolysis and loss of the sulfonic groups to the water portion of the reaction mix.

Catalyst Suppliers

Licensor: RWE/DEA (formerly Deutche Texaco)

Catalyst Deactivation

Loss of acidity by release of acidic components is the main cause of catalyst deactivation. The proprietary catalyst minimizes this loss, but it does occur over an extended period. Ultimately, the catalyst must be replaced.

Process Unit[10]

The reaction system at the temperature and pressure conditions reported (150–170°C, 50–70 bar) consists of liquid water and supercritical hydrocarbon mixture of predominantly n-butene. To avoid costly separation for the feed, butanes are not removed but go through the process unchanged. Once the butenes are converted to alcohol, a cooler condenses the water and hydrocarbon product. The water is removed from the sump of the separator and recycled to the reactor after removing acidic impurities in an anion exchanger. Separation of the organic product occurs in a debutanizer column with the sec-butanol as bottoms product. The mixture of butane and unreacted butene is recycled to the reactor where it is mixed with fresh feed. A purge, of course, is required to remove the inert butane in accordance with the desired material balance.

Turning to the reactor, it consists of a series of beds (e.g., four) in a single shell designed for upflow operation. A portion of the water is added to the combined fresh and recycle C_4 stream. After passing through Bed no. 1, the water is separated from the organic phase with relative ease, because they are essentially immiscible. The organic phase is then sent to the next bed with additional water added to it. This sequence is repeated at each bed, and the removed water is recycled via the anion exchanger.

The recovered sec-butyl alcohol from the debutanizer described previously contains about 1 wt% sec-butyl ether and water. The concentration of sec-butyl alcohol is typically 94–97 wt%, depending on the n-butene content of the feedstock. A high content of n-butene in the feed of 90 wt% or higher assures high reaction rates and smaller equipment when operating at up to 60–70% conversion. A more dilute feed for the same conversion would produce a large drop in butene concentration and lower average reaction rates. Feedstock with less than 90% n-butene should be concentrated prior to use.

By-products include sec-butyl ether, which increases from 1–5% as the run progresses, and operating temperature is increased as catalyst activity declines. It does not affect MEK synthesis from sec-butanol. Isobutene, if present, forms tert-butanol, which does interfere with the MEK production and must be removed if significant in amount.

10.4 ACRYLONITRILE → ACRYLAMIDE

Acrylamide is used to produce polyacrylamide, which is used extensively as a flocculating agent in water treatment plants and for paper and ore processing. It is also used in producing various copolymers for application in paints and resins.[11] Hydration via homogeneous catalysis by sulfuric acid is used, but the more recent plants employ heterogeneous catalysis.

Chemistry

$$H_2C = CHCN + H_2O \rightarrow H_2C = CHCONH_2$$

acrylonitrile   acrylamide

80–120°C
Raney Cu

Catalyst Type and Supplier

Raney copper of a size usually recommended for slurry reactions (30–35 m) or copper chromite (Dow process).

Suppliers

Activated Metals, Grace Davison, Engelhard

Process Licensor or Process Owner

Mitsubishi-Kasel, Cyanamid, Mitsui Toatsu, Dow

Process Unit[11,12]

Acrylonitrile and excess water are added to a stirred reactor along with Raney copper slurry catalyst (30–35 m). After several hours of contact time, the slurry is separated from the catalyst and concentrated, and the catalyst and unreacted acrylonitrile are recycled. The resulting aqueous solution is in a convenient 30–50% acrylamide range. Acrylonitrile conversion is in the range of 50–80% and selectivity to acrylamide is variously reported as 96–99%.[11,12]

REFERENCES (HYDRATION)

1. Chenier, P. J., *Survey of Industrial Chemistry*, 2nd edition, VCH, New York, 1992.
2. Logston, J. E. in *Kirk-Othmer Encyclopedia of Chemical Technology*, 4th edition, Vol. 9, p. 812, Wiley, New York, 1994.
3. Olah, G. A. and Molnar, A., *Hydrocarbon Chemistry*, Wiley, 1995.
4. Taft, R. W., *J. Am. Chem. Soc.,* 74, 5372 (1952).
5. Farkas, A., in *Ullmann's Encyclopedia of Industrial Chemistry*, 5th edition, Vol. A9, VCH, New York, 1987.
6. Gel'bshtein, A. I., Bakshi, M. and Temkin, M. I., *Akad. Nauk, SSSr,* 132, 384 (1960).
7. Logston, J. E. and Loke, R. A., in, *Kirk-Othmer Encyclopedia of Chemical Technology*, 4th edition, Vol. 20, p. 216, Wiley, New York, 1996.
8. Weissermel, K. and Arpe, H. J., *Industrial Organic Chemical Technology*, 3rd edition, VCH, New York, 1997.
9. Billing, E. in *Kirk-Othmer Encyclopedia of Chemical Technology*, 4th edition, Vol. 4, p. 691, Wiley, 1992.
10. Prezel, Jay M., Koog, W. and Dettmer, M., *Hydrocarbon Process.,* p. 75, Nov., 1998.

11. Weissermel, K. and Arpe, H. J., *Industrial Organic Chemistry*, 3rd edition, VCH, New York, 1997.
12. Preschev, G.; Schwind, O. W. and Weiberg, O. in *Ullmann's Encyclopedia of Industrial Chemistry*, 5th edition, Vol. A1, VCH, New York, 1985.

11 Hydrochlorination

11.1 ALKENES

Hydrochlorination of alkenes has been exclusively accomplished in the liquid phase using homogeneous Friedel–Craft catalyst such as $AlCl_3$, but the usual handling and waste disposal problems over the years have encouraged the development of heterogeneous catalysts of particular value in producing ethyl chloride.[2] The liquid-phase process has remained useful, and the choice between gas or liquid depends on the process economics in a particular plant relative to other production units.[3]

11.1.1 ETHYLENE + HYDROGEN CHLORIDE → ETHYL CHLORIDE

The major use of ethyl chloride has been in the production of tetraethyl lead, which is now banned in many areas of the world. Its remaining uses include as a reactant in the production of fine chemicals and ethyl cellulose. It is also used directly as a solvent for natural fragrances and as a blowing agent, refrigerant, and topical anesthetic.[1,5]

Chemistry

$$C_2H_4 + HCl \rightarrow C_2H_5Cl$$

>200°C and 5–15 bar
gas-phase

Catalyst Types

Reported catalysts include copper chloride, promoted zinc chloride-on-alumina, $AlCl_3$-on-alumina, thorium oxychloride-on-silica, rare-earth oxides-on-alumina or silica, and platinum-on-alumina.[1,2,6]

Process Units

Both an adiabatic fixed-bed reactor and a fluidized bed have been used. The reaction is not highly favored above 200°C. However, higher temperatures are necessary to produce acceptable rapid reaction rates. This situation means that high conversions are not attainable, and recycling of unreacted HCl and ethylene is necessary. Reported attainable single-pass conversion is 50 percent.[2] Typical operating conditions are variously reported as 5–15 bar at 150–250°C[1,2,3] or higher, as at 450°C.[3,6] High pressure favors the main reaction over side reactions. Selectivity to ethyl chloride is high (>99%).[1,2,3]

Process Kinetics

A model based on HCl adsorption as the rate determining step and α-alumina as the catalyst has been presented.[8]

$$r_{MC} = \frac{k(P_{HCl} - P_{MC}^2/P_M K)}{1 + K_{HCl}P_{MC}^2/P_M K}$$

where subscripts

HCl = hydrogen chloride
MC = methyl chloride
M = methanol

A power-law model was also presented.

$$r_{MC} = kP_M^m P_{HCl}^n$$

where exponents

m = 0.68
n = 0.54

This power-law form could be usefully applied in the narrow range of conditions for industrial operation. In such a case, the reverse reaction could be included and applicable exponents determined by operation analysis. See Ref. 7 for homogeneous systems that may provide some reasonable analogies.

11.2 ALCOHOLS

11.2.1 METHANOL + HYDROGEN CHLORIDE → METHYL CHLORIDE

Hydrochlorination of methanol is a major route to the production of methyl chloride. This situation exists because both methanol and HCl are relatively low-cost reactants. In fact, HCl is often present in petrochemical operations involving chlorinated compounds. Silicones are produced via a reaction with methyl chloride, which produces HCl as a by-product. Vinyl chloride manufacture and chlorination of various alkanes also produce by-product hydrogen chloride.

Methyl chloride is a valuable solvent, but its major use is in the production of silicones. Additional uses include manufacture of methylcellulose, higher chloromethanes, quaternary amines, and various agriculture chemicals.[1,4]

The reaction can be carried out in the liquid phase with homogenous catalysis or in the vapor phase with a heterogeneous catalyst.

Chemistry

$$CH_3OH + HCl \rightarrow CH_3Cl + H_2O \qquad \Delta H = -8.17 \text{ kcal}$$

300–380°C @ 3-6 bar[1,4]

Side Reactions[8]

$$2CH_3OH \Leftrightarrow (CH_3)_2O + H_2O$$

$$(CH_3)_2O + 2HCl \rightarrow 2CH_3Cl + H_2O$$

The two reaction producing are irreversible while the ether forming reaction is equilibrium limited. The main reaction is dominant, since very little ether is formed.[8]

Catalyst Type and Supplier[1,4]

Alumina gel or gamma alumina (8-12 mesh) supported $CuCl_2$ or $ZnCl_2$ and silica-alumina are described in the patent literature.

Suppliers

United Catalysts, Alcoa, Engelhard, Akzo Nobel

Process Unit

An adiabatic fixed-bed reactor is used. A stoichiometric mixture of methanol and HCl, vaporized and preheated to 200°C, is fed to the reactor where exothermic reaction heat raises the temperature to the 300–380°C range. Recovery of product is accomplished by water scrubbing and condensation of the methyl chloride by refrigerated cooling (bp @ 1 bar = −23.7°C). Conversions of 95% and yields of ~98% are reported.[4]

REFERENCES (HYDROCHLORINATION)

1. Weissermel, H., J. Arpe, *Industrial Organic Chemistry*, 3rd edition, VCH, New York, 1997.
2. Lindel, W. in *Ullmann's Encyclopedia of Industrial Chemistry*, Vol. A6, p. 260, VCH, New York, 1986.
3. Stirling, R. G. in *Encyclopedia of Chemical Processing and Design*, J. J. McKetta, editor, Vol. 20, p. 68, Marcel Dekker, New York, 1984.
4. Holbrook, M. T., in *Kirk-Othmer Encyclopedia of Chemical Technology*, 4th edition, Vol. 5, page 1028, Wiley, New York, 1993.
5. *Merk Index*, 11th edition, Merk & Lumpary, Rabway, NJ, 1989.
6. Olah, G. A. and Molnar, A., *Hydrocarbon Chemistry*, Wiley, New York, 1995.
7. Raghumath, C. and Doraiswamy, L. K., *Chem. Eng. Sci.,* 29 (2) 349 (1974).
8. Becerra, A. M.; CastroLuna, A.E.; Ardissone, D. R. and Ponxi, M.I. *Ind. Eng. Chem. Res.*, 31, (4), 1040, (1992).

12 Hydrogenation

12.1 GENERAL BACKGROUND

Hydrogenation is the most ubiquitous reaction in organic chemistry as well as in the commercial organic chemical industry. Both heterogeneous and homogeneous catalysts are used, but heterogeneous catalysts dominate commercial practice, especially for large-scale production. The reactions are exothermic and reversible so that temperature control is essential. From the array of modern supported catalysts, however, a standard catalyst can usually be selected that has a high activity, permitting operation at mild temperatures that favor high yields of the hydrogenated product.

Although there is a rich literature on hydrogenation catalysis in the field of organic synthesis, a major source of recommendations for practical application is the catalyst manufacturer. It has been observed that the characteristic properties of each hydrogenation catalyst type toward each functional group does not, in general, vary between compounds unless a complex structure could prove to be an impediment to the rate of reaction.[1] This truism enables reasonable selection of a standard catalyst, about which much operating information is known for a new process involving a similar functional group.

12.1.1 CATALYST TYPES AND SUPPLIERS

Hydrogenation catalysts most frequently used in commercial practice are base metals (Ni, Cu, Co, and Ca) and platinum (precious) metals (Pt, Pd, Rh, Ru).

The platinum metals are in most cases more active than the base metals and thus require milder conditions and smaller amounts of catalyst. Nickel is the preferred catalyst for hydrogenating a wide variety of functional groups in the many instances where higher temperature and pressure present no physical or chemical disadvantages. Platinum group metals are more costly but are used in cases where equipment limitations or stability, require operating only under mild conditions.[2]

The platinum metals are very often attractive for small operations for which the capital outlay for catalyst is not excessive. Large-scale production using precious metals requires careful economic analysis involving catalyst costs, which constitute a major outlay, metal recovery costs, and incremental selectivity and yield advantages for the metal. The superior selectivity of precious metal catalysts can prove very attractive in certain reaction systems especially when the catalyst to be used requires only a low precious metal loading (0.5–1.0%) on a support.

Useful Tables

Table 12.1 provides a general overview of catalyst types used for most reactions of commercial interest. The reaction types are listed alphabetically, and the catalysts are designated by the major metallic component. There are, of course, many varieties of a particular active-metal catalyst, differing in such characteristics as type of support and support surface area, pore volume, pore-size distribution, crush strength, size, and form. Of additional importance is the type of promoter

TABLE 12.1
Recommended Initial Catalyst Selection for Process Research and Development Involving Hydrogenation Catalysts

Reactions (Reductions)	Recommended Initial Catalysts							
	Pd	Pt	Rh	Ru	Ni	Co	Cu	CuCr
Acetylenes to								
Cis-olefins	✔							
Alkanes	✔	✔			✔			
Acid chlorides to								
Alcohols	✔	✔		✔				
Aldehydes	✔							
Acids to alcohols				✔				✔
Anhydrides to esters, alcohols, or lactones	✔	✔						✔
Anilines to								
Cyclohexylamines			✔	✔				
Dicyclohexylamines	✔	✔						
Cyclohexanones (reductive hydrolysis)	✔							
Aromatic (carbocyclic) to								
Cycloalkane			✔	✔	✔			
Cyclo-olefin			✔	✔				
Benzyl derivatives to cyclohexyl derivatives		✔	✔	✔				
Carbonyl (unsaturated) to								
Unsaturated alcohols		✔		✔				
Aliphatic aldehydes	✔				✔			
Carbonyls (aliphatic) to alcohols		✔		✔	✔	✔	✔	✔
Carbonyls (aromatic) to								
Aromatic alcohols	✔				✔	✔	✔	✔
Saturated alcohols		✔	✔					
Alkanes	✔							
Dienes to olefins	✔				✔	✔		
Esters to alcohols or acids		✔		✔			✔	✔
Halonitroaromatics to haloanilines		✔	✔	✔				
Heterocyclic ring reductions								
Furans	✔		✔					
Pyrroles		✔	✔					
Pyridines	✔	✔	✔	✔				
Nitriles (aliphatic) to								
Primary amines	✔			✔	✔	✔		
Secondary amines		✔			✔	✔		
Ternary amines	✔	✔			✔			

TABLE 12.1

Recommended Initial Catalyst Selection for Process Research and Development Involving Hydrogenation Catalysts (continued)

Reactions (Reductions)	Recommended Initial Catalysts							
	Pd	Pt	Rh	Ru	Ni	Co	Cu	CuCr
Nitriles (aromatic) to								
Primary benzylamines	✔					✔		
Secondary amines	✔	✔				✔		
Nitriles to aldehydes (reductive hydrolysis)	✔							
Nitro (aliphatic) to amine	✔	✔			✔			
Nitro (aromatic) to								
Anilines	✔	✔			✔		✔	✔
Hydroxylamine		✔						
Amino phenol		✔						
Olefins to alkanes	✔	✔			✔			
Oximes to								
Primary amines	✔		✔		✔	✔		
Secondary amines	✔	✔						
Hydroxylamines		✔						
Phenols to								
Cyclohexanols	✔		✔	✔	✔			
Cyclohexanones	✔		✔					
Reductive alkylation (reductive amination)	✔	✔			✔	✔		✔

Recommendations are based on the active metal. Selection of the optimum support and promoters would follow from existing in-house knowledge or supplier and literature information.

Reprinted by permission: Pressure Chemical Co., *Heterogeneous Catalog,* Pittsburgh, PA (source of commercial and experimental catalysts).

that may be present and the surface area of the main active metals, but details of this type may be proprietary in some cases. Most of the reactions and catalysts summarized in Table 12.1 are addressed in greater detail beginning on page 120, using the most important commercial examples.

Table 12.2 is a useful summary of selective hydrogenations of various functional groups in the presence of another functional group that is not to be hydrogenated. Reactions of this type are important in a number of commercial processes and of special significance in the fine chemicals industry.

Table 12.3 provides a description of each commercial hydrogenation catalyst type, including a listing of suppliers. See "Catalyst Supports" for details on the characteristics of the various types of commercial supports (page 117). Following this "General Background" section, commercial hydrogenation processes are discussed according to reaction types. More specific information is given on the catalysts most commonly used for each hydrogenation process defined by the functional group being hydrogenated.

TABLE 12.2
Selective Hydrogenation of Different Functional Groups

In the Presence of	To Be Hydrogenated								
	$C{<}^{O}_{OH}$	$C{<}^{O}_{OR}$	$^{R}_{R'}{>}{=}O$	$^{R}_{H}{>}{=}O$	${>}{=}{<}$	$-NO_2$	$-CN$	⬡ (O)	$-{\equiv}-$
$-C{<}^{O}_{OH}$	—	Cu-Cr Cu	Cu-Cr	Cu-Cr	Pd Ni	Cu-Cr Pd	Co	Pt Pd	Pd
$-C{<}^{O}_{OR}$		—	Cu-Cr	Cu-Cr	Pd	Cu-Cr Pd	Co	Pt Pd	Pd
$^{R}_{R'}{>}{=}O$		Cu-Cr Cu	—	Cu-Cr Ni	Pd	Cu-Cr Pd	Co Ru	Pd	Pd
$^{R}_{H}{>}{=}O$		Cu-Cr Cu		—	Pd	Cu-Cr Cu	Co		Pd
${>}{=}{<}$			Cu-Cr	Cu-Cr Ni	—	Cu-Cr Cu	Co		Pd
$-NO_2$					Pd	—	Co		Pd
$-CN$	Cu-Cr	Cu-Cr	Cu-Cr	Cu-Cr	Pd	Cu	—	Pt	Pd
⬡ (O)	Cu-Cr	Cu-Cr	Pd Cu-Cr	Pd Cu-Cr	Pd	Cu-Cr Cu	Co	—	Pd
$-{\equiv}-$	Cu-Cr	Cu-Cr	Cu-Cr	Cu-Cr		Cu-Cr Cu	Co		—
$-NH_2$	Cu-Cr	Cu-Cr	Cu-Cr	Cu-Cr	Pd	Cu-Cr	Co	Ni Pt	Pd
$R-O-R'$	Cu-Cr Ni	Cu-Cr Ni	Cu-Cr Ni	Cu-Cr Ni	Ni	Cu-Cr	Co	Pt Ni	Pd
F	Ni	Ni	Ni	Ni	Pd	Cu-Cr Ni	Pd Co	Pd Pt	Pd
Cl, Br, J	Pd Ni	Pd	Pd	Pd	Pd	Pd	Pd	Pt	Pd
$-OH$		Pd	Cu-Cr		Pd	Cu Cu-Cr	Co	Ni Pd	Pd

Reprinted by permission: *General Hydrogenation Catalysts*, Technical Bulletin, Süd-Chemie AG.

TABLE 12.3
Commercial Hydrogenation Catalysts

Catalyst	Metal content, wt%, nominal range	Unreduced chemical state	Typical promoters used	Form	Supports[1]	Suppliers
Base Metals						
Cobalt (supported)	25–60	CoO	Zr, Mn	Fixed bed: • tablets • extrusions Slurry: • powder	• Kieselguhr • silica • silica-alumina • refractory oxide • alumina	Engelhard United Catalysts Synetix Celanese
Cobalt (Raney, sponge metal)	90–99	Co metallic		Slurry: • powder	None	Activated Metals & Chemicals, Inc. Grace Davison
Copper	15–30	CuO		Fixed bed: • tablets • spheres Slurry: • powder	• silica • BaSO$_4$ • alumina	Engelhard United Catalysts Synetix Celanese
Copper (Raney, sponge metal)	93–99	Cu metallic		Slurry: • powder	None	Activated Metals & Chemicals, Inc. Grace Davison Engelhard
Copper chromite Cu: 30–40 Cr: 26–30 Available in Cu/Cr atomic ratio form	CuO:CuCr$_2$O$_4$	Ba, Mn		Fixed bed: • tablets • extrusions Slurry: • powder	None	Engelhard United Catalysts
Nickel	10–68	NiO	Cu, Zr, Mo	Fixed bed: • tablets • spheres • extrudates Slurry: • powder • fat protected powder, flakes, droplets	• Kieselguhr • alumina • silica • various proprietary	Criterion Engelhard United Catalysts Synetix Celanese Acreon
Nickel (Raney, sponge metal)	88–96 44–52 (in fatty amine)	Ni metallic	Fe, Cr, Mo	Slurry • powder • cubes in fatty amine	None	Activated Metals & Chemicals, Inc. Grace Davison Engelhard

TABLE 12.3
Commercial Hydrogenation Catalysts (continued)

Catalyst	Metal content, wt%, nominal range	Unreduced chemical state	Typical promoters used	Form	Supports[1]	Suppliers
Platinum Metals						
Palladium, slurry	1–10 5 is common	PdO		• powder paste (water wet)	• activated carbon • alumina	Engelhard Johnson-Matthey Precious Metals Corp.
Palladium, fixed bed	0.1–1 0.5 is common	PdO		• extrudates • spheres • tablets • granules	• alumina • activated carbon	Acreom/Procatalyse BASF Engelhard Hülls ICI Katacol Synetix Johnson-Matthey Precious Metals Corp. United Catalysts
Platinum, slurry	1–10 5 is common	PtO$_2$ (or as hydrous oxides)		• powder paste (water wet)	• activated carbon • alumina	Engelhard Johnson-Matthey Precious Metals Corp.
Platinum, fixed bed	0.3–0.5 0.5 is common	PtO$_2$ (or as hydrous oxides)		• extrudates • spheres • tablets • granules	• alumina • alumina • alumina • activated carbon	Acreom/Procatalyse Engelhard Johnson-Matthey Precious Metals Corp. United Catalysts
Rhodium, slurry	5	Rh$_2$O$_3$xH$_2$O		• powder paste (water wet)	• activated carbon • alumina	Engelhard Johnson-Matthey Precious Metals Corp.
Rhodium, fixed bed	0.5	Rh$_2$O$_3$xH$_2$O		• extrudates • spheres • tablets	• alumina • alumina • alumina	Engelhard Johnson-Matthey Precious Metals Corp.
Ruthenium, slurry	5	Ru$_2$O$_3$xH$_2$O		• powder paste (water wet)	• activated carbon • alumina	Engelhard Johnson-Matthey Precious Metals Corp.
Ruthenium, fixed bed	0.5	Ru$_2$O$_3$xH$_2$O		• extrudates • spheres	• alumina • alumina • alumina	Engelhard Johnson-Matthey Precious Metals Corp.

12.1.2 GENERAL MECHANISM

Hydrogenation is certainly the most studied reaction in organic chemical catalysis. The Horriuti–Polanyi mechanism has been used extensively for over 60 years.[3] In step 1 of this mechanism, dissociative chemisorption on the catalytic metal produces the active hydrogen species.

$$H_2 \qquad + 2\;* \quad \rightleftharpoons \qquad \overset{2\ H}{\underset{*}{|}} \qquad (1)$$

$$\overset{\diagup}{\underset{\diagdown}{}}C\!=\!C\overset{\diagup}{\underset{\diagdown}{}} \quad + 2\;* \quad \rightleftharpoons \quad \underset{*}{-}\!\overset{|}{\underset{|}{C}}\!-\!\underset{*}{\overset{|}{\underset{|}{C}}}\!- \quad (2)$$

$$-\!\underset{*}{\overset{|}{\underset{|}{C}}}\!-\!\underset{*}{\overset{|}{\underset{|}{C}}}\!- \;+\; \underset{*}{\overset{|}{H}} \quad \rightleftharpoons \quad -\!\underset{*}{\overset{|}{\underset{|}{C}}}\!-\!\underset{H}{\overset{|}{\underset{|}{C}}}\!- \quad (3)$$

$$-\!\underset{*}{\overset{|}{\underset{|}{C}}}\!-\!\underset{H}{\overset{|}{\underset{|}{C}}}\!- \;+\; \underset{*}{\overset{|}{H}} \quad \rightleftharpoons \quad -\!\underset{H}{\overset{|}{\underset{|}{C}}}\!-\!\underset{H}{\overset{|}{\underset{|}{C}}}\!- \quad (4)$$

* = active site

The relative rates of steps 3 and 4 depend on the concentration of adsorbed hydrogen and thus all factors influencing the availability of hydrogen on the catalyst surface.[4]

hydrogen partial pressure	type of support
catalyst surface area	surface modifiers
metal structure	temperature
metal loading	additives
degree of turbulence	diffusion characteristics

Any one or more of these factors can influence the rate-determining and geometry-determining steps.[4] Of course, side reactions can greatly influence outcomes and increase complexity of the reaction network. Such is the situation with the hydrogenation of C_4 and higher olefins.

There are many examples of the effect of the above listed variables on changing the selectivity of hydrogenation reaction systems. Unfortunately, no clear *a priori* guidelines exist for predicting what effect changes might produce. Once a favorable modification in reaction conditions or catalyst characteristics is discovered, it can become useful to apply the knowledge to further optimize the catalyst and/or operating conditions. Of course, the particular catalyst itself plays the major role, and, in many processes, the catalyst provides such a high degree of selectivity that increased selectivity by other means is not needed or expected.[6]

There is no doubt that operating conditions can play a role. Examples of the effects on selectivity of temperature, pressure, additives, pore diffusion, and mass transport have been reported[4,6] and are summarized in Table 12.4. Unfortunately, these observations are generalizations that are not necessarily all applicable to other similar compounds. They do, however, strongly suggest that changes in one or more of these variables may lead to improved selectivity. The catalyst manufacturer can often suggest optimal conditions for the desired selectivity. If a new reaction system, it is not too difficult to explore easily managed variables such as temperature and pressure.

Hydrogen Adsorption and Surface Migration

Because of the small size of hydrogen relative to other atoms, after adsorbing on metal surfaces and dissociating into hydrogen atoms, it can migrate into the bulk metal crystallite and, in some cases, react with the bulk metal to form metal hydrides.[8] The adsorption step is thought to be the

TABLE 12.4
Factors Affecting Selectivity in Catalytic Hydrogenation Other Than Catalyst Type

Variable	Examples	Effect on Selectivity
Hydrogen partial pressure	acetylene → ethylene	Low H_2 pressure when H_2 is present in excess avoids saturation to ethane.
	disubstituted cyclic olefins	Cis isomer favored at increased H_2 pressure.
	alkyl substituted methylene cycolhaxane	Cis isomer decreases at increased H_2 pressure.
	aromatics	Intermediate increases then decreases with increasing H_2 pressure.
	vegetable oils	Both the selectivity of the consecutive reactions and the isomerization reactions are favored at low pressures (low H_2 availability).
Temperature	acetylene → ethylene → ethane	Lower temperature favors ethylene—lowers H_2 availability.
	cyclic ketones and substituted aromatics	Opposite effect to that of pressure.
Additives	acetylene → ethylene Pb added to Pd catalyst Many catalyst systems use additives. (See particular catalyst or reaction of interest.)	Low adsorbed and chemisorbed hydrogen caused by added Pb (CO also reduces H_2 coverage).
Metal loading	2-methylmethylene cyclohexane	Cis isomer favored at low metal loading (large reactant-to-metal ratio makes H_2 readily available to active sites).
	vegetable oils	Trans isomer favored at high loading (low H_2 availability).
Pore diffusion	consecutive reactions $A + H_2 \rightarrow B$ $B + H_2 \rightarrow C$	At some point, selectivity to C increases with increase in H_2 pressure. As chemical rate increases, the rate of transport in the pores begins to influence the global rate and causes a low rate, which increases the opportunity for B to react to C.
Mass transfer in slurry reactors	vegetable oil $A + H_2 \rightarrow B$ $B + H_2 \rightarrow C$	Low agitation reduces H_2 availability and favors trans isomer. If catalyst particles are very small, there is a possibility for particles to penetrate the gas-liquid film enveloping the gas bubble. Reaction can then be rapid and favor production of C.

Based on information from Refs. 4, 6, and 7.

controlling step in this series of steps. Of the transition metals important in catalysis, only palladium forms a stable hydride at low pressures. Nickel and copper form unstable hydrides but only at very high pressures.[9] The unique behavior of Pd ultimately may have been part of an explanation of its successful role in many instances of selective catalysis. As shown in Figure 12.1, a metal surface in which hydrogen has penetrated may be less reactive, since rearrangement of the surface is required to release hydrogen atoms.[10]

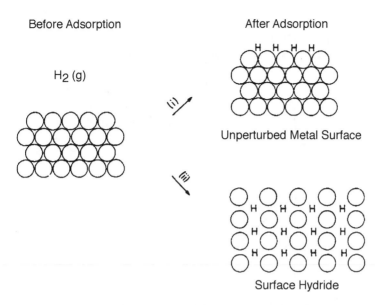

FIGURE 12.1 Possible Extremes in Hydrogen Adsorption on Metal Surfaces. Reprinted from Geus, J. W., *Hydrogen Effects in Catalysis,* Z. Paul and P. G. Menon, eds., p. 196, Marcel Dekker, New York, 1988, by courtesy of Marcel Dekker, Inc.

Surface migration of atomic hydrogen is a phenomenon of great interest, for it appears to have a significant effect on catalysis in some reaction systems. The process has been called *spillover* when it involves surface diffusion from the metallic site to the support (Al_2O_3, SiO_2, O_2, C, etc.). Extensive research has produced some general agreement on the overall phenomenon, but the details remain unsettled. There is solid evidence, however, that supports such as refractory oxides or active carbon become catalytically active themselves. More importantly, the spilled-over hydrogen proves useful in making atomic hydrogen available to the active metal surfaces and for removing catalyst poisons such as sulfur by formation of hydrogen sulfide.[11] The various supports behave differently, and the studies of spillover may lead to some understanding of the often seemingly mysterious but dramatic improvement in selectivity and yield exhibited by a particular support compared to all others tested using the same active metal. In the meantime, guidance is best sought from technical experts at the various catalyst manufacturers.

Mass Transport Effects in Hydrogenation

Many hydrogenation reactions involve a three-phase system of hydrogen gas, the substrate liquid, and the solid catalyst on a porous support. Since the solubility of hydrogen in organic liquids is low, and the substrate concentration is often high and is also the more strongly adsorbed, it is possible to rationalize the various possible rate controlling conditions based solely on hydrogen transport.[12] Figure 12.2 illustrates the series of steps involved.

I. H_2 transport through the gas phase.
II. H_2 transport through the liquid to the catalyst exterior.
III. & IV. H_2 pore diffusion to the catalytic metal in the pore followed by reaction or, depending on the relative rates of reaction and diffusion, concurrent diffusion and reaction throughout the support.

Steps I and II are influenced by the reactor design and operating conditions such as degree of turbulence. Steps III and IV can be influenced most effectively by the design of the catalyst.

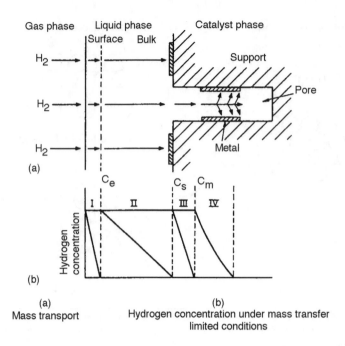

FIGURE 12.2 Hydrogen Transport in Liquid-Phase Hydrogenation. Reprinted from *Chemical Engineer (London),* March 1974, Acres, G. J. K., Bird, A. J. and Davidson, P. J., "Recent Development in Platinum Metal Catalyst Systems," pp. 145–157, copyright 1974, with permission from Elsevier Science.

Assuming a given catalytic metal, the remaining degrees of freedom involve the support characteristics, including surface area, pore size and pore-size distribution, and metal distribution on the catalyst surfaces.

Several interesting studies have been conducted to illustrate the following axioms.[12]

1. If a catalyst operates under conditions of chemical-rate control, thereby utilizing its intrinsic activity, maximum activity and poison resistance will be attained along with the selectivity favored by the active catalytic components.
2. If a catalyst operates under diffusion control, the net reaction rate will be lowered, and the selectivity can be altered. In the case of consecutive reactions (A → B → C), diffusion control would retard the counter diffusion of B and favor the production of C.

These issues have been illustrated by several well conceived experiments. Figure 12.3 presents the behavior of four catalyst preparations of Pd on charcoal in which Pd_E and Pd_I represent, respectively, Pd only on the exterior of the support and Pd within the support. Charcoal of high surface area is indicated by C_H and of low surface area as C_L. The reaction is the hydrogenation of nitrobenzene to aniline. Note the following:

- At constant metal loading, maximum activity is realized with metal deposited on the exterior (egg-shell pattern) of the catalyst.
- Low support surface area lowers the activity.
- Internal metal deposition causes an even greater decline in observed rate as internal diffusion (step III) becomes controlling.

Figure 12.4 shows observed conversions for the hydrogenation of benzoic acid that contains impurities that are catalyst poisons. The product is hexahydrobenzoic acid. Because the poison is

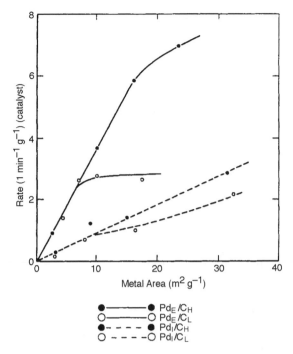

FIGURE 12.3 Effect of Palladium Distribution and Charcoal Surface Area on Reaction Rate. Pd_E = Pd on exterior of support only, Pd_I = Pd within exterior of support, C_H = high-area charcoal support, C_L = low-area charcoal support. Reprinted from *Chemical Engineer (London),* March 1974, Acres, G. J. K., Bird, A. J. and Davidson, P. J., "Recent Development in Platinum Metal Catalyst Systems," pp. 145–157, copyright 1974, with permission from Elsevier Science.

distributed throughout the pores in the case of the preparation that had the Pd deposited internally, more catalyst surface was available for adsorbing the poison. Thus, poison was removed over time toward the inlet of the pores, leaving unpoisoned depths available for reaction. Although the reaction was controlled by diffusion for the internally dispersed metal, it outperformed the reaction-controlled, exterior-loaded catalyst.

An example of a dual-path reaction affected by changes in the controlling step was studied using the hydrogenation of 2, 4 dinitrotoluene with Pd/charcoal in a solution of isopropanol-water.[9]

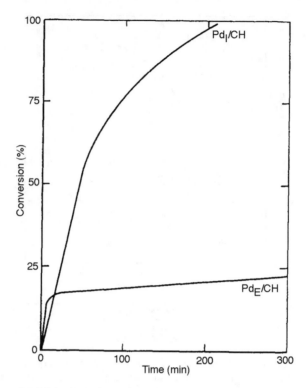

FIGURE 12.4 Effect of Pd Distribution on Poison Resistance in the Hydrogenation of Benzoic Acid. Reprinted from *Chemical Engineer (London),* March 1974, Acres, G. J. K., Bird, A. J. and Davidson, P. J., "Recent Development in Platinum Metal Catalyst Systems," pp. 145–157, copyright 1974, with permission from Elsevier Science.

The quantity of ortho isomer was observed to be low in the product when operating in a diffusion-controlled region, but it increased when conditions were altered so that the intrinsic reaction rate was the controlling step. Furthermore, when the temperature was increased at constant pressure, a point was reached where the ortho-isomer production declined, which corresponds to the onset of the diffusion-controlled region. Thus, favorable selectivity to the para-isomer requires catalyst pore and surface-area design by the catalyst manufacturers conducive to operation in the diffusion-controlling region.

These several examples illustrate the important contribution to be made by the catalyst manufactured in creating the specific catalyst design that will serve the goals of maximum yield and long life. The most useful generalization can be simply stated. A reaction system can be optimized not only by selecting the best catalyst chemistry, but also the best catalyst-metal distribution and support design. To be sure, the careful placement of small amounts of costly precious metal may be more crucial than the disposition of basic metals on catalysts. But the same principles of catalyst design apply to base metals.

12.1.3 CATALYST CHARACTERISTICS

Surface Characteristics

Surface characteristics of hydrogenation catalysts are important not only in fundamental studies but also in comparing various preparation methods and operating procedures. Classical methods involving the determination of chemisorption isotherms continue to be useful, since they do not require expensive spectroscopic equipment and also rest on a foundation of decades of valuable experience. These isotherms enable calculation of the metal surface area and its dispersion and from it, the mean crystallite diameter.[13] Since hydrogen is mainly chemisorbed by the metallic part of the catalyst surface for all hydrogenation-catalyst metals except copper, it is a test case of choice. For the exception, copper, adsorptive decomposition of N_2O is the accepted method.[13]

$$N_2O \text{ (gas)} + 2Cu \rightarrow Cu_2O + N_2 \text{ (gas)}$$

In interpreting results, it is important to realize that, often, the conditions of the metal surface during actual reaction may be different than that after the pretreatment procedure for the metal surface area measurements. These procedures include oxidation, reduction, and evacuation at high temperature. Most methods of surface study, including XPS (x-ray photoelectron spectroscopy), ESCA (electron spectroscopy for chemical analysis), x-ray diffraction, and x-ray spectroscopy, require conditions unlike those in the conduct of practical reactions. Surprisingly, many useful correlations and insights have been gained, but one should be prepared at times to find no correlation between free-metal surface areas and catalytic activity and selectivity. Clearly, metal dispersion (along with many other characteristics of the active components), promoters, and support play a major role in the ultimate behavior of a particular hydrogenation catalyst. Catalyst manufacturers offer numerous types of catalysts with the same active metal, but each one having unique characteristics that favor its use for a particular reaction or class of reaction. For this reason alone, recommendations of a manufacturer's technical expert can be most valuable.

Catalyst Supports

Hydrogenation reactions are conducted in stirred-slurry or fixed-bed reactors. Slurry reactors require powdered catalysts, with activated charcoal or Kieselguhr being the most common support. Fixed-bed reactors employ alumina or Kieselguhr supports as tablets, extrudates, or spheres. Activated charcoal is not used in fixed beds, because it does not have sufficient crushing strength. It can, however, be used as a support in granular form in trickle-bed reactors because of the lower pressure drops in reactors of this type.

Charcoal has a distinct advantage when used as a support for precious metal catalysts because of the ease of recovering the valuable metal from deactivated catalysts. The charcoal can be readily removed by burning.

In selecting a support, characteristics such as average pore size, pore-size distribution, surface area, and active-metal distribution (edge, uniform, and mixed) are important, and catalyst manufacturers offer both standard and special proprietary supports to aid in attaining the desired activity and selectivity.

Catalytic metal and support interaction can vary significantly between various supports. In many such instances, activity and selectivity will be favored by one type of support. These effects have been studied for certain hydrogenations and hypotheses on the mode of action proposed. No clear *a priori* selection criterion seems to have emerged, and the potential user is well advised to seek help from catalyst manufacturers who are custodians of the experience and lore related to the catalysts they offer. In the section on specific commercial reactions, suggestions are given on the recommended support and its effects, where applicable and available.

Role of Support in Hydrogenation

There is no doubt that the support often plays an important role on the degree of dispersion of the active metal. Many examples have been published showing a significant effect of dispersion on activity and selectivity for structurally sensitive reactions, or because the highly dispersed metal-to-support interaction produces a different type of active center.[17]

Other support metal interactions have been demonstrated, including poison resistance, electronic interaction between the metal and support, a strong-metal support interaction, and spillover of species initially adsorbed on the metal on to the support surfaces.[17] These phenomena have been studied intensively in recent years, and interesting observations and insights have been described. No theory has yet been proposed that will permit *a priori* metal/support selection capable of producing a desired result. Catalyst manufacturers, however, have much application experience in these areas and can suggest an optimum support for a given situation that, in some cases, may be a proprietary product of special value.

Catalyst Deactivation

Since many hydrogenation reactions are operated at modest temperatures, sintering of the support, agglomeration of the active metal, and coking are often not as much of a problem, as for higher temperature reactions such as dehydrogenation. But over long periods of on-stream time, effects of larger crystallites and some carbonaceous deposits may be noted by a gradually declining activity. Constant, lower-temperature operation (<200°C) can increase the deactivating effect of catalyst poisons,[14] and thus poisons become a significant concern in hydrogenation systems.

Poisons

Poisons are usually grouped as *permanent* and *temporary* poisons. In common usage, a temporary poison is one that, when removed from the feed, will ultimately desorb from the catalyst as fresh feed is fed, and the activity will return to its original condition. A permanent poison is one that cannot be removed by the previously described maneuver or any other operating adjustment. Some so-called permanent poisons, however, can be removed by an oxidative catalyst regeneration procedure. Some authors,[14] therefore, call the temporary poison described above as an *inhibitor* and the poison removed by regeneration as *temporary*.[14] In this classification system, the poison that cannot be removed off-line by oxidative regeneration is called *permanent*, which indeed it is.

In general, catalyst poisons for hydrogenation reactions include compounds of S, Se, Te, P, As, Hg, N, Si, and Zn, halides, Pb, NH_3, CO.[14,15] Briefly, the Group VIII metals (Fe, Co, Ni, Ru, Rh, Pd, and Pt) plus copper, which constitute the major commercial hydrogenation catalysts, are poisoned by compounds of VA and VIA elements (N, P, As, S, Se, Te) plus Hg, Si, and Cl. Examples are given in Table 12.5 of typical compounds encountered. More information can be found in the following sections on specific processes.

Sulfur compounds are some of the most common hydrogenation catalyst poisons, since sulfur is contained to some extent in all feedstocks derived from raw materials of animal or vegetable origin (petroleum, coal, and natural fats). Generalizations on sulfur poisoning indicate the complexity of the phenomenon shared in many ways with other known poisons. These include the observation that the extent of poisoning depends on the nature of reacting organic compounds undergoing reaction, the type of sulfur compound(s) present, operating conditions, and the catalyst properties (type of support, pore structure, active metal and its dispersion, and catalyst pretreatment).

Sintering

Both the active metal and supports such as alumina are subject to sintering at high temperatures. Since many hydrogenations are conducted at moderate temperatures for thermodynamic reasons, deactivation by sintering is not a major problem unless a runaway reaction occurs or oxidative regeneration is attempted without proper control.

TABLE 12.5
Typical Poisons for Hydrogenation Catalysts[1-3]

Poison Type	Examples	Mode of Action
Sulfur compounds	H_2S Thiols R-SH Organic sulfide R-S-R Organic disulfide RS-SR Order of deactivation intensity: $H_2S \gg RSH \geq R(S_2)R \geq RSSR$ $> RSR$, thiophene Sulfur compounds occur in petroleum and coal liquids before or after processing Carbon oxysulfide COS (present in fluid cracking reaction products)	• Dissociatively adsorbed to form metal sulfide ($Ni + H_2S + H_2 + NiS$) • At low temperatures, characteristic of many hydrogenation reactions, associative adsorption often occurs, and the original sulfur compound is strongly chemisorbed. In such cases, toxicity toward the desired reaction depends on molecular size, structure, and adsorptive strength. • At higher temperatures, dissociative adsorption can occur with the formation of metal sulfides that reduce the absorptive capacity of the catalyst for H_2 and organic reactants. • If water is present, $COS + H_2O \rightarrow CO_2 + H_2S$ in feed, and H_2S is a strong poison. • Regeneration is not usually practical. Although oxidation converts sulfur compounds to less toxic structures, metallic oxides form more rapidly and cover the metal sulfides to prevent oxidation. • Bimetallic Pd catalysts can resist small amounts of H_2S. It is best to remove sulfur from feed prior to entering the reactor.
Metallic and Semimetallic Compounds		
Arsenic	AsH_3 (occurs in light ends from fluid catalytic cracking)	• Forms surface compound (alloy) with catalytic metal blocking catalytic action. • Regeneration is possible, since As_2O_3 formed migrates to the support, and active metal sites are restored. But regeneration is only practical if deactivation causes by one-time feed contamination. If As is continuously in feed, unacceptably short cycle times are required. It is best to use special adsorbent unit prior to main reactor.
Mercury	Hg (occurs in some natural gases)	• Forms amalgam with catalytic metal that has no useful catalytic activity. Heat treatment vaporizes Hg and restores activity, but it is often impractical as described above for arsenic.
Silicon	Silicone antifoaming agents (often present in effluents from coking units)	• Hydrogenolysis to silicon occurs on the metal, and the silicon covers the active metal blocking, thereby, access of reactants to the active catalytic metal.
Lead	Tetraethyl lead contamination (generally not a problem since the phase-out of tetraethyl lead as an anti-knock agent)	• Metallic lead, which is produced by hydrogenolysis of an organic lead compound, deposits on the catalytic metal and tends to form an alloy with it that destroys catalytic activity.
Ammonia Organic nitrogen compounds	NH_3 Piperidine	These compounds act as inhibitors or, selectively, promoters. By adsorbing at a strength less than that of the desired reactant and greater than the undesired reactant, the preferred reaction path is favored.
Carbon monoxide	CO	Carbon monoxide acts as an inhibitor or selectivity promoter. By adsorbing at a strength less than that of the desired reactant and greater than the undesired reactant, the preferred reaction path is favored.
Chlorides	Organic chlorides (original source of chlorine may be salt in crude petroleum)	Chlorides adsorb strongly on active metals because of high electronegativity and compete thereby with reactants. If H_2 is present, HCl is formed that passes out of the reactor. However, HCl can cause stress cracking of reactor. It is best to remove it from the feed.
Phosphorus compounds	Detergents such as used in lube oil, phosphatides in vegetable oils and animal fat	

12.2 HYDROGENATION OF AROMATIC RINGS

12.2.1 ANILINE → CYCLOHEXYLAMINE AND DICYCLOHEXYLAMINE

Although the major use of aniline is in producing methylene diphenylene isocyanate for reacting with polyols to form rigid polyurethane foams and elastomers, a significant quantity of aniline is hydrogenated to cyclohexylamine and dicyclohexylamine.[1,5]

Cyclohexylamine has a variety of uses as a component in making vulcanizing accelerators, plasticizers, emulsifiers, antioxidants, and artificial sweeteners[1,2] (cyclohexyl sulfanates, cyclamates), corrosion inhibitors, and a hardener for epoxy resins.[5]

Dicyclohexylamine and its salts are anticorrosion agents. It is also used in producing vulcanization accelerators, pesticides, and corrosion inhibitors.[5]

Chemistry

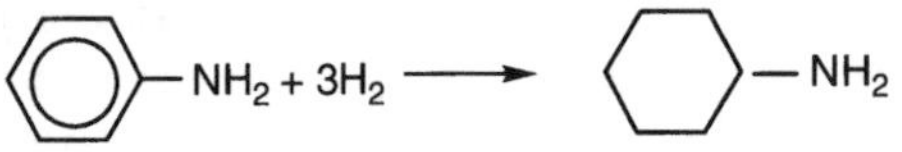

Mechanism

Unless careful choice of catalyst and operating conditions are made, both the cyclohexylamine and dicyclohexylamine are formed.[2,3]

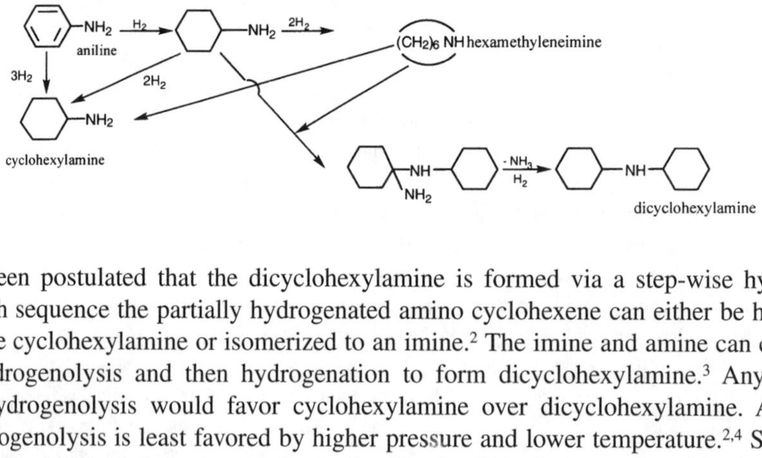

It has been postulated that the dicyclohexylamine is formed via a step-wise hydrogenation during which sequence the partially hydrogenated amino cyclohexene can either be hydrogenated further to the cyclohexylamine or isomerized to an imine.[2] The imine and amine can combine and undergo hydrogenolysis and then hydrogenation to form dicyclohexylamine.[3] Any action that decreases hydrogenolysis would favor cyclohexylamine over dicyclohexylamine. As has been shown, hydrogenolysis is least favored by higher pressure and lower temperature.[2,4] Solvents with high dielectric constants (e.g., water) favor the diamine.[2] In the case of noble-metal catalysts, ruthenium is preferred, because it is the least active for hydrogenolysis.[3,3] Everything else being equal, palladium favors hydrogenolysis, and thus dicyclohexylamine production, especially at higher temperatures.

A more recent study using laboratory-prepared Co/Al$_2$O$_3$ suggests a slightly different mechanism as shown in Figure 12.5.[7] The intermediate in the path to dicyclohexylamine is suggested to be N-phenylcyclohexylamine. The primary product over Co and Ni/Al$_2$O$_3$ is said to be cyclohexylamine. The high activity of Ni/Al$_2$O$_3$, however, favors further reaction to dicyclohexylamine by both pathways. In contrast, the lower-activity Co/Al$_2$O$_3$ tends to be more selective to cyclohexylamine. It is important to remember, however, that proprietary promoters may alter the performance of both of these catalysts.

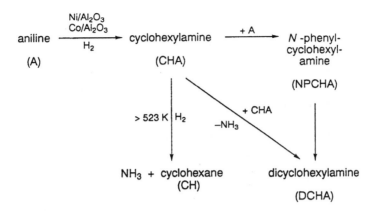

FIGURE 12.5   Alternative Reaction Scheme for Cyclohexylamine and Dicyclohexylamine. Reprinted by permission: Narayenan, S. and Urimkrishman, R. P., *Journal Chemical Society, Faraday Transactions,* 93 (10), 2009 (1997), The Royal Society of Chemistry, Cambridge, England.

Catalysts and Reactor Types

Probably the most used commercial catalyst for manufacturing cyclohexylamine is a cobalt (46%) on silica with a basic promoter such as a Mn or Ca oxide, which reduces the tendency to form dicyclohexylamine, as does the presence of ammonia. The reaction is carried out in the range of 40 to 60 bar and a temperature around 230°C.[1,6] A liquid-phase slurry reaction is favored with a catalyst loading at 4 to 5 wt%. Most of the operating conditions in the literature are obtained from patents.

Ni-on-Kieselguhr (65% Ni), Raney cobalt treated with CaO, and noble metals are also used. Nickel is more active than cobalt and could prove more efficient for a continuous fixed-bed, vapor-phase process. Of the noble metals, Rh and Ru are the least active in hydrogenolysis and therefore produce less dicyclohexylamine than Pd or Pt catalysts.[2] Ruthenium requires higher pressures (20–50 bar) than rhodium but lower temperatures than cobalt or nickel (160–180°C).

Alkyl anilines can be hydrogenated to alkyl cyclohexylamines using any one of these same catalysts. N-methylcyclohexylamine is used in rubber vulcanization accelerators. N, N, dimethyl-cyclohexylamine is a catalyst for polyurethane production, and N-ethylcyclohexylamine is used in the manufacture of herbicides.

Dicyclohexylamine is most effectively produced by hydrogenation of an equal mixture of cyclohexanone and cyclohexylamine using Pd-on-carbon at lower pressures, 4 bar.[5] Low pressure, high temperature, and carbon-supported Pd all favor the dicyclohexylamine.[2]

Clearly, catalyst type, additives and operating conditions play a major role in selectivity toward the monoamine. In some processes where mixtures of both the monoamine and the diamine (in lesser quantity) are produced, recycle of diamine to the hydrogenator inhibits additional diamine formation. Operating conditions are perhaps such that equilibrium is reached in such cases.

Catalyst Suppliers

See Table 12.3.

Other Anilines

Other aniline compounds also can be hydrogenated to the cyclohexyl form in good yield in a similar fashion, provided no easily hydrogenated functional groups are present that must be preserved. Hydrogenation of methyl aniline yields N-methylcyclohexylamine (used as a vulcanization accel-erator), dimethylaniline yields N,N-dimethylhexylamine (used as a catalyst for polyurethane pro-duction), and ethylaniline yields N-ethylcyclohexylamine (used as a herbicide).

12.2.2 Benzene → Cyclohexane

Benzene hydrogenation to cyclohexane is a major petrochemical process. Most of the cyclohexane (96–98%) is used to produce a cyclohexanol/cyclohexanone mixture by catalyzed air oxidation. The product, called *KA oil* (ketone and aldehyde), is then converted to adipic acid by nitric acid oxidation. It is postulated that the cyclohexanol in the mixture is converted to cyclohexanone, and the total cyclohexanone forms adipic acid by one of three possible pathways.[1] Adipic acid is a comonomer with hexamethylenediamine in Nylon 6,6. The KA oil can also be thermally dehydrogenated in the vapor state to all cyclohexanone, which is then used to make cyclohexanone oxine by reaction with hydroxylanine. Beckmann's rearrangement of the oxine yields caprolactame monomer for use in producing Nylon 6.

Chemistry

Benzene is readily hydrogenated to cyclohexane using supported nickel or platinum in fixed beds. One process uses Raney nickel in a continuous-stirred-tank reactor or a homogeneous catalyst (see "Catalyst Types and Suppliers" and "Process Units" sections that follow). The reaction has been proposed to proceed through an intermediate cyclohexene,[5,9] which is not as strongly adsorbed as cyclohexadiene. Other mechanistic routes have been reviewed,[5] but the overall result may be depicted as follows. The system is no doubt complex. For example, three different forms of bezene adsorption have been observed.[6]

	Temp., K	DH, kcal	K_p
Reaction #1	400	−50.62	17.87
	500	−51.62	0.876
Reaction #2	400	−46.62	7.447×10^7
	500	−47.64	538

Equilibrium favors the hydroisomerization product, methylcyclopentane, but, fortunately, its rate of reaction is low, and short contact time assures minimal production of this unwanted isomer.

Cyclohexane for nylon production must be very pure, and the close boiling points of cyclohexane and benzene make separation difficult and costly. It is, therefore, essential to react all the benzene to cyclohexone. Specifications for maximum benzene content of the product are often in the range of 50 ppm and aliphatic content of no more than 5000 ppm, making a purity of 99.5 percent.[2,3] In recent years, purity specifications have increased in many cases to 99.8% or even higher.

To accomplish such high conversions, lower temperatures at the outlet section must be maintained in the range of 177–205°C where equilibrium is favorable for cyclohexane.[2,3] The reaction is highly exothermic, and clever process design that assures controlled removal of this heat of reaction is essential. At the same time, catalyst activity must be high enough to allow short contact time so as to prevent side reactions. The effect of pressure on equilibrium is shown on Figure 12.6. Excess hydrogen and pressure favor high conversion.

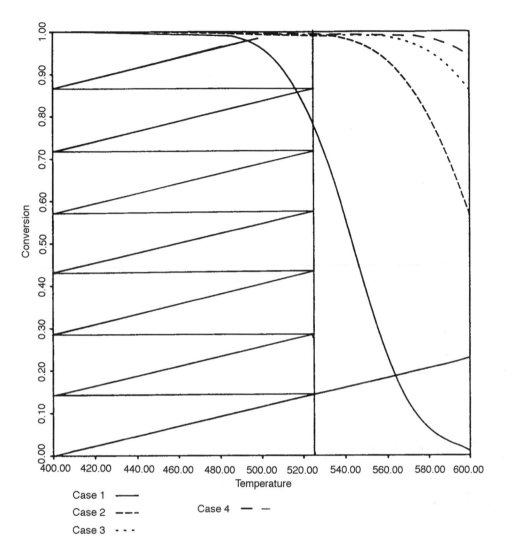

FIGURE 12.6   Cyclohexane Equilibrium and Adiabatic Reaction Temperature (H_2/benzene = 4). Case 1 = 1 atm, Case 2 = 5 atm, Case 3 = 10 atm, and Case 4 = 15 atm. Maximum allowable temperature is 525 K. Example stages at 10 atm. Fewer stages are possible with higher H_2/benzene mole ratio in feed to each stage. Temperature in Kelvins. Reprinted by permission from Rase, H. F., *Fixed-Bed Reactor Design and Diagnostics,* Butterworths, Boston, 1990. Copyright held by author.

If the reaction temperature gets too high (above 260°C), not only will the reverse reaction become significant, but highly exothermic hydrocracking of cyclohexane can occur, causing a runaway reaction that requires emergency quenching. Judicious use of recycle and lower activity catalyst in earlier beds aid greatly in temperature control. Alternatively, a multitubular reactor can be used.

Catalyst Types and Suppliers

Nickel on a refractory support is used in a number of fixed-bed processes both mixed-phase and liquid phase. Platinum-on-Al_2O_3 is also used in several processes. Although more expensive, it has an advantage of low-temperature operation (300°F/150°C or less). Raney nickel is used in one major process. Acidic characteristics in the support must be strictly avoided, since isomerization to methylcyclopentane is acid catalyzed.[1]

Supplier	Ni-on-Al$_2$O$_3$ or Proprietary Support	Pt-on-Al$_2$O$_3$ or Proprietary Support	Sponge or Raney Nickel
United Catalysts	✔		
Celanese	✔		
Engelhard	✔	✔	
Johnson-Matthey		✔	
Activated Metals & Chemicals			✔
Grace Davison			✔
Precious Metals Corp.		✔	
Synetix	✔		
UOP LLC		✔	

Licensors	Description
ABB Lummus Global	• Fixed-bed
IFP	• Raney nickel catalyst or proprietary homogeneous catalyst in CSTR. Finishing reactor has standard Ni-on-alumina.
UOP LLC	• Fixed-bed, liquid-phase process
CDTECH	• Catalytic distillation

Catalyst Deactivation

Both nickel and platinum are subject to poisoning by sulfur compounds, as well as P and As compounds. A 1 ppm, sulfur specification is usually set for the benzene feed. Suppliers offer formulations with some resistance to sulfur poisoning by means of extensive metal surface areas. This characteristic is most valuable in the event of some upset that caused higher feed-benzene sulfur content. Halides are also poisons that can enter the system from another upstream unit. Metal chlorides can form that are volatile. Ammonia complexes with nickel and deactivates the site. Carbon monoxide, which can be present in some hydrogen streams depending on their source, forms metal carbonyls[4] that destroy activity of the catalyst metal content.

Sintering of active metal with reduction in activation can become a problem, as can coke formation, if temperatures exceed the normal operating range excessively and over an extended period. Temperature excesses are usually guarded against by careful temperature control (see "Process Units" section), not only because of these deactivating effects but also because of the unfavorable equilibrium conditions thus created.

Process Units

Hydrogenation of benzene represents a classical example of a highly exothermic equilibrium-limited reaction systems that can be handled by a variety of reactor configurations. All are based on effective reaction temperature control.

Adiabatic Reactors in Series

A series of three to four adiabatic fixed-bed reactors can be operated with cooling between beds and recycling of a portion of the condensed product, along with introduction of cool benzene feed at each stage. The entire operation can be conducted in the vapor phase within each reactor if the pressure is maintained at 10 to 15 atmospheres. In this mode, temperature control is more tedious but can be improved by using a lower activity catalyst in the first bed.

Alternatively, the process can be conducted in a mixed-phase environment in which a portion of the cooled and condensed product from the reaction train is recycled along with benzene feed and hydrogen to each reactor inlet. Adiabatic temperature is controlled in each not only by means of the inlet temperature but also by partial vaporization of the liquid-phase (benzene and cyclohexene). This direct contact release of latent heat is a highly efficient means of heat transfer. As in the all-vapor phase process, reactor outlet temperatures are lowered by interstage heat exchangers prior to entering the next stage using boiler feedwater as the coolant, which generates useful high-pressure steam.

Multiple fixed-bed reactors have also been used in an all liquid-phase reactor system with benzene introduced at the inlet of each stage. Operating pressures must be higher (20–30 bar).

Nonadiabatic Multitubular Reactor

A single multitubular reactor has been used in several processes operating in either a vapor phase or mixed-phase regime. Heat of reaction is removed by boiler feedwater at a temperature set by the back pressure on the associated steam drum from which water is circulated by thermosiphon action to the shell side of the reactor. As in the adiabatic case, mixed-phase operation provides greater operating flexibility because of the latent heat of vaporization of benzene and cyclohexane, which serves as a direct contact means of finessing temperature control.

In the early 1970s, there were at least ten processes for licensing, but the patents on many of these have expired. Each one was characterized by clever ideas for controlling temperature and saving energy. One very different process was introduced by IFP in recent years.[17] It used a continuous stirred-tank reactor with Raney nickel catalyst. The liquid reactants are circulated through a boiler feedwater exchanger and back to the reactor as means of maintaining complete mixing and aiding in temperature control while producing steam for plant use. Operating conditions are such that the liquid in the reactor is maintained at its boiling temperature. Most of the conversion occurs in this CSTR reactor, and the cyclohexane and the small amount of unreacted benzene vapor generated continuously passes to a vapor-phase adiabatic reactor with a supported Ni catalyst, where essentially complete conversion is reached. Continuous vaporization of the product not only provides a powerful means for maintaining constant temperature but also favors the forward reaction by removing product at a steady rate. A newer version of the process uses a homogeneous catalyst in the CSTR, thus avoiding a problem that might occur if circulation is impaired and the Raney nickel would settle out.

In all processes, product is cooled and flashed to separate hydrogen for recycle. A portion of the condensate is used as described in many of the processes as a diluent for temperature control. The remainder goes to a stabilizer for removal of light ends. Cyclohexane and small amounts of benzene and methylcyclopentane constitute the final product. It is important, therefore, to avoid higher temperatures so that significant amounts of methylcyclopentane are not formed and so that benzene is essentially totally consumed. For adiabatic reactors in series, the final reactor serves as a finishing reactor with feed to it entering at conversions in the high nineties, under which conditions the outlet temperature is easily controlled. In the case of the multitubular reactor, the outlet temperature will approach very closely that of the boiling feedwater in the shell. Thus, this temperature must be controlled to a favorable equilibrium range for cyclohexane formation and allow essentially complete conversion.

The catalyst support must be strong for mixed-phase processes to withstand vaporization of liquid in both on the catalyst outer surface and within pores. Special alumina or other refractory supports are used.

Process Kinetics

A commercial process with a dominant single reaction is an obvious choice for a thorough kinetic study, and many have been done. A rather thorough study of all three supported catalysts of general interest (Pt, Pd, and Ni) has been reported.[5] As determined by many other studies, the reaction is zero-order in benzene.[6,8] The rate is expressed in the Langmuir form[4,5] as follows. See Ref. 5 for a detailed review.

$$r = k\frac{K_H P_H}{1 + K_H P_H}$$

where k = the rate constant

       K_H = the adsorption coefficient for molecular hydrogen

       P_H = the partial pressure of hydrogen

The activation energy on all three catalysts was the same (14 ± 1 kcal/mole). For process modeling, a simpler first-order equation has proved useful for both liquid and gas-phase units. The partial pressure of hydrogen is proportional to the hydrogen content of the liquid phase.

$$r = kP_H$$

12.2.3 Benzoic Acid → Cyclohexane Carboxylic Acid

Benzoic acid, which is produced commercially primarily by oxidation of toluene, can be hydrogenated over palladium catalyst to cyclohexane carboxylic acid.[1] This hydrogenation is used as a route to $\in$ −caprolactam for Nylon 6. After purification, the carboxylic acid is reacted with nitroxyl-sulfuric acid to produce the desired $\in$ −caprolactam.

Chemistry and Catalyst

The reaction is similar to hydrogenation of benzene, but palladium-on-carbon is used instead of nickel catalyst, because the carboxylic acid side chain attacks the nickel and renders it inactive.

    The reaction is conducted with the organic compounds in the liquid phase at 170°C and hydrogen pressure of 90–200 psig.[1–3] A typical slurry-phase Pd-on-carbon catalyst is used, having a 5% palladium content.

Catalyst Suppliers

See Table 12.3.

Process Unit

Hydrogenation is carried out in three continuous-stirred reactors connected in series and equipped with cooling oils.[2] Additional temperature control is maintained by recycle of a portion of the cooled product.[3] With effective temperature control, essentially complete conversion can be realized. Catalyst is separated from the product by centrifuging, and the recovered catalyst is returned to the first reactor along with make-up catalyst.[2] Flashing of product enables complete separation of any remaining catalyst.

12.2.4 Naphthalene → Tetralin → Decalin

Tetralin (1,2,3,4-tetrahydronapthalene) and decalin (decahydronapthalene) are both valuable solvents in manufacturing paints, fats, resins, lacquers, varnishes, shoe creams, and floor waxes. The high flash points and low vapor pressures make them ideal solvents and cleaners.[1,2] Neither is highly toxic.

Chemistry and Catalysts

Hydrogenation is accomplished with ease using nickel sulfide or nickel molybdenum catalysts at 20 to 60 atm and 400°C in the liquid phase using a fixed-bed reactor.[4] Higher pressures drive the reaction to decalin. The yield of tetralin is reported to be greater than 97 percent.

naphthalene   tetralin   decalin

Higher yields are possible with palladium catalyst, since it does not promote hydrogenation of the second ring.[4] Lower operating temperatures and pressures are also possible with palladium. Naphthalene feedstock must undergo desulfurization to avoid adsorption of sulfur compounds. It is possible to manage sulfur content of feedstock when using nickel-molybdenum catalyst, since it is usually operated in the sulfide form and actually serves as a desulfurization catalysts (see "Hydrotreating" section). Excessive amounts of H$_2$S in the feed or produced in the reactor and organo-sulfur compounds compete successfully for active sites, and removal of major concentrations of sulfur compounds from naphthalene feedstocks remains necessary.

Catalyst Suppliers

See Table 12.3.

Mechanism

It has been proposed that tetralin is formed in a two-step reaction that is suggested to be at equilibrium.[5]

naphthalene   tetralin

Process Kinetics

Studies on reactions of this type have been used as model reactions for those that occur in hydroprocessing of heavy petroleum feedstocks or coal liquids. The rate of conversion of methyl-naphthalene to tetralin and decalin products was found to be first-order.[5] By analogy, the rate for naphthalene conversion to tetralin and decalin should also be first-order. An additional study on naphthalene hydrogenation invoked the more complex rate form of Langmuir–Hinshelwood.[6]

Other Polynuclear Aromatics → Total or Partial Saturation

In addition to naphthalene many more complex polynuclear aromatics are often hydrogenated primarily in hydroprocessing of heavy petroleum feedstocks or coal liquids. See the section on "Hydrotreating." Nickel-molybdenum and cobalt molybdenum catalysts in their sulfided forms are most effective.

12.2.5 Phenol → Cyclohexanone*

Cyclohexanone can be used directly to form caprolactam by reaction with hydroxylamine using acid catalysis. Phenol provides a convenient direct route to cyclohexanone by hydrogenation.[1,2]

Chemistry

The reaction is exothermic but operates at modest temperatures and pressures using a palladium catalyst:

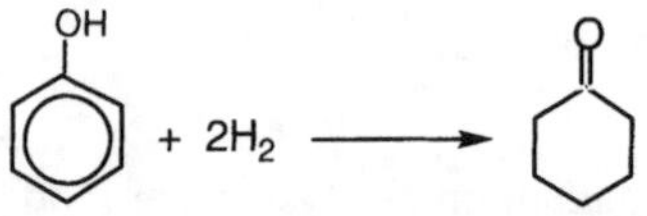

phenol cyclohexanone

Temp., K	ΔH, kcal	K_p
400	−33.14	2.133×10^4
500	−34.04	2.183

One can speculate that the stepwise hydrogenation of the aromatic ring proceeds to the second stage, at which point cyclohexanone, the keto tautomer, is formed. Basic promoters improve selectivity.

Catalyst and Processes

Both gas-phase and liquid-phase processes are practiced. In the gas-phase process, palladium of 0.2–0.5% on zeolite or alumina is used in a fixed-bed reactor operating at 140–170°C and slightly above atmospheric pressure.[1] The liquid-phase process operates at 175°C and 13 atm using Pd-on-carbon and is reported to be more selective to cyclohexanone.[3] However, yields above 90% are reported for both processes. Although phenol is more costly than cyclohexane, the simplicity of this process makes it attractive for caprolactam producers who have reliable supplies of phenol.

Catalyst Suppliers

See Table 12.3. There is ample evidence to conclude that the support must include some additive such as CaO to neutralize the acidity of the support and prevent coke formation.

12.2.6 Phenol → Cyclohexanol

Almost pure cyclohexanol can be produced from phenol by hydrogenation using the proper catalyst system, nickel on silica or alumina.[1] Cyclohexanol from cyclohexane, however, is the favored route because of the lower cost of cyclohexane.

Chemistry

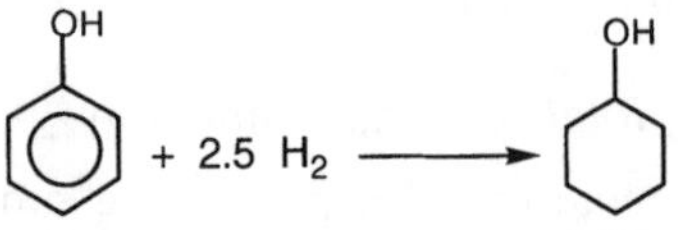

phenol cyclahexanol

* See also next reaction to cyclohexanol.

Temp., K	ΔH, kcal	K_p
400	−48.77	19.36×10^4
500	−49.81	3.4995

Catalyst and Processes

A vapor-phrase process with a fixed-bed of nickel-on-silica or alumina is reported.[1] Operation is at 120–200°C and 20 atmospheres. Yields are close to 100 percent.[1] One might assume that the nickel hydrogenation characteristics, along with the neutral support, favor selective hydrogenation of the ring to cyclohexanol.

Catalyst Suppliers

See Table 12.3. A basic promoter is reported to be used.

Process Kinetics and Mechanisms

The industrial process goal is to produce either cyclohexanone or cyclohexanol in high purity. This goal has been accomplished as already noted by using a palladium catalyst to selectively produce cyclohexanone and a nickel catalyst to selectively produce cyclohexanol. In each case, the desired result is accomplished by the action of the catalyst in shifting the equilibrium steps of adsorption and surface reaction.[6]

It has been suggested that adsorbed phenol in coplanar orientation favors complete hydrogenation to cyclohexanol,[7] as is the case for nickel catalyst. Non-planar orientation postulated for palladium favors partial hydrogenation to cyclohexanone. On palladium the non-planar form reacts in a stepwise fashion with the dissociatively adsorbed hydrogen atoms.[5]

By contrast, the coplanar orientation of phenol has suggested a series of steps involving an intermediate cyclohexanone that is rapidly reacted to cyclohexanol. These steps include adsorption surface reaction and desorption.

Kinetics for both catalysts expressed as rate of phenol conversion have been presented in identical forms[4,6] using a Langmuir–Hinshelwood form based on phenol and hydrogen adsorbed on active sites.

$$r = \frac{k K_P K_H P_H}{(1 + K_P P_P + K_H P_H)^2}$$

where subscripts P and H represent, respectively, phenol and hydrogen.

If phenol is strongly adsorbed ($K_P P_P \gg 1 + K_H P_H$), the equation reduces to a power-law equation.

$$r = k' P_P^m P_H^n$$

Depending on the strength of adsorption of phenol and hydrogen, the orders m and n may vary.

A completely basic support such as MgO strongly adsorbs phenol, and Pd/MgO catalyst produces a value of m = −1 and of n = 1. On Pd/alumina, the orders were 1 and 2, respectively, indicating a less strongly adsorbed phenol.[5,7] Temperature also affects the order, particularly of hydrogen.

For the Ni/SiO$_2$ catalyst, the order for phenol was found to be one and unaffected by temperature. The order for hydrogen, however, increases with increasing temperature, because of the resulting lower loss of surface active hydrogen. In such a situation, the hydrogen partial pressure has a more important effect on the rate and can be represented by a higher order.[6]

Based on these studies, it maybe reasonably suggested that the power-law equation can be applied commercially, with m and n determined experimentally or from plant operation. In such a situation, the catalyst would be set and the operating range narrowly defined.

12.3 HYDROGENATION OF HETEROCYCLIC COMPOUNDS

Tetrahydrofuran is a valuable solvent for polymers such as polyvinylchloride. It is used as a reaction medium for Grignard reactions and in producing special packaging materials. It is miscible in both water and organics.

Heterocyclic rings are scattered throughout nature in complex molecules. Coal-tar light oil is rich in six-member ring heterocyclics such as pyridine; but the coal-tar production is no longer adequate to supply the demand, and pyridine is now synthesized from the reaction of aldehydes and ammonia. Furfural produced from pentosans derived from bagasse or corn cobs is a renewable source for five-membered heterocyclic rings. Clever synthesis of complex molecules using either six-members or five-members, or both, has produced a large variety of dyes. Hydrogenation plays a role in some cases.

12.3.1 FURAN → TETRAHYDROFURAN

Furan, which is produced by decarboxylating furfural, can be hydrogenated to tetrahydrofuran. Similarly, furan derivatives such as alkyl furans can also be hydrogenated to corresponding alkyl-tetrahydrofurans. The route to tetrahydrofuran represents a process based on a renewable resource. There are a number of other processes based on petroleum sources.

Mechanism

The reaction has been described[3] as a stepwise hydrogenation via the intermediate, dihydrofuran (one double bond).

Catalyst and Process Unit

The furan ring is more easily hydrogenated than the benzene ring, but moderate temperatures must be used to avoid hydrogenolysis.[4] The noble metals (Pd, Pt, Rh, and Ru) all accomplish the hydrogenation, as do nickel-based catalysts. Apparently, nickel catalysts (Ni/Kieselguhr) required the controlled addition of an amine to prevent unwanted side reactions,[1] but nickel is reported to be preferred for industrial processes,[6] perhaps for economic reasons. Ruthenium is not as active as palladium or rhodium and requires higher operating pressure. The reaction can be carried out in the vapor phase using a fixed-bed reactor or in the liquid phase at higher pressures. Operating temperature must be below 150°C to avoid hydrogenolysis. Typical conditions reported for Pd/C are 100°C and 20 bar.[3]

12.3.2 PYRIDINE → PIPERIDINE

Pyridine, which is contained in the coal tar light-oil fraction, is mainly produced synthetically from formaldehyde, acetaldehyde, and ammonia. It is highly reactive and used extensively in producing specialty chemicals. Hydrogenation yields piperidine which is used to form vulcanization accelerators[2] such as dipentamethylenethiuran tetrasulfide.

Chemistry

pyridine $+3H_2 \longrightarrow$ piperidine

If the pyridine is recovered from coal tar, it will contain organic sulfur compounds that must be removed by hydrotreating prior to effective high-yield hydrogenation to piperidine by a Ni/Al_2O_3 catalyst using a fixed-bed reactor. Synthetic pyridine, which contains no sulfur, is usually hydrogenated over ruthenium or rhodium. Ruthenium requires higher pressures. Palladium and platinum have also been used.

12.3.3 PYRROLE → PYRROLIDINE

Although pyrrole can be obtained by destructive distillation of coal, bone, or animal proteins, it is primarily produced synthetically by reacting furan with ammonia. Pyrrolidine can readily be produced in good yield by hydrogenation of pyrrole over platinum or rhodium, but reaction of tetrahydrofuran with ammonia is apparently a more economical route. Pyrrolidine is used as a component in the production of pharmaceuticals and antibiotics.

Chemistry

pyrrole $+ 2H_2 \longrightarrow$ pyrrolidine

12.4 HYDROGENATION OF ALIPHATIC UNSATURATES

12.4.1 GENERAL[3-5]

Industrial hydrogenation of aliphatic unsaturated compounds and unsaturated portions of more complex organics is a widely used reaction that covers many products. In most cases, a high degree of selectivity is required so that only undesired unsaturated entities or compounds are hydrogenated to the desired state. It is in this arena that the art of catalysis has been applied in the most impressive and dramatic fashion.

Selectivity is important in almost all aliphatic hydrogenations. Saturation of olefins invariably involves cis and trans isomer production for butenes and higher olefins. Selectivity in hydrogenation is often required so as to favor partial hydrogenation of dienes and alkynes to mono-olefins. Careful addition of limited amounts of a mild poison that will prevent complete hydrogenation has been practiced. Such methods have been replaced in many processes by specially promoted catalysts that accomplish the feat more effectively and reliably.

Catalyst Deactivation

The most common poisons in petrochemical plant hydrogen streams are CO and H_2S. Carbon monoxide has an inhibiting effect that can improve selectivity if reduced hydrogen availability is favorable for the desired reaction path. Hydrogen sulfide is a progressive poison. Thiophenes, mercaptans, and disulfides present in the hydrocarbon feed can cause rapid poisoning of double-bond saturation reactions.

Carbonaceous deposits are to be expected especially with C_2 and higher unsaturates. These can be removed by regeneration with air or by treatment at elevated temperatures with steam followed by hydrogen. Such treatment is not as effective if the deposit has aged and become graphitic in character.

Polymerization of olefins and dienes can occur at conditions ideal for hydrogenation. Depending on the operating conditions and molecular weight of hydrocarbons being processed, the polymers formed can range from solid deposits to heavy oils. The notorious "green oil," which is produced during selective hydrogenation of ethylene product from a steam cracking unit, also covers the catalyst. Both types of deposits tend to cause plugging. The diene polymers cause gum formation that can quickly produce bed plugging. When possible, liquid phrase operation can be advantageous, since the flowing liquid hydrocarbon acts to wash out newly deposited polymer.

Catalyst Types

The catalysts most frequently used are supported Pd or supported nickel. Platinum is useful in some cases, and Raney or sponge nickel is also used. Table 12.6 provides a summary. The order of hydrogenation activity of the platinum metals is Pd > Rh, Ru Pt > Os > Ir.[6] Raney or sponge nickel compares to Pd in activity. The level of Pt activity is ideal for avoiding isomerization.[7]

TABLE 12.6
Catalysts for Hydrogenation of Aliphatic Unsaturates

Reaction Type	Catalysts	Active Metal Percentages, wt%
Alkene → alkane	Pd/Al_2O_3	Pd = 0.03–0.4
	Ni/Al_2O_3	Ni = 12–60, also on Kieselguhr
	Pt/Al_2O_3	Pt = 0.1–2 preferred when cis-trans isomerization is to be avoided
	Ni-W (sulfided)	various compositions (sulfur resistant)
	Ni	Sponge nickel, Raney nickel
Alkadienes and higher acetylenes → alkenes (selective)	Pd/Al_2O_3 Ni/Al_2O_3	Pd = 0.3 to 0.5
Acetylene → ethylene (selective)	Pd/Al_2O_3	*Pd = 0.02 → 0.05 promoted
	Ni/Al_2O_3 (front-end hydrogenation)	Ni = 0.5–3.0 Co and Cr promoted, partially sulfided
Methylacetylene and propyldiene → propylene	Pd/Al_2O_3	Pd = 0.01 → 0.03 promoted
Vinylacetylene in butdrene → butadiene	promoted alumina	(proprietary promoted alumina)[†]

Suppliers: United Catalysts, Engelhard, Johnson-Matthey (noble metals), Activated Metals (sponge nickel), Grace Davison (Raney nickel), Precious Metals Corp. (noble metals), Criterion (noble metals), BASF, Synetix UOP LLC, Beiring Research Inst.

Licensors: CDTECH, [†]UOP LLC (KLP Process), [**]UOP LLC, and Hülls

Supports

The catalyst supports used are often low-acid Al_2O_3, and some have added basic compounds. The purpose is to prevent acid-catalyzed polymerization by the support. Charcoal is selected for noble-metal powdered catalysts in slurry reactions involving production of fine custom chemicals requiring selective hydrogenation of certain multiple bonds. Metal content is higher in the range of five percent. Charcoal cannot be regenerated, but the metal is more easily recovered from charcoal. Fortunately, in many cases, the metal loading is so low on Al_2O_3 supported catalysts used in these systems that recovery is not warranted.

Process Units

Since selectivity, conversion, and safety depend on good temperature control, adiabatic reactors with intermediate cooling or nonadiabatic multitubular reactors are required. When the amount of a given unsaturate to be hydrogenated is small, one adiabatic reactor is indicated. If fouling is expected an additional reactor for standby service is used. When increased amounts of reactant require better control, two adiabatic reactors in series may be used with intermediate cooling along with a standby reactor. At even higher concentrations three beds in series may be used with intermediate cooling which can include cool hydrogen or cooled product recycle. If a multitubular reactor is used, it may be necessary to complete the conversion in a single adiabatic bed.

For liquid systems, a recirculated reactor with exterior cooling of recycle can be useful, especially if some vaporization occurs in the reactor, thereby providing direct-contact cooling. Trickle-bed reactors are also used. A single bed is adequate for smaller amounts of unsaturated reactants. Multiple beds with intermediate cooling are used for better temperature control when larger conversions are expected. The trickle bed requires careful planning so that good distribution of liquid and hydrogen gas is attained. Pressure drop and mass velocity must be adequate to effect good distribution and thorough contact of reactants and catalyst throughout the bed cross-section. Adequate distributors must be provided at the beginning of each bed. Liquid operation has the advantage of providing continuous removal of oligomers from the catalyst surfaces.

Fine chemicals and specialities are often produced in small quantities for which batch operation in stirred reactors with cooling coils and/or cooling jackets is ideal for the precise control essential for high yields and attainment of rigid quality standards.

12.4.2 ALKENES → ALKANES

Olefins such as ethylene and propylene are major starting compounds for a host of products including polymers and a variety of key chemical intermediates. The focus, therefore, is on maximizing production of these olefins rather than hydrogenating them. However, there are many instances where saturation of olefins or olefinic functional groups is essential. In the petrochemical industry, steam cracking products can include significant components in the gasoline boiling range when naphtha is the primary feedstock. If this pyrolysis gasoline cut is to be used as a gasoline blending stock, dienes and acetylenic compounds must be removed, but the olefins must be retained to produce a high octane stock (see below). If, however, the pyrolysis gasoline cut is to be used as a source of aromatics, the olefins must be hydrogenated after the diolefins and acetylinic compounds have been converted to olefins.[1] Nickel is often the catalyst of choice in this second-stage process. Moderate operating conditions and partial sulfiding of the nickel prevent hydrogenation of the aromatics. Although palladium is more active, it is also a more costly and cannot easily be regenerated by removing polymer that might coat the catalyst over time.

There are many speciality and fine chemicals that require selective hydrogenation of olefinic functional groups. In this arena, palladium is invariably the choice because of its unique selectivity characteristics, which permit hydrogenation of a double bond without hydrogenating other functional groups. Table 12.2 provides examples for hydrogenation of olefinic entities in the presence of other groups that can also be hydrogenated at more vigorous conditions or with a less selective catalyst.

Platinum, although less active than palladium and nickel, is advantageous when double-bond migration is to be avoided.[2] Ruthenium provides an unusual selectivity for monosubstituted olefins in mixtures with di- and trisubstituted olefins. This characteristic is associated with the very slow double-bond migration that occurs with ruthenium.[2]

Chemistry, Thermodynamics, and Mechanism

The overall reaction of saturation of a double bond is seemingly quite simple, but, beginning with butene, formation of isomers becomes possible and occurs. The Horiuti–Polanyi mechanism involves dissociative adsorption of hydrogen and associative adsorption of the alkene followed by stepwise addition of hydrogen. The intermediate half-hydrogenated step is reversible and accounts for isomerization of the alkane,[7] which can cause the formation in the case of butene of both butane and isobutane.

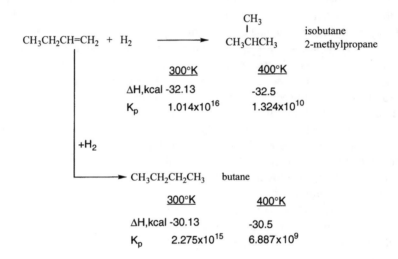

Obviously, from an overall-reaction perspective, both are favored, but the reversibility of the intermediate reaction allows for isomerization of butene-1 to cis and trans butenes. The comparative isomerization rates have been reported on several catalysts:[6,7]

    cis-2->trans-2->1-butene on ruthenium and osmium
    cis-2->1-butene>trans-2-butene on platinum and iridium

Palladium catalyst exhibits the highest activity for isomerization, and platinum one of the lowest. In fact, platinum is often used when isomerization is not wanted. Thus, catalyst selection should be guided by the desired outcome.

Other forms of surface species have been suggested, as shown in Figure 12.7, in explaining complex reactions involving various side reactions. Many more details have been adroitly reviewed in Ref. 7. Numerous mechanisms have been proposed. One such mechanism for 2-butene isomerization shown in Figure 12.8 is exemplary.

Process Kinetics

The rate of hydrogenation of most olefins, at the modest temperatures (150–170°C) used to favor high equilibrium conversion, can usually be expressed in a simple form for a gas-phase operation.[8]

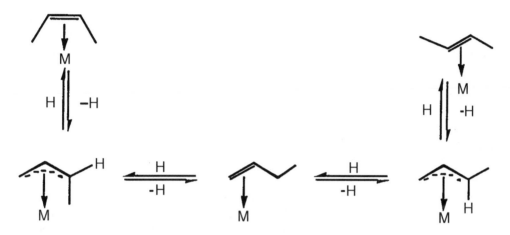

FIGURE 12.7 Other Types of Surface Species. 1. π-adsorbed intermediate, 2. dissociatively adsorbed alkene, 3. σ-allyl, 4. π-allyl. Reprinted by permission of John Wiley & Sons, Inc.: Olah, G. A. and Molnar, A., *Hydrocarbon Chemistry*, p. 449, Wiley, New York, copyright © 1995.

FIGURE 12.8 Cis-trans Isomerization via Double-Bond Migration Involving π-Allyl Intermediates. Reprinted by permission of John Wiley & Sons, Inc.: Olah, G. A. and Molnar, A., *Hydrocarbon Chemistry*, p. 449, Wiley, New York, copyright © 1995.

$$r = kP_O^m P_{H_2}^n$$

where subscript O represents the olefin. In many cases, $m = 1$ and $n = 0$ or a negative value.

12.4.3 ALKADIENE → ALKENES

This selective reaction is an important process for both the petroleum refining and petrochemical industry. Fluid catalytic cracking products include highly unsaturated C_4 and C_5 cuts that are separated by distillation and contain butenes and butadiene and amylenes, pentadienes, and cyclopentadienes, respectively. These dienes can be selectively hydrogenated to the corresponding olefins. The preferred catalyst is Ni/low-acid Al_2O_3, but Pd/Al_2O_3 is used with feedstocks having low amounts of insoluble gums.[1] By converting the butadiene selectively to butene, acid consumption is reduced in the subsequent alkylation unit. The pentadienes and cyclopentadienes in the C_5 cut can be selectively hydrogenated to eliminate these dienes. Hydroisomerization occurs simultaneously to increase the yield of isoamylenes used in production of TAME (tertiary-amylmethylether).

Another major use of selective hydrogenation of dienes is an important part of the production of propylene by steam cracking. The propylene produced will contain small amounts of methyl acetylene and propyldiene that must be converted to propylene. The propylene thus formed adds to the total propylene production; removal of methyl acetylene by hydrogenation to propylene occurs simultaneously and is necessary for polymerization grade propylene.

Distillation of the liquid products from steam cracking of naphtha produces a cut in the gasoline boiling range (C_{5+}) that can contain as much as 15–20 wt% of diolefins, cyclodiolefins, and alkenyl aromatics,[1] all of which polymerize in storage creating copious quantities of gum. Selective hydrogenation is used to convert these compounds to olefins, cycloolefins, and alkyl aromatics. If the aromatics are to be separated into benzene, toluene, and xylenes, the aromatic boiling range cut is removed by distillation. It is subjected to hydrotreating to saturate the olefins so that the aromatics can be separated by extraction (see section on "Hydrotreating").

Heavy gasolines produced by coking or visbreaking require similar treatment for removal of dienes. These gasolines, however, have high mercaptan content, and special sulfur resistant catalysts are used.[1] Steam cracking of naphtha also produces significant amounts of butenes, butadiene, and various acetylenics. The acetylenics are removed selectively first. Then, the valuable butadiene can be extracted. Alternatively, the selective hydrogenation can be carried out to convert both acetylenics and butadiene to butenes.

Mechanism

Selective hydrogenations of dienes are most complex, and it has proved difficult to develop well documented mechanisms. Reaction schemes have been presented such as that depicted in Figure 12.9 for 1,3 butadiene.

Process Kinetics

For selective hydrogenations, power-law models usually give better fits than Langmuir–Hinshelwood models. Extensive studies at industrial conditions have shown that the rate equations with the best fit were first order in hydrogen and zero order in diolefin.[9,10] (See also "Alkynes → Alkenes" for additional comments.)

Any effort to include all the complexity of the various side reactions and catalyst deactivation requires the implementation of the theory of complex reactions. One such thorough study has been presented in considerable detail.[13]

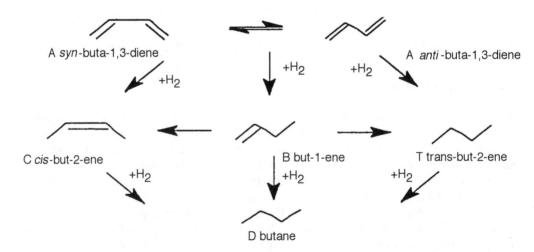

FIGURE 12.9   Reaction Scheme for Selective Hydrogenation of 1,3 Butadiene. Reprinted from *Chemical Engineering Science* 51 (11), 2879 (1996). Goetz, J., Murain, D. Y., Vlischenko, M. and Touroude, R., "Kinetics of Buta-1,3-diene Hydrogenation over Platinum Catalysts," pp. 2879–2884, copyright 1996 with permission from Elsevier Science.

12.4.4 ALKYNES → ALKENES

ACETYLENE → ETHYLENE

Reacting acetylene selectively to ethylene is an important operation in every ethylene plant. Acetylene is strongly adsorbed on many catalysts, and production of many products from ethylene can be adversely affected by the presence of acetylene in the ethylene feed. Polyethylene units, both catalytic and noncatalytic are subject to lower production and poor molecular weight and quality control when acetylene is present in the ethylene feed. Low-density polyethylene is produced via a high-pressure free-radical process, and acetylene present in the feed acts as a chain-transfer agent that inhibits chain growth and thus reduces molecular weight of the product. Other polyethylene processes, such as Ziegler and Ziegler-Natta, employ heterogeneous catalysts to produce linear polyethylene. These catalysts are poisoned by acetylene, which causes a decline in production and molecular weight.

Catalyst Types

The catalyst of choice for selective hydrogenation of acetylene in an ethylene stream is palladium-on-alumina of low loadings in the range of 0.015–0.05 wt% palladium. Since acetylene is adsorbed more strongly than ethylene on the palladium, selectivity is enhanced at mild conditions, low Pd loadings, and low ratios of H_2-to-acetylene (2 or less). Selectivity is also favored by high rates of internal mass transfer. Mass transfer resistance increases as coke builds up in the pores and causes increased ethylene hydrogenation. Shell-type catalysts are used to overcome this problem to some degree.[11] Earlier, Pd catalysts were used with the continuous addition of carbon monoxide, which has a deactivating effect and makes possible hydrogenation of acetylene to ethylene without significant loss of ethylene to ethane. New catalysts, however, are now available that accomplish the same goal more effectively. These are proprietary catalysts that contain small amounts of a promoter such as gold. Carbon monoxide is not required with these catalysts and, in fact, must be removed to avoid deactivation. All palladium catalysts are poisoned by sulfur compounds, but usually H_2S has been removed in the demethanizer overhead. The Pd/Al_2O_3 catalysts are used for the ethylene-rich streams after the demethanizer. Such units are called *back-end hydrogenation* and are now the most common.

Front-end hydrogenation, however, is employed in some older plants, and is operated prior to the demethanizer, and contains all the hydrogen and methane in the pyrolysis gas. For these units, cobalt and chromium-promoted nickel on a refractory alumina support is the catalyst of choice. Unlike palladium catalyst, promoted nickel has some tolerance for sulfur compounds (H_2S or mercaptans) in the feed. In fact, the desired selectivity is attained by maintaining nickel in the sulfided form. The catalyst is presulfided during start-up, and the sulfur in the feed, along with H_2S addition, is used to maintain optimum sulfur content. There is an operating temperature range in which H_2S addition is effectively used to control activity and selectively. The catalyst nickel content (0.5–2.0 wt%) is selected, based on the hydrogen content and partial pressure. Promoter content is in the range of 0.05–0.2 wt%. At higher hydrogen partial pressures (60–70 psia), a low nickel content is indicated to prevent hydrogenation of excessive amounts of ethylene to ethane. Also, higher sulfur content in the feed is required (>20 ppm) to maintain selectivity.

Process Kinetics

Although selective hydrogenation of steam-cracking product usually is accomplished after ethylene and propylene are separated, an interesting study of the mixture prior to separation provides useful insights on appropriate rate forms.[9] Again, power-law models proved best.

$$\overset{H_2}{C_2H_2 \rightarrow} C_2H_4 \overset{H_2}{\rightarrow} C_2H_6 \tag{1}$$

$$\overset{H_2}{C_3H_4 \rightarrow} C_3H_5 \overset{H_2}{\rightarrow} C_3H_8 \tag{2}$$

$$(-r_A) = k_1 P_H$$

$$r_E = k_2 P_H^3$$

$$(-r_{MAPD}) = k_3 P_H$$

$$r_p = k_4 P_H^3$$

where subscripts A, E, MAPD, and P refer, respectively, to acetylene, ethane, methylacetylene + propadiene, and propane.

The higher exponent on the saturation steps gives a quantitative representation for the value of avoiding a high hydrogen-to-reactant ratio. If the C_2 and C_3 streams are selectively hydrogenated separately, the first two rate equations applying to Reaction 1 would be used for the C_2 stream and the second two rates for the C_3 stream.

Langmuir–Hinshelwood equations have also been presented,[11] but the power-law equations are recommended. Simplified Langmuir–Hinshelwood equations were successfully used for a front-end hydrogenation model.[12]

METHYLACETYLENE AND PROPYLDIENE → PROPYLENE

These compounds are usually selectively hydrogenated in the propylene-rich stream from the depropanizer of an ethylene/propylene producing plant. Catalysts similar to those used for acetylene are applied. The promoter and pore size of the support may be different. Modestly, higher concentrations of hydrogen are acceptable.

12.4.5 ALKYNES → ALKANES

In cases where complete hydrogenation is desired, and hydrogenation of other unsaturates in an aliphatic stream, Raney or sponge nickel catalysts are well suited, as are nickel catalysts.

12.5 HYDROGENATION OF NITRILES TO AMINES

12.5.1 GENERAL

Hydrogenation of nitriles to amines is one of a number of ways to produce industrially important amines. Commercial amines are generally designated in three main groupings, lower aliphatic amines (chain lengths of 1 to 8 carbon atoms), fatty amines (aliphatic amines with chain lengths of 9–24 carbon atoms), and aromatic amines. Several of the most frequently used processes are:[4]

- Ammonolysis by alkylation of alcohols
- Reaction of NH_3 and H_2 with an aldehyde or ketone
- Hydrogenation of nitriles
- Hydrogenation of nitroaromatics

Selection of a given process depends primarily on availability and cost of the feedstock, the energy costs, and the ease of the transformation.

The lower aliphatic amines are most frequently prepared by the alcohol route, with the nitrile route being used for special cases. Natural fatty-acid amines are formed via the nitrile, but synthetic fatty amines are more conveniently produced via the oxo-alcohol route. Aromatic amines can be produced by direct hydrogenation of the nitroaromatic (see "Hydrogenation Nitroaromatics"), which is preferable in most cases because of the direct procedures for producing a wide variety of nitroaromatics. The aliphatic nitriles are readily produced, and hydrogenation to aliphatic amines is an economic procedure.[4] This section on hydrogenation of nitriles focuses, therefore, on aliphatic nitriles, although several special cases where aromatic nitriles are hydrogenated to desired products are illustrated.

Uses

Amines have basic characteristics and are reactive entities that find wide use in synthetic chemistry. They are used as the amine itself or as intermediates in the production of a long list of important commercial products including agricultural chemicals, pharmaceuticals, surfactants, fabric softeners, plastics, paper and textile processing agents, corrosion protection agents, and absorbent liquids for removing acid-gas impurities (e.g., H_2S, CO_2) from process vapor streams.

Chemistry

The chemical reaction scheme is quite complex and tends to produce a mixture of primary, secondary, and tertiary amines. As shown in the following abbreviated scheme, the intermediate imine permits routes to all three amines.[1,2]

$$RCH_2C \equiv N \xrightarrow{H_2} RCH_2CH=NH \xrightarrow{H_2} RCH_2CH_2NH_2 \qquad \text{primary}$$

$$RCH_2CH=NH + RCH_2CH_2NH_2 \longrightarrow RCH_2\underset{\underset{NH_2}{|}}{C}HNHCH_2CH_2R \quad \text{secondary}$$

$$\downarrow H_2$$

$$(RCH_2CH_2)_2NH + NH_3 \quad \text{tertiary}$$

Mechanism

A more detailed representation is given in Figure 12.10 that approaches a mechanism but without surface chemistry details. The scheme presented begins with the aldimine, the intermediate in the formation of the primary amine, and details a scheme for the formation of secondary and tertiary amines. Secondary amines are formed as shown by nucleophilic addition of the primary amine to form the aminodialkylamine, elimination of ammonia, and subsequent hydrogenation of the C=N group.[3] The tertiary amine is formed via an enamine since there is no hydrogen on the beta carbon atom.[3] References 2, 6, and 8 provide additional mechanistic insights.

Effect of Operating Variables on Selectivity[1-3]

Temperature

Industrial operating temperatures are selected through experience to avoid equilibrium limitations due to excessive temperatures that would favor reverse reactions for this exothermic system. Temperature within the equilibrium-favorable range, however, can have an effect on selectivity caused by differences between activation energies for hydrogenation and condensation. Such differences would be a function of the catalyst selected for use. A rough generalization suggests that an increase in temperature can favor secondary and tertiary amines, and a lower temperature will favor the primary.

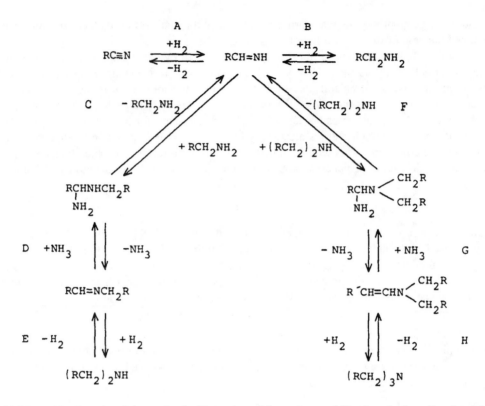

FIGURE 12.10   Reaction Scheme for the Formation of Secondary and Tertriary Amines. Reprinted from Volf, J. and Pasek, J., in *Catalytic Hydrogenation: Studies in Surface Science,* Vol. 27, p. 110, copyright 1986, with permission from Elsevier Science. Note: The intermediates produced by Reactions D and C are, respectively, alkylidenealkylamine and enamine.

Addition of Ammonia

Ammonia is added to the reaction feed to combine with the intermediate to aid in preventing formation of secondary and tertiary amines. Obviously, such addition must be controlled to reach a concentration of intermediate just sufficient for the initial step to the primary amine. In the context of Figure 12.10, the addition of ammonia tends to reverse the reaction to alkylideneamine and enamine and reduce the rate of formation of secondary and tertiary amine.

Hydrogen Pressure

An increase in hydrogen pressure increases the rate of the overall reaction when the hydrogenation step is rate controlling. Increased pressure increases hydrogen concentration in the liquid phase or the hydrogen partial pressure for a gas-phase system. There are, however, important differences between catalysts that are not well understood but that cause significant changes in selectivity. An example of the effect of pressure on hydrogenation of valeronitrile to amines at lower and higher pressures has been reported for several different catalysts.[3–5]

	"High" Pressure	Lower Pressure
Rhodium	monopentylamine	dipentylamine
Raney nickel or Ni/Al$_2$O$_3$	dipentylamine	monopentylamine
Cobalt	insensitive	

Specific Reaction Characteristics

The following general observations apply:

- Aromatic nitriles (e.g., benzonitrile) are hydrogenated to primary and secondary amines only. Tertiary amines are not formed, because the intermediate enamine cannot be formed.
- Rate of hydrogenation decreases with an increase in chain length.
- Other functional groups close to the cyano group are often hydrogenated as well, causing low selectivity.
- Groups that are not hydrogenated as readily as the cyano group are usually left untouched.
- Hydrogenation of olefinic or aromatic bonds can be accomplished by high-activity hydrogenation catalysts.

Effect of Solvent in Liquid-Phase Hydrogenation

- Solubility of hydrogen is affected by the type solvent. Hydrocarbons, for example, are the best solvents and, when used, provide the highest activity for hydrogenation.

Addition of Water

Addition of water to liquid-phase systems does affect selectivities, but its effect varies among catalyst types. In some cases, selectivity to primary amines is improved by addition of water and in others, selectivity to secondary amines is favored. Totally anhydrous solvents prevent the reaction from going to completion.[3,6] Adding water is worth a try to see if it improves in the direction desired. Catalyst suppliers can be helpful relative to their particular catalysts.

Catalyst Types and Suppliers

Table 12.7 summarizes information on commercial catalysts used in nitrile hydrogenation. Cobalt, nickel, and Raney or sponge nickel appear to be the most frequently used. The precious metals, in many cases, require less severe operating conditions and are often used in fine chemical synthesis where convenience and small scale can prove attractive for such high-cost catalysts. Cobalt is less likely to cause undesired coupling reactions.

Catalyst Poisons

Compounds of sulfur, phosphorous, chlorine, and nitrogen are poisons that could be encountered, especially with natural fatty nitriles.[13] Upsets in the refining process may cause significant deactivating effects due to contaminated recycle.

Process Units

Lighter, lower boiling nitriles are hydrogenated in fixed-bed reactors in the vapor phase while the higher boiling fatty amines are produced in gas-liquid-solid systems. In such cases, stirred slurry reactors, batch, or continuous are used. A pump-around (loop) continuous reactor is useful for large production at elevated pressures since it eliminates sealing problems with agitator shaft and provides excellent heat transfer for temperature control by means of a heat-exchanger in the pump-around loop.

12.5.2 Hydrogenation of Lower Aliphatic Nitriles

In general, aliphatic amines with one to eight carbon atoms in the named alkyl group are produced commercially by ammonolysis of alcohols. Production from nitriles is only employed when that route is more economical. The hydrogenation of adiponitrile is a typical example.

TABLE 12.7
Catalyst Types for Hydrogenation of Nitriles

Catalyst	Typical Compositions, wt%	Form	Special Characteristic
Nickel (supported)	50–60% ni on Kieselguhr or silica-alumina	tablets powder	Best for primary amine selectivity when ammonia is added as a diluent
Raney nickel (sponge)	Promoters: 2% Fe, 2.5% Cr	powder	Best for primary amine selectivity when ammonia is added as a diluent
Cobalt	40–54% Co, Zr promoted on Kieselguhr	tablets powder extrusions	Best for dinitriles
Raney cobalt (sponge)	<0.5% Fe	powder	Best for dinitriles
Copper-chromite	40% Ni, 2.5% Cr	tablets or powder	Best for use when other unsaturated groups present are not to be saturated
Palladium	1–10% Pd-on-carbon	powder	Primary or tertiary
Platinum	1–10% Pt-on-carbon	powder	Tertiary
Rhodium	5% Rh-on-carbon	powder	Primary or secondary

Suppliers: Ni, Co, and Cu-chromite: United Catalyst, Engelhard, Synetix, Celanese. Raney or sponge, Ni, and Co: Activated Metals, Grace Davison. Pd, Pt, Rh: Engelhard, Johnson-Matthey, Precious Metals Corp.

Source: manufacturers' literature.

ADIPONITRILE → HEXAMETHYLENEDIAMINE

Hexamethylenediamine is a major chemical product because it, copolymerized with adipic acid, yields nylon-66. Adiponitrile can be made by dimerizing acrylnitrile or from butadiene by reaction with HCN followed by isomerization of the intermediate product. The butadine process accounts for roughly 70% of adiponitrile production. In either case, the hydrogenation of adiponitrile to the diamine is identical.

Chemistry

$$N \equiv C \; CH_2CH_2CH_2CH_2C \equiv N + 4H_2 \rightarrow NH_2CH_2CH_2CH_2CH_2CH_2CH_2NH_2$$

Adiponitrile   Hexamethylenediame
(hexanedinitrile)   (1,6 hexanediamine)

Catalyst Types

Usual catalysts are Zr-promoted Ni or Co on Kieselguhr for a high-pressure process and Raney or sponge nickel for a low-pressure process. See Table 12.3 for a list of suppliers.

Mechanism

The reaction, as with other nitriles, proceeds via a series of steps involving intermediate imines and, inevitably, side reactions. All possible reactions have been documented some years ago[8] and have been summarized more recently.[7] The commercial goal is to develop catalysts and processes that are highly selective. For practical purposes the following reaction scheme is often used:[14]

$$NC-(CH_2)_4-CN+2H_2 \rightarrow NH_2-(CH_2)_5-CN$$
$$ADN \qquad\qquad\qquad ACN$$

$$(1)$$

$$NH_2-(CH_2)_5-CN+2H_2 \rightarrow NH_2-(CH_2)_6-NH_2$$
$$ACN \qquad\qquad\qquad HMD$$

$$(2)$$

where   ADN = adiponitrile
      ACN = aminocapronitrile
      HMD = hexylmethylenediamine

The more detailed representation is shown for the general case in Figure 12.11. By-products of significance, other than secondary and tertiary amines, include cyclic compounds. The cyclic compounds observed have been described by reaction studies using Raney nickel catalysts as follows:[1,5]

- Reductive coupling leading to 1,2 diaminocycloalkenes (DCA)
- Base-catalyzed reaction causing further hydrogenation to aminomethyl cycloalkylamine (AMCA)
- Nucleophilic addition of a primary amine to a primary imine leading, by loss of NH_3, to azacycloalkenes which are hydrogenated to azacycloalkanes

$$N\equiv C-CH_2CH_2CH_2CH_2-C\equiv N + H_2 \longrightarrow HN=CHCH_2CH_2CH_2CH_2CN$$
ADN   Iminocapronitrile   **(1)**

$$HN=CHCH_2CH_2CH_2CH_2CN + H_2 \longrightarrow H_2NCH_2CH_2CH_2CH_2CN \quad (2)$$
Iminocapronitrile   Aminocapronitrile

$$H_2NCHCH_2CH_2CH_2CN + H_2 \longrightarrow H_2NCH_2CH_2CH_2CH_2CH_2CH=NH \quad (3)$$
Aminocapronitrile   "Aminoaldimine"

$$H_2NCH_2CH_2CH_2CH_2CH_2CH=NH + H_2 \longrightarrow H_2NCH_2CH_2CH_2CH_2CH_2CH_2NH_2$$
"Aminoaldimine"   Hexamethylenediamine (HMI)   **(4)**

FIGURE 12.11 Proposed Main Reaction Sequence for Hydrogenation of Adiponitrile to Hexamethylenediamine. Side reactions have also been described that, in some cases, produce ammonia, and ammonia added to the feed tends to suppress such reactions. Reprinted from Luedeke, V. D. in *Encyclopedia of Chemical Processing and Design*, J. J. McKetta, ed., Vol. 26, p. 226, Marcel Dekker, New York, 1987, by courtesy of Marcel Dekker, Inc.

DCA formation can be prevented by poisoning the sites involved using sodium carboxylates so as to cover 5% of nickel surface without loss of activity for the main reaction.[15] AMCA and AZACA, however, initially reduce selectivity followed by catalyst deactivation requiring the addition of fresh Raney or sponge catalyst.

Side Reactions

Of course, even in the most advanced processes side reactions occur. Some of the side reactions involve coupling between the amine entities of the diamines. The by-products must be removed by a combination of azeotropic and vacuum distillation. A number of the by-products have value on the commercial market or they can be recycled to the reactor section.[7] Some, such as 1,2-diaminocyclohexane, cause color in the final product, which is unacceptable for nylon-66 production.[7] It is usually kept below 50 ppm in the product from the purification section.

Production of unsaturated by-products and intermediates must be controlled in the reactor system, since these compounds can combine to form polymers on the catalyst surface, thereby lowering the activity of the catalyst precipitously.

Process Units

The reaction is conducted in a gas-liquid-solid catalyst regime. Two major types of processes are in use, a high- and a low-pressure design. The two differ in the catalyst used and the additives to the reaction liquid.

High-Pressure Processes

These processes generally employ supported promoted nickel or cobalt catalysts in an adiabatic fixed-bed reactor. Ammonia and hydrogen in large excess are fed to the reactor along with the adiponitrile. The ammonia, which serves to prevent condensation reactions, is recovered by distillation and recycled to the reactors as is unreacted hydrogen. This large recycle serves as a means for moderating reaction temperature increases by absorbing reaction heat (~75 kcal/mole).

The high-pressure process operates in the range of 200–300 bar and 90–150°C.[7] By careful control of temperature, yields in the range of 98–99 percent are attained. High-pressure design requires high levels of capital spending. Once the unit is paid off, however, economical production is possible over an extended period, particularly because the catalyst can be reactivated by steam to remove polymer coatings.

Low-Pressure Process

The low-pressure process uses a promoted Raney-nickel powdered catalyst and operates at 60–100°C and 20–50 atmospheres[7,9,10] in a slurry reactor. Liquid feed of adiponitrile, along with fresh and recycled hydrogen, methanol, caustic, and water are fed into the reactor. Pump-around of reactor contents enables temperature control by external heat exchange and also provides turbulence in the reactor to assure good contact with the catalyst. One such system employes an inner tube within a tube.[9] This arrangement promotes rapid circulation of the fine catalyst and reaction mix up the inner tube and back down the larger outer tube. Product is withdrawn, and hydrogen is separated and recycled except for necessary purging to remove excess accumulated inerts. Catalyst is separated and a portion returned to the reactor along with fresh catalyst to maintain a required average catalyst activity.

The lower-pressure process obviously affords savings in capital costs, but catalyst costs are thought to be higher because deactivated catalyst is not regenerated. Catalyst consumption, however, can be reduced by removing impurities from the adiponitrile feed. The presence of caustic, water, and methanol serves to improve selectivity of the Raney-nickel catalyst and thus assures good yields of hexamethylenediamine.

Deactivation and Poisons

Impurities in the adiponitrile such as organic chloride, sulfur, and phosphorous compounds are poisons for the catalysts in both processes.[7] In addition, isomers of adiponitrile and other reaction products of adiponitrile product tend to polymerize on the catalyst surfaces, causing rapid activity decline. Great care in purifying the adiponitrile feed is essential.

Process Kinetics

A Langmuir–Hinshelwood model has been proposed for the liquid-phase Raney nickel catalyzed process. It is based on competitive adsorption.[14]

$$r_{H_2} = \frac{2(k_1 C_{ADN} + k_2 \alpha_{ACN} C_{ACN}) \alpha_H (CH)^{0.5}}{[C_{ADN} + \alpha_{ACN} C_{ACN} + \alpha_H (CH)^{0.5}]^2}$$

with

$$\alpha_{ACN} = \frac{K_{ACN}}{K_{ADN}}, \qquad \alpha_H = \frac{\sqrt{K_H}}{K_{ADN}}$$

The first term in the parenthesis accounts for Reaction 1, and the second term for Reaction 2,

where
C = indicated component concentration in bulk liquid
k_1 and k_2 = rate constants for Reactions 1 and 2
K_i = adsorption constants
ADN, ACN, H = adiponitrile, aminocapronitrile, and hydrogen, respectively

A detailed process model has been presented by the same authors based on a mini-plant with a continuously operated Raney-nickel catalyst slurry in a draft-tube bubble column. Both intrinsic kinetics and fluid-flow characteristics were studied.[16] Model equations based on rates of Reactions 1 and 2 are as shown below.

$$\frac{dC_{HMD}}{dt} = \frac{-dC_{ADN}}{dt} = \phi r_1 \frac{m}{(1 - \epsilon_g)V_R} + \frac{Q}{(1 - \epsilon_g)V_R}(C_{ADN_0} - C_{ADN})$$

where
f = deactivation function
ϵ_g = gas hold up
m = catalyst mass
V_R = reactor volume
Q = volumetric flow rate
C_{ADN_0} = concentration of ADN at reactor inlet

3-AMINOPROPIONITRILE → 1,3-DIAMINOPROPANE

Certain other diamines are routinely produced via the nitrile intermediate.

The nitrile is produced by reacting ammonia with acrylonitrile in a first-stage reactor, and then the nitrile is hydrogenated in the second reactor to yield the diamine. Operating conditions reported are 100–300 bar at 60–120°C in the hydrogenation step.[10,11] A fixed-bed adiabatic reactor is used with temperature control by cooled recycle. Nickel or cobalt catalysts supported on Kieselguhr are used. Batch reactors are also used.

$$H_2H\ CH_2CH_2C \equiv N + 2H_2 \rightarrow H_2N\ CH_2CH_2CH_2NH_2$$

3-aminopropionitrile   1, 3 diaminopropane
(3-aminopropanenitrile)   (1,3 propanediamine)

As with hexamethylenidiamine, an excess of ammonia or methanol and water favors the formation of the diamine. A major side reaction produces bis(aminopropyl) amine, which also has commercial value. The diamine is used in producing dyes, textile and paper treating agents, epoxy curing agents, and corrosion inhibitors.

Various alkyl-substituted aminopropionitriles can be prepared in a similar two-stage process.[12] A primary amine is reacted with acrylonitrile to yield the alkyl-substituted aminopropylamine, which can then be hydrogenated to the corresponding diamine. Most such compounds are used in epoxy resin formulations.

12.5.3 HYDROGENATION OF LONGER-CHAIN NITRILES

LONGER-CHAIN DIAMINES

Although hydrogenation of the corresponding dinitrile is an alternative route to 1,4 diaminobutane and 1,5 diaminopentane, it is a major process for diamines with seven or more carbon atoms. Both batch and continuous processes similar to those described for hexamethylenediamine are used. Some of these longer-chain diamines have been used in producing special polyamides.

POLYAMINES

Polyamines, which are used in treatment of fabric and paper for water repellency,[10] can be produced from the reaction of a diaminoethane or a diaminopropane with acrylonitrile.[10] The resulting product is then hydrogenated and can consist of various polyamides of higher carbon numbers and amine content.

For example,

$$H_2NCH_2CH_2CH_2NH_2 + CH_2 = CH - C \equiv N \rightarrow$$

1,3 diaminopropane   acyrlonitrile

$$H_2NCH_2CH_2CH_2NHCH_2CH_2C \equiv N \xrightarrow{2H_2} H_2NCH_2CH_2CH_2NH\ CH_2CH_2CH_2NH_2$$

dipropylene triamine
(3-aminopropyl-amino propane)

Higher polyamines can be formed as the intermediates react with additional diamine. Mixtures are marketed for special purpose uses, or separation following reaction permits the sale of pure compounds.

FATTY NITRILES → FATTY AMINES

Fatty amines are straight-chain amines variously defined as C_8-C_{24} or $C_{10}-C_{22}$ amines originally produced from natural fats. That source continues to be an important segment of the business and depends on converting fatty acids to nitriles which are subsequently hydrogenated to amines. In recent years, synthetic oxy alcohols and carbonyls provide an alternate route of direct amination to the desired amine, but natural fat continues to be a major source of important higher molecular weight amines.

Primary, secondary and tertiary amines are the major products, but some derivatives are marketed for special purposes. The amines and derivatives are widely used in many essential products such as cationic surfactants, fabric softeners, corrosion inhibitors, bactercides, ore flotation agents, thickeners, pesticides, antistatic agents, and dispersants for pigments.[10,11]

Chemistry

Natural fats are composed of fatty-acid triglycerides that must be cleaved by hydrolysis to release the fatty acids.

$$
\begin{array}{c}
H_2C\text{-}O\text{-}\overset{\displaystyle O}{\overset{\|}{C}}\text{-}R_1 \\
| \\
| \quad\quad O \\
| \quad\quad \| \\
H\,C\text{-}O\text{-}C\text{-}R_2 \\
| \\
| \quad\quad O \\
| \quad\quad \| \\
H_2C\text{-}O\text{-}C\text{-}R_3
\end{array}
\quad \underset{}{\overset{+\,H_2O}{\rightleftarrows}} \quad
\begin{array}{c}
H_2\,C\text{-}OH \;+\; R_1COOH \\
| \\
| \\
H\,C\text{-}OH \;+\; R_2COOH \\
| \\
| \\
H_2\,C\text{-}OH \;+\; R_3COOH
\end{array}
$$

The acids thus released vary with the natural oil used. For example, tallow oil consists primarily of C_{16} and C_{18} acids, while coconut oil yields primarily C_{12} and C_{14} acids.

The released acids are then reacted with ammonia to produce the nitriles.

$$RCOOH + NH_3 \rightarrow RCN + 2H_2O$$

The nitrile is hydrogenated in a series of reactions that can yield a mixture of primary, secondary, and tertiary amines; or under, certain conditions and catalysts, produce any one of these selectively:[3]

Primary $RC \equiv N \xrightarrow{+H_2} RCH = NH \xrightarrow{+H_2} RCH_2NH_2$

Secondary $RCH = NH + RCH_2NH_2 \longrightarrow \underset{\underset{NH_2}{|}}{RCHNHCH_2R} \xrightarrow{-NH_3} RCH = NCH_2R \xrightarrow{+H_2} (RCH_2)_2NH$

Tertiary $RCH = NH + R(CH_2)_2NH \longrightarrow \underset{\underset{NH_2}{|}}{RCHN(CH_2R)_2} \xrightarrow{-NH_3} RCH = N(CH_2R)_2 \xrightarrow{+H_2} (RCH_2)_3N$

A more detailed representation is given in Figure 12.10.

Amine mixtures are marketed and identified according to the natural fat from which they are derived (e.g., tallow, soya, etc.). Pure fatty amines are also produced by separating the individual fatty acids prior to nitrile production or separating the nitrile product mix, using vacuum distillation in each case.

Catalyst Types

Most of these catalysts are available prereduced and stabilized. Depending on the desired amine (primary, secondary, or tertiary), the type of catalyst and promoter play an important role. For example, Raney (sponge) nickel is often best for primary amine production, whereas supported catalysts are known to produce secondary amines preferentially in many cases.[17] As an example of promoter action, copper-chromite catalyst promoted with barium favors a one-step synthesis of tertiary fatty amines.[18]

	wt%					
	Ni	Cu	Cr	Co	Form	Support
Nickel	50–65				powder	Kieselguhr or proprietary
Nickel	50–55				powder	alumina
Nickel	20–25				droplets in hardened tallow oil	proprietary
Nickel	15–50				extrusions	silica/alumina, alumina
Cobalt				54 (Zr promoted)	powder or tablets	silica
Cobalt				50 10–20	extrusions	silica/alumina, alumina
Copper-chromite		40	25		tablets or powder	
Copper-chromite		35	31	(2% Ba, 2.5% Mn)	tablets	
Raney or sponge nickel	44–52				powder	
Raney or sponge Nickel	44–52 (1% Mo)				powder	

Suppliers: United Catalysts, Engelhard, Celanese, Synetix, Activated Metals & Chemicals (sponge nickel), Grace Davison (Raney nickel)

Catalyst Poisons

Poisons that can be introduced by the natural fats include organo-sulfur, phosphorous, chlorine, and nitrogen compounds.[12,13] Care must be exercised in the upstream refining process.

Process Units

The hydrogenation can be done in batch or continuous mode. The open literature suggests that most units operate in the liquid phase with hydrogen and ammonia in the gas phase. For batch reactors, powdered catalyst is used with good agitation. Continuous reactors are usually packed beds of extrudates or tablets with pump-around through heat exchangers to control reaction temperature. Alternatively, a continuous slurry reactor may be used with sufficient gas and liquid flow to provide good suspension of the catalyst powder. Reported operating conditions are summarized in Table 12.8.

12.5.4 HYDROGENATION OF AROMATIC NITRILES

BENZONITRILE → BENZYLAMINE

Benzylamine, which is used as a corrosion inhibitor and in synthesizing various pharmaceuticals, can be manufactured by hydrogenating benzonitrile.

TABLE 12.8
Catalysts and Conditions for Fatty Amines from Fatty Nitriles

Product	Temp., °C	Pressure, bar	Catalysts	Special Conditions
Primary amines	80–150	10–550	nickel Raney nickel cobalt	Ammonia added to feed to suppress secondary and tertiary amine production
Secondary amines	150–20	50–200	nickel cobalt	Ammonia must be removed (see reaction sequence) by purging with hydrogen
Tertiary amines	160–230	7–14	nickel cobalt	Secondary amine used as feed; hydrogen purge necessary to remove ammonia
Unsaturated amines	Similar conditions as above		copper chromite	Similar conditions as above

Sources: Refs. 10–12 and manufacturers' literature.

Chemistry

benzonitrile + 2H₂ → benzylamine

Dibenzylamine can also be formed as shown in Figure 12.11.

Catalysts and conditions that will not cause hydrogenation of the aromatic ring are required. Also, catalyst and reaction conditions affect the selectivity to the primary or secondary amine. Alkyl substituted amines can be formed in a similar manner.

Catalyst Types

	wt%					
	Pd	Pt	Co	Cu	Cr	Form
Pd/C	5					powder
Pt/C		5				powder
Co/silica			40			powder or tablets
Cu-chromite				40	26	tablets

Suppliers: Engelhard, Johnson Matthey (Pd, Pt only), United Catalysts, Celanese, Precious Metals Corp. (Pd, Pt only)

Conditions

- Use Pd in an alcohol solution with 1–2 equivalents of HCl for benzylamine.
- Use Pt in neutral solution for dibenzylamine.
- Use Co or copper chromite for benzylamine with dibenzylamine in feed to prevent additional secondary amine formation.
- Use 50–100°C and 1–10 bar.

Mechanism

A reaction scheme involving an imine intermediate has been proposed. Promoters play an important role. With Pt catalysts a small amount of tin increases the hydrogenation rate, but large amounts poison the platinum.[19]

Process Kinetics

The rate of reaction of benzonitrile in the liquid phase over Pt has been well represented by a Power-Law equation based solely on the concentration of benzonitrile to the power of 0.37–0.43.[19]

12.6   HYDROGENATION OF NITROAROMATICS

In contrast to aliphatic compounds from which the nitrile is easier to produce, aromatic compounds can be more easily directly nitrated to nitroaromatics. Thus, most aromatic amines are produced by hydrogenating the corresponding nitro compound. Because water is one of the products, the heat of reaction is quite large (150 kcal/mg nitrobenzene to aniline), and special care must be exercised in the design of reaction cooling and control systems.

12.6.1   NITROBENZENE → ANILINE

Aniline has been a major chemical product for many decades, although its uses have changed. Its major use today is in the production of 4,4' methylenediphenyl isocyanate (60–65%), which is copolymerized with polyhydric alcohols such as polypropylene glycol to form a rigid polyurethane. Other uses of aniline all fall in the category of derivatives used as intermediates in production of dyes, herbicides, fungicides, defoliants, rubber chemicals, pharmaceuticals, photographic developers, amino resins, and corrosion inhibitors.[1,2]

Although aniline is also produced by ammonolysis of phenol, and direct amination of benzene, the nitrobenzene route continues to dominate and give the highest selectivity.

Chemistry

$$NO_2 \text{ (nitrobenzene)} + 3H_2 \longrightarrow NH_2 \text{ (aniline)} + 2H_2O$$

See Table 12.9 for operating conditions.

Mechanism

Haber's mechanism (Figure 12.12), proposed in 1898, continues to be the accepted description of the reaction path and of the intermediates that are formed, and it is applied to all nitro compounds.[1,2,3]

Process Units

Most modern industrial aniline processes employ continuous vapor-phase operation using either a fixed-bed or fluidized-bed reactor. Liquid-phase slurry reactor systems are often operated as batch reactors for producing specialized aromatic amines to be used in dyes, pharmaceuticals, and agricultural products.

Because of the high exothermic heat of reaction and ease of hydrogenation, much care in controlling temperature must be exercised in order to prevent runaway incidents of explosive nature. Such stringent condition require effective cooling and moderately active catalysts. See Table 12.9 for typical operating conditions.

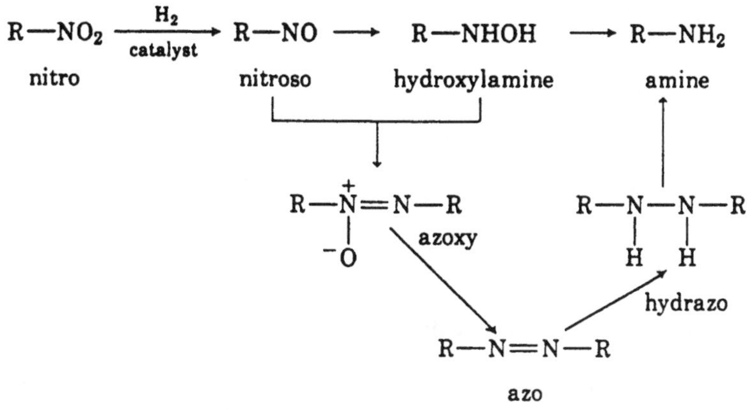

FIGURE 12.12   Haber's Mechanism for Hydrogenation of Nitroaromatics. Reprinted by permission of John Wiley & Sons, Inc. Schilling, S. L., in *Kirk-Othmer Encyclopedia of Chemical Technology*, 4th ed., Vol. 2, p. 482, © 1992, Wiley, New York, 1992. Copyright © 1992.

TABLE 12.9
Catalysts and Operating Conditions for Vapor-Phase Hydrogenation of Nitrobenzene to Aniline

Reactor System	Catalyst	Temp., °C[*]	P, bar	Form[†]	Comments	Catalyst Suppliers[§]
Fixed bed (gas phase)	52–65% Ni on Kieselguhr or silica-alumina	300–475	1–5	T or E	Must be presulfided to moderate activity	1, 4, 5
	Cu/Cr, 35% Cu, 31% Cr, 2% Ba, 2.5% Mn	150–250	1–5	T	Ba and Mn stabilize condition of the copper	1, 4, 5
	0.2–0.5% Pd-on-alumina	275–400	1–5	E, T, S		1, 2, 3, 4
	30% Cu on 66% ZnO	200–300	1–10	T		1, 4, 5
	30% Cu on silica	200–300	1–10	S, T		1, 4
	Cu, Mn, Fe oxides on pumice	250–300	1–10			
Fluidized bed	Cu 15% promoted with 0.3% Cr, Ba, and Zn	250	1–5	P	Supported on a silica hydrogel mixed with catalytic components and spray dried to make a tough silica particle	1, 4

Source: Based on information from suppliers' literature and Refs. 2, 6, and 8.

[*]The higher temperature is a peak or maximum temperature.
[†]T = tablets, E = extrusions, P = powder, S = spheres.
[§]1 = Engelhard, 2 = Johnson Matthey, 3 = Precious Metals Corp., 4 = United Catalysts, 5 = Celanese.

Fixed-Bed Reactors

Multitubular reactors with heat-transfer liquid on the shell side seem to be preferred. Boiling boiler feedwater can be used up to about 300°C, but, above that temperature, steam pressures become excessive. Molten salts can be used at higher temperatures. Hot spots are usually monitored by thermocouples in a single representative tube. Means for emergency shutdown can include increased flow of cold hydrogen or no flow of nitrobenzene. Also, temperature control during regular operation can be facilitated by varying the excess hydrogen flow.

If adiabatic fixed-bed reactors are used, multiple reactors or multiple beds with intermediate cooling will be required.

Fluidized Bed Reactors

These reactors have the advantage of rapid heat transfer between the fluidized particles and the internal heat exchanger bundles inserted in the reactor. A hot particle resides briefly on the tubular surface, and, because of the large ΔT and good contact with the metal surface, heat transfers rapidly. Obviously, the fluidized catalyst must be attrition resistant. Emergency measures must be available if flow is interrupted and fluidization seizes causing a rapid temperature increase.

Typical Catalyst Types and Operating Conditions

Many patents and catalyst have been described, but commercial practice suggests that Pd, Ni, Cu, and Cu/Cr catalysts are the most frequently used. Table 12.9 summarizes information on catalysts used for aniline vapor-phase, continuous processes.

Process Kinetics

A very thorough kinetic study of the hydrogenation of nitrobenzene to aniline has been presented with the goal of producing a model for industrial use. Intrinsic kinetics was well represented by the following equation for the rate of hydrogenation over an industrial copper catalyst:[7]

$$r = \frac{k_1 P_{nb} P_H}{1 + k_2 P_{nb}}$$

where nb and H represent nitrobenzene and hydrogen, respectively.

Since the catalyst deactivates over a period of several months due to carbonaceous deposits, the authors developed and tested an equation that included the deactivation with time. Experimental observation indicated that hydrogen impeded the deactivation.

$$r = \frac{k_1 P_{nb} P_H}{1 + k_2 P_{nb}} \left(1 - k_1^* \left[\left(\frac{P_H}{P_{nb}} \right)_\tau - \left(\frac{P_H}{P_{nb}} \right)_{\tau 0} \right] \right)$$

A third formulation also was developed for the effect of diffusion on the reaction rate.

Power-law equations have also been proposed in earlier work. Laboratory kinetic studies using precious-metal catalysts and nickel with nitrobenzene in the liquid phase suggest first-order with respect to both hydrogen and nitrobenzene, provided agitation is adequate to eliminate mass-transfer controlling resistances.[2,4] Studies in the vapor phase using copper chromite catalysts indicated half-order dependence on both nitrobenzene and hydrogen partial pressures.[5] At high ratios of hydrogen to nitrobenzene, the hydrogen partial pressure term becomes a constant.

12.6.2   ANILINE DERIVATIVES

Although so named, most aniline derivatives obtained by hydrogenation are really not produced from aniline. Rather, they are made by hydrogenating the corresponding substituted nitrobenzene. Most of these hydrogenations are conducted in the liquid phase, which allows independent variations of temperature and pressure[6] and the use of a solvent to improve selectivity and temperature control. Methanol is a frequently used solvent in amounts such that a solution of 15–35%[6] is produced, thus lowering the reaction rate. This condition is particularly important for nitro compounds with two or more nitro groups, for which case the heat of reaction increases in direct proportion to the number of nitro groups. If a solvent is not used, the reaction must be controlled by careful addition over time of the nitro compound in a batch process.

Generally, these liquid-phase hydrogenations are conducted in the range of 100–170°C. Above 170°C, the tendency to hydrogenate the ring becomes significant. Pressures are in the range of 10–200 bar. Lower ranges of pressure and temperature are used for compounds with certain chains subject to removal or cleavage.[6]

A summary of some typical compounds is presented in Table 12.10, including suggested catalysts and suppliers. Standard stirred reactors or pump-around slurry reactors are used for batch processes, which are usually favored when producing smaller quantities of valuable products.[2,6] Continuous slurry reactors in series with substantial amounts of cooled recycle are used in larger production units. These reactors usually have internal cooling coils or bundles. Agitation is created by the upflow of feed and recycled products.

12.7 HYDROGENATION OF HALOAROMATICS

Haloaromatics impose a significant selectivity problem to avoid dehalogenation by hydrogenolysis. Selective inhibitors for Pt catalyst include morpholine and very small amounts of bases such as MgO or $Ca(OH)_2$. Platinum catalyst can be properly inhibited by sulfiding, and palladium performs well with an added alkyl or phenol.[1] Also, high concentration of hydrogen at the catalyst surface improves selectivity. Such a condition is favored by vigorous agitation, higher pressure, and low catalyst loadings.[1]

12.8 HYDROGENATION OF CARBONYL COMPOUNDS

12.8.1 GENERAL

Hydrogenation of aldehydes and ketones to alcohols is a major set of reactions of significant commercial importance. Hundreds of valuable compounds are produced, and only processes for major commodity products and examples of significant speciality products are described. A significant portion of aldehydes that are hydrogenated to alcohols is produced via the oxo process (hydroformylation).

Catalyst Types (General Comments)

Merchant catalysts most commonly associated with hydrogenation of carbonyls are copper chromite, nickel, Raney nickel, cobalt, and noble metals (Pt, Pd, Os, Ru). Each has particular advantages, depending on the results desired. Catalyst costs decrease in the following order: noble metals > cobalt > copper chromite > nickel. Selectivity declines in the order: noble metals > copper chromite> cobalt> nickel. Of course, other catalysts have been used. These include CuO-ZnO-on-alumina (similar to a shift conversion catalyst) and Cu/SiO_2.

Table 12.11 summarizes some general characteristics of these catalyst types with reference to carbonyl hydrogenation. More specific detail is presented in the sections that follow.

Catalyst Deactivation

All the catalysts listed are susceptible to poisoning by sulfur compounds and chlorine. Since feedstocks for a hydroformylation unit are cuts from petroleum refining streams, sulfur and chlorine contamination is a possibility, if only during operational upsets at the refinery. Depending on local conditions, a guard reactor may be required.

TABLE 12.10
Typical Examples of Hydrogenation of Other Nitrobenzene, Toluene, and Xylene Compounds

Reactant	Product	Operating Conditions	Typical Catalyst	Suppliers	Uses
m-nitroaniline	m-phenylenediamine (1,2-benzenediamine)		• 0.1–0.3% Pd/C	1, 2, 3, 6	Production of dyes, fungicides; also ortho and para diemines
3,4-dichlorolnitrobenzene	3,4-dichlorolaminebenzene (3,4 dichloroaniline)	35 bars 360–389 K	• 0.3–0.5% Pd/C with inhibitors added to prevent dehalogenation • 5% Pt/C sulfided to prevent dehalogenation	1, 2, 3, 6	Production of herbicides, fungicides, and dye intermediates
o-nitrotoluene	o-toluidine (2, methylbenzamine)	Batch: 20–50 bar Continuous: 100+ bar	• 0.3–0.5% Pd/C • 35% Cu, 31% Cr, 2% Ba, 2.5% Mn • 30% Cu/silica • 50% Ni/Kieselguhr	1, 2, 3, 6 1, 2, 3 1, 2, 3 1, 2, 3	Production of herbicides, antioxidants, pigments
2,4-dinitrotoluene (1-methyl-2,4-dinitrobenzene) (Commercial product is 80/20% mixture of 2,4 and 2,6.)	toluenediamine (toluene-2,4-diamine)	Continuous process-loop reactor or cascade of multiple reactors, 150°C max. and 70 bar (Dinitrotoluene concentration held at low value, recycle and catalyst recovery practical, methanol used as solvent)	• Raney nickel • 5% Pd/C	4, 5 1, 2, 3, 4, 6	Manufacture of toluene diisocyanate (TDI) by recycling with phosgene; TDI polymerized with propylene glycol to produce polyurethane foams

TABLE 12.10
Typical Examples of Hydrogenation of Other Nitrobenzene, Toluene, and Xylene Compounds (continued)

Reactant	Product	Operating Conditions	Typical Catalyst	Suppliers	Uses
2,4-dinitroxylene	2,4-xylidine		• 35% Cu, 31% Cr, 2% Mn	1, 2	Production of dyes
			• 30% Cu/Silica	1, 2	
			• 0.3–0.5% Pd/C	1, 2, 3, 6	
			• 35% Cu, 31% Cr, 2% Ba, 2.5% Mn	1, 2	
3,4-dinitroxylene	3,4-xylidine		• 30% Cu/silica	1, 2	Production of synthetic riboflavin (vitamin B_2)
			• 0.3–0.5% Pd/C	1, 2, 3, 6	

Source: Based on information from suppliers' literature and Refs. 2, 3, and 6.

Suppliers: 1 = Engelhard, 2 = United Catalysts, 3 = Johnson-Matthey, 4 = Activated Metals & Chemicals, 5 = Grace Davison, 6 = Precious Metals Corp.

TABLE 12.11
Characteristics and Suppliers of Catalysts for Carbonyl Hydrogenation

Catalyst Type	Description	Major Characteristics	Application
Nickel	40–50% nickel on Kieselguhr refractory or high surface area silica-alumina 10–30% nickel-on-alumina	• High activity, useful at lower temperatures and pressures • More susceptible to side reaction including ether and hydrocarbon formation	Saturated aldehydes and ketones Useful if saturation of double bonds is desired
Nickel (Raney, sponge metal)	Powder (shipped in alkaline water slurry)	• High activity special handling (pyrpopheric)	Slurry processes very active
Cu chromite	Copper (35–40%), chromium (20–45%) Promoted with O_2 for improved stability and resistance to crystallization of copper	• High selectivity for carbonyl hydrogenation • Lower activity than nickel	Saturated and unsaturated aldehydes and ketones, both aliphatic and aromatic
Cobalt	40% cobalt on high-strength silica support 10–20% cobalt on high-strength inert alumina	• Activity and selectivity intermediate between nickel and copper	Saturated aliphatic aldehydes and ketones
Noble metals	Powder 3–10%	• Highly selective	Useful particularly in fine chemical synthesis where catalyst cost is not a major issue but high selectivity is essential

Note: See individual reaction examples for more detail.

Catalyst suppliers: Activated Metals & Chemicals and Grace Davison (Raney/sponge nickel), Engelhard, Johnson-Matthey and Precious Metals Corp. (noble metals only), United Catalysts, Synetix.

Carbon monoxide is a poison for both nickel and cobalt catalysts, which form gaseous carbonyls. Carbon monoxide should be less than 20 ppm in hydrogen feed and 5 ppm in the aldehyde feed. Copper chromite catalysts are not affected. Organic acid impurities can cause rapid growth of copper crystallites and loss of activity. Barium-promoted copper chromite provides resistance to such action as well as providing thermal stability.

Cobalt and cobalt oxides can be carried over from the oxo reactors to the hydrogenation reactor system and cause deactivation of the copper-chromite catalyst in the first reactor.[23]

As with so many other organic reactions, coking can occur on any of these catalysts. Over time, activity may decline, but, in some cases, selectivity may be improved because the less-active sites may not become coked but may often be the more selective sites.[5]

Process Units

Carbonyl hydrogenation is practiced in both vapor-phase and liquid-phase operations, with liquid phase being most common for higher alcohols and for batch processes involving stirred vessels. Adiabatic fixed-bed reactors in series with intermediate cooling or multitubular heat exchange reactors are used for vapor-phase systems, while trickle adiabatic beds are extensively employed for liquid carbonyl feed and hydrogen gas fed concurrently. The liquid phase reactors are operated at higher pressures to maximize the dissolved H_2 in the reacting liquid.

Process Kinetics

Rate expressions are presented when available for the various individual reactions following this section. It should be recognized that many of these reaction systems can involve numerous side reactions, and kinetic models for all simultaneous and consecutive reactions are most difficult to develop and confirm. It is best to employ the best possible selective catalyst so that most side reactions can be minimized.

Mass transfer to and from the catalyst surface in a trickle bed may lead to lower apparent rate constants than those obtained by studies in a well mixed reactor. The flow velocity of the liquid in a trickle bed can be low, and the film thickness around the catalyst excessive. If higher pressure drops are practical, the use of smaller catalyst particles may prove useful. Special catalyst shapes such as "clover" designs that present increased exterior area per unit mass of catalyst could also be considered favorably, since significant pressure drop increases would not be required.

Selectivity in Carbonyl Hydrogenation Reactions

Selectivity plays an important role in catalyst selection. Intrinsic catalyst properties that favor the kinetically controlled reaction rates are, of course, paramount. Usually, these characteristics have been determined empirically after years of successful applications. There are, in addition, physical characteristics such as pore size, pore size distribution, and catalyst particle size that can influence outcomes and, at times, lead to an extra selectivity level so important to product economics.

Since many aldehydes and ketones that are hydrogenated commercially to alcohols are large molecules or highly branched molecules, diffusion in and out of catalyst pores may influence the net competing reaction rates. These phenomena could be significant in the hydrogenation of unsaturated aldehydes or ketones.[6]

In such a case, for example, the net reaction rate for the desired intermediate is influenced by the rate of transport into the pore by the unsaturated aldehyde and the rate of transport out of the pore by the unsaturated alcohol. The opportunity for the intermediate to react further to the final states will increase when the transport of the intermediate is slow and saturated. Certainly, saturated-aldehyde catalyst design related to pores and pore structure can be an important consideration in improving catalyst performance. Catalyst suppliers can be most helpful in selecting the ideal catalysts in their product lines.

For a given catalyst, higher temperatures and/or higher H_2 concentrations within the range of favorable equilibrium limitations will cause an increase in the chemical reaction rate. At some point, the transport in and out of the pores will begin to affect the net rates of reaction. Accordingly, the selectivity for the production of an intermediate will decline.[6] In effect, the intermediate will reside longer in the pore and is more likely to react to the saturated form. It is conceivable that the intrinsic chemical kinetic rates of both reactions could be increased by higher hydrogen pressures to such an extent that the transport in and out of the pores could become controlling. In such a situation, the net rate of formation of intermediate and saturated products will be equal.

There is no truly *a priori* procedure for predicting such outcomes of this type. Empirical conclusions based on experience with a given process as provided by catalyst suppliers are most valuable and can be verified by appropriate tests.

One should not forget, however, that catalyst chemistry can invariably play the dominate role in producing the desired selectivity (see the section on "Hydrogenation of Crotonaldehyde").

12.8.2   ALIPHATIC OXO ALDEHYDES → ALIPHATIC ALCOHOLS

A major source of aliphatic aldehydes is the oxo process (hydroformylation), which involves reacting an olefin with CO and H_2 to produce an aldehyde with one carbon more than the original olefin.

$$\overset{\text{cat.}}{RCH = CH_2 + CO + H_2 \rightarrow RCH_2CH_2CH=O}$$

Homogenous organo-metallic complexes serve as catalysts.[1] Cobalt hydrocarbonyl [HCO(CO)$_4$] is used for producing C$_6$ and higher linear and branched aldehydes with about 80% linear and the remainder primarily 2-methyl branched isomers. A process using rhodium hydrocarbonyl [4Rh(CO)$_4$] yields an equal amount of linear and branched isomers. A third process using a liquid-modified rhodium catalyst [HRh(CO)$_2$L$_2$ or HRH(CO)L$_3$], where L is a tertiary phosphine, provides excellent yields to the linear aldehyde of 92%.[1] This phosphine modified rhodium catalyst is attractive for producing n-butraldehyde in high yields.

For completeness, it should be stated that a low-pressure process using a cobalt-carbonylorganophosphine complex, which is a strong hydrogenation catalyst, simultaneously hydrogenates the aldehyde to the alcohol. However, approximately 10% of the olefin feed is hydrogenated to paraffins.[1]

In the several processes that produce aldehydes, hydrogenation to the corresponding alcohols must follow the oxo reaction system. Isomeric aldehydes can be separated by distillation and then hydrogenated or the mixture can be hydrogenated and the resulting alcohols separated.[2]

Chemistry of Hydrogenation of Aldehydes

The general hydrogenation reaction is simply shown as follows:

$$RCH_2CH_2CHO + H_2 \rightarrow RCH_2CH_2CH_2OH$$

The actual reaction system is subject to many side reactions (Figure 12.13) that must be minimized by careful control of operating conditions and, above all, proper choice of a selective catalyst. The complexity of the reaction system shown in Figure 12.13 emphasizes the value of recommendations by the catalyst manufacturer in selecting the best catalyst for a given aldehyde.

12.8.3   MAJOR OXO ALDEHYDES → OXO ALCOHOLS

N-BUTYRALDEHYDE → N-BUTYL ALCOHOL (1-BUTANOL) AND
ISOBUTYRALDEHYDE → ISO-BUTYL ALCOHOL (2-METHYL-1-PROPANOL)

Most oxo plants using a feedstock of n-butene produce a high yield of the more desired 1-butanol. The aldehydes may be separated prior to hydrogenation or the alcohols separated after hydrogenation.

Over 50 percent of all oxo aldehyde production is involved in producing derivatives for use as plasticizers, and n-butyraldehyde is the major aldehyde in this category. It is used to produce 2-ethyl-2-hexanol via aldol condensation and then hydrogenated to 2-ethyl-1-1 hexanol, which is reacted with various acids or anhydrides to form esters for plasticizers (see 2-ethyl-2-hexanol).

The n-butyl alcohol is an important product obtained from n-butyraldehyde via hydrogenation. It is used as a solvent in paint and lacquer formulations and also in the formulation of pharmaceuticals, waxes and resins.[3] Its largest use, other than in the production of 2-ethyl-1-hexanol, is in the production of n-butylacrylate and methacrylate, which are used in making emulsion polymers for latex paints. Another large use of n-butyl alcohol is in producing butyl glycol ethers, essential ingredients in vinyl and acrylic paints, lacquers, and varnishes.[3] Butyl acetate, which is made from acetic acid and butyl alcohol, is an important solvent for quick-drying paints. Iso-butyl alcohol has limited uses, mainly as a specialty solvent.

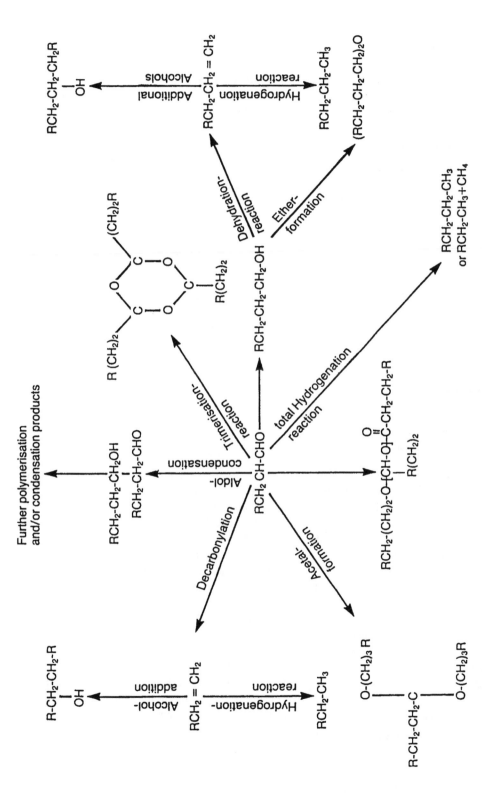

FIGURE 12.13 Possible Side Reactions in the Hydrogenation of Oxo-Aldehydes to Alcohols. Reprinted by permission: "Aldehyde and Ketone Hydrogenation," technical bulletin, Süd-Chemie, AG,

Catalyst Types

Copper chromite catalysts, barium promoted, have excellent selectivity and allow higher temperature operation where economically justified. They are less active but more selective than nickel catalysts. It can be advantageous for adiabatic beds in series to have copper chromite in the first reactor, where the rates are higher, and then follow with the more active nickel catalyst in subsequent beds, where a lower driving force occurs as equilibrium is approached and can be mitigated by the lower operating temperature possible with nickel that will minimize the reverse reaction. The equilibrium constant for the reaction at 400 K (127°C) is 22.59, while that at 500 K (227°C) is 18.53.

Typical catalysts include the following (numbers are weight percentages):

	Cu	Cr	Ba
Copper chromite	23	22	5
	34	30	8
	37	12	10
	Ni	**Cu**	**Support**
Supported nickel	50	0	Kieselguhr or silica-alumina
Raney or sponge Ni	40	2	Kieselguhr

Forms include tablets and extrudates. The copper promoted nickel is more selective. Suppliers are United Catalysts, Engelhard, and Synetix, plus Activated Metals & Chemicals and Grace Davison (Raney/sponge nickel).

Catalyst Poisons

See general summary for carbonyl compounds.

Process Units

Butyraldehyde is usually hydrogenated in the vapor phase using adiabatic reactors in series with intermediate cooling and/or recirculation of cooled hydrogen. Alternatively, multiple beds, with intermediate cooling, in a single reactor shell or a nonadiabatic reactor, can be used. The adiabatic beds make possible the use of different catalyst in lead and final beds as discussed under "Catalyst Types." Operating conditions reported are 110–180°C and 5–50 bar.[23] Selection of operating pressure depends considerably on the energy costs and steam pressures at the existing plant location. Recovered heat of reaction is only truly valuable if it is generated at a temperature that will produce usable saturated steam. Furthermore, the selected pressure must be based on optimum recycle compression costs and plant hydrogen pressure.

The most efficient heat recovery is attainable with a well designed multitubular heat-transfer reactor. The cooling medium is boiler feed water on the shell-side of the reactor. The back pressure on the steam separator drum connected to the shell sets the temperature of the boiling water and the generated steam. The heat of reaction is transferred to the water if the boiling water operates at roughly the inlet temperature of the aldehyde and hydrogen feed. Then the increase in temperature due to the reaction in the tubes provides the driving force for the transfer of reaction heat to the boiling water. An orderly optimization of operating conditions of a heat-transfer reactor for butyraldehyde hydrogenation has been described.[8]

Process Kinetics

A typical Langmuir–Hinshelwood equation has been applied successfully.[8]

$$r = \frac{k\left(P_H P_{BAL} - \dfrac{P_{BOH}}{K}\right)}{(1 + K_1 P_{BAL} + K_2 P_{BOH} + K_3 P_H)^2}$$

where
k = rate constant
K = equilibrium constant
K_1, K_2, K_3 = "adsorption" constants
BAL = butyaldehyde
BOH = butyl alcohol
H = hydrogen

The equation and constants, although not reported, are said to apply for both n-butyl aldehyde and iso-butyl alcohol. Equations for the rate of production of impurities were written as simple power-law equations.

PENTANALS → PENTANOLS

Hydroformylation of butenes yields pentanals that can be hydrogenated to pentanols. Butenes from catalytic cracking and steam cracking are used as feed olefins after the removal of butadiene by extractive distillation and isobutene by reaction with methanol to produce the gasoline additive methyl tert-butyl ether. Because of the high reactivity of isobutenes, the other butenes remain unreacted and can be used for hydroformylation. These are the isomers 1-butene, 2 cis-2-butene, and 2 trans-2-butene. The products after hydrogenation of the aldehydes are 1-pentanol and 2 methyl-l-butanol. The cobalt catalyst complex for the oxo process followed by hydrogenation yields 70% 1-pentanol and 30% 2-methyl-l-butanol, while the rhodium catalyst complex produces respectively, 90 and 10 percent.[4]

Both alcohols are used extensively as solvents and extracting agents. A major role is played by 1-pentanol in synthetic organic chemistry, including the production of dyes, flavorings, pharmaceuticals, and pesticides.[4]

Chemistry

$$CH_3CH_2CH_2CH_2CHO + H_2 \rightarrow CH_3CH_2CH_2CH_2CH_2OH$$

1-pentanol

$$\underset{\qquad\quad |}{CH_3CH_2CHCHO} + H_2 \rightarrow \underset{\qquad\qquad |}{CH_3CH_2CHCH_2OH}$$
$$\underset{CH_3}{\quad} \qquad\qquad \underset{CH_3}{\qquad}$$

2-methyl-l-butanol

Catalysts

Copper chromite and nickel catalysts as described for butanal hydrogenation are used, and reactor design requires means for adequate cooling as described for butanol production.

12.8.4   HIGHER ALIPHATIC OXO ALCOHOLS

It is common practice to refer to the C_6–C_{18} alcohols as *higher alcohols,* with the C_6–C_{11} range called *plasticizer alcohols* and the C_{12}–C_{18} range named *detergent-range alcohols.* These designations are based on their use. Olefins for producing the desired oxo aldehyde are obtained mainly from petroleum refinery product streams.

Propylene and butenes are polymerized using phosphoric acid to yield a mixture of branched olefins. These are usually used for producing alcohols in the plasticizer range. Linear olefins for linear detergent-range alcohols are obtained from higher molecular-weight refinery paraffin cuts. The paraffins are separated by molecular sieves and dehydrogenated. The aldehydes thus produced are linear. Also, ethylene may be polymerized to larger linear alpha olefins for detergent-range alcohol production. In all cases, the oxo-aldehydes are hydrogenated to the corresponding alcohols.

PLASTICIZER ALCOHOLS

Over 50% of oxo alcohols are consumed in producing derivatives for use as plasticizers. The major alcohol in this category is 2-ethyl-l-hexanol. It is synthesized from propylene via the oxo aldehyde, butyraldehyde, which is subjected to aldol condensation.[1,9]

BUTYRALDEHYDE → 2-ETHYL-2-HEXENAL → 2-ETHYL-L-HEXANOL

Aldol Condensation

$$2CH_3CH_2CH_2CHO \xrightarrow{\text{NaOH}} \underset{\text{2-ethyl-2-hexenal}}{CH_3CH_2CH_2CH = \overset{\overset{\displaystyle C_2H_5}{|}}{C}CHO}$$

n-butyraldehyde   2-ethyl-2-hexenal

Hydrogenation

$$CH_3CH_2CH_2CH = \overset{\overset{\displaystyle C_2H_5}{|}}{C}CHO + 2H_2 \xrightarrow{\text{Cat.}} CH_3CH_2CH_2CH_2\overset{\overset{\displaystyle C_2H_5}{|}}{C}HCH_2OH$$

2-ethyl-2-hexenal   2-ethyl-1-hexanol 230°C, 50-200 bar

Many side reactions are possible, but clever processes have been developed to maximize yield of the desired ethyl-hexanol. A rather complete discussion has been published.[9]

Uses

The 2-ethyl-l-hexanol is reacted with various acids or anhydrides to produce plasticizers, of which di-2 ethylhexylphthalate is the most common. It is used as a plasticizer for polyvinyl chloride. A number of other oxo alcohols in the range of C_4 to C_{13} serve as intermediates to other special purpose plasticizers, primarily for polyvinyl chloride.[2] Many of these plasticizers are based on nonlinear alcohols, although linear-based products have special applications.

Catalysts

The major catalysts used commercially are copper-chromite and nickel.

The aldehyde feedstock may contain residual iso-butyraldehyde and other contaminates such as acids, acetals, ethers, esters, and formates. If these are not removed by distillation, higher temperatures are required, for which copper chromite is more suitable. Nickel at higher temperatures causes the formation of C_1–C_4 paraffins and also ethers. If sufficient ether is formed, it can form azeotropes with the alcohol and cause costly difficulties in producing pure product. Supplier recommendations for special nickel catalysts that will avoid these problems should be observed.

	Wt%				
	Ni	**Cu**	**Cr**	**Ba**	**Support**
Copper chromite[*]		23	22	5	
		34	30	8	
		37	12	10	
Nickel[*]	50				Kieselguhr
	40	2			Kieselguhr
Sponge (Raney) nickel[†]					

[*]Engelhard, United Catalyst, Synetix
[†]Activated Metals, Grace Davison

Process Units

As in other carbonyl hydrogenations, two adiabatic beds in series with intermediate cooling are common. Operation in the vapor or liquid phase are both practiced. One unit has been described in which the first bed is operated in the vapor phase, followed by an intermediate distillation for removing heavy impurities. The resulting pure condensed overhead is fed to a second reactor operating as a trickle bed, and the output is fed to a series of distillation columns for final purification and recycle of useful fractions.[9] The use of the intermediate column between hydrogenation stages is said to lead to a higher quality final product.[9] It would be logical to assume that the second bed (trickle bed) could use nickel catalyst and obtain the advantage of higher activity at lower temperature.

Operating Conditions

As with other carbonyl hydrogenations, suggested operating conditions vary. Much depends on the particular catalyst characteristic and the state of the feed. Higher pressures may be indicated for liquid-phase operation to increase the rate of hydrogen transfer to the catalyst surface.

Process Kinetics

A reaction scheme based on two parallel paths has been proposed, and it provided the best fit for a copper-chromite catalyst.[24]

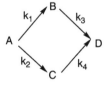

where   A = 2-ethylhexenal
        B = 2-ethylhexenol
        C = 2-ethylhexanal
        D = 2-ethylhexanol

First-order rate forms gave satisfactory fits to the experimental data.

$$\frac{dA}{dt} = k_1 A + k_2 A$$

$$\frac{dB}{dt} = k_1 A - k_3 B$$

$$\frac{dC}{dt} = k_2A - k_4C$$

$$\frac{dD}{dt} = k_3B + d_4C$$

Values of the rate constants at various temperatures, as well as activation energies, for the several reactions are reported.

OTHER PLASTICIZER-RANGE ALCOHOLS (C_6–C_{11})

Table 12.12 summarizes other alcohols in this range that are produced in a similar fashion to 2-ethyl-1-hexanol. Only those involving hydrogenation of oxy aldehydes are listed. Most of the linear alcohols are produced by the Ziegler process and do not involve hydrogenation of aldehydes. Process units and catalysts will be similar to those just described for 2-ethyl-1-hexanol.

DETERGENT-RANGE ALCOHOLS (C_{12}–C_{18})

Detergent-range alcohols are used to produce a wide range of surfactants including anionic, cationic, and nonionic. Examples are given in Table 12.13 along with direct uses of the alcohols in products and processes.

12.8.5   POLYHYDRIC ALCOHOLS

Polyhydric alcohols, alcohols with two or more hydroxyl groups, are produced by a variety of methods, depending on the cost of the feedstock and the most convenient and economic synthesis route. One synthesis path directly related to hydrogenation of aldehyde groups involves the alcohol addition of formaldehyde to a higher aldehyde with the desired number of carbon atoms and structure.[12,16] Catalytic hydrogenation of the hydroxyaldehyde yields the desired alcohol.

R^1 and R = H, alkyl, or CH_2OH

Catalytic hydrogenation is viable only if the alcohol addition operating temperature can be high enough to produce practical rates, while not too high to cause a low equilibrium conversion to the hydroxyaldehyde that must be hydrogenated in the second stage. A highly successful process, especially because of an adequate supply of the aldehydes used, illustrates this genre.

HYDROXYPIVALDEHYDE → NEOPENTYL GLYCOL

Aldol addition of isobutyraldehyde and formaldehyde yields hydroxypivaldehyde, which is then hydrogenated in the liquid or gaseous phase using copper chromite, nickel, or cobalt catalysts.[16]

hydroxypivaldehyde   neopentyl glycol
(3-hydroxy-2,2-dimethyl propanal)   (2,2 dimethyl-1,3 propanediol)

TABLE 12.12
Other Plasticizer-Range Alcohols

Alcohol	Preparation	Uses
C$_6$ 2-methyl-l-pentanol CH$_3$CH$_2$CH$_2$CH(CH$_3$)CH$_2$OH	Aldol condensation of propionaldehyde followed by hydrogenation of the methyl pentanal	Solvent
2-ethyl-l-butanol	Aldol condensation of acetalydehyde and 1-butanol followed by hydrogenation of the ethyl butanal	Component of penetrating oils, paints (flow improver), and corrosion inhibitor Solvent, extracting agent, cleaner for printed circuits Phthalate esters as plasticizers Nitric acid ester as cetane number enhancer
C$_6$ alcohol mixtures, mainly 1-hexanol and 2-methyl-i-pentanol	Hydrogenation of C$_6$ aldehydes	Plasticizer, solvent, flow improver
C$_7$ C$_7$ mixture	Hydroformylation of isohexane, subsequent hydrogenation of the hexanals, and esterification of the alcohol Major product: di-isoheptyl phthalate	Important plasticizer for PVC floor products
C$_8$ 2-ethyl, 1-l-hexanol C$_8$ mixtures	See text Hydroformylation of heptene mixtures and hydrogenation of branched aldehydes so produced to yield branched alcohols called *isooctyl alcohols*	Plasticizer for PVC Solvent, foam suppressants, extracting agent, emulsifier, herbicide production Major use: production of plasticizers such as diisooctyl phthalate for PVC
C$_9$ 2,6-dimethyl-4-heptanol [(CH$_3$)$_2$ CHCH$_2$]$_2$ CHOH	Aldo condensation of acetone and hydrogenation of the ketone thus formed	Esters used as defoaming agents and extraction solvents
C$_9$ mixtures	Hydroformylation of butene dimers and hydrogenation of the aldehydes Product: isononyl alcohol, a mixture of alkyl-branched alcohols (e.g., 3,5,5-trimethyl-l-hexanol and dimethyl-l-heptanols)	Phalate esters as plasticizers Acrylates for lubricating oil additives Hydraulic fluids
C$_{10}$ mixtures, trimethylheptanols, dimethyloctanols	Hydroformylation of tripropylene and hydrogenation of the aldehyde	Phthalate esters as plasticizers Esters of trichlorophenoloxyacetic acid as herbicides Other derivatives used in detergents, pharmaceuticals, and cosmetics

Source: Based on information from Ref. 10.

TABLE 12.13
Application of Major Detergent-Range Alcohols

Alcohol	Preparation	Example Uses
Linear C_{12}–C_{18} alcohol mixtures	Most are produced directly by the Ziegler process, but some are produced by modified oxo processes that yield up to 80% of n-aldehydes, which are converted to n-alcohols by hydrogenation.	**Surfactants, anionic** Linear alkylbenzene sulfonates, $R\text{-}C_6H_4\text{-}SO_3Na$, R = 10–13, used in laundry detergents, powdered and liquid Alkyl sulfates, $R\text{-}CH_2\text{-}O\text{-}SO_3Na$; R = 12–15, used in toothpaste, shampoos, hand dishwashing detergents; R = 16–18, heavy-duty detergents Alkylether sulfates, $RO(CH_2\text{-}CH_2\text{-}O)_n\text{-}SO_3Na$, R = 12–14, made from polyetholxylated alcohols, used in easy-care detergents, foam baths, shampoos, hand dishwashing liquids **Surfactants, nonionic** Alkyl poly(ethylene glycol) ethers, $RO(CH_2\text{-}Ch_2\text{-}O)_n\,H$, used in laundry detergents; very good for synthetic textiles and low-temperature washing Specialties: (a) acrylic esters polymerized and used as viscosity index improvers for lube oil, (b) many detergents used as emulsifiers in various industries and products
C12–C18 alcohols used as alcohols	Hydrogenation of oxy aldehydes	Used as or in emollient, drilling mud, plastic mold release agent, foam control, flotation agent
Hexadecanol, octadecenol		Cosmetics and pharmaceuticals
Isotridecyl alcohol	Hydrogenation of aldehyde prepared by oxo process from tetrapropylene	Used in paints, plasticizers (phthalate ester) for PVC wire insulation
Isohaxadecyl and isooctadecyl alcohols	Aldo condensation of iso octyladehyde and isononyaldehydes, respectively, and subsequent hydrogenation	Used in lubricants and for evaporation prevention

Source: Based on information from Refs. 1, 10, and 11.

Various processes have been described with reaction temperatures reported between 100 and 175°C.[16]

Neopentyl glycol is used extensively as the glycol to react with terephtalic acid or isocyanate to produce polyesters and polyurethanes of excellent chemical stability.

12.8.6   SUGAR ALCOHOLS

Sugar alcohols are also classified as polyhydric alcohols.

GLUCOSE → SORBITOL

Sorbitol is the most widely used sugar alcohol. It is produced by hydrogenating glucose.

αD-(+)-GLUCOSE   open chain form   SORBITOL

The reaction is conducted commercially as a batch or continuous process. The reaction temperature must be kept below 150°C (typically 140°C) to avoid by-product formation. Operating pressures reported vary between 30 and 125 atmospheres. Solutions of 50% dextrose in water constitute the reactor feed. This feedstock is pretreated by deionization to remove ionic impurities. The pH must be kept on the moderate acid side of 5 to 6 to avoid unwanted side reactions[13,14,15] and to protect the activity of the nickel.

Catalysts Used

Type	Suppliers	Ni	Ru	Mo	Al	Form
Raney or sponge nickel	Activated Metals and Chemicals, Grace Davison	88–94		2	4–10	30–35 μ powder
Ni-on-Kieselguhr	United Catalysts, Engelhard, Synetix	50				powder
Ruthenium	Engelhard, Johnson-Matthey, Precious Metals Corp.		5			powder

Ruthenium has been suggested for situations where the feedstock contains some polysaccharides that would lead to impurities in the product. Ruthenium is also resistant to acid conditions.

Process Units

Both batch operation and continuous slurry-bed operation have been reported. After hydrogenation, the product is purified by ion exchange and activated carbon. The purified product is evaporated to a 10% solution for marketing.[14,15]

Uses

Sorbitol is used to produce ascorbic acid (vitamin C), alkyd resins (polyol content), poly (oxy propylene) derivatives of sorbitol for polyol content of polyurethanes, and adhesives. Sorbitol is widely used in its pure form as an FDA-approved additive to foods as texture, body, and shelf-life improver. It is used as a humectant and softener for candies and shredded coconut. Sorbitol is also used in pharmaceuticals to give body to cough syrups and other liquid medicines and provide stability to suspensions. Finally, sorbitol is used as a component in wash-and-wear cotton treatment agents and as a plasticizer for soil release agents.[15]

Process Kinetics

The reaction of glucose to sorbitol has been reported to be first-order in glucose and catalyst and half-order in hydrogen pressure.[29,30]

OTHER SUGAR ALCOHOLS: FRUCTOSE → MANNITOL, XYLOSE → XYLITOL

Both of these alcohols are manufactured by hydrogenation using catalysts and processes similar to that for sorbitol. Their use in products is not as large and diverse as that for sorbitol, but they fulfill important niches. The primary use of mannitol is in pharmaceuticals, mainly as a primary component in producing chewable tablets of medicines and vitamins. Xylitol is used as a humectant and as a source of parenteral nutrition.[14] Mannitol is often produced by hydrogenating invert sugar (fructose + glucose), in which case both sorbitol and mannitol are produced. The reaction has been shown to produce an intermediate gluconic acid formed by dehydrogenation of glucose. The gluconic acid is rapidly consumed by the adsorbed fructose forming sorbitol and mannitol.[25]

Process Kinetics

The xylose reaction has been found to fit a simple first-order expression in xylose concentration.[31]

12.8.7   UNSATURATED ALIPHATIC ALDEHYDES

The double bond is more readily hydrogenated than the carbonyl group. Palladium has the unique quality of stopping hydrogenation with the formation of the saturated aldehyde. Total hydrogenation to the saturated aldehyde can be accomplished with active nickel catalysis. If the unsaturated alcohol is wanted, more selective catalysts are used under mild conditions and a low hydrogen-to-aldehyde unsaturated feed ratio. The following are typical examples.

ACROLEIN → N-PROPANOL OR PROPANAL OR ALLYL ALCOHOL

Acrolein, which is produced by the catalytic oxidation of propylene, is a highly reactive compound used in several commercially important synthesis such as the production of the amino acid methionine. Hydrogenation yields n-propylalcohol, propionaldehyde, and allyl alcohol. Selectivity to any one of these products requires careful choice of catalyst.

Chemistry

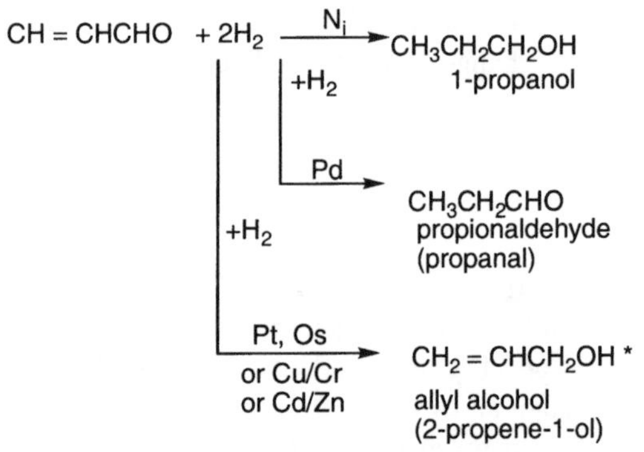

$$CH = CHCHO \ + 2H_2 \ \xrightarrow{\ N_i\ } CH_3CH_2CH_2OH$$

1-propanol

$+H_2$

$\xrightarrow{\ Pd\ }$ CH_3CH_2CHO
propionaldehyde
(propanal)

$+H_2$

$\xrightarrow[\text{or Cu/Cr}]{\text{Pt, Os}}$
or Cd/Zn $CH_2 = CHCH_2OH$ *

allyl alcohol
(2-propene-1-ol)

* Alkaline hydrolysis of allyl chloride is a preferred process for allyl alcohol.

Catalysts

Catalyst	Suppliers
Ni: Raney nickel or sponge nickel, or high-nickel catalysts on Kieselguhr (e.g., 50% nickel)	United Catalysts, Engelhard, Synetix (Ni-on-Kieselguhr); Activated Metals & Chemicals, Grace Davison (sponge and Raney nickel)
Pd: 5%-on-carbon (slurry phase) or 0.5% on Al_2O_3 (fixed-bed)	Engelhard, Johnson-Matthey, Precious Metals Corp.
Pt-on carbon modified with $FeCl_2$ and $Zn(OAc)_2$	Engelhard, Johnson-Matthey
Os-on-carbon	Engelhard, Johnson-Matthey
Copper/cadmium or copper chromite	Engelhard, United Catalysts, Celanese

Propionaldehyde is now largely supplied by oxo synthesis from ethylene, and 1-propanol is the result of hydrogenating this aldehyde. Acrolein is a source for these compounds in some instances. The n-propyl alcohol (1-propanol) is used as a solvent for resins and cellulose derivatives. Allyl alcohol is an intermediate in the production of allylethers, glycerol, resins, and plasticizers.

CROTONALDEHYDE → N-BUTANAL, 2-BUTEN-1-OL, N-BUTANOL

Crotonaldehyde, which is produced by aldolization of acetaldehyde, can be hydrogenated to 2-butene-1-ol, and n-butanol.

$$CH_3 - CH = CH - CHO + H_2 \xrightarrow{Pd} CH_3 - CH_2 - CH_2 - CHO$$

crotonaldehyde   butyraldehyde
2-butenal   n-butanal

$+2H_2 \xrightarrow{Ni} CH_3CH_2CH_2CH_2OH$
n-butanol

$+H_2$

$\xrightarrow[\text{Pt, Os}]{Cu/Cr,\ Cu/Cd} CH_3CH = CHCH_2OH$
crotyl alcohol
2-buten-1-ol

See "Acrolein" for catalyst descriptions and suppliers.

The crotonaldehyde route to butyraldehyde and n-butanol provides an alternative to that of the oxy process followed by hydrogenation of the aldehyde to butanol. The basic raw material for crotonaldehyde is ethylene (ethylene → acetaldehyde → crotonaldehyde), while that of the oxy route is propylene (propylene → butyraldehyde). The economics have become more favorable for the oxy route in recent years.

Process Kinetics

A Langmuir–Hinshelwood equation has been proposed, based on a study using copper chromite as the catalyst for the formation of 2-butene-1-ol (crotyl alcohol). The rate determining step was postulated as the reaction of a second hydrogen with the partially hydrogenated crotonaldehyde in the absorbed state.[28]

$$C_3H_6CHO * + H * C_3H_6CHOH * + *$$

where * represents the active site.

$$r = \frac{kK_HK_CK^*P_HP_C}{\left(1 + K_H^{1/2}P_H^{1/2} + K_CP_C + K_H^{1/2}K_CK^*P_H^{1/2}P_C\right)^2}$$

where K_H and K_C = equilibrium adsorption constants

K^* = equilibrium constant for first surface reaction with hydrogen

subscripts H and C = hydrogen and crotonaldehyde, respectively

It is possible to simplify the equation in restricted temperature regions and pressures such that hydrogen exhibits first-order behavior and crotonaldehyde zero-order. In such a situation, the rate may simply be expressed as follows:

$$r = k^1P_H$$

where k^1 is an apparent rate constant.

More recently, a rather thorough study of the hydrogenation of crotonaldehyde over a 5% Cu/Al_2O_3 catalyst has suggested a reaction scheme involving six reactions and all three possible products.[32] Because hydrogen is fed in excess, the six reactions can be represented as pseudo-first-order. Separate experiments using butanal, crotyl alcohol, and crotonaldehyde provided the data necessary to obtain best fit values for the six constants. Separate studies on the effect of thiophene in the feed demonstrated a decrease in the rate of hydrogenation butanal and an increase in the rate of crotyl alcohol formation. Conversely, feeds free of thiophene showed very low selectivity to crotyl alcohol and high selectivity to butanol.[32]

12.8.8   HYDROGENATION OF ALIPHATIC KETONES

ACETONE → DIACETONE ALCOHOL → HEXYLENE GLYCOL

Catalysts and operating conditions are in general similar for both aliphatic aldehydes and ketones. Since ketones are more sterically hindered, they require slightly more severe conditions.

<pre>
            OH   OH   OH
            | | |
CH₃ CCH₂ COCH₃ + H₂ → CH₃ CCH₂ CHCH₃
            | |
            CH₃   CH₃

Diacetone   2-methyl-2,4-pentanediol
(4-hydroxy-4-methyl-2-pentanone)
</pre>

Diacetone alcohol is made by base-catalyzed condensation of acetone and is an important commercial solvent. Hydrogenation of the ketone group produces hexylene glycol, which is used in cosmetics and as a component of castor-oil based brake fluid in which it serves as a coupling agent.

OTHER ALIPHATIC KETONES

Other aliphatic ketones in addition to being valuable commercial products are sources of useful derivatives, including alcohols. Some examples are summarized in Table 12.14.

TABLE 12.14
Other Aliphatic Ketone Hydrogenation Examples

Ketone	Product	Catalyst	Uses
Diisobutyl ketone (2,6-dimethyl-4-heptanone) $(CH_3)_2 CHCH_2COCH_2CH(CH_3)_2$	diisobutylcarbinol (2,6-dimethyl-4-heptanol) $(CH_3)_2CHCH_2CH(OH)CH_2CH(CH_3)_2$		
Isophonene (3,5,5-trimethyl-2-cyclohexene)	3,3,5-trimethylcyclohexanol m-homomenthol		Fragrance and synthesis of vasodilator
Ethyl amyl ketone 3-octanone $CH_3CH_2CO(CH_2)_4CH_3$	Ethanyl amyl alcohol (3-octanol) $CH_3CH_2CH(OH)(CH_2)_4CH_3$		Component of lavender fragrance, also mushroom fragrance
Methyl isobutyl ketone $CH_3COCH_2CH(CH_3)_2$ 4-methyl-2-pentenone	Methyl-isobutylcarbinol $CH_3,CHCH_2,CH(OH)CH_3$ 4-methyl-2-pentanol	Ni	

12.8.9 HYDROGENATION OF AROMATIC ALDEHYDES AND KETONES

There are three distinct possible products: aromatic alcohol, alcyclic alcohol, and alkylaromatic.

Obviously, catalyst selectivity, temperature, and hydrogen-to-carbonyl ratio must play an important role. For good selectivity to the alcohol, excess hydrogen should be avoided and mild conditions used. Although palladium is active for aldehyde hydrogenation, it is an excellent catalyst for selective reduction of aromatic ketones to aromatic alcohols. Hydrogenolysis to the hydrocarbon is avoided by neutral or slightly basic conditions, stoichiometric ratio of H_2-to-carbonyl, and moderate temperatures and pressure. Excessive temperature can bring on the reduction of the aromatic ring. Copper chromite catalysts are used extensively for hydrogenation of both aromatic aldehydes and ketones to alcohols. Palladium is the most active catalyst for hydrogenation of aromatic aldehydes and ketones to alcohols and is used at mild conditions in many fine chemical operations.

ACETOPHENONE → METHYLPHENYLCARBINOL

Acetophenone occurs in many natural products, imparting a sweet "orange blossom" odor. It is an industrial by-product of phenol production from cumene and of the oxirane process, which produces propylene oxide by oxidation of ethylbenzene along with styrene and acetophenone.[18,19] The acetophenone can be hydrogenated to methyl phenylcarbinol and subsequently dehydrated to styrene. It is estimated that about 17% of styrene produced is accounted for by total production of styrene by the oxirane process.

Chemistry

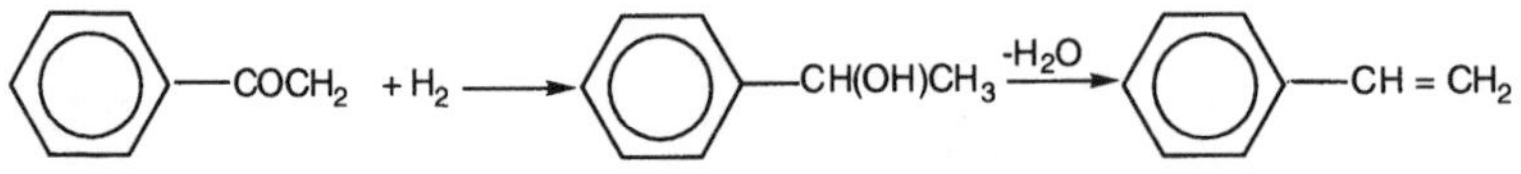

acetophonone
(phenyl methyl ketone)
(1-phenylethanone)

methylphenylcarbinol
1-phenylethyl alcohol

styrene

Typical Catalysts

	Wt%					
	Cu	ZnO	Cr	Ba	Mn	Form
Copper chromite	40		26			Tablets
	35		31	2	2.5	Tablets
Copper zinc	30	65				Tablets or extrudates

Suppliers: United Catalysts, Engelhard, Celanese
Operating conditions:[18] 70–100°C, 70 bar

BENZALDEHYDE → BENZYL ALCOHOL

Benzyl alcohol is a very important solvent with minimal odor. It is not carcinogenic and is used as a preservative in small amounts in pharmaceuticals, perfumes, soaps, and detergents. Ballpoint pen ink contains a small amount as a flow improver. In the dyeing of various synthetic textiles, benzyl alcohol is an important dye additive. Of course, a major use is its role as the alcohol in forming many valuable esters.

Chemistry

Benzaldehyde is readily hydrogenated to benzyl alcohol in good yield using nickel or platinum catalysts. A platinum/Al_2O_3/Li oxide catalyst has been reported as an excellent commercial catalyst for fixed-bed operation.[20]

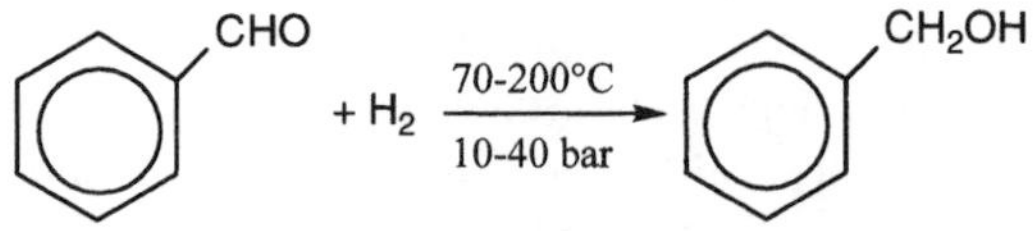

benzylaldehyde   benzyl alcohol

Two side reactions are feasible thermodynamically.[26]

$$C_7H_6O + H_2 \rightarrow C_7H_8O + H_2O$$

$$C_7H_8 + H_2 \rightarrow C_6H_6 + CH_4$$

Catalyst Types

Pt-on-alumina/Li oxide. Suppliers are Engelhard, Johnson-Matthey, and Precious Metals Corp.

Economics

When benzaldehyde is available as a side reaction product at low cost, its hydrogenation to the alcohol can be most economically attractive. If, however, its cost is high, the chlorination of toluene to benzyl chloride followed by hydrolysis to the alcohol in basic solution will be the dominant process.

Process Unit

The direct hydrogenation reaction is carried out in a slurry or trickle-bed reactor.

Process Kinetics

A Langmiur–Hinshelwood equation with the surface reaction being controlling has been proposed.[26]

$$r = \frac{kK_BC_B}{1 + K_BC_B} \times \frac{K_HP_H}{[1 + (K_HP_H)^{0.5}]^2}$$

where C_B = benzaldehyde concentration
$\quad\quad\quad$ P_H = hydrogen pressure
$\quad\quad\quad$ K_B = adsorption constant for benzaldehyde
$\quad\quad\quad$ K_H = adsorption constant for hydrogent
$\quad\quad\quad$ k = rate constant

The equation was tested on both a trickle-bed and slurry reactors and gave good agreement for both mass transfer of hydrogen to the catalyst surface and internally. The internal gradient was expressed via an effectiveness factor. The effectiveness factor was very low and the gas-liquid mass-transfer gradient was also significant. But the liquid-solid gradient was negligible.

A slurry reactor produced rates 50 times those of the trickle-bed because of the small catalyst size, and the effective contacting of catalyst and reactants due to the agitation of the suspended slurry. Although this study is based on nickel catalyst, the effects of mass transfer should be similar for any practical catalyst.

Platinum catalysts (Pt/SiO_2, Pt/TiO_2, and Pt/Al_2O_3) have also been studied.[27] The Pt/TiO_2 catalyst maintained the highest selectivity to benzyl alcohol. A Langmuir–Hinshelwood form was presented based on the presumed rate limiting step of the addition of the second hydrogen atom.

$$C_6H_5CH_2O* + H* \rightarrow C_6H_5CH_2OH* + *$$

where * indicates an active site

$$r = \frac{kK_1K_HK_BP_HP_B}{\left(1 + K_BP_B + K_H^{1/2}P_H^{1/2}\right)^2}$$

where K_1 = equilibrium constant for the reaction between adsorbed benzaldehyde and
                  the first hydrogen
        K_B and K_H = adsorption equilibrium constants for benzaldehyde and hydrogen,
                  respectively

and subscripts H and B refer, respectively, to hydrogen and benzaldehyde.

12.8.10   AROMATIC UNSATURATED ALDEHYDES

Although the carbonyl group of aromatic aldehydes is more readily hydrogenated than that for
aliphatic aldehydes, the unsaturation form remains the preferred path. Selective catalysts such as
copper-chromite, modified platinum, or osmium are used. See the following example.

CINNAMALDEHYDE → CINNAMIC ALCOHOL

Cinnamaldehyde is produced from the alcohol condensation of benzaldehyde and acetaldehyde.[18]
Hydrogenation of the aldehyde thus formed yields cinnamic alcohol.

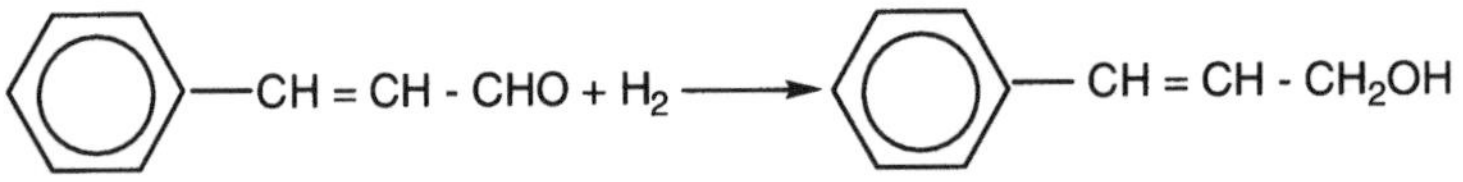

cinnamaldehyde   cinnamic alcohol

3-phenyl-2-propenal   3-phenyl-2-propen-1-ol

The alcohol exists esterfied in nature in cinnamon leaves, hyacinth, and other sources. It is
produced synthetically as shown above and is used in perfume and deodorant preparations. It is
also used in synthesizing the antibiotic chloramphenicol, in various flavor systems as a modifier,
and in inks and animal repellents.

Catalysts

	Typical types, Wt%					
	Pt	Cu	Cr	Ba	Form	Reactor
Copper chromite		40	26		Tablets	Adiabatic, fixed bed
		60	10		Tablets	Adiabatic, fixed bed
		33	30	8	Tablets	Adiabatic, fixed bed
Platinum	5				On-carbon Powder	Slurry

Suppliers: United Catalysts, Engelhard, Johnson-Matthey, Precious Metals Corp.

CINNAMALDEHYDE → HYDROCINNAMIC ALCOHOL

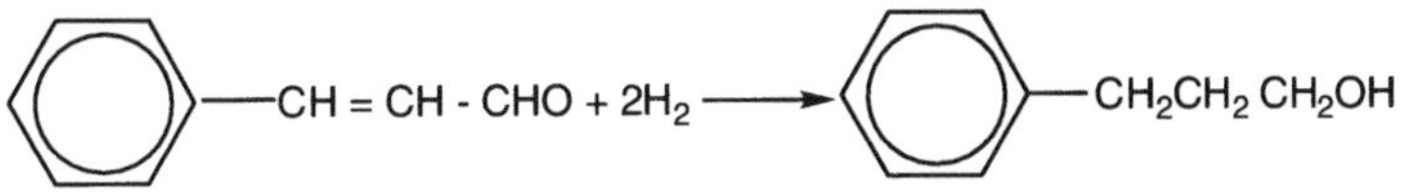

$$\text{—CH} = \text{CH - CHO} + 2H_2 \longrightarrow \text{—CH}_2\text{CH}_2\text{ CH}_2\text{OH}$$

3-phenylpropanol

This reaction carries the hydrogenation one step further by saturating the side chain in addition to converting the carbonyl to the alcohol. The resulting alcohol is used as a perfume ingredient. No special selectivity is required other than the prevention of hydrogenolysis of the side chain to alkyl hydrocarbon.

If a saturated aromatic aldehyde is desired, a modified palladium catalyst has been described using ferrous sulfate.[21]

Catalysts

0.5% Pd-on-charcoal (avoid acid conditions to prevent hydrogenolysis).

Suppliers

Engelhard, Johnson Matthey, Precious Metals Corp.

12.8.11   HYDROGENATION OF HETEROCYCLIC ALDEHYDE

FURFURAL → FURFUROL

Furfural, for many years, has been and continues to be a major chemical derived from renewable agriculture products (bagasse, corn cobs, brans, and cereal straws). It is hydrogenated to furfurol (furfuryl alcohol), which is an important solvent and a component in the synthesis of an ulcer drug, and it forms commercially important resins with urea and formaldehyde. The resins are used in chemical resistant construction and for plywood glues.

Chemistry

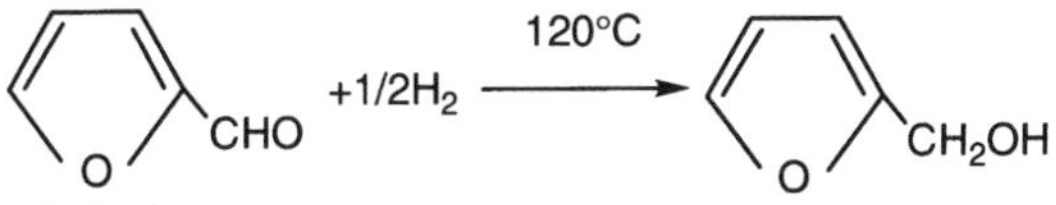

$$\overset{120°C}{\underset{+1/2H_2}{\longrightarrow}}$$

furfural
2-furancarboxaldehyde
2-furaldehyde

furfurol
furfuryl alcohol
2-furanmethanol

Both the furan ring and the carbonyl side chain are subject to chemical transformation. As shown in Figure 12.14, many side reactions are possible, and it is essential that the most selective catalyst and proper operating conditions be selected. Barium promoted copper chromite is the usual catalyst of choice. The barium retards excess hydrogenations such as that of the ring and hydrogenolysis of the side chain.[22]

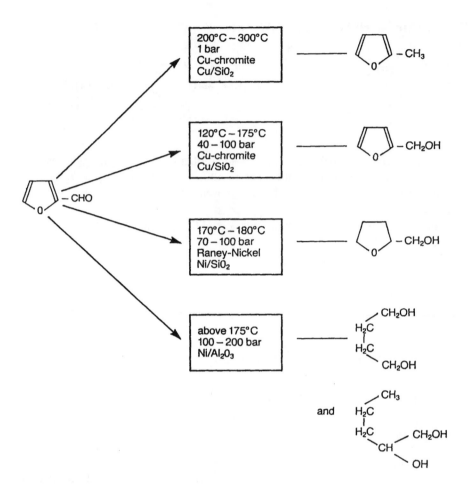

FIGURE 12.14 Effect of Operating Conditions and Catalyst Type on Furfural Hydrogenation. Note: Low-pressure operation at lower temperatures in a fixed-bed reaction will give furfural in good yield. The high-pressure examples are for slurry liquid-phase systems. Reprinted by permission, *Furfural Hydrogenation,* technical bulletin, Süd-Chemie AG.

Catalyst Types

	Wt%				
	Cu	Cr	Ba	Mn	Form
Copper chromite	22	17	5		Tablet
Copper chromite	35	31	2	2.5	Powder

Supplier: United Catalysts, Engelhard

Catalyst Poisoning, Coking, and Sintering

Table 12.15 provides a list of possible poisons along with a description of coke formation and sintering (crystal growth). It is necessary that these poisons, if present in the feed, be reduced to low levels. The catalyst supplier will provide recommended allowable maxima for poisons in the feed.

TABLE 12.15
Coking, Poisoning, and Sintering of Furfural Hydrogenation Catalyst

Type of Poison	Poisoning Mechanism	Decontamination/Regeneration
Polymer formation mostly (-O-$(CH_2)_4$-O-$(CH_2)_4$-)$_n$, probably via tetrahydrofuran and ring opening via hydrogenolysis	Blocking of active sites, pore filling, carbon deposition	Regeneration by "burning off" the polymers; activity of tablets is partly restored
Sulfur compounds	Blocking of active copper sites by formation of copper sulfide	Not possible
Chlorine	Chlorine compounds from hydrochloric acid, which attacks the active copper surface and can destroy the surface texture of the catalyst	Not possible
Water	Can accelerate, in combination with organic and/or inorganic acids, the growth of copper crystallites and so diminish the activity	Not possible
Organic acids	Can dissolve part of the catalyst and so destroy the catalyst structure; along with water, acids can accelerate the growth of copper crystallites	Not possible
High temperatures	At temperatures higher than 300°C, the copper crystallite growth remarkably accelerated	Not possible

Source: Reprinted by permission, Furfural Hydrogenation, technical bulletin, Süd-Chemie AG.

Catalyst Regeneration

Regeneration of the coked catalyst is accomplished by purging the reactor with nitrogen, cooling to near ambient temperature, and then adding air to produce a low oxygen content of 1%. The temperature is increased over a period of hours until it reaches some maximum recommended value (e.g., 250°C). After regeneration is complete, the reactor is again purged with nitrogen and the catalyst reduced with hydrogen in preparation for another cycle of several months. Catalyst manufacturers supply detailed directions for all phases of operation related to the catalyst.[22]

Process Units

The tableted catalyst is used for vapor-phase operation in fixed-bed reactors, many of which are multitubular nonadiabatic heat-transfer reactors. For batch slurry-phase, the powdered catalyst used must be sturdy so as to permit filtration of the catalyst after each cycle. Also, high surface area is desirable to reduce batch times. High pressure assures high concentrations of hydrogen in the liquid.

Approximate Operating Conditions[22]

Fixed bed	1–2 bar, 100–140°C
Batch, slurry-phase	150 bar, 160–180°C

A moderate-size multitubular reactor has been reported with 6500 tubes, 25 mm in diameter, and 1.5 m in length.[18] Operating conditions were said to be 120°C and 1.5–2.5 bar for this vapor-phase reactor. Exothermic heat is removed by shell-side circulation of boiler feedwater.

Process Kinetics

A simplified Langmuir–Hinshelwood equation has been proposed for copper chromite catalyst with vapor-phase operation.[28] The rate-determining step was assumed to be the addition of the second hydrogen to the adsorbed partially hydrogenated furfural.

$$C_4H_3OCH_2O^* + H^* \rightarrow C_4H_3OCH_2OH^*$$

where * = active site

$$r = \frac{kP_HP_F}{(1 + K_H^{1/2}P_H^{1/2} + K_FP_F)^2}$$

where k = an apparent rate constant

and subscripts H and F refer to hydrogen and furfural, respectively.

12.9   HYDROGENATION OF RESINS, ROSINS, AND WAXES

The terms *resins, rosins,* and *waxes* have been used hundreds of years to describe certain high-molecular-weight polymers obtained from natural sources. Natural resins are high-molecular-weight compounds that form films. They can be molded into useful objects or made into threads. Rosin is a hard and brittle resin that has many uses (see below). The harvesting of these natural products is highly labor intensive. Consider, for example the resin, called *shellac,* which is obtained from the secretion of larvae on the twigs of various host trees in India and China and adjacent countries in the Far East. It is estimated that 1.5 million insects produce one pound of shellac, which must be tediously removed from the gathered twigs and washed, cleaned, and filtered.[1] Most other natural resins are obtained from plants by equally labor-intensive methods. It is not surprising, therefore, that less costly synthetic polymers have replaced natural resins in many applications. It is understandable that these polymers were often called *synthetic resins* when used to replace natural resins. But the synthetic polymers usually are totally different molecular structures, and the use of the word *resin* for such polymers is fading in practice among polymer chemists and engineers.

Hydrogenation of natural resins is practiced in instances where the product must be color free and free of destabilizing or undesirable entities. Color, which is usually yellow or brown, is caused by conjugated double bonds, and hydrogenation eliminates the color by saturating the double bonds. Color can also be caused by sulfur and oxygen compounds, which can also be removed by hydrogenation. Great care in selecting a catalyst is necessary. Excessive activity and acid supports can cause hydrogenolysis and hydrogenation of other entities in the polymer such as aromatics.[4] When this situation occurs, the desired characteristics of the resin may have been destroyed.

Reactors

Resins are often hydrogenated in stirred batch reactors or continuous pump-around slurry reactors. Adiabatic fixed-bed reactors are used when high production is required.

Poisons

Natural products often contain sulfur and chlorine and removal may be necessary prior to hydrogenation. Often selection of a resistant catalyst eliminates the need for pretreatment. Catalyst suppliers should be consulted in such situations.

Some examples of hydrogenation of these large, high-molecular weight compounds are presented here to illustrate typical catalyst applications and operating conditions.

12.9.1 Rosins → Stable, Pale-Color Rosins

Rosins can be obtained from gum flowing from cuts in pine trees (gum rosin), pine-tree stumps (wood rosin), or from digested wood chips in paper manufacture (tall-oil rosin). The rosin fraction is obtained by a different procedure for each source. In the case of gum, the rosin is separated as a residue from the crude turpentine. Wood rosin is isolated by a solvent process, and tall-oil rosin is released by a caustic digestion process.[1] Rosins consist of a mixture monocarboxylic acids of alkylated hydrophenanthrene nuclei. The dominant acid is abietic acid. The other acids, in general, have different positions of the double bonds.

Chemistry

Hydrogenation eliminates the possibility of conjugation by saturating one or both double bonds.[1]

abietic acid   dihydroabietic acids   tetrahydroabietic acid

This hydrogenation eliminates aging and oxidative degradation and produces a product relatively free of dark color bodies so that it has a stable pale color. Acid polymerization accomplishes similar results.

Uses

The major use of rosin is for paper sizing made by partial saponification of the hydrogenated rosin. The sizing gives the paper water resistance and gloss and prevents yellowing. Formulation of other products involving rosins or rosin derivatives includes printing inks, pressure sensitive adhesives, hot-melt adhesives, chewing-gum modifiers, sealing wax, and polishes.

Typical Catalysts

| | Wt% | | | |
	Ni	Zr	Pd	Form
Nickel/Kieselguhr	20–65			Tablets, powder, extrudates
Nickel/silica-alumina	50			Extrudates
Palladium/carbon			0.1–0.5	Granules

Suppliers: Celanese, United Catalysts, Engelhard, Johnson-Matthey (Pd), Precious Metals Corp. (Pd)

Note: Zirconium promotion allows lower-temperature operation, thereby reducing hydrogenoloysis.[4]

12.9.2 WAXES

The major commercial and industrial waxes from natural sources are petroleum and coal based. There are, of course, many other sources of wax, including minerals, vegetables, animals, and insects.[2] The only common characteristics are water repellency, solid condition at room temperature, and deformability under applied pressure, insolubility in water, combustibility, and capability of producing a glossy film on various surfaces. In general, waxes are blends of organic esters, alcohols, esters, and hydrocarbons; but the natural waxes vary all over the map in composition. Petroleum waxes are primarily hydrocarbon waxes, whereas many of the vegetable and animal waxes are primarily esters.

PETROLEUM WAXES

Petroleum waxes (paraffin and microcrystalline) are produced in the refining of crude oil in quantities of about 1 million tons per year in the U.S.A. alone. A majority of this production is devoted to paraffin wax, which is a by-product of lubricating oil manufacture.

Paraffin Wax

Lubricating oil fractions obtained from vacuum distillation of the heavy portions of crude oil are subjected to further solvent extraction to remove aromatic compounds because of their poor lubricating properties. The raffinate contains mainly normal alkanes, which include the wax components that are removed by cooling and crystallizing of the wax or, more frequently, by solvent dewaxing using toluene to associate with the lube oil components and methyl ethyl ketone, which acts as the wax precipitating agent.[3]

The wax product contains some unsaturation and slight amounts of aromatics. These impurities cause off colors, odor, and instability. For many years, adsorption was practiced using activated earth to decolorize the wax, but hydrogenation (hydrofinishing) is more commonly employed. It not only saturates color bodies but also reduces traces of aromatics and other possible carcinogenic compounds.[3] Food-grade waxes require such treatment and, in many cases, all paraffin wax is so treated regardless of final use. The hydrofinishing is carried out in an adiabatic, fixed-bed reactor.

Uses[3]

Paraffin wax continues to be used for candles, since there is a significant demand for decorative candles. But the major use today is for coating paper board for drinking cups and cartons for milk, juices, and other foods. Waxed paper continues as an important product in the home for microwave use and in pastry preparation. It is also used for treating paper bags. Wax is applied directly to many product items to preserve freshness and reduce water loss. Finally, more costly natural waxes such as carnauba wax is often blended with paraffin and microcrystalline wax to produce an excellent liquid or solid polish.

Typical Catalysts

Nickel-on-Kieselguhr or silica-alumina (50 wt% Ni) extrudates
Nickel-on-alumina (10–30% Ni) extrudates
Suppliers: Engelhard, United Catalysts, Celanese
Operating Conditions: 250–350°C, 70–125 bar

Microcrystalline Wax

Microcrystalline wax is obtained from crude-storage tank bottoms and vacuum residual by extraction and dewaxing. In each case, the crude material is deasphalted prior to extraction and dewaxing.

The very small crystals of wax attract oil, and the oily product must be deoiled, although some oil will remain.

Hydrofinishing to remove color is accomplished at conditions similar to that used for paraffin wax. The lower melting and softer grade is used for laminations in paper manufacture, and the harder and higher melting wax is used in producing coatings and polishes.[1]

Typical Catalysts

If the residues from which the microcrystalline waxes are produced contained sulfur compounds, a nickel-tungsten catalyst is suggested because of its sulfur resistance. If organosulfur compounds or H_2S is not present, use same catalyst as for paraffin wax.

SYNTHETIC WAXES

The major synthetic wax product requiring hydrogenation is that produced by the Fischer–Tropsch process, mainly in South Africa. Carbon monoxide and hydrogen obtained from coal are reacted to produce various synthetic fuels and oils as well as a wax that basically is a polymethylene polymer. The heavy wax fraction requires decolorization by hydrogenation. It is used in adhesives and hot-melt coatings in conjunction with other waxes to increase the melting point and hardness.

A nickel catalyst, as described for paraffinic wax, is used. Operating conditions are reported 200–250°C and 50–70 bar.[3]

Lower-molecular-weight polyethylenes (>10,000) are often used as a synthetic wax in admixture with paraffin wax[2] to produce desired properties for regular paraffin-wax products. A major use is as a mold and extrusion aid and color dispersant for plastics.[1] Hydrogenation is usually not required.

12.9.3 PETROLEUM RESINS

Steam cracking units for ethylene or propylene manufacture produce a heavy residue usually highly aromatic in character. Depending on plant economics this material can be polymerized by acid catalysis and then hydrogenated to eliminate color bodies caused by double bonds in the aliphatic portions of the molecule. The resulting product has been used in the manufacture of floor tiles. Similarly, catalytic-cracking residue can also be so employed where economics so indicate an advantage over alternate applications.

Typical Catalysts and Conditions

	Wt%, Ni	Form	Application	Condition
Nickel/Kieselguhr	50	Powder	Batch or continuous slurry reactor	200–150°C, 70–130 bar
Nickel/silica-alumina	50	Extrudates	Fixed-bed reactor	200–300°C, 30–150 bar

Suppliers: Engelhard, United Catalysts

12.9.4 SYNTHETIC RESINS

The name *synthetic resin* is sometimes used to refer to a variety of synthetic polymers in the molecular-weight range of natural resins in which they are admixed or totally replaced. Most of these so-called synthetic resins usually have little or no color problem and do not require hydrogenation, because very few double bonds are contained in the complex polymeric matrix.

12.10   SELECTIVE HYDROGENATION OF FATS AND OILS

Hydrogenation of fats and oils is one of the first commercial hydrogenation processes. Not many years after Sabtier demonstrated that double bonds in light hydrocarbons could be hydrogenated in the vapor phase using nickel or noble metal catalysts, W. Normann patented (1902) a liquid phase hydrogenation process for fats and oils. A plant was built in England in 1907, and Proctor and Gamble obtained rights to the Normann patent in 1911. Over the years, the production of edible fats and oils has soared, with vegetable sources now dominating the field formally held by butter and lard.

12.10.1   GENERAL BACKGROUND

The chemistry of the production of margarine, shortening, confectionery fat, and cooking oil is highly complex. Like crude petroleum and petroleum products, fats and oils are mixtures of compounds. Natural fats are mixtures of triglycerides, which are esters of fatty acids and glycerine.

$$R_1COOCH_2$$
$$|$$
$$R_2COOCH$$
$$|$$
$$R_3COOCH_2$$

where R_1C, R_2C, and R_3C represent the carbon chain of the original fatty acid. These are primarily straight chains with even numbers of carbon atoms.

As with other natural sources such as petroleum, the natural fat or oil is a complex mixture with varying chain lengths and degree of saturation (saturated and mono, di and polyunsaturated). Chain lengths vary between 4 and 24 carbon atoms, with lengths of 16 and 18 being the most common. Table 12.16 provides a summary of fatty acid compositions of various vegetable fats and oils.

TABLE 12.16
Typical Fatty Acid Compositions of Some Edible Plant Fats and Oils

Fatty Acid	Coconut	Palm Kernel	Palm	Cottonseed	Sunflower	Corn	Soybean	Canola	Rapeseed
Saturates									
6–0	1								
8–0	8	3							
10–0	6	4							
12–0	47	48							
14–0	18	16	1	1					
16–0	9	8	45	21	7	11	11	4	3
18–0	3	3	4	3	5	2	4	2	1
20–0									
Monounsaturates									
16–1				1					
18–1	6	16	40	19	19	28	24	61	13
20–1									7
22–1								1	52
Polyunsaturates									
18–2	2	2	10	54	68	58	54	22	14
18–3				1	1	1	7	10	10

Reprinted by permission: *Fats and Oils Hydrogenation Manual,* Engelhard Corp., Iselin, NJ., 1992.

TABLE 12.17
C-18 Fatty Acids

Fatty Acid	Structure	MP, °C	BP, °C
Stearic (octadecanoic)	$CH_3(CH_2)_{16}COOH$	69–70	383
Elaidic (trans-9-octadecenoic)	$CH_3(CH_2)_6CH_2$ and $CH_2(CH_2)_6COOH$ (trans double bond)	44–95	234
Oleic (cis-9-octadecenoic)	$CH_3(CH_2)_6CH_2$ and $CH_2(CH_2)_6COOH$ (cis double bond)	4 (pure) 16 (commercial)	286 (pure)
Linoleic (cis,cis 9,12-octadecadienoic)	$CH_3(CH_2)_4$ … $(CH_2)_7COOH$ (two cis double bonds with central CH_2)	–5	202
Linolenic (cis,cis,cis-9-12-15-octadectrienoic)	CH_3CH_2 … CH_2 … CH_2 … $(CH_2)_7(COOH)$ (three cis double bonds)	–11	230

12.10.2 PURPOSE OF HYDROGENATION OF VEGETABLE OILS

Hydrogenation of vegetable oils accomplishes two major purposes:

1. Increase the melting point so as to produce products such as margarine and shortening
2. Increase the stability to prevent rancidity

The C-18 fatty acids are the dominant components of the triglycerides in soybean, corn, and sunflower oils. Table 12.17 lists these acids, along with common and systematic names and structural formulas. Table 12.18 lists the iodine value and melting point of C-18 fatty acids and corresponding triglycerides.

TABLE 12.18
Properties of C-18 Fatty Acids

			Free Acids		Triglycerides	
Acid	Symbols	C=C	IV	MP, °C	IV	MP, °C
Linolenic	LN $C_{18\text{-}3}$	3	273	−11	261.6	−24
Linoleic	LO $C_{18\text{-}2}$	2	181	− 5	173.2	−13
Oleic	O $C_{18\text{-}1}$	1	90	16	86.0	5
Elaidic	E $C_{18\text{-}1}$	1	90	44	86.0	42
Stearic	S $C_{18\text{-}0}$	0	0	70	0	73

Source: Fats and Oils Hydrogenation Manual, Engelhard Corp., Iselin, NJ 1992, reprinted by permission.

Fatty acids of natural triglycerides with unsaturation all occur in the cis isomeric form, and multiple double bonds are isolated on either of side of a methylene group. However, hydrogenation can cause isomerization to a trans configuration and also conjugated double bonds. In addition, hydrogenation saturates double bonds selectively when operating conditions are controlled and a selective catalyst is used.

Table 12.18 illustrates the effect of saturation on iodine value and melting point. Since iodine value is the grams of iodine that react with one gram of fat so as to saturate all existing double bonds, it is direct measure of unsaturation at any point in the hydrogenation process. And, of course, it is a measure of the unsaturation of the virgin oil or its components. It is clear from Table 12.18 that the melting point increases with increased saturation (lower IV) and also with isomerization of cis to trans configuration (oleic → elaidic). Thus, by proper choice of catalyst and operating conditions, it is possible to arrive at the desired properties for the production of a given product.

Hydrogenation or partial hydrogenation also stabilizes the resulting product so as to avoid rancidity due to oxidative reactions during storage. These reactions break the hydrocarbon chain, producing volatile unsaturated aldehydes that cause unpleasant flavors and odors.[1] Since a methylene group between a double bond is reactive, it is thought that these are sites for the oxidation to occur. Since linolenic acid has three double bonds and two related methylene groups, it seems logical that this fatty acid is the main culprit, and indeed it is. However, there are additional hypotheses involving other factors.[2] But much-improved shelf life is attained by lowering the linolenic acid content through hydrogenation. The difference in relative oxidation rates of the several C_{18} fatty acids[1] confirms this empirical conclusion (stearic = 1, oleic = 10, linoleic = 100, linolenic = 200).

12.10.3 PROCESS DETAILS

Chemistry

The chemical reactions that occur in edible oil hydrogenation are profoundly affected by transport processes commonly associated with gas-liquid-solid catalyst systems. These processes must be recognized to understand the ultimate results of the reaction process. Seven categories of physical and chemical steps have been suggested as a means of defining the reaction system.[3]

1. Transfer of Reactants to Outer Surface of Catalyst
   - Hydrogen transfers from gas phase to liquid, with a portion of it dissolving in the oil. In commercial batch hydrogenation, the rate of hydrogen dissolution is the rate-controlling step. This rate increases with the degree of agitation and with an increase in hydrogen pressure.
   - The oil phase molecules (triglycerides) transfer from the liquid to the catalyst surface, as does the dissolved hydrogen. Hydrogen, being a small molecule, transfers rapidly, but the net rate is hindered by the characteristically low concentration of dissolved hydrogen. By contrast, various triglyceride molecules may be present in fairly high concentrations, but the large size and shape can impose a hindering effect.

2. Transfer in Catalyst Pores
   - Again, hydrogen diffuses readily in the catalyst pores, but the net rate is impeded by the low concentration of dissolved hydrogen. Pore structure and shape play an important role in the diffusion of the various triglycerides in the pores. Molecular size and shape effect diffusion and can affect selectivity. It is known that narrow-pore catalyst reduces geometrical isomerization,[8] and trans isomers are believed to diffuse more easily than cis isomers.[3]

3. Chemisorption and Reaction on the Active Catalyst Surface
   - Hydrogen, as in other hydrogenations, experiences dissociative chemisorption and then reacts with a double bond of the fatty acid portion of the triglyceride, which is chemisorbed on adjacent vacant nickel atoms. The configuration of nickel is such that the distance between nickel atoms is approximately the same as the two carbon atoms of the unsaturated entity. Hydrogenation occurs via the half-hydrogenated state as generally accepted for most hydrogenations.

The reaction system is complicated by the fact that the natural triglycerides are mixtures of glycerine—fatty acid esters even within a single carbon number. The large fatty acid portion of the molecule can be saturated, monounsaturated, or polyunsaturated (conjugated or unconjugated), and a single molecule can have all unsaturated chains, all saturated, or mixtures of the two, as illustrated in Figure 12.15 for soybean oil. Numerous studies have been conducted through the years to better understand this complex system. In more recent years, methyl esters of individual fatty acids have been studied as a means of isolating the hydrogenation behavior of acid types. These studies, along with those of natural triglyceride mixtures, have led to some useful conclusions:[3,4,5]

- Polyunsaturated portions of triglycerides are preferentially adsorbed.
- If hydrogen concentration at the catalyst surface is high, adsorbed monounsaturated groups will be hydrogenated rapidly.
- If hydrogen concentration at the catalyst surface is low, a polyunsaturated group can displace the mono unsaturate, and it will be selectively hydrogenated.

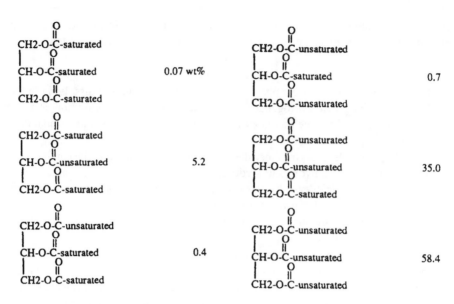

FIGURE 12.15   Typical Distribution of Triglycerides in Soybean Oil (weight percent). Reprinted by permission: *Fats and Oils Hydrogenation Manual,* Engelhard Corporation, Iselin, NJ, 1992.

- Double bonds on fatty-acid chains of triglycerides in the α position (exterior) are more readily hydrogenated than chains in the ß (interior) position.[6,7]
- In addition to hydrogenation, isomerization occurs. The half-hydrogenated state can proceed by further reaction to the fully-hydrogenated state at higher concentrations of adsorbed hydrogen. But, at lower hydrogen concentrations, the half-hydrogenated state is more apt to decompose in a manner that produces both geometrical and positional isomers. Figures 12.16 and 12.17 represent these several reaction paths based on the classical Horiuti and Polanyi general theory of the half-hydrogenated state.[9]
- After hydrogenation of a double bond, the strong attraction of the triglyceride at that region of the molecule is greatly limited, and desorption occurs. The molecule then travels through the pores to the bulk oil. The shape of the molecule changes upon hydrogenation, and it is reasonable to conclude that saturated molecules diffuse through the pores more easily than unsaturated molecules. Differences in shape between various isomers may also affect transport in the pores.[3]

Catalyst Types and Suppliers

Although fat and oil hydrogenation has been practiced in fixed beds, most plants use suspended catalyst systems (slurries) operated continuously or batchwise. Batch slurry reactors are, by far, the most widely used.

Nickel is the catalyst of choice by most producers because of its low cost, activity, and relatively lower sensitivity to many poisons compared to copper or palladium catalysts.[2] Nickel is most often supported on Kieselguhr, but various proprietary supports are also offered. Silica-alumina, for example, provides acid sites that would favor isomerization reactions.

The supported catalyst particles are in the range of 1–50 μm, with a major portion in the 10 μm (100,000 Å) range.[2,10] Nickel crystallite sizes distributed throughout the catalyst are 50–100 Å in size, and total particle surface area is 200–600 m²/g with 20–30% as active nickel.[10] The catalyst must be filtered from the product, and the particles must be in a size range that will favor efficient filterability.

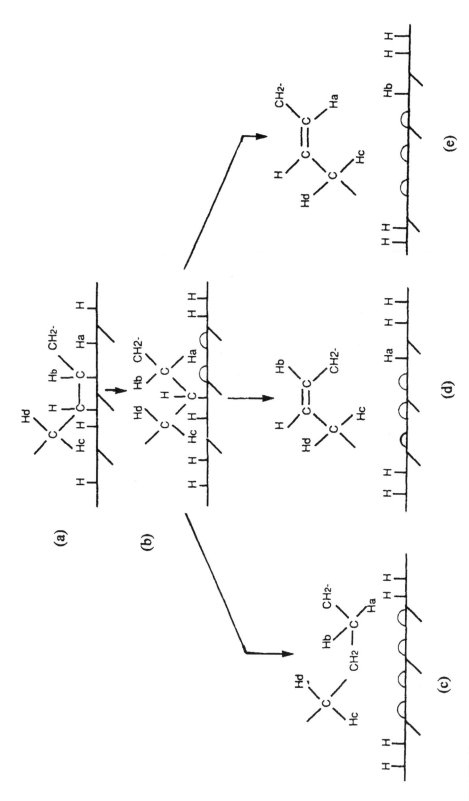

FIGURE 12.16 Geometric Isomerization (half circles represent vacated nickel sites). (a) chemisorbed fatty-acid group at double bond, (b) half-hydrogenated state, (c) complete hydrogenated state with addition of another hydrogen from the surface (d) description of original cis-molecule by reattachment of hydrogen, H_a, (e) cis-to-trans isomerization by rotation of the molecule and loss of H_b to the surface, followed by desorption. Adapted by permission, *Fats and Oils Hydrogenation Manual*, Engelhard Corporation, Iselin, NJ, 1992.

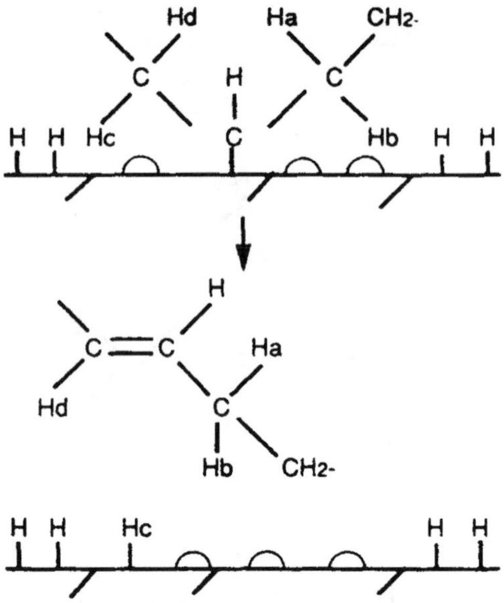

FIGURE 12.17   Positional Isomerization (half circles represent vacated nickel sites). Half-hydrogenated state losses, either Hc or Hd, to the active surface causing a shift in the double bond from its initial position. Reprinted by permission: *Fats and Oils Hydrogenation Manual,* Engelhard Corporation, Iselin, NJ, 1992.

Typical Catalysts

   Nickel content: 18–25%
   Support: 5–25% (Kieselguhr or proprietary)
   Hardened oil: remainder

Suppliers

   United Catalysts, Engelhard, Celanese, Synetix

   The catalyst is dry reduced by the manufacturer and encapsulated in a protective hardened oil (50°C mp). Several forms are available (droplets, pastilles, flakes or pellets), all with the protective coating. When introduced into the reactor, the coating melts and releases the activated catalyst particles. Stabilized catalyst without hardened-oil protectant is available when there is a need to avoid the inclusion of hardened oil, though small in quantity, in the final product. Such stabilized catalysts have been only superficially oxidized after reduction so as to eliminate the pyrophoric property of the reduced surface nickel.[10]

   Catalyst manufacturers tailor their catalysts for particular uses and should be consulted for recommendations. An example of a special-purpose catalyst is a partially sulfided product that reduces hydrogenation activity, thereby encouraging isomerization of cis-to-trans isomers. High trans-isomer content is important in margarine blend stocks and confectionery fats where a rapid decline in solid content with an increase in temperature is desired, so that the product melts rapidly in the mouth. The partially sulfided catalyst is referred to as a *sulfur-promoted catalyst* even though its activity per unit surface area has been reduced. This loss of activity on a unit area or mass basis is easily compensated for by increasing the catalyst loading to the reactor. Other promoters in amounts as high as 2% have been studied, including palladium, copper, chromium, and platinum, all of which enhanced the activity.

   Pore size and nickel distribution are also given special considerations in catalyst design. As can be seen in Figure 12.18, a pore much larger than the triglyceride molecule size of 15–20 Å

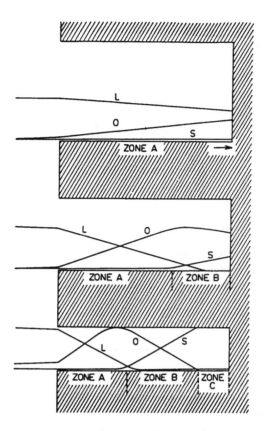

FIGURE 12.18 Qualitative Representation of Reactant Concentrations in Pores of Different Widths (large medium, and small). L = linoleic group, O = oleic group, S = stearic group. Reprinted by permission: Conerien J. W. E., *J. Amer. Oil Chemists Soc.* 53, 386 (1976), American Oil Chemists Society. Zone A constitutes the entire length of large pores and smaller portions of smaller pores, favors hydrogenation or isomerization of one double bond. In Zone B, nearly all of the polyunsaturates have been hydrogenated to oleic groups, and stearic groups are formed by hydrogenation of the oleic groups. Zone C illustrates characteristics of narrow pores, a region of essentially no reaction.

will produce, at most, the hydrogenation of only one double bond. The smaller the pores, the greater opportunity for further hydrogenation. Although a typical catalyst will have pore size distributions ranging from less than 20–200 Å in size, manufacturers are able, by proprietary means, to change the distribution to fit the desired process result.

Catalyst Deactivation

Catalyst poisons that can be present in vegetable oils include organic sulfur and phosphorous compounds, halogenated compounds, free fatty acids, fatty-acid oxides, soaps, and water.[2] Hydrogen must be free of other potential poisons such as H_2S and CO that result from malfunctions in the hydrogen production unit. As is the case for all nickel-based hydrogenations, sulfur compounds can interact strongly with nickel and significantly reduce total nickel activity. If nickel is highly dispersed on the support, the catalyst will be less sensitive to sulfur poisoning. Although some sites will be deactivated, many sites will remain unaffected.

Phosphatides (lecithins), which are mixtures of diglycerides of stearic, palmitic, and oleic acids connected to the choline ester of phosphoric acid, are natural constituents of both animal and vegetable oils. These phosphatides are normally removed in the bleaching step of the oil refining process, but small amounts can at times break through due to faulty operation. In such cases, pore mouths and exterior surfaces of the catalyst can be rapidly deactivated as the phosphatides cover these surfaces. The deactivation of the exterior and pore-mouth surfaces causes more action within the pores, resulting in higher saturation rates, higher melting points, and less isomerization.[10]

Other possible constituents of refined oil feed to the hydrogenator due to upsets in the refining process include soaps, free fatty acids, water, and chlorinated organics soaps. Fatty acids and fatty-acid oxides tend to be adsorbed at the pore mouths and can ultimately seriously inhibit the flow of reactants and products. Their effect is temporary, since removal can restore lost activity. Of course, chlorine, nitrogen, and arsenic compounds are permanent poisons when they degrade and become strongly attached to the nickel surfaces. Finally, steam formed from water in the feed can destroy activity by oxidizing nickel to nickel oxide. If water content is kept low, this situation need not be a major problem.

Catalyst Regeneration

The catalyst is often reused several times. In fact, if successful filtration devoid of air leakage is accomplished, start-up activity on the second use may be better than when the catalyst is fresh. However, continued use ultimately causes diminishing activity and selectivity.[2] Although sulfur poisoning over several batches may improve selectivity while reducing activity, continued reuse will finally require new catalyst. Regeneration is possible at this point, but the extra facility costs are often not justifiable.

Three routines for catalyst replacement and addition have been reported.[10]

1. Use the minimum amount of fresh catalyst each batch and then discard it.
2. Reuse catalyst a number of times and discard 10–20% of used catalyst, replacing it with fresh catalyst each time.
3. Use a flexible program of new and used catalyst, depending on performance of each previous batch.

The choice of procedure depends on the type of product and stringency of product specifications.

Process Units

The total process consists of a series of operations consisting of oil extraction, caustic treatment to remove gum, and then bleaching, followed by hydrogenation.

The industry is dominated by dead-end batch hydrogenators with a 2:1 or 3:1 liquid level to diameter and a top "dead" space for hydrogen accumulation. Mixing is accomplished by two impellers for the 2:1 designs and three for the 3:1 design. The tanks are made of carbon steel with stainless-steel heating and cooling coils. The top impeller is located below but close to the operating gas-liquid surface. It is a standard axial-flow impeller that creates a vortex action in the top-liquid area so as to entrain hydrogen from the dead space back into the liquid. The next one or two impellers can be radial-flow turbine agitators operating in a baffled tank with baffles extending toward the top liquid but below the liquid level so that a vortex can form in that region. Alternatively, specially designed high-efficiency impellers with broad blades and a low angle of attack can be used in place of the turbine agitators. They combine good shearing effect and high capacity for oil flows.

Although the dead-end hydrogenator just described dominates the industry, other reactors are offered as licensed processes. These include loop reactors by Buss and Lurgi, each of which employs a mixing jet that uses hydrogen to propel the oil into the reactor while mixing the liquid and circulating it through an external heat exchangers. Praxair offers a reactor design with high internal circulation downward through a draft tube.[10] These reactors are better mixed and provide better heat transfer by means of the external heat exchangers, but they are more difficult to start up and are more likely to produce catalyst fines if not properly operated.[2]

Process Kinetics

Many studies on this complex reaction system have been published, but practical working equations based on first-order, irreversible kinetics have been the most useful approach. Using the sequence of hydrogenation reactions as follows, useful rate equations, integrated equations, and selectivities can be expressed.[3,11]

$$
\begin{array}{ccccccc}
 & k_1 & & k_2 & & k_3 & \\
\text{UUU} & \rightarrow & \text{UU} & \rightarrow & \text{U} & \rightarrow & \text{S}
\end{array}
$$

| Triunsaturated fatty acid group Linolenic | Diunsaturated fatty acid group Linoleic | Monosaturated fatty acid group Oleic | Saturated fatty acid group Stearic |

$$\frac{dL1}{dt} = k_1(L1)$$

$$\frac{dL}{dt} = k_2(L)$$

$$\frac{dO1}{dt} = k_1(O1)$$

These equations can be solved for the compositions of acid groups at any time, and the iodine value (IV) can be calculated accordingly.[3,11]

$$L1 - L1_0 e^{-k_1 t} L1 = L1_0 e^{-k_1 t}$$

$$L = L1_0 \left(\frac{k_1}{k_2 - k_1}\right)(e^{-k_1 t} - e^{-k_2 t}) + L_0 e^{-k_2 t}$$

$$O1 = L1_0 \left(\frac{k_1}{k_2 - k_1}\right)\left(\frac{k_2}{k_3 - k_1}\right)(e^{-k_1 t} - e^{-k_2 t}) - L1_0 \left(\frac{k_1}{k_2 - k_1}\right)\left(\frac{k_2}{k_3 - k_2}\right)(e^{-k_2 t} - e^{-k_3 t})$$
$$+ L_0 \left(\frac{k_2}{k_3 - k_2}\right)(e^{-k_2 t} - e^{-k_3 t}) + O1_0 e^{-k_3 t}$$

$$\text{I.V.} = 2.6161\ L1 + 1.7321\ L + 0.8601\ O1$$

where Ll, L, and O1 are the mole percentages of linolenic linoleic and oleic, respectively.

IV is the iodine value, the number of grams of iodine that will react with 100 g of fat. Iodine reacts with double bonds, and thus the IV is a measure of the unsaturation of the oils. Refractive index measurements of reactor contents can be correlated with the iodine value.

Another useful relation expresses the overall rate of hydrogenation in terms of the change in iodine value with time.[3]

$$\frac{d(IV)}{dt} = -k(IV)$$

or integrating,

$$\ln[(IV)_0/(IV)_t] = kt$$

where $(IV)_0$ is the initial IV and $(IV)_t$ is the value at time t.

The integrated equation suggests that a semi-logarithmic plot of (IV), versus t is a straight line with a slope of -k. Actually, operation at lower temperatures tends to exhibit induction periods while the fresh catalyst is in the process of being activated.[3] At a higher temperature, complete activation is rapid, and a straight line results beginning essentially at zero (Figure 12.19). Reused catalysts do not exhibit induction periods. Other curve shapes have been discussed and explained.[3] Clearly, these plots are useful empirical tools for catalyst studies and operation analysis.

Various selectivity ratios have been defined in terms of rate constants as follows:

$$S_I = \frac{k_2}{k_3}, \text{ linoleic selectivity (preference for linoleic over oleic hydrogenation)}$$

$$S_{II} = \frac{k_1}{k_2}, \text{ linolenic selectivity (preference for linolenic over linoleic hydrogenation)}$$

$$S_i = \frac{\text{number of trans double bonds formed}}{\text{number of hydrogenated double bonds}}$$

The S_I selectivity ratio and S_i isomerization ratio are frequently used in quantifying the effects of catalyst characteristics and operating conditions on product characteristics.[3] Figure 12.20 shows the theoretical effect of various S_I values on linoleic, oleic, and stearic acid compositions at various levels of hydrogenation expressed in terms of iodine value reduction. At low values of S_I, stearic formation starts immediately. By contrast, at high values, stearic formation does not become significant until the linoleic is essentially all reacted.[12] Nickel catalysts usually have S_I values in the range of 5–100, depending on catalyst characteristics and operating conditions.

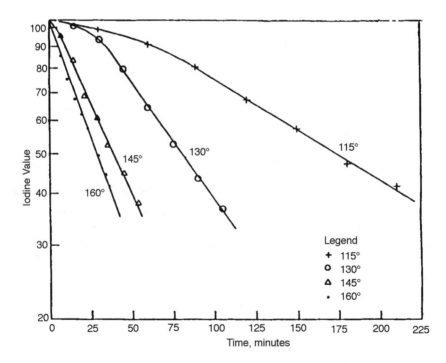

FIGURE 12.19 Effect of Temperature on Hydrogenation Rate: 60 psig, 0.07% Ni, 1175 rpm. Reprinted by permission: Albright, L. F., *Hydrogenation Proceedings of an AOCS Colloquium,* p. 16, 1987, American Oil Chemists Society.

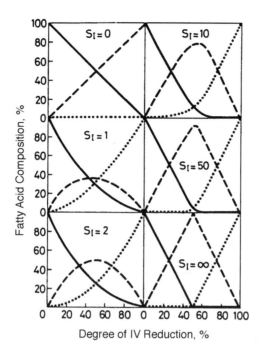

FIGURE 12.20 Theoretical Curves for Hydrogenation of Linoleic Esters at Various Values of S_I. Solid line = linoleic, dashed line = oleic, dotted line = stearic, and IV = iodine value. Reprinted by permission: Coenen, W. E., *J. Am. Oil Chemists Soc.*, 53, 384 (1976), American Oil Chemists Society.

The reaction system is strongly affected by mass transfer of hydrogen to the catalyst surface which, in turn, can alter selectivity, including that of trans isomerization. Table 12.19 summarizes the factors that influence selectivity by process conditions that affect dissolved hydrogen concentration at the catalyst surface.

An important goal in producing cooking oils and margarines is to reduce a large proportion of linolenic acid groups to assure oxidative stability while retaining as much linoleic as possible. Linoleic acid in the natural cis form is an essential fatty acid in human nutrition. Although the body makes important essential chemicals (e.g., prostaglandions from linoleic acid), it must obtain the necessary linoleic acid by ingesting fats containing linoleic acid groups.

Polyunsaturated fats have been recommended in the diet as a means for reducing serum cholesterol levels in the blood, believed to be a major cause of heart disease. More recently, cholesterol associated with high-density lipoproteins (HDLs) has been determined to be a favorable form of cholesterol, but the low-density lipoproteins (LDLs) have been targeted as the ones that can cause heart disease and thrombosis. Ingestion of saturated fats has been implicated as a major contributor to a higher proportion of LDL in the blood, although saturated fats vary in this effect[12] with dairy products (high), beef fat (less), and cocoa butter (slightly).

Conflicting conclusions have resulted from studies on trans fatty acids. These isomers formed in hydrogenation of fatty oils have in some studies been shown to have similar effects as saturated fats. Other studies suggest that the trans acids increase both LDL and HDL levels.[12] Some health organizations are recommending reduction in trans fatty acids in the diet, which might suggest using more liquid margarine and oils that have been hydrogenated only lightly.

Polyunsaturated fats in the diet, such as linoleic, have also been recommended for lowering cholesterol levels in the blood. More recently, the ratio of low-density lipids (LDL) to high-density

TABLE 12.19
Effects of Process Factors that Influence Hydrogen
Concentration at Catalyst Surface and Selectivities

		Effect on		
	Increase of	H_2 Conc.	S_1	S_i
Increases supply	Pressure	+	–	–
	Stirring intensity	+	–	–
Increases demand	Temperature	–	+	+
	Catalyst amount	–	+	+
	Catalyst activity	–	+	+
	Unsaturation of oil	–	+	+

Note: S_1 = preference of hydrogen of linoleic acid over oleic acid; S_i = number of trans double bonds formed per double-bond hydrogenation.

Source: Coenen, J. W. E., *J. American Oil Chem. Soc*. 53, 385 (1976). Reprinted by permission.

lipids (HDL) has been a much-used measure of healthy (less than a 3.5:5 ratio) serum cholesterol or a signal of the possibility of the ultimate development of clogging deposits in the circulatory system (high-value).

12.11   MISCELLANEOUS HYDROGENATIONS

Hydrogenation is the most used reaction in the synthesis of thousands of commodities and speciALITIES. Table 12.20 summarizes some typical significant products for which hydrogenation constitutes one of several steps in the synthesis route. Two important large-scale reactions, the production of 1, 4-butanediol and tetrahydrofuran from maleic acid or malate ester, and MIBK from acetone, are considered in more detail.

MALEIC ACID OR MALATE ESTER → 1,4-BUTANEDIOL AND TETRAHYDROFURAN

Tetrahydrofuran (THF) and its precursor, 1,4-butanediol, are important commercial products. THF is a valuable solvent, especially useful for poly(vinyl chloride). The diol, 1,4-butanediol is used to produce poly(butylene terephthalate) and certain cross-linked polyurethanes.[2]

The maleic route has several different reaction schemes, all of which involve hydrogenation.

Chemistry[1,3]

Method 1 (Davy Technology)

Maleic anhydride is esterfied with methanol to yield dimethyl maleate, which undergoes hydrogenation as follows:

TABLE 12.20
Miscellaneous Hydrogenations

cyclododecatrine 	cyclododecane	1,12-dodecanedioc precursor to polyamides and polyester	Hydrogenation to cyclododecane is done in liquid phase over Ni catalyst at 200°C and 10–15 bar. This is followed by oxidation to alcohol/ketone mixture, which is oxidatively cleaved to form the acid.
$\xrightarrow[\text{Ni}]{\text{H}_2}$	$C_{10}H_{20} \xrightarrow{O_2}$	$HOOC(CH_2)_{10}COOH$	

dimethyl terephthalate 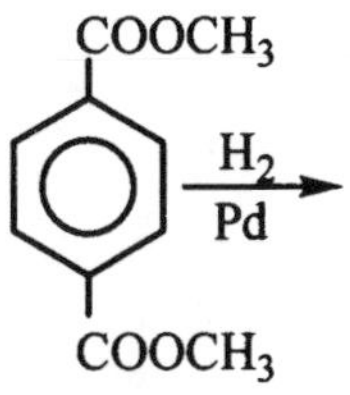	cyclohexane-1,4-dicarboxylic acid 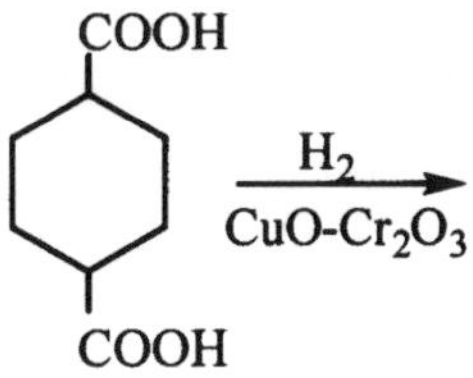	1,4-dimethylol cyclohexane 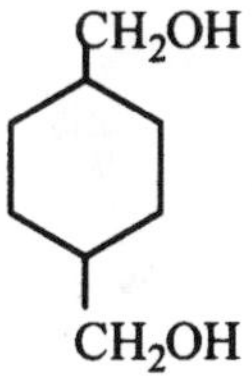	High heat of reaction of first step requires circulation of liquid reactant and product through exterior coolers. Uses: diol in production of polyurethanes, polyesters, and polycarbonates.

mesityl oxide	methyl isobutyl ketone (MIBK)	Methyl isobutyl carbinol is a by-product, the amount of which can be controlled. The Pd catalyst is selective to MIBK and is used in a reactive distillation. See also methyl isobutyl ketone.	This is an old but still used process for MIBK. The hydrogenation step is preceded by two homogeneous catalyzed reactions, aldol condensation o acetone to diacetone alcohol followed by its dehydration to mesityl oxide.
$(CH_3)_2C{=}CHCCH$ $\|$ O	$\longrightarrow$ $(CH_2)_2CHCH_2CCH_3$ $\|$ O		

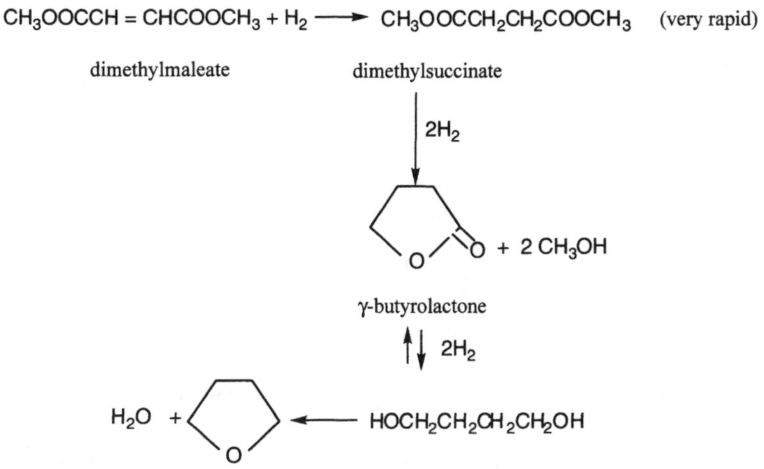

150–170°C, depending on the process
1–3 bar for vapor phase Cu, Ni, or Pd catalyst.

$$CH_3OOCCH = CHCOOCH_3 + H_2 \longrightarrow CH_3OOCCH_2CH_2COOCH_3 \quad \text{(very rapid)}$$

dimethylmaleate   dimethylsuccinate

$\downarrow$ 2H$_2$

$+ \ 2 \ CH_3OH$

γ-butyrolactone

$\updownarrow$ 2H$_2$

$H_2O \ + $ $\longleftarrow HOCH_2CH_2CH_2CH_2OH$

Method 2 (Lurgi/BP)

$$HOOCCH = CHCOOH + 5H_2 \rightarrow HOCH_2CH_2CH_2CH_2OH + H_2O$$

$$\text{maleic acid} \qquad\qquad\qquad \text{1,4-butanediol}$$

As above, 1,4-butanediol can be dehydrated to tetrahydrofuran.

Method 3 (duPont)

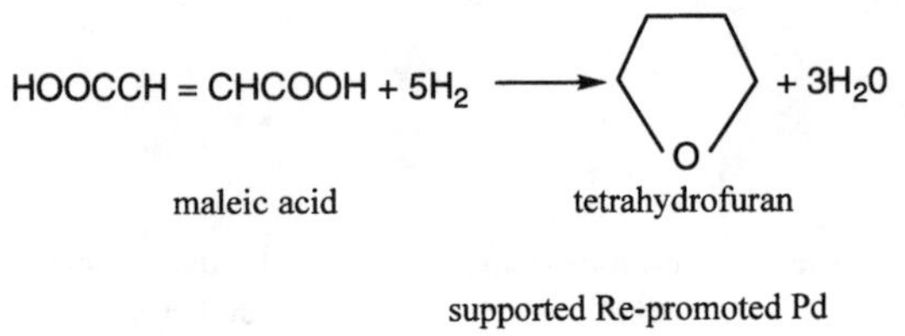

$$HOOCCH = CHCOOH + 5H_2 \longrightarrow \text{[ring structure]} + 3H_2O$$

maleic acid   tetrahydrofuran

supported Re-promoted Pd

This is a single-step process. Details are not published.

Process Unit: Davy Technology[1,3]

Maleic anhydride and excess methanol are mixed together at modest temperatures to produce the ester methyl maleate. This monoester is then esterfied further, using an acidic ion-exchange resin, to the di-ester. After purifying the dimethyl maleate by removing water and excess methanol, the dimethyl maleate stream is purified more completely by distillation.

The purified dimethylmaleate is then heated, and the vapor, along with hydrogen, is passed to a fixed-bed adiabatic reactor containing a supported copper hydrogenation catalyst. The reaction proceeds in rapid steps. Since the butyrolactone-to-butanediol reaction is at equilibrium, recovered butyrolactone can be recycled to extinction and the butanediol production increased. Selectivity to butanediol or tetrahydrofuran can be manipulated to some extent by changing operating conditions. Unwanted by-products are reported to be less than one percent.

Process Unit: Lurgi/BP

Maleic acid produced from maleic anhydride via a water scrubber is passed to a fixed-bed hydrogenation reactor. The overhead gases from the scrubber are incinerated. The crude hydrogenation product is then cooled and passed to a series of distillation units to produce commercial grade butanediol with a yield of 94 percent.[4]

ACETONE DIACETONE → METHYL ISOBUTYL KETONE

This is one of those reaction systems that certainly celebrate the ingenuity of catalyst development chemists and chemical engineers. The original process required three independent steps (see also Table 12.20). The first two of these involved homogeneous catalysis.

1. Aldol condensation of acetone to diacetone alcohol, catalyzed by an alkali
2. Acid catalyzed (H_2SO_4 or H_3PO_4) dehydration to mesityl oxide
3. Hydrogenation of olefinic bond in mesityl oxide to produce methyl isobutylketone

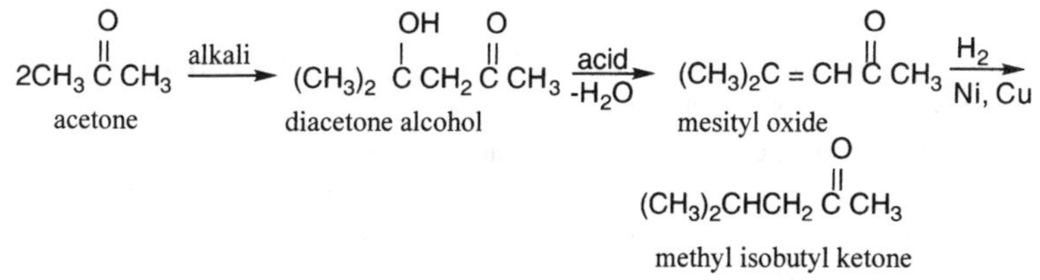

$$2CH_3 \overset{O}{\underset{\|}{C}} CH_3 \xrightarrow{\text{alkali}} (CH_3)_2 \overset{OH}{\underset{|}{C}} CH_2 \overset{O}{\underset{\|}{C}} CH_3 \xrightarrow[-H_2O]{\text{acid}} (CH_3)_2C = CH \overset{O}{\underset{\|}{C}} CH_3 \xrightarrow{\underset{\text{Ni, Cu}}{H_2}}$$

acetone   diacetone alcohol   mesityl oxide

$$(CH_3)_2CHCH_2 \overset{O}{\underset{\|}{C}} CH_3$$

methyl isobutyl ketone

This original process is still in use at some locations, but a relatively new one-step process employing a multifunctional catalyst has been widely adopted. It offers lower investment and operating costs and avoids undesired secondary products.

Chemistry and Catalyst (One-Step Process)

The condensation, dehydration, and hydrogenation steps shown above occur simultaneously on a new type of catalyst. It is described as an acidic cation-exchange macroporous resin to which palladium is fixed. The acidic sites catalyze both the condensation of acetone to diacetone alcohol (also catalyzed by acid) and the dehydration of the diacetone alcohol to mesityl oxide. The palladium component catalyzes the hydrogeneration of the oxide to methyl isobutyl ketone. Because of the intimate contact of the several catalytic sites, the reactions occur almost simultaneously.[5,6]

Since the several steps are not carried out in separate reactions but, rather, simultaneously, the formation of undesired side products from intermediates does not occur. Since methyl isobutyl ketone (MIBK) is less reactive, it is possible to continue condensation of mesityl oxide up to its equilibrium value, and it is removed rapidly via hydrogenation to MIBK. The cycle continues beyond what would occur if each step were brought to equilibrium separately. The result is a higher yield for the one-step process.[6]

Catalyst Type and Suppliers

The catalyst and processes are proprietary. Reported details and licensors or owners are:[2,3,6]

- RWE-DEA (was Deutsche Texaco):
  acidic ion-exchange resin impregnated with palladium
- Veba-Chemie:
  acidic ion-exchange resin impregnated with 0.05% palladium.
- Tokiuyama Soda:
  zirconium phosphate impregnated with 0.1–0.5% palladium
- Sumitamo Chemical:
  niobium and palladium presumably on ion-exchange resin
- Hülls AG
- Edeleanu

Process Unit (RWE-DEA Process)[2,5,6]

Acetone and hydrogen are fed concurrently in a trickle-phase multitubular reactor operating at 130–140°C and 30 bar. The reactor is temperature-controlled via circulating hot water. Hydrogen is fed in excess (~2:1 H_2-to-acetone). Once-through acetone conversion is 40 percent, and selectivity for the several processes is in the range of 92–95 percent. Side reaction products include small amounts of hexanes and diisobutyl ketone.

The raw product, containing about 35% MIBK, is distilled to remove light ends. Unreacted acetone is recycled. The MIBK and H_2O bottoms is separated into an aqueous phase and an organic phase. The aqueous phase is distilled as an azeotrope. The azeotrope is mixed with the organic phase and cooled to yield a water phase and a MIBK phase. The MIBK thus obtained is then distilled to remove high-boiling impurities.

As is the case with multitubular reactors in general, careful catalyst loading of each tube and pressure test confirmation of equal average bed void fraction are necessary. The issue is crucial in this particular case, since good contacting of reactants and products is essential for the success of this multi-simultaneous reaction system.

Process Kinetics and Reaction Scheme

A reaction scheme has been suggested for the bifunctional catalyst (Pt-HZSM-5) used in the one-step process as shown in Figure 12.21.[8] No detailed mechanism has been determined, although an early one-step catalyst study involving Pd on zirconium phosphate suggested the nature of the acidic active site.[9]

Initial activity of the Pt-HZSM-5 catalyst increases with Pt loading up to a certain value and then remains constant, at which point the limiting reaction step becomes the acid-catalyzed formation of the intermediate mesityl oxide.[8] One might expect that a useful commercial rate form might be a first-order expression in terms of acetone.

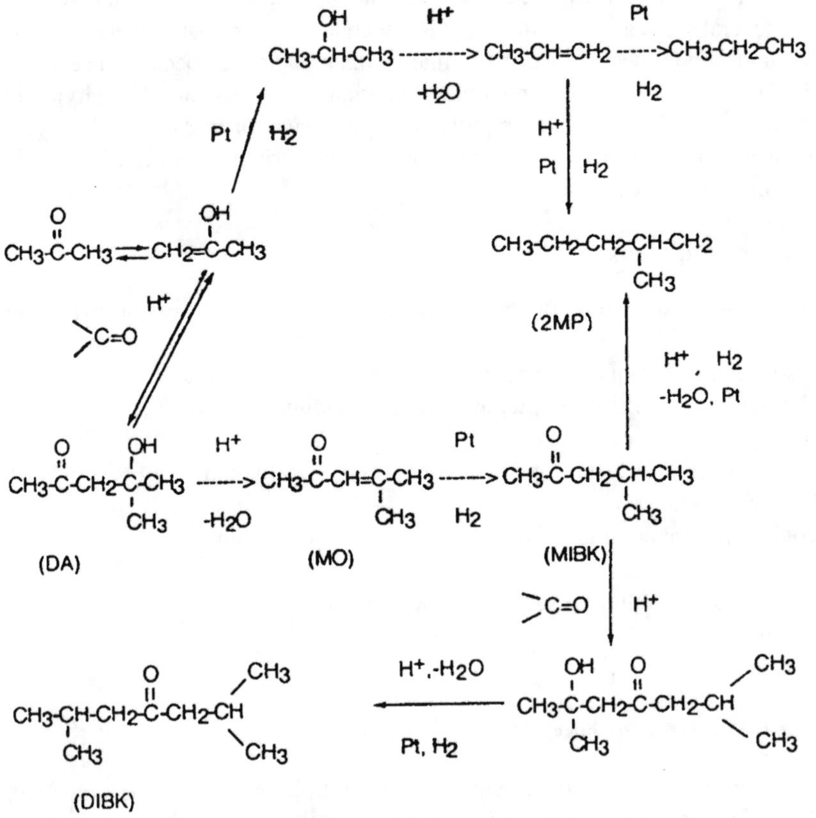

FIGURE 12.21 Proposed Reaction Scheme for One-Step MIBK Process. Reprinted from Melo, L.; Romb, E.; Dominguez, J. M.; Magnoux, P.; and Guisnet, M., in *Heterogeneous Datalysis and Fine Chemicals III: Studies in Surface Science,* Vol. 78, M. Guisnet et al., eds., p. 701, copyright 1993, with permission Elsevier Science. 2MP = 2-methylpentane, DA = doacetone alcohol, MO = mesityl oxide, MIBK = methylisobutylketone, DIBK = diisobutylketone.

REFERENCES, SECTION 12.1 (GENERAL BACKGROUND)

1. Rylander, P. N., in *Ullmann's Encyclopedia of Industrial Chemistry,* 5th ed., Vol. A13, p. 195, VCH, New York, 1989.
2. Rylander, P. N. in *Catalysis*, Vol. 4, p. 2, edited by J. R. Anderson and M. Boudart, Springer Verlag, Berlin, 1983.
3. Horriuti, J. and Polanyi, M., *Trans. Faraday Soc.*, 30, 663, 1164 (1934).
4. Molnar, A. and Smith, G. V. in *Hydrogen Effects in Catalysis,* edited by Z. Paal and P. G. Menon, p. 499, Marcel Dekker, New York, 1988.
5. Kieboom, A. P. G. and van Rantwijh, F. *Hydrogenation and Hydrogenalysis in Synthetic Organic Chemistry,* Delft University Press, 1977.
6. Schoon, Nils-Herman in *Hydrogen Effects in Catalysis,* Z. Paal and P. G. Menon, eds., p. 622, Marcel Dekker, Inc., New York, 1988.
7. Aper, E., Wichtendahl, R. and Deckwer, W. D. *Chem. Eng. Sci.,* 35, 217 (1980).
8. Christmann, K. R. in *Hydrogen Effects in Catalysis*, Z. Paal and P. G. Menon, p. 3, Marcel Dekker Inc., New York, 1988.
9. Geus, J. W. in *Hydrogen Effects in Catalysis,* edited by Z. Paal and P. G. Menon, p. 85, Marcel Dekker, Inc., New York, 1988.
10. Geus, J. W. in *Hydrogen Effects in Catalysis,* edited by Z. Paal and P. G. Menon, p. 196, Marcel Dekker, Inc., New York, 1988.
11. Conner, N. C., in *Hydrogen Effects in Catalysis,* edited by Z. Paal and P. G. Menon, p. 311, Marcel Dekker, Inc., New York, 1988.
12. Acres, G. J. K., Bird, A. J. and Davidson, P. J., *Chem. Eng.* (London), 145, March, 1974.
13. Scholten, J. J. F., Pijpers, A. P. and Hustings, A. M. L., *Catal. Rev.-Sci. Eng.,* 27 (1), 151–206, (1985).
14. Boitiaux, J. P., Cosyns, J. and Verna, F. in "Catalyst Deactivation, 1987," edited by E. Delman and G. F. Froment, *Studies in Surface Science and Catalysis,* 34, 105 (1987).
15. Bartholomew, C. H. in "Catalyst Deactivation, 1987," edited by E. Delman and G. F. Froment, "*Studies in Surface Science and Catalysis,* 34, 81 (1987).
16. Bartholomew, C. H., and Agrawal, P. K., *Advances in Catalysis*, 31, 135 (1982).
17. Pajon, G. M. and Telchner, S. J., in "Catalytic Hydrogenation," edited by L. Cerveny, *Studies in Surface Science and Catalysis,* 27, 277 (1986).

REFERENCES, SECTION 12.2 (HYDROGENATION OF AROMATICS)

1. Franck, H. G. and Stadelhofer, J. W., *Industrial Aromatic Chemistry,* Springer Verlag, New York, 1988.
2. Rylander, P. N., in *Catalysis: Science and Technology,* F. R. Anderson and M. Bonrdart, eds., Vol. 4, Springer Verlag, New York, 1983.
3. Greenfield, H., *J. Org. Chem.*, 29, 3082 (1964).
4. Rylander, P. N., Hasbrouck, L., and Karpenko, I., *Ann. N.Y. Acad. Sci.* 214, 100 (1973).
5. Mercker, H. J. in *Ullmann's Encyclopedia Industrial Chemistry,* 5th ed., Vol. A2, p. 10, VCH, New York, 1985.
6. Amini, B. in *Kirk-Othmer Encyclopedia of Chemical Technology,* 4th ed., Vol. 2, p. 432, Wiley, New York, 1992.
7. Narayanan, S. and Unnikrishman, R. P., *J. Chem. Soc. Faraday Transaction,* 93 (10), 2009, (1997).

REFERENCES, SECTION 12.2.2 (HYDROGENATION OF BENZENE)

1. Van Asselt and Van Krevelen, W. *Chem. Eng. Sci.,* 18, 471 (1963).
2. Franck, H. G. and Stadelhofer, J. W., "Industrial Organic Chemistry," Springer-Verlag, New York, 1988.
3. Eastman, A. D. and Mears, D. in *Kirk-Othmer Encyclopedia of Chemical Technology,* Vol. 13, p. 832, Wiley, New York, 1995.

4. Aben, P. L., Plateeuw, J. C. and Stouthamer, B., *Proc. 4th Intl. Congr. Catal.,* Vol. 1, p. 395, Akademia, Kiado, Budapest, 1971.
5. Stanislaus, A. and Cooper, B. H., *Catal. Rev. Sci. Eng.* 36, (1) 75 (1994).
6. Van Meertten, R. Z. C., Verhaak, A. C. M. and Coenen, J. W. E., *J. Catal.* 44, 217 (1976).
7. *Hydrocarbon Proc.,* "Petrochemical Processes–95," March, 1995.
8. Kehoe, J. P. and Butt, J. B., *J. Appl. Chem. Biotechnol.,* 22, 23 (1972).
9. Kalechits, I.V., Lipovich, V.G. and Vykhovanets, V.V., *Dokl. AN SSSR,* 138, 381 (1961) as reported by Ref. 5.

REFERENCES, SECTION 12.2.3 (HYDROGENATION OF BENZOIC ACID)

1. Franck, H. G. and Stadelhofer, J. N., *Industrial Organic Chemistry,* Springer-Verlag, New York, 1988.
2. Luekeke, V. D. in *Encyclopedia of Chemical Processing and Design,* J. J. McKetta, Exec. Ed., Vol. 1, p. 84, Marcel Dekker, New York, 1978.
3. Rylander, P. N., in *Catalysis: Science and Technology,* J. R. Anderson and M. Bonrdart, eds., Vol. 4, p. 9, Springer-Verlag, New York, 1983.

REFERENCES, SECTION 12.2.4 (HYDROGENATION OF NAPHTHALENE)

1. Mason, R. I. in *Kirk-Othmer Encyclopedia of Chemical Technology,* 4th ed., Vol. 16, p. 963, Wiley, New York, 1995.
2. Colin, S., in *Ullmann's Encyclopedia of Industrial Chemistry,* 5th ed., Vol. A17, p. 6, VCH, New York, 1991.
3. Franck, H. G. and Stadelhofer, J. W., *Industrial Organic Chemistry,* Springer-Verlag, New York, 1988.
4. Rylander, R. N., in *Catalysis: Science and Technology,* J. R. Anderson and M. Bourdant, eds., Vol. 4, p. 9, Springer-Verlag, New York, 1983.
5. Patzer, J. F. II, Farraudo, R. N. and Montagna, A. A., *Ind. Eng. Chem. Res. Dev.,* 18 (4), 625 (1979).
6. Sundaram, K. M., Katzer, J. R. and Bischoff, K. B., *Chem. Eng. Commun.,* 71, 53 (1988).

REFERENCES, SECTION 12.2.5 (HYDROGENATION OF PHENOL)

1. Franck, H. G. and Stadelhofer, J. W., *Industrial Aromatic Chemistry,* Springer-Verlag, New York, 1988.
2. Chenier, P. J., *Survey of Industrial Chemistry,* 2nd ed., VCH, New York, 1992.
3. Fisher, W. B. and Van Peppen, J. R. in *Kirk-Othmer Encyclopedia of Chemical Technology,* 4th ed., Vol. 7, p. 853, Wiley, New York, 1993.
4. Galvagno, S., Donato, A., Neri, G. and R. Pietropaolo, R. *J. Chem. Tech. Biotechnol.,* 51, 145 (1991).
5. Mahata, N. and Vishwanathan, V., *J. of Molecular Catalysis Chem.,* 120 (1–3), 267 (1997).
6. Shin, E.-J. and Keane, M. A., *J. of Catalysis,* 173, 450 (1998).
7. Neri, G., Visco, A. M., Donato, A., Milone, C., Malentacchi, M., and Gubitosa, G., *Appl. Catal. A.,* 110, 49 (1994).

REFERENCES, SECTION 12.3 (HYDROGENATION OF HETEROCYCLIC COMPOUNDS)

1. Manley, D. G., U.S. Patent 3,021,342 to Quaker Oats.
2. Franck, H. G. and Stadelhofer, J. M., *Industrial Organic Chemistry,* Springer-Verlag, New York, 1988.
3. Godawa, C., Gaset, A., Kalck, P. and Maire, Y., *J. Molecular Catalysis* 34, 199 (1986).

4. *Engelhard Catalysts,* Engelhard publication, 1977.
5. *General Hydrogenation Catalysts,* Chemie Publication, undated.
6. McKillip, W. J., Collin, G. et al. in *Ullmann's Encyclopedia of Industrial Chemistry,* 5th ed., Vol. A12, p. 119, VCH, New York, 1989.

REFERENCES, SECTION 12.4 (HYDROGENATION OF ALIPHATIC UNSATURATES)

1. LePage, J. F. et al, *Applied Heterogeneous Catalysis,* Technip, Paris, 1987.
2. *Engelhard Catalysts and Precious Metal Chemicals Catalog,* Engelhard Corporation, Newark, N.J. (1985).
3. United Catalysts literature.
4. Knid, L., Winter, O., and Stork, K., *Ethylene,* Marcel Dekker, New York, 1980.
5. Royer, D. J. in *Ulmann's Encyclopedia of Industrial Chemistry,* 5th ed., Vol. A10, p. 83, VCH, New York, 1987.
6. Nishimura, S., Sakamoto, H. and Ozawa, T., *Chem. Lett.,* 855 (1973).
7. Olah, G. M. and Molnar, A., *Hydrocarbon Chemistry,* Wiley, New York, 1995.
8. Zelinsky, N. D. in *Studies is Surface Science and Catalysis,* Vol. 27, p. 1, Elsevier, New York, 1986.
9. Godinez, C., Cabanes, A. L. and Villora, G., *Can J. Chem. Eng.* 74 (1), 84 (1996).
10. Furlong, B. K., Hightower, J. W., Chan, T. Y. L., Sarkany, A., and Guczi, L., *Appl. Catal.* A 117, 41 (1994).
11. Asplund, S., *J. Catal.* 158, 267 (1986).
12. Schbib, N. S., Garcia, M. A., Gigola, C. E. and Errazu, A. F., *Ind. Eng. Chem. Res.,* 35, 1496 (1996).
13. Goetz, J., Murzin, D. Y. and Touroude, R. A., *Ind. Eng. Chem. Res.* 35, 703 (1996).

REFERENCES, SECTION 12.5 (HYDROGENATION OF NITRILES)

1. Rylander, P. N. in *Catalysis: Science and Technology,* J. R. Anderson and M. Boudart, eds., Vol. 4, p. 1 Springer Verlag, Berlin, 1983.
2. Wolf, U. and Pasek, J. in "Catalytic Hydrogenation," *Studies in Surface Science and Catalysis,* L. Cervany, ed., Vol. 27, p. 105, Elsevier, New York, 1986.
3. de Bellefon, C. and Fouilloux, P., *Catal. Rev.: Sci. Engr.,* 36, 459 (1994).
4. Tricotte, M. G. and Johnson, T. A. in *Kirk-Othmer Encyclopedia of Chemical Technology,* 4th ed., Vol. 2, p. 369, Wiley, New York, 1992.
5. Schilling, S. L. in *Kirk-Othmer Encyclopedia of Chemical Technology,* 4th ed., Vol. 2, p. 482, Wiley, New York, 1992.
6. Greenfield, H., *I & EC Prod. Res. Dev.,* 4, 143 (1967).
7. Luedeke, V. D. in *Encyclopedia of Chemical Processing and Design,* J. J. McKetta, ed., Vol. 26, p. 222, Marcel Dekker, New York, 1987.
8. Friedlin, L. Kh. and Sladkova, T. A., *Russ. Chem. Rev.,* 33, 6 (1964).
9. Bartalini, to Monsanto Chemical Co., U.S. Patent 3831305, June, 1974.
10. Frank, D. and Reck, R. A., *Ullmann's Encyclopedia of Industrial Chemistry,* 5th ed., Vol. A2, p. 22, VCH, New York, 1985.
11. Watts, W. L., Brennen, M. E. and Yeakey, E. L., *Encyclopedia of Chemical Processing and Design,* J. J. McKetta, ed., Vol. 3, p. 144, Marcel Dekker, New York, 1977.
12. Casey, J. P., in *Kirk-Othmer Encyclopedia of Chemical Technology,* 4th ed., Vol. 2, p. 405, Wiley, New York, 1992.
13. Klimmek, H., *J. Am. Oil Chem. Soc.,* 61, 200 (1984).
14. Joly-Vurllemin, C., Gavory, D., Cordier, G., De Bellefon, C. and Delmas, H., *Chem. Eng. Sci,* 49 (24A) 4839 (1994).

15. Philippe, M. in "Heterogeneous Catalysis and Fine Chemicals: Studies in Surface Science and Catalysis," Vol. 78, 291, Elsevier, New York, 1993.
16. Gavroy, D., Joly-Vuillemin, C., Cordier, G., Fouilloux, P. and Delmas, H., *Catal. Today*, 24 (1–2), 103 (1995).
17. Friedli, F. E. and Gilbert, R. M., *J. Amer. Oil Chemists Soc.*, 67 (1) 48 (1990).
18. Barrault, J., Brunet, S., Suppo-Esseyem, N., Piccirilli, A. and Quimon, C., *J. Amer. Oil Chemists Soc.*, 71 (11), 1231 (1994).
19. Galvagno, S. *J. Mol. Catal.*, 58, 215 (1990).

REFERENCES, SECTION 12.6 (HYDROGENATION OF NITROAROMATICS)

1. Haber, F., *Z. Electrochem.*, 4, 506 (1898).
2. Stratz, A. M. in "Catalysts of Organic Reactions," J. R. Kosak, ed., p. 335, Marcel Dekker, New York, 1984.
3. Schilling, S. L. in *Kirk-Othmer Encyclopedia of Chemical Technology*, 4th ed., Vol. 2, p. 482, Wiley, New York, 1992.
4. Yao, H. C. and Emmet, P. H., *J. Am. Chem. Soc.*, 83, 796, 799 (1961).
5. Gharda, X. H. and Sliepcevich, C. M., *Ind. Eng. Chem.*, 52, 417 (1960).
6. Vogt, P. F. in *Ullmann's Encyclopedia of Industrial Chemistry*, 5th ed., Vol. A2, p. 44, VCH, New York, 1985.
7. Petrov, L., Kumbilieva, K. and Kirkov, N., *Appl. Catal.*, 59 (1), 30 (1990).

REFERENCE, SECTION 12.7 (HYDROGENATION OF HALOAROMATICS)

1. Rylander, P.N. in *Catalysis–Science and Technology*, J.R. Anderson and M. Boudart, eds., Vol. 4, p.2, Springer-Verlag, New York, 1983.

REFERENCES, SECTION 12.8 (HYDROGENATION OF CARBONYL COMPOUNDS)

1. Wagner, J. D., Lappin, G. R., and Zietz, J. R., *Kirk-Othmer Encyclopedia of Chemical Technology*, 4th ed., Vol. 1, p. 893, Wiley, New York, 1991.
2. Allen, P. N., Pruett, R. L. and Wickson, E. J., *Encyclopedia of Chemical Processing and Design*, J. J. McKetta, ed., Vol. 33, p. 46, Marcel Dekker, Inc., New York., 1990.
3. Billig, E., *Kirk-Othmer Encyclopedia of Chemical Technology*, 4th ed., Vol. 4, p.691, Wiley, New York., 1992.
4. Lappe, P. and Hofnann, T., *Ullmann's Encyclopedia of Industrial Chemistry*, 5th ed., Vol. A19, p. 49, VCH, Weinheim, Germany, 1991.
5. Anderson, P., Esk, M. and Wrammerfors, A., Catalyst deactivation, 1991, C. H. Bartholomew and J. B. Butt, eds., *Studies in Surface Science* 68, 61, Elsevier, New York, 1991.
6. Menon, F. C. in *Hydrogen Effects in Catalysis*, Z. Paal, Z. and P.G. Menon, eds., p. 611, Marcel Dekker, Inc., 1988.
7. Loefler, C. E., Stantzenberger, L. and Unruh, J. D. in *Encyclopedia of Chemical Processing and Design*, J. J. McKetta, ed., Vol. 1, p. 387, Marcel Dekker, Inc., New York, 1976.
8. Cropley, J. B., Burgess, L. M. and Luke, R. A., *Chemtech*, p. 374, June, 1984.
9. Comils, B. and Mullen, A., *Hydrocarbon Proc.*, p. 93, Nov., 1990.
10. Falloe, J., Lipps, W. and Grubler, I. in *Ullmann's Encyclopedia of Industrial Chemistry*, 5th ed., Vol. A1 p. 283, Weinheim, Germany, 1985.

11. Jakob, G. and Albrecht, L., in *Ullmann's Encyclopedia of Industrial Chemistry,* 5th ed., Vol. A8, p. 338, VCH, Weinheim, Germany, 1987.
12. Schossig, J. in *Ullmann's Encyclopedia of Industrial Chemistry,* 5th ed., Vol. A1 p. 305, VCH, Weiheim, Germany, 1985.
13. Rylander, P. in *Catalysis,* J. R. Anderson and M. Bondart, eds., Vol. 4, p.1, Springer-Verlag, New York, 1983.
14. Fedor, W. S., Millar, J. and Accola, A. J., *Ind. Eng. Chem.* 52, 282 (1960).
15. Benson, F. H., in *Kirk-Othmer Encyclopedia of Chemical Technology,* 3rd ed., 765, Wiley, New York, 1978.
16. Merger, F. and Paetsch, J. in *Ullmann's Encyclopedia Industrial Chemistry,* 5th ed., Vol. A1, p. 307, VCH, Weiheim, Germany, 1985.
17. Schmidt, K., *Chem. Ind.* 18, (4), 204 (1966).
18. Frenck, H. G. and Stadelhofer, J. W., *Industrial Organic Chemistry.*
19. Chenier, P. J., *Survey of Industrial Chemistry.*
20. Bruhue, F. in *Ullmann's Encyclopedia of Industrial Chemistry,* 5th ed., Vol. A4, p. 1, VCH, New York, 1985.
21. *Engelhard Technical Bulletin,* 4, 131 (1964).
22. *Furfural Hydrogenation,* Süd-Chemie, A.G., Technical Bulletin.
23. Süd-Chemie, AG, *Catalysts of Süd-Chemie AG: Aldehyde and Ketone Hydrogenation,* Germany.
24. Bel chikova, G.M., Kashina, V.V., Pilyaevskii, V.P., Gurevich, G.S., Pritsker, A.A. and Fuks, I.S., *Journal of Applied Chem.,* USSR (English translation) 51 (2), 458 (1978).
25. Van Hengstum, A.J., Kieboom, A.P.G. and Van Bekkum, H., *Starch/Staerke,* 36 (9), 317 (1984).
26. Herskowitz, M., *Studies in Surface Science: Heterogeneous Catalysis and Fine Chemicals,* Vol. 59, p. 105, Elsevier, Amsterdam, 1991.
27. Vannice, M.A., and Poondi, D., *J. Catal.* 169 (1), 166 (1997).
28. Rao, R., Dandekar, A., Baker, T.K. and Vannice, M.A., *J. Catal.,* 171, 406 (1997).
29. Brahme, P.H., Pai, M.U. and Narshiman, G., *Brit Chem. Eng.,* 9, 984 (1964).
30. Froment, G.F. and de Groof, W. *Recil. Conf. Colloq. Pharm-Ind.,* 9, 36 (1966).
31. Wisniak, J., Hershkowitz, M., Lebowitz, R. and Stein, S. *Ind. Eng. Chem. Prod. Res. Develop.* 13 (1), 75 (1974).
32. Hutchings, G.J., King, F., Okoye, H.P., Padley, M.B. and Rochester, C.H., *J. Catal.,* 148, 453 (1994).

REFERENCES, SECTION 12.9 (HYDROGENATION OF RESINS, ROSINS, AND WAXES)

1. Class, J. A., *Encyclopedia of Polymer Science and Engineering,* 2nd ed., Vol. 14, p. 438, Wiley-Interscience, New York, 1988.
2. Letcher, C. S. in *Kirk-Othmer Encyclopedia of Chemical Technology,* 3rd ed., Vol. 24, p. 466, Wiley, Interscience, New York, 1984.
3. *The Petroleum Handbook,* 6th ed., Royal Dutch Shell Companies, Elsevier, New York, 1983.
4. Süd-Chemie, *Technical Bulletin: Resins and High-Molecular Feed Hydrogenation.*

REFERENCES, SECTION 12.10 (FATS AND OILS)

1. Bechmann, H. J., *J. Am. Oil Chem. Soc.,* 60, 282 (1983).
2. Gran, R. J., Casano, A. F., and Baltanas, M. A., *Catal. Rev. Sci. Eng.,* 30 (1) 1(1988).
3. Albright, L. F. *Hydrogenation: Proceedings of an AOCS Colloquium,* R. Hastert, ed., Amer. Oil Chemists Society, Champaign, IL, 1987.
4. Heertje, I. and Bierma, H. I. *J. Catalysis,* 21, 20 (1971).
5. Heertje, I., Koch, G. K., and Watson, W. J., *J. Catalysis.,* 32, 337 (1974).
6. Drozdowski, B., *J. Am. Oil Chem. Soc.,* 54, 600 (1977).
7. Kalmal, T. M. B. and Lakshiminarayana, A., *J. Am. Oil Chem. Soc.,* 56, 578 (1979).

8. Van der Plant, P. J., *J. Catalysis*, 26, 42 (1972).
9. Horiuti, J. and Polanyi, M. *Trans. Faraday Soc.*, 30, 1164 (1934).
10. Patteron, H. B. N., *Hydrogenation of Fats and Oils: Theory and Practice,* AOCS Press, Champaign, IL, 1994.
11. Albright, L. F., *J. Am. Oil Chem. Soc.*, 42, 250 (1965).
12. Coenen, J. W. E., *J. Am. Oil Chem. Soc*, 53, 382 (1976).

REFERENCES, SECTION 12.11 (MISCELLANEOUS HYDROGENATIONS)

1. Harris, N. and Tuck, M. W., *Hydrocarbon Proc.,* p. 79, May, 1990.
2. Weissermel, L and Arpe, J.J., *Industrial Organic Chemistry,* 3rd ed., VCH, New York, 1997.
3. Olah, G. A. and Molnar, A., *Hydrocarbon Chemistry,* Wiley, New York, 1993.
4. Petrochemical Processes 97, *Hydrocarbon Proc.,* p. 118, March, 1997.
5. Braithwaite, J. in *Kirk-Othmer Encyclopedia of Chemical Technology,* 4th ed., Vol. 14, p. 990, Wiley, New York, 1995.
6. Lange, P. M., Martinola, F. B. and Oeckl, S., *Hydrocarbon Proc.,* p. 51, December, 1985.
7. *Hydrocarbon Proc.,* p. 184, November, 1977.
8. Melo, L., Rombi, E, Dominguez, Magnoux, P. and Guisnet, M. in *Heterogeneous Catalysts and Fine Chemicals: Studies in Surface Science and Catalysis,* Guisnet, M. et al. eds., Vol. 78, p. 701, Elsevier, New York, 1993.
9. Onoue, Y., Mizutani, Y., Akiyama, S., Izumi, Y. and Watanabe, Y., *Chemtech,* p. 52, Jan., 1997.

13 Hydrogenolysis

Hydrogenolysis is defined as the splitting of a bond with simultaneous addition of a hydrogen atom to each of the fragments. Based on commercial production, the major uses of this reaction occur in the petroleum and natural fatty-oil industries. Hydrotreating of petroleum stocks includes among other reactions the hydrogenolysis of organo-sulfur, -nitrogen, -oxygen, and -metallic compounds. (See "Hydrotreating" section.) Hydrogenolysis of fatty acids and fatty esters produced from natural oils is a significant pathway to fatty alcohols.

13.1 NATURAL FATTY ACIDS AND FATTY ESTERS → FATTY ALCOHOLS

Fatty alcohols, which are aliphatic alcohols of carbon chain lengths from C_6–C_{22}, are manufactured from natural fats and oils or synthetically by Ziegler polymerization of ethylene with subsequent oxidation and hydrolysis. In each case, linear alcohols with an even number of carbon atoms are produced. Natural fats and oils can contain some unsaturation, which can prove helpful for syntheses of other useful compounds. If unsaturation is desired, only natural sources can be used.

Linear fatty alcohols are used as intermediates in producing nonionic detergents, anionic detergents, and various specialized detergents. The fatty alcohols are also used directly as emulsifying agents, body enhancers in cosmetic creams, and lubricants in polymer processing.[6]

For special applications, unsaturated fatty alcohols have certain desirable properties such as lower melting points than saturated alcohols of the same carbon number. The configuration of the double bond can also affect these and other physical properties. They are used in heavy-duty detergents, cosmetic creams, textile and leather processing, plasticizers, and lubricating-oil additives.[6] Unsaturated alcohols are valuable intermediates for producing various products because of the two reactive species double bonds and the hydroxyl group. The most widely used monounsaturated fatty alcohol is oleyl alcohol ($C_{18}H_{36}O$) from tallow oil. It can be sold pure or in a C_{16}–C_{18} mixture, the major component of which is oleyl alcohol.

Chemistry

The various paths to fatty alcohols illustrated in Figure 13.1 conform to the several processes utilized. All involve the first step of splitting out the fatty acids from the natural oil by hydrolysis or by transesterification using methanol to form a methyl ester directly. If the first step is hydrolysis, then the second step involves the formation of a fatty acid ester (methyl or long-chain ester). The third and final step in all cases is the hydrogenolysis of the ester to two moles of alcohol. The hydrogenolysis step, the subject of this section, is often referred to in the natural fats industry as *hydrogenation,* but *hydrogenolysis* is the more exact descriptive term.

If methanol is used to form a methyl ester, one of the alcohols formed by hydrogenolysis will be methanol and the other the fatty alcohol. By contrast, if a fatty alcohol is used to form the ester,

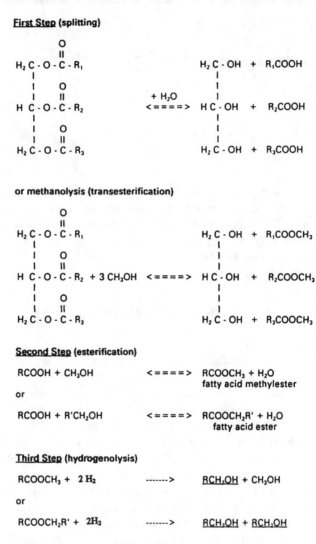

First Step (splitting)

$$
\begin{array}{llll}
& \text{O} & & \\
& \parallel & & \\
\text{H}_2\text{C - O - C - R}_1 & & & \text{H}_2\text{C - OH} \quad + \quad \text{R}_1\text{COOH} \\
\mid & & & \mid \\
\mid & \text{O} & & \mid \\
\mid & \parallel & + \text{H}_2\text{O} & \mid \\
\text{H C - O - C - R}_2 & \quad <====> & \text{H C - OH} \quad + \quad \text{R}_2\text{COOH} \\
\mid & & & \mid \\
\mid & \text{O} & & \mid \\
\mid & \parallel & & \mid \\
\text{H}_2\text{C - O - C - R}_3 & & & \text{H}_2\text{C - OH} \quad + \quad \text{R}_3\text{COOH}
\end{array}
$$

or methanolysis (transesterification)

$$
\begin{array}{llll}
& \text{O} & & \\
& \parallel & & \\
\text{H}_2\text{C - O - C - R}_1 & & & \text{H}_2\text{C - OH} \quad + \quad \text{R}_1\text{COOCH}_3 \\
\mid & & & \mid \\
\mid & \text{O} & & \mid \\
\mid & \parallel & & \mid \\
\text{H C - O - C - R}_2 + 3\,\text{CH}_3\text{OH} & <====> & \text{H C - OH} \quad + \quad \text{R}_2\text{COOCH}_3 \\
\mid & & & \mid \\
\mid & \text{O} & & \mid \\
\mid & \parallel & & \mid \\
\text{H}_2\text{C - O - C - R}_3 & & & \text{H}_2\text{C - OH} \quad + \quad \text{R}_3\text{COOCH}_3
\end{array}
$$

Second Step (esterification)

$$\text{RCOOH} + \text{CH}_3\text{OH} \quad <====> \quad \text{RCOOCH}_3 + \text{H}_2\text{O}$$
fatty acid methylester

or

$$\text{RCOOH} + \text{R'CH}_2\text{OH} \quad <====> \quad \text{RCOOCH}_2\text{R'} + \text{H}_2\text{O}$$
fatty acid ester

Third Step (hydrogenolysis)

$$\text{RCOOCH}_3 + 2\,\text{H}_2 \quad ------> \quad \underline{\text{RCH}_2\text{OH}} + \text{CH}_3\text{OH}$$

or

$$\text{RCOOCH}_2\text{R'} + 2\text{H}_2 \quad ------> \quad \underline{\text{RCH}_2\text{OH}} + \underline{\text{RCH}_2\text{OH}}$$

FIGURE 13.1 Steps in the Production of Fatty Acids from Natural Fats. Reprinted by permission: *Fatty Acid and Fatty-Acid Ester Hydrogenation,* technical bulletin, Süd-Chemie AG.

two fatty alcohols will be produced that can be identical if the initial alcohol is so chosen. These several reaction pathways provide the opportunity to choose from various alternative processes described later in this section.

Only a few studies have been done on the possible mechanistic pathways. Early on, it was suggested that fatty-acid ester hydrogenolysis proceeded through a hemiacetal intermediate.[1]

$$
\text{RCOOR}' + \text{H}_2 \longrightarrow \text{RC} \overset{\displaystyle \text{OH}}{\underset{\displaystyle \text{H}}{|}} \text{OR}'
$$

Undesired side reactions are the hydrogenation of the alcohol to a hydrocarbon and the saturation of unsaturated alcohols when the actual goal is to produce unsaturated alcohols.[3]

$$RCH_2OH + H_2 \rightarrow RCH_3 + H_2O$$

$$RCH = CHR'CH_2OH + H_2 \rightarrow RCH_2 - CH_2R'CH_2OH$$

Both these reactions are thermodynamically favored, but they can be kinetically controlled by proper choice of catalyst.

Thermodynamics

Equilibrium constants for liquid-phase hydrogenolysis have been determined for the mixed esters of coconut oil (C_8 to C_{18}) and tallow oil (C_{12} to C_{18}) and reported as follows.[4]

$$K = 9.65 \times 10^{-8} \exp\left\{\frac{4900}{T}\right\} \text{ for coconut oil}$$

$$K = atm^{-2}$$
$$T = °K$$

$$K = 16.8 \times 10^{-8} \exp\left\{\frac{4700}{T}\right\} \text{ for tallow oil}$$

Pure alcohols have higher equilibrium constants than exhibited by the mixtures, since the distribution of chain lengths for both alcohols and esters in the mixture effectively reduces the equilibrium conversion.[5]

A comparison of equilibrium constants for both the hydrogenolysis and the side reactions at 25 and 300°C shows that hydrogenolysis varies from 1.9×10^3 to 3.5×10^{-1}, hydrocarbon formation from 2.1×10^{16} to 4.3×10^3, and saturation from 4.1×10^{18} to 1.6×10^4.[3] Clearly, the high values for the equilibrium constants for the side reactions suggest careful choice of selective catalysts so that the rates of such reactions will be negligible. Conversely, the rapidly declining equilibrium constants with temperature of the main reaction suggest equilibrium control and thus careful choice of reaction conditions.

Catalyst Types and Suppliers

The most frequently used commercial catalysts are summarized in Table 13.1 along with typical applications. The exact nature and goals of the particular process used along with typical feed analysis is essential for the supplier to provide an optimum catalyst selection.

Various improvements of the copper-chromite and zinc-chromite catalysts have been made by the several suppliers. Specific details are, of course, proprietary; but patent literature and journal articles suggest the use of zinc as a means of enhancing the activity of copper chromite catalysts and improvement of zinc chromite performance by the addition of alumina.[3] Ruthenium-tin and ruthenium-tin boride have been reported, respectively, to give improved activity for hydrogenolysis and selective hydrogenolysis.[3]

Catalyst Deactivation

The most usual poisons for the copper catalyst that can be present in fatty esters or fatty acids are compounds of sulfur, chlorine, and phosphorous.[7,8] These compounds can occur in varying amounts, depending on the source of the natural fatty acid and prior process steps.

The active component of the catalyst is a nonstoichiometric Cu(O), a partially reduced copper, which reacts with sulfur and phosphorous species to produce poisoned (inactive) sites.[9] Barium in

TABLE 13.1
Fatty Ester Hydrogenolysis Catalysts

General Type	Typical Atomic Weight Percentages					Application*
	Cu	Cr	Ba	Mn	Zn	
Unpromoted copper	40	26				Resistant to free fatty acids
chromite	38	30				
Ba-promoted copper	37	32	11			Ba provides better thermal
chromite	33	30	8			resistance and hinders agglomeration of Cu
	34	27	7			
Ba and Mn promoted copper	37	32	3	3		Mn retards complete reduction
chromite						of CuO assuring active double-bond
						isomerization and hydrogenation.
Mn–promoted copper	39	37		3		Mn retards complete reduction
chromite	28	34		2.5		of CuO assuring active double-bond
	36	33		3		isomerization and hydrogenation.
Unpromoted copper-chromite	low ratio of Cu/Cr ≈ 0.7					Lower activity but higher resistance to acid attack
Zinc chromite		23			48	Highly selective. Will not hydrogenate double bonds, but less active
Raney/sponge copper	(1–6% residual aluminum)					Used in slurry systems (30–35 μ)

*Application comments based on "Fatty Acid and Fatty Acid Ester Hydrogenation," Süd Chemie, AG, Munich

Suppliers:	Activated Metals & Chemicals (sponge copper)	Catalyst Forms (varies with catalyst type)
	Grace Davison (Raney copper)	tablets
	Engelhard	powder
	United Catalysts	extrusions

sufficient quantity as caustic sites will adsorb sulfur compounds and prevent, to some extent, the deactivation of the copper sites.[8] Chlorine is highly toxic to copper catalysts and is thought to promote the recrystallization of inactive metallic copper.[7]

Small sizes of catalysts, when used in a fixed bed, will adsorb poisons more readily because of the shorter diffusion path. Accordingly, poisons will be adsorbed in the upper portion of the bed while the remainder of the bed remains free from poisons.

The partially reduced Cu(O) is subject to agglomeration at high temperature and loss of activity. Barium inhibits such agglomeration. Manganese retards excessive reduction of CuO and thereby assures hydrogenation of double bonds.

Regeneration

The catalyst, when properly selected and with careful control of temperature and feedstock characteristics, will deactivate gradually and ultimately must be replaced. Regeneration is therefore not practiced. Catalyst consumption varies between 0.3 and 0.7 percent.[6]

Process Units[6,7,10]

Three major fatty-ester hydrogenolysis reactor systems are being used. Gas-phase and trickle-bed, both of which use fixed-bed catalytic reactors, operate with a feedstock of the methyl esters of fatty acids from the original oil. Transesterification of the oil with methanol seems to be the predominate route to the methyl esters, although esterification of the fatty acids released by the splitting process is an alternative route (see "Chemistry"). Suspension hydrogenolysis uses either methyl esters or a mixture of fatty acids and fatty alcohols in a backmixed reactor.

The gas-phase process is used with feeds that can be vaporized in the presence of sufficient hydrogen or a combination of hydrogen and inerts or added methanol. Operating conditions are in the range of 200–250 bar (2900–3590 psi) and 230–250°C. Yields are 99% alcohol with 0.3% catalyst usage [(mass of catalyst deactivated/mass of fatty alcohol produced) × 100]. Hydrogen recycle rates are high (~600 moles H_2 per mole of ester) to maintain gaseous conditions.

Trickle-bed operation allows for much less hydrogen recycle since the esters undergoing hydrogenolysis remain in the liquid phase and flow downflow concurrently with the hydrogen in a ratio of 100 moles H_2 per mole of esters. Since the reaction is diffusion limited in the liquid phase, the rate is sensitive to hydrogen pressure. Higher operating pressures than in the gas-phase design are reported. Reported reaction conditions are 200–300 bar (2900–4351 psi) and 250°C. A supported copper-chromite on a silicon dioxide is sometimes used to provide a more porous structure for better liquid diffusion characteristics.

Two types of suspension processes are in operation. One uses the methyl ester with hydrogen fed upflow through a vertical unpacked reactor. Operating conditions are reported to be 250–300°C at 250 bar. The feed stream is mixed with copper chromite catalyst in the form of a powder. The high ratio of hydrogen-to-ester (~50:1) that is used not only favors the equilibrium but also serves to agitate the contents of the reactor. Of course, hydrogen is recycled after separation from the product stream. Methanol is removed by distillation and the fatty alcohol separated from the catalyst by filtration. Catalyst can be recycled, but overall usage of catalyst is higher (0.5–0.7%) than in the fixed-bed processes. However, the suspension process offers a simple means for adding fresh catalyst without the necessity of a shutdown for catalyst addition or replacement in the case of fixed-bed units.

Although it is possible to produce fatty alcohols by direct hydrogenolysis of fatty acids, such an approach has proved impractical because of the corrosive characteristic of the acids and the necessity for a more costly acid-resistant catalyst. More recently (1980s) Lurgi developed a suspension process that enables direct addition of fatty acid into the reactor, which contains a large excess of fatty alcohol (accomplished at start-up by direct addition). The reactor is a loop-type reactor to which hydrogen, fatty alcohol (250:1 alcohol/fatty acid), and copper chromite catalyst are added. The copper chromite rapidly catalyzes the esterification of the fatty acid by the fatty alcohol. This reaction yields a fatty ester plus water.

The copper chromite then catalyzes the slower hydrogenolysis of the ester to two moles of alcohol per mole of ester reacted. Recycle of a portion of this product is accomplished to maintain the same excess of alcohol in the reactor. Reported typical operating conditions are 300 bar (296 atm, 4350 psi), at 280–300°C. Catalyst consumption is higher than the fixed-bed processes (0.5–0.7%).

Hydrogen and low-boiling fatty acids in the reactor product are separated at a lower pressure, and catalyst is segregated by centrifugation to produce an alcohol slurry phase. A portion of this slurry can be discarded and replaced by fresh catalyst to be charged to the reactor with the remaining slurry. This arrangement permits easy control of catalyst activity.

Suspension processes are reported to produce an alcohol product with 2–5% unconverted ester, which can be removed by an alkali wash. In some designs, overhydrogenation can occur and produce up to 2–3% hydrocarbons, compared to 1% for fixed-bed operation. But careful choice of catalyst, with the help of catalyst suppliers, along with optimum operating conditions, can minimize this impurity. With copper chromite catalyst systems, unsaturated fatty esters will be saturated in each of these processes. If unsaturation is desired to be retained, zinc-chromite catalyst systems are indicated.

Process Kinetics

Kinetics have been determined for copper chromite catalysts using coconut oil and tallow-oil esters in the range of 270–320°C and pressures of 47 and 165 atmospheres.[3,4]

$$r = k_1 C_{ester}(P_H)^2 - k_2(C_{alcohol})^2$$

where r = hr^{-1}, C = wt% of indicated component
          P_H = hydrogen partial pressure in atm
          k_1 = hr^{-1} atm^{-1}
          k_2 = hr^{-1}

Water was found to inhibit the rate of hydrogenolysis, and data at 285°C were found to fit the following expression.

$$k_1 = \frac{k_{10}}{1 + 14C_{H_2O}}$$

$$k_2 = \frac{k_{20}}{1 + 14C_{H_2O}}$$

where C_{H2O} = wt% H_2O
          k_{10} and k_{20} = forward and reverse rate constants for zero percent H_2O

Mean rate constant values for hydrogenolysis of coconut and tallow esters on a water-free basis are:

$$k_{10} = 10^3 \exp\left\{\frac{-8600}{T/K}\right\}$$

$$k_{20} = 3.109 \exp\left\{\frac{-13100}{T/K}\right\}$$

It is postulated that water is formed during reduction of CuO and reduces catalyst activity. Activation energies have been reported for the several reactions of interest.[3,4,5]

saturated alcohol formation	16.7 kcal/mole
hydrocarbon formation	26.3 kcal/mole
unsaturated alcohol formation	24.6 kcal/mole

The unsaturated case requires a different catalyst. As in most commercial situations, the equations presented here are useful forms with which to construct a process model. The actual values reported could be quite different, depending on the characteristics of the catalyst being considered.

13.2 DIMETHYL TEREPHTHALATE →
1,4 DIMETHYLOLCYCLOHEXANE

Dimethyl terephthalate (DMT) is an alternative to terephthalic acid for use in polyester manufacture, but it also has an additional use. The product of a two-step hydrogenation-hydrogenolysis of DMT, 1,4 dimethylolcyclohexane acts as the diol in producing amorphous polyesters with high clarity for products such as films and bottles.[1] The reaction demonstrates a clever use of two catalysts and operating conditions to realize the desired result. The first step involves selective hydrogenation of the aromatic ring, and the second step is a standard ester hydrogenolysis.

dimethyl terephthalate   dimethyl hexahydrophthalate 1,4 dimethylolcyclohexane

Process Unit

The reaction is conducted in two fixed-bed reactors in series at 160–180°C and 300–400 bar.[2] High pressure is used in the first reactor to facilitate high reaction rate. The second reaction is equilibrium limited at high temperature and low pressures. So the high pressure is maintained to force the equilibrium to the right.[2] With all the hydrogenation going on, the heat of reaction is high (47.3 kcal/mole), and intermediate cooled product is recycled to control temperature.[3]

Palladium is an excellent hydrogenation catalyst for the aromatic ring at moderate temperatures, and copper chromite is ideal for the hydrogenolysis of the ester groups to alcohols.

Catalyst Type	Form
Copper chromite Promoted	Tablets
Palladium 0.5% kPd-on-carbon	

Catalyst Suppliers

United Catalysts, Engelhard, for both catalysts; and Johnson-Matthey and Precious Metals Corp. for Pd only.

13.3 TOLUENE → BENZENE (HYDRODEALKYLATION)

The major building blocks of aromatic chemistry (*benzene, toluene, ethylbenzene,* and the *xylenes*) are obtained almost exclusively from catalytic reformate and pyrolysis. In the steam cracking of naphtha, there often is produced an oversupply of toluene and an undersupply of benzene. In that situation, hydrodealkylation is an economical process for rectifying the imbalance. When the opposite situation occurs, the hydrodealkylation plant can be idled for the duration.

The major uses of toluene (solvent and production of toluene diisocyanate, benzoic acid, benzaldhyde, chlorinated toluene derivatives, nitrotoluenes, and benzyl alcohol) are almost overshadowed in the U.S.A., where production of benzene consumes over half of the toluene supply.[1] Ethylbenzene production alone consumes about one-half of the available benzene. The other major products from benzene are cumene and cyclohexane.

Chemistry

Two practical routes for hydrodealkylation of toluene are both used commercially. Thermal free-radical hydrodealkylation was the initial process used and continues to be used. It requires a higher temperature than the catalytic process.

The reaction is highly exothermic and temperature must be controlled by injection of recycled cold hydrogen. Without proper control, temperature runaway is possible. Operating temperatures for the catalytic process are in the range of 550–650°C. Pressures used are reported to be from 35 to 70 bar[1, 2, 3] and hydrogen-to-toluene feed ratios of 4–8 to 1 toluene are typical.

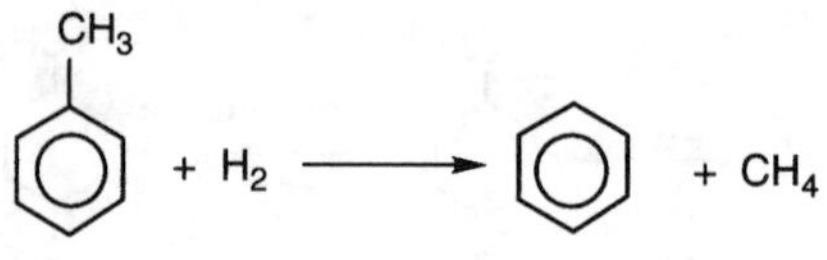

Thermodynamics

At the conditions given, the equilibrium constant is very favorable (191.42 @ 900 K and 1176.7 @ 700 K). Lower temperatures are avoided, since hydrogenation of the aromatic ring becomes favorable (K = 68.6 at 500 K compared to 2.27×10^{-6} at 700 K).

Catalysts

The catalysts used include Cr_2O_3 Al_2O_3, Mo_2O_3/alumina, or CoO/alumina. The chromium-on-alumina catalyst (3–12% Cr_2O_3, spheres or tablets) has been the most widely used.

Suppliers

United Catalyst, Engelhard

Form

Spheres, tablets

Catalyst Deactivation

The severe conditions cause a rapid coke laydown that must be removed periodically by regeneration. Long catalyst life is reported when careful control of regeneration temperatures is practiced. Excessive temperatures, about 650°C, can convert Cr_2O_3 to an inactive form and reduce the surface area of alumina.[4] Like many other oxide catalysts, it is resistant to poisons, but water vapor is a temporary poison.

Process Unit

There are several proprietary processes that are available for license. In general, they consist of two fixed-bed reactors in series with intermediate cooling using cold hydrogen (recycled plus fresh). Since one mole of methane is produced for each mole of benzene produced, it is necessary to separate the methane from the hydrogen that results as off gas from the partial condensation of product. The liquid product is separated by distillation and unreacted toluene recycled. The methane in the H_2–CH_4 stream may be removed by cryogenic separation and the methane diverted to a steam reformer to convert it to H_2 and CO.[2] Alternatively, the methane-hydrogen mixture may be added to the fuel gas system.

The same reactor system can be used for mixtures of toluene, benzene, C-8 aromatics, paraffins; and even some thiophene. These catalysts, which are resistant to H_2S to a reasonable degree, will convert thiophene to H_2S, saturate olefins, and dealkylate the various alkyl benzenes along with the toluene to benzene.

13.4 METHYL AND DIMETHYL NAPHTHALENE

Conversion to naphthalene can be accomplished by the same processes as used for toluene hydrolealkylation.

REFERENCES, SECTION 13.1 (FATTY ACIDS AND ESTERS)

1. Normann, W. *Angew. Chem.* 44, 714 (1931).
2. Yan, T. Y.; Albright, L. F. and Case, L. F., *Ind. Eng. Chem. Prod. Res. Dev.,* 4, 101 (1965).
3. Turek, T. and Trimm, D. *L., Catal. Rev.* 36 (4), 646 (1994).
4. Mutzall, K. M. K. and van den Berg, P. J., *Proc. 4th Eur. Sym. Chem. React. Eng.,* Pergamon Press, pp. 277–285, 1968.
5. Coenen, J. W. E., *Serfen-,Oele-, Fette-, Wachse* 14, 341 (1958).
6. Noweck, K. and Ridder, H., in *Ullmann's Encyclopedia of Industrial Chemistry,* 5th edition, Vol. Al0, p. 277, VCH, Weinheim, Germany, 1987.
7. Voeste, T. and Buchold, H., *J. Am. Oil Chem. Soc.,* 61, 350 (1984).
8. *Fatty Acid and Fatty Acid Ester Hydrogenation,* Süd Chemie, A. G., Munich, Germany.
9. Hughes, R., *Deactivation of Catalysts*, Academic Press, London, 1984.
10. Xreutzer, U. R., *J. Am. Oil. Chem. Soc.,* 61, 343 (1984).

REFERENCES, SECTION 13.2 (DIMETHYLHEXADYDROPHTHALATE)

1. Jadhav, I. Y. and Kantor, S. W., in *Concise Encyclopedia of Polymer Science and Engineering,* J. I. Kroschwitz, ed., p. 814, Wiley, New York (1990).
2. Rylander, P. N. in *Catalysis: Science and Technology,* J. R. Anderson and M. Boudart, ed., Vol. 4, p. 2, Springer Verlag, New York, 1983.
3. Rylander, P. N. in *Ullmann's Encyclopedia of Industrial Chemistry,* 5th edition, Vol. A13, p. 441, VCH, Weinheim, Germany (1989).

REFERENCES, SECTION 13.3 (HYDRODEALKYLATION OF TOLUENE)

1. Franck, H.G. and Stadelhefer, J.W., *Industrial Aromatic Chemistry,* Springer Verlag, New York, 1988.
2. Weissermel, K, and Arpe, H.J., *Industrial Organic Chemistry,* 3rd edition, VCH, New York, 1997.
3. Folkins, H.O., in *Ullmann's Encyclopedia of Industrial Chemistry,* 5th edition, Vol. A3, p. 484, VCH, Weinheim, Germany, 1985.
4. Thomas, C.L., *Catalytic Processes and Proven Catalysts,* Academic Press, New York, 1970.

14 Isomerization*

14.1 META-XYLENE → PARA- AND ORTHO-XYLENE

The C_8 portion of catalytic reformate (ortho-, meta-, and para-xylenes and ethylbenzene) is a valuable feedstock for petrochemicals. Often p-xylene is in most demand, and an effort to maximize its production is common. Ortho-xylene is next in demand, while meta-xylene has only a few uses.

The separation of the C_8 aromatics is complex because of their close boiling points (normal bp in °C: o-xylene 144.4, m-xylene 139.1, p-xylene 138.4, ethylbenzene 136.2).

Ethylbenzene can be removed by distillation, but higher energy consumption results because of the 2°C difference in boiling points between p-xylene and ethylbenzene. A column of 300–360 trays and a reflux of 80:1 to 120:1 has been reported.[1] Alternatively, the ethylbenzene may be left in the mixture and altered chemically to yield an easily separated product. Obviously, distillation requires high energy consumption and is no longer favored, although units are still in operation.

Separation of o-xylene from the other xylenes and ethylbenzene is challenging but not as difficult as ethylbenzene separation. A tower with 120–150 trays and a reflux ratio of 10:1 to 15:1 has been reported.[1]

The m-xylene and p-xylene fraction can then be separated by fractional crystallization or selective adsorption on molecular sieves. The crystallization process yields a liquid mixture of about 85% m-xylene and about 15% para-xylene. The crystallized product is 99.5 wt% p-xylene.[3] The mother liquor contains the aforementioned meta and para xylenes plus ethylbenzene, unless it was previously removed by distillation.

The adsorption process uses molecular sieves and produces a rather sharp separation yielding a p-xylene product of 99.9 wt% and 97 wt% recovery per pass. The raffinate is essentially all m-xylene with about 1 wt% p-xylene.[3]

In both cases, the rich m-xylene stream can be isomerized to an equilibrium mixture of the three isomers. This mixture can then be recycled through the separation section to produce additional p-xylene. It is the isomerization process that is of interest here, and basically the isomerization of m-xylene to an equilibrium mixture of the three xylene isomers and the isomerization of ethylbenzene to o-xylene or to benzene via dealkylation.[3]

Chemistry

In most cases, dual-function catalysts are used. The acid function provides the protons for carbocations that are the intermediates in the isomerization. A hydrogenation function, usually Pt, is provided for various purposes, depending on the particular process.

---

* See also Chapter 18, "Petroleum Refining."

Xylene Isomerization

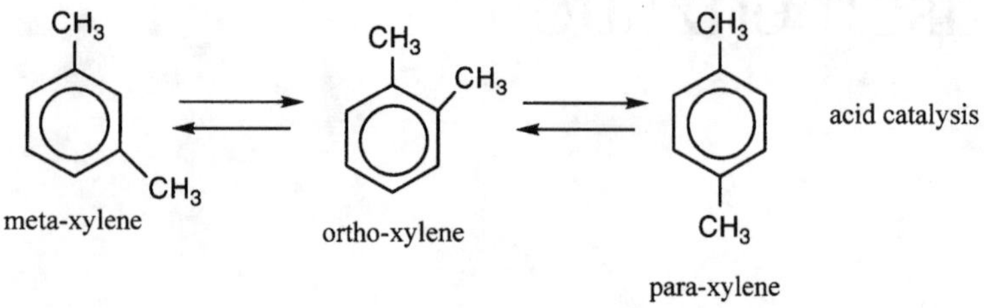

meta-xylene   ortho-xylene   para-xylene   acid catalysis

A contrary view of the reaction scheme for isomerization has been documented by several authors, namely, that a direct interconversion between para-xylene and ortho-xylene does not occur.[9]

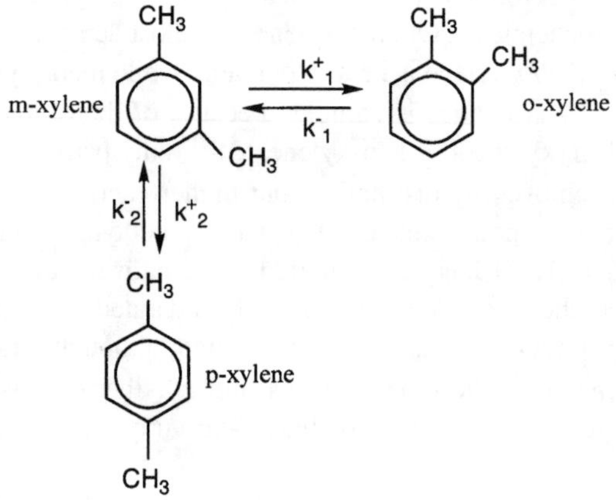

Ethylbenzene Isomerization

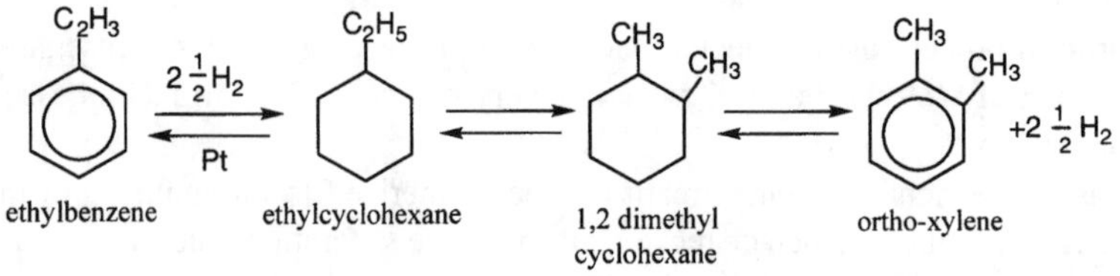

ethylbenzene   ethylcyclohexane   1,2 dimethyl   ortho-xylene
                                                cyclohexane

Ethylbenzene Dealkylation:

Ethylbenzene Dealkylation

ethylbenzene   benzene

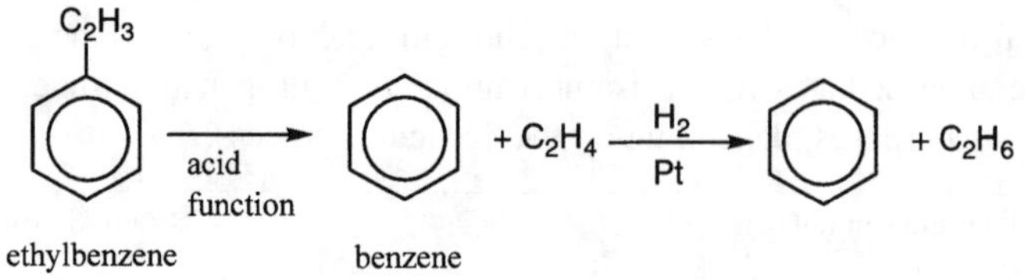

This reaction is not equilibrium limited, and conversion to benzene is high (70%).[3] The purpose of the hydrogenation is to prevent coke formation via ethylene oligomerization. The theoretical equilibrium compositions for the C_8 aromatics are shown in Figure 14.1.

Catalyst Types and Suppliers[1-4]

The major licensors and suppliers of proprietary catalysts are shown in Table 14.1

Effect of Zeolite Void Structure

Acid zeolites with ten- or-twelve-membered rings have been studied to highlight the differences in pore shapes and dimensions relative to isomerization and disproportionation of m-xylene.[7] Ten-membered ring zeolites with crystals larger than 1 μm exhibit a high para/ortho selectivity, while selectivity for disproportionation versus isomerization is low.[7] The selectivity to isomerization is caused by steric hindrance due to the pore walls that prevents formation of transition-state complexes that lead to disproportionation.

Deactivation[3]

The catalysts, although of various types, would be deactivated by the poisons listed in Table 14.2, which is based on the Isomar catalyst.[3] In addition, coke formation will occur over time, and regeneration is necessary. Maintaining the licensor's recommended H_2/CH ratio will prolong periods between regeneration. As catalyst activity declines operating temperature is raised over time to a recommended maximum, at which point the catalyst is regenerated by careful burning off of the coke.

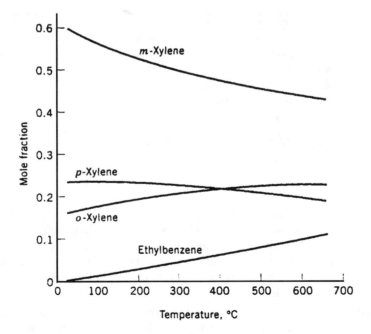

FIGURE 14.1 Equilibrium Mole Fractions for C_8 Aromatics. Reprinted by permission of John Wiley & Sons, Inc., Cammella, W. J. in *Kirk–Othmer Encyclopedia of Chemical Technology*, Suppl., 4th edition, p. 835, Wiley, New York, copyright © 1998.

TABLE 14.1
Catalyst Types and Suppliers[1-4]

Licensor/ Suppliers	Process	Catalyst	Conditions
Engelhard/ARCO also IFP		Octifining 0.5 wt% Pt on silica-alumina or on alumina and H-mordenite	425–480°C 11–25 bar $H_2/HC = 4$–6:1 Ethylbenzene converted to xylene
UOP LLC	Isomar™	Pt-on-alumina	400°C, 12.5 bar Ethylbenzene converted to xylene
UOP LLC	Isomer™	Pt-on-zeolite	Ethylbenzene converted to benzene
Mobil	MHTI™	Acidic ZSM-5 zeolite partially exchanged with Pt	427–460°C 15–18 bar $H_2/HC = 1.5$–2 Ethylbenzenes converted to benzene
Mobil	MHAI™	Two beds #1 Pt-on-ZSM-5 converts EB to benzene #2 Pt-on-ZSM-5, low crystal size, isomerizes xylenes	400–480°C 4.5–29 bar $H_2/Hc = 1$ to 5:1

TABLE 14.2
List of Poisons for Typical Platinum-Based Xylene Isomerization Catalysts and Allowable Limits (Isomar Process)

Contaminant	Effect	Limit
Water	Deactivates catalyst, promotes corrosion. Irreversible.	200 ppm, max.
Total chloride	Increases acid function, increases cracking. Reversible.	2 ppm, max.
Total nitrogen	Neutralizes acid sites, deactivates catalyst. Irreversible.	1 ppm, max.
Total sulfur	Attenuates Pt activity, increases cracking. Reversible.	1 ppm, max.
Lead	Poisons acid and Pt sites. Irreversible.	20 ppb, max.
Copper	Poisons acid and Pt sites. Irreversible.	20 ppb, max.
Arsenic	Poisons acid and Pt sites. Irreversible.	2 ppb, max.

Reprinted by permission: Jeanneret, J. J. in *Handbook of Petroleum Refining Processes,* 2nd edition, R. A. Myers, editor, p. 2.37, copyright McGraw-Hill Companies, New York, 1996.

Process Units[1-5]

The isomerization reactor is part of a rather complex larger unit consisting of a distillation section for isolating aromatics from the catalytic reforming or steam cracking product, the separation of C_8 aromatics, the removal of ortho-xylene by distillation, and the separation of para-xylene from the meta-xylene and ethylbenzene by crystallization or selective adsorption. The raffinate or mother liquid is passed to the isomerization section. This feed to the isomerization section is rich in m-xylene and low in para- and ortho-xylene content.

The isomerization feed is heated along with H_2 in a direct-fired heater. The heated vapor-phase mixture then passes to a fixed-bed reactor where equilibrium is established. Referring to Figure 14.1 at 400°C, this composition at equilibrium is approximately 8% ethylbenzene, 22% ortho, 22% para-xylene, and 48% m-xylene.[4] The reactor product undergoes condensation and separation of H_2, which is recycled and joined by make-up H_2 to the reactor inlet. The condensed product is then passed to a distillation column for removal of light-end by-products. The C_8 ortho and para enriched bottoms is recycled to the ortho and para recovery systems described above.

Depending on the catalyst and process selected, the ethylbenzene content is either converted to xylenes or benzene and ethane. Either procedure eliminates the difficult problem of separating ethylbenzene from the xylenes. If the xylene forming catalyst is used, recycling of unreacted alkylcyclohexanes back to the isomerization reactor is necessary to limit formation of excessive amounts of alkylcyclohexanes. However, conversion of ethylbenzene to xylenes ultimately increases p-xylene yield. Catalysts that convert ethylbenzene to benzene and ethane eliminate the recycle of intermediates and reduce the required equipment size, but also reduce p-xylene production since the ethylbenzene will not be converted to xylenes.

Although p-xylene is most in demand, o-xylene is also required at an increasing pace. These processes also produce considerable amounts of o-xylene which, as noted previously, is removed by distillation with purities of 99%.

Process Kinetics

It is rare indeed to find a detailed description of a successful commercial model that embodies the results of a number of independent studies of reaction kinetics, catalyst deactivation (coking) intra-particle diffusion, and side reactions of disproportionation. Such a study has been published, the strategy of which is shown in Figure 14.2. Equations first-order in each xylene were used for the isomerization reactions and second-order for the disproportionation reactions.[6] At the time, the catalyst of choice was amorphous silica-alumina. More recently, zeolites have the great advantage of a new and valuable dimension of different crystallographic structures with various shapes and intracrystalline cavities that have a more profound effect on selectivity. The detailed studies described for the older silica-alumina, however, should be useful as a guide to a successful model for other acidic type catalysts such as zeolites.

A kinetic model for the liquid-phase isomerization over a shape-selective ZSM-5 zeolite has been presented[8] based on the following scheme, including both chemical reaction and mass transport in the pores.

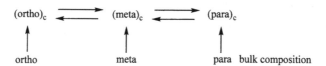

A rather complex series of four rate equations involving twelve adjustable constants was proposed. The reader is referred to this article for insights on the issues of pore sizes and characteristics.

Kinetics based on the alternate scheme involving no direct interconversion between ortho and para-xylene has been presented for galiosilicates with ZSM-5 structure[9] See "Chemistry" section for reaction scheme.

$$r_m = -(k_1^+ + k_2^+)C_m + k_1^-C_o + k_2^-C_p$$

$$r_o = (k_1^+C_m - k_1^-C_o)$$

$$r_p = (k_2^+C_m - k_2^-C_p)$$

where subscripts m, o, and p represent, respectively, meta-xylene, ortho-xylene, and para-xylene.

Over the past several decades, evidence has accumulated suggesting that direct conversion between p-xylene and ortho-xylene does not occur.[9–12]

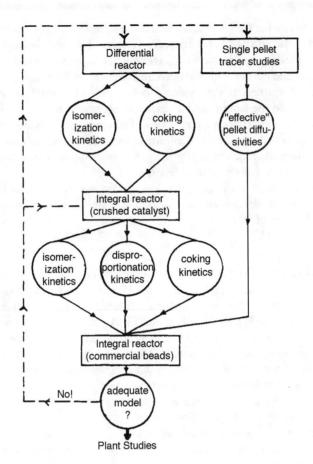

FIGURE 14.2 Strategy for Model Development. Reprinted with permission from Orr, N. H., Cresswell, D. L., and Edwards, D. E., *Ind. Eng. Chem. Des. Dev.*, 22, 135 (1983). Copyright 1983, American Chemical Society.

REFERENCES (XYLENE ISOMERIZATION)

1. Veba, J. F., Graeser, U., and Sims, T. A. in *Ullmann's Encyclopedia of Industrial Chemistry*, 5th edition, Vol. A28, p. 423, VCH, New York, 1996.
2. Weissermel, K. and Arpe, H. J., *Industrial Organic Chemistry*, 3rd edition, VCH, New York, 1997.
3. Jeanneret, J. J. in *Handbook of Petroleum Refining Processes*, R. A. Meyers, editor, 2nd edition, p. 2.37, McGraw-Hill, New York, 1997.
4. Cannella, W. J. in *Kirk-Othmer Encyclopedia of Chemical Technology*, 4th edition, supplement, p. 831, Wiley, New York, 1998.
5. *Hydrocarbon Proc.: Petrochemical Process.* '97, p. 166, March, 1997.
6. Orr, N. H., Cresswell, D. L, and Edwards, D., *Ind. Eng. Chem. Process Des. Dev.*, 22, 135 (1983).
7. Martens, J.A., Perez-Parientes, J., Sastre, E., Corma, A, and Jacobs, P.A., *Appl. Catal.* 45 (1), 85 (1988).
8. Cappellazzo, O., Cao, G., Messina, G. and Morbidelli, M., *Ind. Eng. Chem. Res.* 30, 2280 (1991).
9. Richter, M., Kosslick, H., Tuan, V.A., Richter-Mendau, Parlitz, D., Vorbeck, G. and Szulzewsky, K., *Ber. Bunsenges. Phys. Chem.*, 96 (4), 586 (1992).
10. Hanson, K.L. and Engel, A.J., *AI. Che. J.*, 13, 260 (1967).
11. Cortes, A. and Corma, A., *J. Catal.* 51, 338 (1978).
12. Kaeding, W.W., Chu, C., Young, L.B., Weinstein, B. and Butter, S., *J. Catal.* 67, 159 (1981).

15 Oxidation (Inorganic)

Although there are only a few inorganics produced via catalytic oxidation, they are major commodity chemicals: sulfuric acid, nitric acid, and sulfur from H_2S by oxidation. The catalysts are quite different in each case.

15.1 SULFUR DIOXIDE → SULFUR TRIOXIDE → SULFURIC ACID

Sulfuric acid is the number one chemical, based on tonnage. It is so important to the operation of a healthy economy that trends in its total production have been used as an indication of trends in the economy itself. In more recent decades, although it has remained number one, it has lost some of its sales to hydrochloric acid, particularly in metal cleaning operations. Hydrochloric acid is a by-product of so many hydrocarbon chlorinations that it has tended to decline in price.

The major use of sulfuric acid is in the production of phosphate fertilizers, particularly in the U.S.A., where agriculture products have become one of our major exports. Sulfuric acid is reacted with phosphate rock, a complex calcium phosphate, to produce phosphoric acid. The phosphoric acid is then reacted with ammonia to produce ammonium phosphate, the leading phosphate source for fertilizers. Phosphoric acid is also used to make other phosphates, such as detergent builders, which aid in complexing ions in the wash water that interfere with dirt removal, prevent redeposition of dirt, buffer the wash water, and act as an antibacterial and anticorrosive agent.

Sulfuric acid is also an important component in the production of dyes, paper, plastics, explosives, textiles, petroleum products, and in the processing of ores.[1,2]

The so-called *contact process* has been used commercially since the 1890s, initially replacing the chamber process when high-strength acid began to be required by a fledgling dye industry. The chamber process produced 65% acid, which could be concentrated to 78 percent.[3] Over the years, the chamber process was totally replaced by the contact process. At first, platinum was used as the direct oxidation catalyst of sulfur dioxide to sulfur trioxide. Although the high activity of platinum permitted operation at modest temperatures where the equilibrium of this exothermic reaction was most favorable, its high cost and susceptibility to poisoning (by many metallic compounds, halogens, and other trace materials of varying nature peculiar to each sulfur source) made the search for an alternative catalyst a high priority.[3] By 1920, the first commercial plant using vanadium catalysts was constructed and successfully operated. It was soon followed by a rapidly growing number, and in the short span of 8 years, vanadium catalysts replaced platinum essentially all over the world.

Chemistry

The reactions for the entire process involve the burning of sulfur to sulfur dioxide, the catalytic oxidation of sulfur dioxide to sulfur trioxide, and the dissolution of SO_3 into sulfuric acid (98.5%). Water is added and acid removed to maintain the circulating acid at 98–99 percent.

1. Sulfur Burning

$$S + 1/2O_2 \rightarrow SO_2 \qquad \Delta H = -70.94 \text{ kcal/gmole (Ref. 2)}$$

$$983 \rightarrow 1150°C$$

2. Sulfur Dioxide Oxidation (Vapor Phase)

$$SO_2 + 1/2\,O_2 \rightarrow SO_3 \qquad \Delta H = -23.64 \text{ kcal/gmole}$$

Catalyst: V_2O_5/Kieselguhr
693 K(in) = 875 K (max)
1.2–1.4 bar inlet pressure (absolute)

3. Sulfur Trioxide Absorption (countercurrent flow)

$$SO_3 + H_2O \rightarrow H_2SO_4 \qquad \Delta H = -31.65 \text{ kcal/gmole (Ref. 2)}$$

Sulfur dioxide oxidation is a classic example of an equilibrium-limited reaction further burdened by a temperature below which the activity of the catalyst is inadequate to meet the needs of essentially complete conversion necessary to avoid problems with unreacted SO_2 recovery. This minimum allowable temperature is reported to be 390°C (693 K). Figure 15.1 shows equilibrium conversion versus temperature (in kelvins) for a feed of 8% SO_2 in air. Quite clearly, temperatures below 700 K would be most attractive especially in the region of high conversion.

Mechanism of Sulfur Dioxide Oxidation

Because of the prime importance of sulfuric acid production, the reaction and the vanadium catalyst have been the focus of numerous investigations for more than four decades. Many excellent studies have been published (see reviews 4–6).

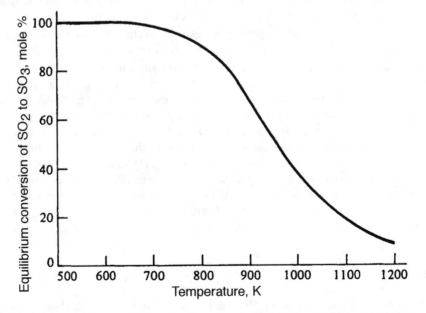

FIGURE 15.1 Equilibrium Conversion versus Temperature in Oxidation of Sulfur Dioxide. Reprinted by permission: Saterfield, C. N., *Heterogeneous Catalysis in Industrial Practice,* 2nd edition, McGraw-Hill, New York, 1991. Copyright held by C. N. Saterfield.

The fact that the catalyst, potassium-promoted vanadium oxide, actually exists as a melt at operating conditions has added complexity to the task of defining a mechanism. In addition, diffusion of gas components into the pores, and thence into the catalyst melt, makes it especially difficult to use rate measurements as an aid to defining a mechanism. The existence of significant interparticle mass transfer resistance was demonstrated early on by comparing SO_2 reaction rates on spherical pellets of varying sizes[5] as shown on Figure 15.2.

Suffice it to say that there is no general agreement on a mechanism, but there are some useful descriptions that have proved helpful in reasoning about reaction behavior. An early and popular mechanism describes a two-step reaction.[8]

$$SO_2 + 2V^{5+} + O^{2-} \Leftrightarrow SO_3 + 2V^{4+}$$

$$1/2 \, O_2 + 2V^{4+} \Leftrightarrow 2V^{5+} + O^{2-}$$

This mechanism accounts, in a simple manner, for the oxidation of SO_2 followed by the reoxidation of the vanadium.

More recently, studies with molten salts without the support have made possible the identification of the vanadium complexes that might be involved.[4-6] Work with melts have suggested many mechanisms of which the following are examples.

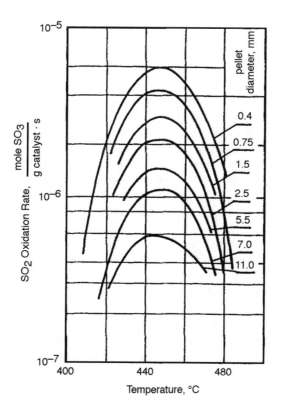

FIGURE 15.2 Sulfur Dioxide Oxidation Rate versus Temperature for Spherical Catalyst Pellets of Various Sizes for Conversion of 95%. Reprinted from Urbanek, A. and Trela, M., *Catalysis Reviews: Science & Engineering*, 21 (1), 73 (1980), by courtesy of Marcel Dekker, Inc. Based on data of Schytil, F. and Schwalb, R., *Chem. Eng. Sci.* 14, 367 (1961).

A. Oxidation of an SO_2 complex which then gives SO_3 and the V^{4+} is reoxidized to V^{5+}. No step is rate controlling.[9]

$$V_2O_5 \cdot SO_3 + SO_2 \Leftrightarrow (VOSO_4)_2 \tag{1}$$

$$(VOSO_4)_2 \Leftrightarrow V_2O_4 \cdot SO_2 + SO_3 + 1/2\, O_2 \tag{2}$$

$$V_2O_4 \cdot 1/2\, O_2 \Leftrightarrow V_2O_4 \cdot SO_3 \tag{3}$$

B. The reaction is independent of the vanadium valence state and may involve a bimolecular complex with two vanadium atoms such as a potassium sulfur vanadate, designated as $[V_2^{5+}]$. This pathway labeled (A) is said to dominate the oxidation route to SO_3 at high SO_3 concentrations and is independent of the vanadium valence state. At low concentrations the redox mechanism is thought to be predominant.[10]

$$[V_2^{5+}] \xrightarrow{+SO_2} [V_2^{5+}]SO_2 \xrightarrow{+SO_2} [V_2^{5+}]2SO_2 \xrightarrow{+O_2} [V_2^{5+}]+2SO_3 \quad (A)$$

$$\downarrow{+O^{2-}} [V_2^{4+}]SO_3 \xrightarrow{+1/2\,O_2} [V_2^{5+}]+O^{2-}+SO_3 \quad (B)$$

C. The catalytic reaction species has been proposed in another study to be $VO_2(SO_4)_2{}^{3-}$, a product of the following equilibrium reaction.[11–13]

$$VO(SO_4)_3{}^{3-} + SO_4{}^{2-} \Leftrightarrow VO_2(SO_4)_2{}^{3-} + S_2O_7{}^{2-}$$

Studies in the melt phrase are advantageous because they eliminate the complexities of mass and heat transfer in the catalyst and the effect of the presence of inerts. The absence of such data makes rate expressions and detailed mechanisms uncertain. In fact, this conclusion made in 1980 essentially applies today. The large amount of work done on mechanism has not been done in vain. It led to a focus on the major technical problem, the need to develop a catalyst that will operate in or below the range of 400–460°C, where valuable compounds with melting points are crystallized out of the melt with the attendant rapid decline in activity. Lower operating temperatures increase equilibrium conversion to SO_3 for the highly exothermic reaction if the active complexes remain in the melt.

A thorough study of a typical catalyst melt was designed so that complexes that precipitate out at lower temperatures could be separated and identified.[7] Precipitation was accompanied by the usual sudden increase in activation and thus a significant reduction in rate. A number of different promoter compositions were studied as summarized in Table 15.1.[7]

Various crystalline compounds that were isolated and identified in the range of 350–480°C had the following formulae, depending on the promoter added: $NaVO(SO_4)_2$ $NaV(SO_4)_2$ $K(VO)_3(SO_4)_5$, $KV(SO_4)_2$, $Cs_2(VO)_2(SO_4)_3$, and $CsV(SO_4)_2$. The break in the Arrhenius plot occurs at lower temperatures with increasing atomic number of the alkali promoter.[7] Thus, as can be seen in Table 15.1, cesium lowers the melt point and allows operation at lower temperature.

The reader is referred to excellent review articles for further comment on mechanism.[3,4,6,9,10]

Catalyst

Catalyst manufacturers offer a variety of shapes and compositions. Typically, a standard catalyst contains 6–9% V_2O_5 and alkali-metal sulfates (potassium sulfate) in a 2–4 K/V atomic ratio. In

TABLE 15.1
Melt Point Variations with Atomic Weight and Quantity of Promoter*

Catalyst Melt	Temperature of Precipitation °C
Na/V = 3	>470
Na/V = 4.7	>460
Na/V = 10	>460
K/V = 3	>475
K/V = 4.7	439
K/V = 10	404
Cs/V = 3	430
Cs/V = 4.7	416
Cs/V = 10	388

*Partial reproduction of a table.

Reprinted by permission: Boghosian, S., Fehrmann, R., Bierrum, N.J. and Paptheodorou, G.N., *J. Catal.* 119, 121 (1989). Academic Press, Inc.

addition, a small amount of sodium as sodium sulfate is added to produce about 1% sodium content. The support is a silica, usually diatomaceous earth (Kieselguhr), along with a biodisperse pore structure. The Kieselguhr provides larger pores for transport together with average size pores of such a size that they will not be filled by melt but offer wall coating so that significant reactant contact area with the melt will be realized.[1,5] Producing optimal pore structures, whatever they may be, is a proprietary procedure of catalyst manufactures.

These standard catalysts have been supplemented by others having particular properties especially useful in certain situations. For example, catalyst with higher vanadium concentrations and with appropriate lowering of the K/Na ratio is more active and can increase production when used in the final stages of the reaction.[12] A low-temperature catalyst that provides a complex with a lower melting point (by virtue of the addition of cesium in place of some of the potassium in the studied recipe) allows operating the last bed, for example, at a lower temperature (below 400°C) where equilibrium conversion is higher. Improved yield without an intermediate absorber is thus possible.

Another catalyst option is a more rugged extrudate, made so by replacing some of the Kieselguhr with colloidal silica that hardens the catalyst[13] and makes it resistant to abrasion. This property is very important in the periodic screening that is required.

Catalyst Shapes and Sizes

Most catalysts for sulfur dioxide oxidation are extrudates with a pore volume of 50%.

Shapes	Size
Cylindrical	6, 8 mm
Rings	10 × 10 × 3.5 mm and 20 × 20 × 7 mm
Ribbed rings	12 × 12 × 4 mm

Ring catalyst produces a lower pressure drop than the solid cylinders, and ribbed rings reduce the ΔP even further. The larger rings and ribbed rings are often used on the top portion of the first bed so that dust can be accumulated without large losses in pressure that would tend to occur with smaller extrudates. Only a small amount of dust is needed to plug the passageway between touching small extrudates.

Ring and ribbed rings also provide additional exterior surface area, which is important since a major portion of the reaction occurs at or near the exterior surface.[14]

Suppliers

Haldor Topsoe
United Catalysts

Process Licensors

Enviro-Chem Systems

Catalyst Deactivation and Loss

Partial Physical Destruction

The V_2O_5 sulfur dioxide oxidation catalyst has a remarkably long life—typically 10 years when handled correctly from catalyst producer to storage to loading in the reactor and subsequent operation. Since some free sulfur trioxide is present in both fresh and used catalyst, it is most important to avoid the presence of moisture. Water vapor that makes contact with the catalyst surface will hydrate the SO_3 to sulfuric acid, causing alteration of the surface species that form crystallites, which can cause significant losses of catalysts solids during handling.[14]

Although the catalyst life is long, the catalysts in the reactor (particularly the first bed) must be screened yearly at a scheduled turnaround to eliminate accumulated dust and fragments of catalyst particles. Lower bed-pressure drop and improved flow distribution will result. Some fresh catalyst must be added due to the losses from attrition in the bed and during the screening process. Attrition in the first bed is more apt to occur because of the higher operating temperatures. When all beds are dumped and screened, the added fresh catalyst is placed in the last bed to permit lower temperature operation in the final bed and thus higher SO_2 conversion.

So-called *dusting* is caused by fine particles of dust or scale carried over in the SO_2 feed stream and by self-attrition of the catalyst during operation. The dust in the feed stream may be caused by contaminated sulfur, failure of molten-sulfur filters, spalling of combustion chamber brick, and equipment and duct scale.[14,15] Modern sulfur-burning plants with sources of high-purity sulfur have eliminated much of the carry-over of dust in the feed stream. Other SO_2 sources, such as waste acid and smelter streams, require elaborate clean-up approaches.

High-Temperature Deactivation

Deactivation of the sulfur-dioxide oxidation catalyst can occur after long-term operation significantly above 600°C, but relatively short high-temperature excursions apparently do not normally affect catalyst activity.[14] Longer-term operation at elevated temperatures (650–700°C) can cause significant deactivation, which is thought to be due to reaction of SiO_2 with potassium that interacts with vanadium and produces an inert vanadium phase.[14]

Poisons

The sulfuric-acid catalyst is rugged, and very few possible contaminants are poisons in the usually accepted definition of compounds that deactivate catalysts even in trace amounts. Because of the long and successful history of sulfuric acid plant operations, there is much valuable lore related to contaminants, many of which are not a problem at the very low levels so often associated with catalyst poisons. Arsenic and fluorides can be said to be the exception. They are detrimental to catalyst activity even at modestly low levels.[14] The effects of feed gas contaminants have been conveniently assembled and reproduced here as Table 15.2.[14] Many of these contaminants are present in unacceptable quantities in SO_2 produced from metal ore smelters but are lowered drastically by a standard cleaning process.

Regeneration

Regeneration of deactivated catalyst has not been a major priority, because of the comparative cost of fresh catalyst. In recent years, regeneration has received more attention as various methods have been developed.[13] One of these methods is based on the concept of an *in situ* reoxidation with a nitrogen stream of very low oxygen content at temperatures below 600°C. Since a goodly portion of the catalyst in a reactor can last up to 10 to 20 years, one would have to look very closely at the economics of any regeneration scheme. New catalyst formulations that appear from time to time may often prove more economical, because of increased production and higher conversion, than any regeneration scheme. The alternative to regeneration is metals recovery by a jobber that yields a low-value ore-like mixture high in vanadium.

Process Units

Sulfur Source

Elemental sulfur delivered in molten form to the user continues to be the preferred raw material for sulfuric acid manufacture. It is quite pure, and the sulfur dioxide produced from it in acid plants does not require elaborate purification and cleaning systems. For many years, elemental sulfur was primarily produced from underground deposits associated with salt domes by means of the Frasch process. This source, although still important, has declined significantly since restrictions of sulfur content of both liquid and gaseous fuels have been enforced. Refineries and natural gas plants now remove sulfur as H_2S from their products by absorption and then convert the H_2S to elemental sulfur by the Claus process (see page 241). Since so much new natural gas and oil discoveries contain significant amounts of sulfur compounds, Claus process elemental sulfur has become the major source of sulfur for sulfuric acid manufacture via sulfur dioxide.

Other sources of sulfur dioxide include waste acid from alkylation, methacrylate, and alcohol plants.[16] The used acid is decomposed to SO_2 and water, and after clean-up and drying it is ready to be converted to sulfuric acid. Metal-sulfide smelting plants produce SO_2 in the metal smelting process. The SO_2 from the smelter has a number of impurities that are removed.

Sulfur Burning to SO_2

Plants using elemental sulfur receive it and store it in the molten state at 270–290°F (132–143°C).[17] The molten sulfur is pumped to a furnace where special burners atomize the melt, which burns in prefiltered and dried air released around the burner. The furnace temperature is controlled by the amount of excess air supplied to the burners.[17] Temperatures above 2000°F (1893°C) are avoided to prevent the formation of nitrogen oxides which, even in small amounts, can increase acid-mist concentrations leaving the absorption tower.[17]

Typically, the SO_2 concentration of the air-SO_2 feed to the reactor (converter) is in the range of 5 to 13%, with a higher range common for sulfur-burning plants and the lower for smelters.[5]

Reactor Details

Sulfur dioxide oxidation reactors are adiabatic fixed-bed reactors with intermediate cooling between beds. The following general characteristics apply:

Diameter: 7–13 m
Single bed height: 0.5–1.0 m
Numbers of beds: 4

TABLE 15.2
Effects of Gas Feed Contaminants

Contaminant	Effect on Vanadium Catalysts
H_2O	Does not have an effect at temperatures above typical dew points for sulfuric acid (150–200°C). At lower temperatures, there may be degradation of the catalyst with loss of activity and mechanical strength, depending on the extent of condensation. Catalyst can usually be regenerated by careful heating.
As_2O_3	At temperatures significantly below 600°C, the catalyst is saturated with arsenic and a reduced plateau of catalytic activity is reached, which apparently does not change appreciably with further exposure. The decrease of activity appears to be connected with blocking of the catalyst surface by arsenic pentoxide (As_2O_5). At temperatures near 600°C, the volatile compound $V_2O_3 \cdot As_2O_5$ can be formed, and some long-term loss of activity may be noted because of vanadium losses.
AsH_3	Because of its easy oxidizability it has the same effect as As_2O_3,
Se	Harmful effect only at temperatures below 400°C; initial activity is restored after heating.
C_nH_m(hydrocarbons)	Harmless in small concentrations. In individual cases, there has been catalyst activity loss as a result of surface deposition of carbon formed by incomplete oxidation of hydrocarbons. The amount of carbon produced is dependent on properties and concentration of the hydrocarbon, the concentration of oxygen, and temperate. In large amounts, heat release from oxidation can be a serious problem.
SiF_4, HF	Sharply reduced activity, but extremely low levels act relatively slowly. HF reacts with silica supports forming volatile SiF_4, and deposition of silica gel on the catalyst surface has also been noted.
$FeSO_4$	Mechanically covers the surface of the catalyst and causes a loss of activity and pressure drop increase. When catalyst beds contaminated with iron are observed in a cold condition, hard crusts between pellets are noted. The crusts contain appreciable potassium plus vanadium, and it is evident that substantial migration of molten-salt actives occurs.
S, CS_2, H_2S	Are not objectionable in small amounts if there is sufficient oxygen to permit oxidation. Large amounts can produce harmful heat release or block the catalyst surface with sulfur deposits.
H_2	May cause loss of catalyst activity by reducing vanadium pentoxide to a lower oxidation state. Heat release may be a problem.
NH_3	Harmless in reasonable quantities. Can be oxidized with objectionable heat release when present in large amounts.
NO, CO_2	Are not objectionable at reasonably low concentrations. (Note that NO may be troublesome in acid plants, because it can contaminate product acids and/or cause formation of submicrometer acidic mists.)
CO	According to most workers, does not harm catalyst but, in the presence of large quantities of CO, it is theoretically possible for the reaction to be inhibited owing to reduction of SO_3 (SO_3 + CO $\Leftrightarrow$ SO_2 + CO_2). Heat release from oxidation can be troublesome, and there is some evidence of vanadium reduction at low temperatures (below approximately 450 to 475°C).
Cl_2, HCl	Do not cause significant problems in low concentrations. If there is extended exposure, losses of vanadium from the catalyst occur as a result of volatile $VOCl_3$ formation.
Pb, Hg	Information is limited, but analyses of spent catalyst from a number of acid plants indicate that compounds of these elements are readily deposited from very low concentrations in gases. Significant catalyst activity loss then occurs. Where *elemental* mercury is present in small concentrations, its volatility is apparently sufficient to prevent deposition on catalyst at operating temperatures.

Reprinted by permission: Donovan, J.R., Stolk, R.D., and Unland, M.L., *Applied Industrial Catalysis*, Vol. 2, p. 275, Academic Press, San Diego, 1993.

Intercooling between beds: Exit gases from each of the first three beds are cooled by
exchanging heat with lower-temperature saturated stream to produce superheated steam.
Each reaction gas stream is then returned to the reactor to enter the bed following that
from which it came.

Catalyst distribution: In general, each subsequent bed consists of a larger catalyst charge than
the previous. A reported distribution for one plant is given in the order Bed 1 to 4, as 19.4,
25.0, 26.7, and 28.9 percent.[17]

Materials of construction: Stainless steel has replaced carbon steel and cast iron.[18]

Absorption and Drying

Concentrated H_2SO_4 in the range of 98.5% is used to dry the combustion air for the sulfur burning
furnace and for the absorption towers (intermediate and final). The towers are typically packed
with 12–15 ft of 3-inch acid-resistant packing supported on an acid-proof brick or ceramic support.[17]
The towers are constructed of steel and lined with an acid resistant membrane and acid-proof
brick.[17] The acid, as it absorbs SO_3, increases in temperature, and that increase is generally restricted
to no more than 63°C to prevent problems with downstream equipment. Water is also added to the
circulating absorption acid to maintain a constant acid composition as product SO_3 is absorbed.
Since the drying tower effluent contains a higher water content, it can be blended with the circulating
absorber acid. Finally, product acid is withdrawn from the circulating stream of the final absorber
and sent to storage for dilution to desired strength or fortification with additional SO_3 to produce
oleum. The final absorber overhead containing no more than 300 ppm of SO_2 exits the absorber at
the top after passing through special mist eliminators as a means of preventing SO_3 mist from
entering the atmosphere.

By-Products

The major by-product is the valuable energy produced by burning SO_2, followed by its oxidation
to sulfur trioxide. There is much effort spent to recover a large portion of the energy so produced.
Sulfur burning produces the highest-temperature gases, and these are passed through a wasteheat
boiler that generates high-temperature steam, which means high-pressure steam that is attractive
for power generation. Large stand-alone plants often generate sufficient superheated steam that it
can be used to drive a large turbogenerator which can be hooked to the local power grid and provide
an attractive income return. Sulfuric acid plants capture all usable heat energy, which includes that
produced by the two oxidations and the heats of absorption.

Reactor Design and Operating Strategy

The reactor is a classic example of a design for an equilibrium reaction for which the goal is
essentially complete conversion of the key reactant (e.g., SO_2). Because of environmental regulations, the release of unreacted SO_2 into the atmosphere is rigidly limited so that a total conversion
of SO_2 of 99.5% or above is required in many jurisdictions.[17]

To attain high conversion, it is necessary to stage an adiabatic reactor and provide for intermediate cooling between beds. This requirement is clearly shown in comparing Figs. 15.3A and
15.3B. Figure 15.3A shows a typical equilibrium plot as SO_2 conversion versus temperature for a
single bed. Figure 15.3B is a plot for multiple beds with intermediate cooling. The diagonal lines
represent the adiabatic temperature increase in a designated bed and the horizontal lines cooling
curves between beds. A high temperature rise in the first bed is permissible, because the outlet
temperature is somewhat below the equilibrium temperature (i.e., equilibrium approach ΔT is
substantial). To reach higher conversions, the effluent from the first bed must be cooled before
entering the second bed. As shown in Figure 15.3B, this same procedure is repeated following
each successive bed.

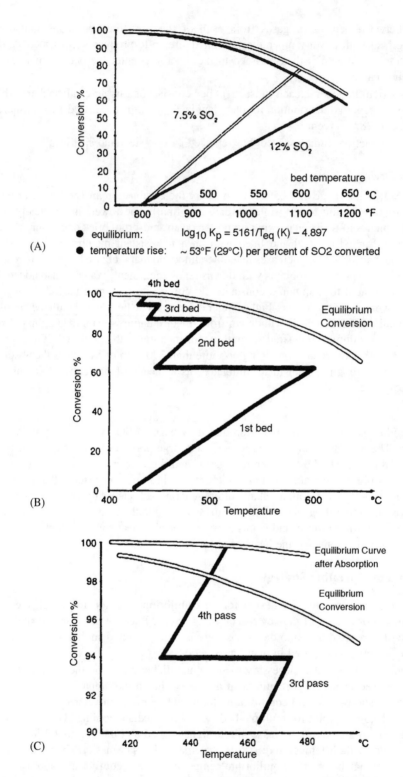

FIGURE 15.3 Equilibrium Conversion—Adiabatic Temperature Plots. Reprinted by permission: Jensen–Holm, H., ASIAFAB, March 1993, British Sulphur Publishing. (A) single-catalyst bed, (B) four-catalyst beds with intermediate cooling, and (C) effect of interstage absorption of SO_3 prior to final bed.

As the average bed temperature is reduced in each bed, the equilibrium approach ΔT declines, as does the rate constant. Thus, the average reaction rate in each successive bed declines, and more catalyst is required. Figure 15.4, a plot of reaction rates versus temperature at various SO_2 conversion levels, illustrates the issue clearly.

This four-bed design with intermediate cooling does very well in producing a 98.5% SO_2 conversion at the reactor outlet. In many locations, this is not enough to meet SO_2 vent regulations. The reactor is followed by heat recovery units and then an absorber that absorbs the SO_3 but not the sulfur dioxide. Thus, any unreacted SO_2 has to be vented. Newer regulations make the vent gas from a 98.5% conversion unit excessive for many locations. Instead, a conversion of 99.5% or above is required. This goal can and has been met. An intermediate absorber must be added to remove SO_3 from the discharge of the third bed. In so doing, the equilibrium line for the final bed is, in effect, moved upward as shown in Figure 15.3C, where the added conversion is illustrated.

An alternative or added improvement will continue to be improved catalysts that maintain activity below 400°C. At these lower temperatures, significantly improved rates will routinely produce higher conversions in the last bed.

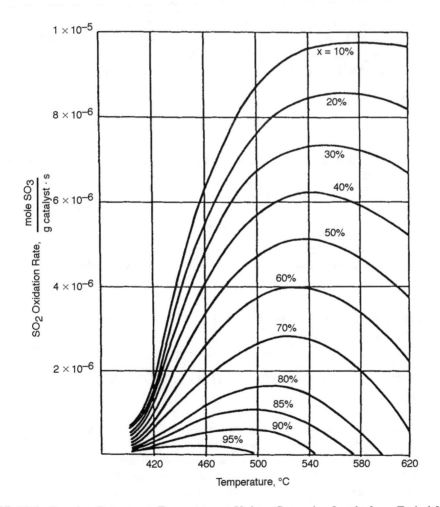

FIGURE 15.4 Reaction Rate versus Temperature at Various Conversion Levels for a Typical Vanadium Catalyst. Reprinted from Urbanek, A. and Trela, M., *Catalysis Reviews: Science & Engineering* 21 (1), 73 (1980), by courtesy of Marcel Dekker, Inc.

Process Kinetics

Possibly no other industrial reaction has been studied so long by so many in an attempt to develop useful rate relations and reasonable and valid mechanisms. An excellent review of the kinetic studies on this reaction has been published,[5] and the authors conclude, in effect, that attempts to develop equations useful for commercial reactors have not been successful.[5]

This single-reaction system is extremely complex, involving mass-transfer limitations of a difficult-to-define system with pores capable of being wetted in different ways that are quite unpredictable, as is the nature of the melt at different temperature levels. Although some equations have been tested on several scales (bench, pilot, and plant), the agreement was good only on the bench scale. No agreement was achieved for the plant-scale reactor.[5] Apparently, some seasoned investigators familiar with plant operations have developed factors with which to multiply calculated bed depths to obtain the required depth, with Bed 1 requiring the largest multiplier, 4.[19] Of course, the required bed depth depends greatly on the particular catalyst used. Others have reported calculated values consistent with a plant reactor in Beds 2, 3, and 4[20] and in Beds 1 and 4.[21]

One major problem related to large commercial reactors with relatively shallow but large diameter beds is the degree of nonuniform flow due to varying gas velocities over the cross-section of the bed that result in significant temperature and composition profiles over the bed area. This problem is better solved by packing the bed with care and providing low ΔP distributor devices that tend to throw the gas over the entire cross-section than by trying to calculate flow distributions.

For industrial use, the best suggestion is to use a Power-Law equation such as the following:

$$r = P_o\left(\frac{P_{SD}}{P_{ST}}\right)^n\left(1 - \frac{P_{ST}}{K_p P_{SD} P_o^{0.5}}\right)$$

at equilibrium $r = 0$, and then the equation reduces to the correct equilibrium expression when solving for K_p at $r = 0$.

$$K_p = \frac{P_{ST}}{P_{SD} P_o^{0.5}}$$

where P_O, P_{SD}, P_{ST} represent, respectively partial pressures of oxygen, sulfur dioxide, and sulfur trioxide.

The reported values of n range from 0.4–0.8. Detailed reviews of a variety of many rate equations in the literature are available.[5,22]

15.2 AMMONIA $\rightarrow$ NITRIC OXIDE $\rightarrow$ NITROGEN DIOXIDE $\rightarrow$ NITRIC ACID

Nitric acid ranks third in production volume on the list of three major inorganic acids: sulfuric, phosphoric, and nitric. Its largest use is in the production of ammonium nitrate for use in fertilizers and explosives. Approximately 65% of nitric acid is used to make fertilizer (mainly ammonium nitrate) and 25% is used in producing explosives (also mainly ammonium nitrate). The remaining 10% is essential in a variety of reactions with organic compounds to yield commercially important compounds such as nitrobenzene, dinitrotoluene (used to make toluene diisocyanate), silver nitrate, nitroglycerin, nitrocellulose, trinitrotoluene (TNT), and also as an oxidizer in the production of such important compounds as adipic acid.[1,2]

Its most important reaction product, ammonium nitrate, decomposes at temperatures above 210°C, but this decomposition can occur at lower temperatures in the presence of metal oxides or powdered metals. Although ammonium nitrate has an excess of oxidizing power, it requires the presence of more sensitive organic explosives that have an excess of reducing power such as TNT (trinitrotoluene) to produce an effective explosive.[3] However, hot aqueous solutions of concentrations above 50% can decompose explosively under hot and confined conditions. Yet its use in fertilizers poses no threat of dangerous decomposition.

Chemistry

The oxidation of NH_3 to NO_2 takes place via two reactions: a catalyzed reaction over platinum-rhodium screens and a homogenous reaction of NO (nitric oxide) with O_2 to produce NO_2, nitrogen dioxide. It is customary in the literature to write the stoichiometric equations in terms of whole numbers.

$$4NH_3 + 5O_2 \rightarrow 4NO + 6H_2O$$

$$2NO + O_2 \rightarrow 2NO_3$$

From an engineering standpoint, it is less likely to cause confusion when dealing with the basis for the heat of reaction, for it is often expressed as kcal/mole when it may actually be kcal/4 moles of ammonia. The following equation will avoid this ambiguity.

1. $NH_3 + 1.25O_2 \rightarrow NO + 1.5 H_2O$ (vapor phase)
   $\Delta H = 54.1$ kcal/mole NH_3 @ 1000 K
   $K = 1.87 \times 10^{14}$ @ 1000 K
   Catalyst: Pt/Rh/Pd
   Conditions: 800–940°C @ 3–6 bar or 7–12 bar

2. $NO + 0.5 O_2 \rightarrow NO_2$ (vapor phase)
   $\Delta H = -13.75$ kcal/mole of NO @ 1000 K
   $K = 0.188$
   $\Delta H = -13.66$ kcal/mole of NO @ 400 K
   $K = -3.483 \times 10^3$
   Homogeneous reaction
   Conditions: 50°C @ 3–6 bar or 7–12 bar[4]

Absorption of NO_2 and its dimer (N_2O_4), with which it equilibrates, produces nitric acid. One concept of this absorption is based on the N_2O_4 as the reactive species formed in the previous vapor-phase oxidation.[6]

$$2NO + O_2 \rightarrow 2NO_2 \Leftrightarrow N_2O_4 \text{ vapor phase}$$

$$N_2O_4 + H_2O \rightarrow HNO_3 + HNO_2 \text{ liquid phase}$$

$$HNO_2 \rightarrow 0.33 \ HNO_3 + 0.33 \ H_2O + 0.667 \ NO \text{ liquid phase}$$

The NO released is oxidized to NO_2 and N_2O_4 by oxygen in the liquid phase and from the vapor phase.

Reaction 1 is irreversible at any temperature, and LeChatelier's principle is not applicable when assessing pressure effects on this main reaction. Undesirable side reactions can occur in the catalyzed ammonia oxidation unit. The following data are at 1000 K except as noted.

	ΔH kcal/mole NH_3*	K_p
3. $NH_3 + 0.75\ O_2 \rightarrow 0.5N_2 + 1.5\ H_2O$	-75.75	2.18×10^{18}
4. $NO \rightarrow 0.5\ N_2 + 0.5\ O_2$	-21.64	1.641×10^3
5. $NH_3 + 1.5NO \rightarrow 1.25\ N_2 + 1.5\ H_2O$	-108.21	3.09×10^{14}
6. $NH_3 \rightarrow 1/2\ N_2 + 1.5\ H_2$	13.11	1.757×10^3
	$(11.43\ @\ 400\ K)$	$(0.178\ @\ 400\ K)$
7. $NH_3 + O_2 \rightarrow 0.5\ N_2O + 1.5\ H_2O$	-65.855	1.85×10^{14}

*Except for Reaction 4, which is given per mole of NO.

Mechanism

Never has a reaction system been studied so long with ultimately so little consensus on a mechanistic explanation. The reason for this situation is clear when one considers the unusual nature of the reactions involved. Oxidation of NH_3 to NO by Pt gauze is one of the highest-temperature and shortest-contact-time commercial processes known. Contact time in the gauze pack area is variously reported as 10^{-3} to 10^{-4} seconds. The estimated actual reaction time is 10^{-11} seconds. With these short residence and reaction times, it was not possible to experimentally detect short-lived intermediates for many years. Although new techniques, such as laser-induced fluorescence, may be successfully applied, there is not much incentive to do so. Much of the lore for successful operation has already been gained over the decades.

Studies using AES and LEED under very low pressure and with single crystals of platinum and platinum wire have culminated in some hypothesis of value[4,7-10]

- At commercial operating conditions the overall rate is mass-transfer controlled. Selectivity in ammonia oxidation is determined by competition between NH_3 and O_2 molecules for active sites.
    (1) Below 200°C, complete coverage by N atoms
    (2) 200–500°C, coverage by NH_3 molecules
    (3) 500–1000°C, coverage by O_2
    (4) Above 1000°C, NH_3 decomposition
    Range 3 falls in the operating range of commercial reactors
- At the conditions in range 3, optimum selectivity to NO is realized at pressures 3–6 bar and around 850°C. At higher pressure, the temperature must be higher, 920–940°C.
- Mass-transfer calculations for ammonia provide an estimate for a maximum mole fraction of 0.05 to 0.06 on the surface of the gauze. These low mole fractions on the gauze surface based on mass transfer along with a feed ratio of NH_3/O_2 of less than stoichiometric (0.3 or less versus 0.8 stoichiometric) assures low NH_3 concentration on the catalyst surface during actual reaction. Such conditions reduce the rate of side reaction 3, 5, and 7. These reactions are thought to be higher order in ammonia than the main reaction. The NO decomposition reaction is, of course, reduced by the short contact times, but higher-pressure operation slightly affects the yield of NO, because the rate of Reaction 4 is apparently second order in NO partial pressure.

- In Range 4, the reaction system becomes dominated by Reaction 6, the decomposition of NH_3 to N_2 and H_2. The hydrogen is rapidly oxidized to water. As noted above, Reaction 6 is endothermic and becomes thermodynamically favorable at elevated temperatures.

Catalyst Type

The catalyst consists of 80 × 80 woven wire gauze composed of 0.003 in. (0.076 mm) diameter platinum (90%) and rhodium (10%) wire or platinum 90%, rhodium (5%), and palladium (5%). Since palladium is less expensive than rhodium, this latter combination is preferred and apparently performs very well. Slightly thinner wire (0.06 mm) is reported to be used for low-pressure units operating near atmospheric pressure.[11]

The gauzes, which handle much like cloth, are packed snugly against each other in numbers varying from 4 to 50, depending on the operating pressure, with low-pressure reactors using fewer gauzes. Gauze diameter is reported to vary from 0.5 to 3 m, depending on plant size and pressure. High-pressure plants (8–10 bar) would use a gauze-pack diameter in the range 0.5 to 1.5 m, depending on plant size. Low-pressure plants would be in the higher range of 2 to 3 m. In all cases, the catalyst gauze-pack diameter-to-height ratio is very large and requires clever reactor design to ensure even distribution of reactants through the pack.

Rhodium content improves the activity of the catalyst as well as its strength. During initial operation, Rh content of the wire surface is enriched, and activity improves.[4] The improvement of catalyst performance upon alloying with rhodium was first reported in 1934[13] and is credited with the success of modern nitric acid plants.[13] Not only did the alloy catalyst improve conversion and selectivity, it reduced Pt loss by oxidation and then vaporization of PtO_2.[14] Rhodium content above 10% was found to be undesirable in that it reduced catalyst activity.[5] The subsequent substitution of Pd for a portion of the Rh (90% Pt, 5% Rh, 5% Pd) produced a catalyst system that performed as well as the original Pt (90%)/Rh (10%) with no loss of activity or selectivity.[15] It has also been claimed that the Pt-Rh-Pd gauze performs with significantly lower precious-metal loss.[9]

Over the years, various improvements have been introduced by catalyst suppliers. These include preactivated gauzes, warp-knitted gauzes, mixed gauze packs with smaller diameter wire in down-stream gauzes, and gauze packs for pressure operation with a nickel-chromium or other high-temperature alloy porous pad or gauzes that substitute for some of the Pt alloy gauzes toward the outlet.[5,12] Plant experience has led to the discovery that some of the downstream gauzes in the pack primarily improve operations by providing sufficient added ΔP to assure more even distribution of flow across the entire reactor diameter and also better mixing.[16] Of course, as is the case in most fixed-bed reactors, extra catalyst has an important role to provide needed activity when the earlier portions of the bed become deactivated. Thus, replacement of a portion of the gauzes near the outlet with an inert back-pressure pad must be done judiciously. Warp-knitted gauzes are stronger yet more flexible. They also have higher surface area per unit area and greater void volume, which means lower ΔP and less tendency to retention of fine solid particles.

Experience has shown that Pt (95%) Rh (5%) is also used successfully for certain operating conditions and reactor designs. Less rhodium, either alone with platinum or palladium, forms less inactive rhodium oxides.

Catalyst Suppliers

Engelhard
Johnson–Matthey
Degussa

Customers can order the desired number of gauzes and arrange and install them at the plant, or they may order a prepacked unit that can be installed straightaway in the reactor.

Catalyst Deactivation

Poisoning

The gauze pack is poisoned or deactivated by almost any solid impurity such as rust (iron oxide), dust, and other impurities carried in the air and ammonia streams. A rather sophisticated filter system is used for both air and liquid ammonia to remove 99.9% of all particles larger than 0.5 mm using plastic or glass fibers for air and Teflon and sintered metals for liquid ammonia.[5] A final filter for the air-gaseous ammonia mixture is necessary to remove, primarily, particles of rust that may have come from corrosion in the equipment prior to the reactor. Although great care is exercised in selecting materials of construction, some corrosion or erosion occurs, especially in the high-temperature region immediately upstream from the reactor. The mixed-gas filter is therefore placed between the mixer and the reactor.

Overloading or failure of filter elements can cause rapid increases in pressure drop and poor distribution of reactants. Even relatively very small particles can cause plugging in 80-mesh gauzes.

Iron oxide is an ammonia decomposition catalyst and not only can cause activity decline by covering portions of the surface but can decompose ammonia and reduce selectivity. Sulfur compounds are also catalyst poisons. The source of sulfur can be organic sulfur compounds from lubricating oil or SO_2 from polluted air.[9] Much care is exercised in cleaning process air and in avoiding lubricating oil carryover from air compression.

Platinum Loss

Start-up of an ammonia oxidation converter requires a heat-up period, which is common to many exothermic adiabatic reactors. In the case of Pt, the fresh gauze is not active enough to produce a self-sustaining reaction. The catalyst must first reach a higher temperature, referred to as an *ignition temperature*. A hydrogen torch installed in the reactor and directed at the gauzes is used to ignite the catalyst.

After the catalyst ignites, a period of increasing conversion occurs as the gauze wire surface begins to roughen as metal migrates to the surface and forms nodules, thereby creating more surface area. When the activity reaches its peak, platinum loss becomes significant and occurs at constant rate.[9] The loss of Pt causes a gradual decline in catalyst activity. The loss process occurs as Pt becomes oxidized producing platinum oxide in gaseous form.[16]

$$Pt(s) + O_2(g) \rightarrow PtO_2(g)$$

The rate of loss increases with operating temperature. There is a tenfold increase from 820–920°C.[4] Platinum is lost preferentially, but some Rh migrates to the surface as Rh_2O_3, which is inactive in the mixed-metal catalyst.[18] Catalyst cycle time (life) has been reported to vary from 2 to 3 months for high-pressure units and 8–12 months for low-pressure plants. At the end of a run, the top several inactive gauzes can be removed and the same number of fresh gauzes placed at the bottom of the pack. For operations in which made-up gauze-pack units purchased from the vender are used, the entire pack unit is easily removed and returned to the manufacturer for service.

Platinum Recovery

Because of the intrinsic value of platinum as a very useful rare metal, units of gauzes called getters are installed just below the reactor gauzes. The getter consists of Pd gauzes. About 70–80% of the lost Pt is recovered. The getter can then be returned to the supplier for Pt recovery and credit. Each sheet of Pd gauze is separated by a base metal (e.g., Ni) screen to improve rigidity of the gauze pack. Some designs employ progressively gauzes of higher surface area per unit mass to better adsorb PtO_2 vapors of lower concentration.

It is possible, particularly during the unsteady ignition period, that some NH_3 bypasses the reactor gauze and reaches the getter where it can react with the NO catalyzed by palladium. This reaction (No. 5 on page 234) is highly exothermic and can melt the Pd getter gauze(s) destroying its effectiveness.[19] Great care and experience are, as in so many start-ups, essential for unit success.

Process Units[1-5, 11]

Three types of plants that differ primarily in operating pressure (which affects other operating variables) are shown in Table 15.3 for typical ammonia-oxidation reactor operation. Lower-pressure reactors have the advantage of lower energy costs, better selectivity, and then higher yields. But low-pressure operation of the absorption system allows excessive amounts of nitrogen oxides to remain in the tail gas. High-pressure plants are much smaller in equipment sizes and are less costly but produce lower yields. High-pressure plants do require higher energy costs due to compressor power requirements, but more efficient removal of nitrogen oxides is realized in the absorption section. The result is that low-pressure plants are no longer built because of excessive emissions of nitrogen oxide. In Europe, medium-pressure plants and medium/high-pressure plants (medium-pressure reactor operation and high-pressure absorption system) are common. Medium/high-pressure plants are preferred in Europe because of the higher cost of ammonia and energy. The combination produces low emissions and higher yields than high-pressure plants, and less platinum loss. The savings in energy and ammonia costs due to higher yields can overcome higher cost of the larger equipment. In the U.S.A., high-pressure operation is used for both reaction and absorption. For both the medium/high-pressure process and the high-pressure process, air must be compressed to the reactor operating pressure, but in the medium/high-pressure process NO_x products from the reactor must be compressed for absorption. Nitrogen oxides (NO_x) are not only poisonous but also create explosion hazards because of their strong oxidizing properties.

TABLE 15.3
Typical Design Data for Ammonia Oxidation Reactors

Reactor Type	Pressure Bar	Number of Gauzes	Gas Velocity m/s	Reaction Temperature °C	Catalyst Loss, g/t HNO_3
Low pressure	1–2	3–5	0.4–1.0	840–850	0.05–0.10
Medium pressure	3–7	6–10	1–3	880–900	0.15–0.20
High pressure	8–12	20–50	2–4	900–950	0.25–0.50

Reprinted by permission: Thiemann, M., Scheibler, E. and Wiegand, K.W. in *Ullmann's Encyclopedia of Industrial Chemistry*, 5th edition, Vol. A17, p. 293, Wiley/VCH, Weinheim, Germany, 1991.

High-pressure plants of the type built in the U.S.A. are reported to be 10–14% lower in cost than the dual-pressure plants.[4] In effect, each area of the world will build either high or dual-pressure plants, depending on economic analysis based on raw material and equipment costs. In the monopressure high-pressure process, the lower yield and additional Pt loss must be weighed against lower plant equipment, construction, and maintenance costs.

To maximize production, high-pressure plants must be operated at higher temperatures, but this condition, along with the higher pressure, favors energy recovery.[4]

Yields as a function of operating pressure are reported as 97–98% at 1 bar, 95–96% at 5 bar, and 94% at 8–10 bar.[11] Reaction temperature ranges are given in Table 15.3.

Ammonia Oxidation Reactor

Filtered air is compressed to the desired operating pressure and mixed with ammonia vapor produced in a vaporizer from the liquid and then filtered. Heaters are used to raise the temperature of each stream to the desired inlet temperature prior to mixing. The mixture consists of 11 mole % NH_3 in high-pressure plants and slightly higher for medium-pressure units. The explosive limit is 12.4 mole %, and the stoichiometric ratio gives a value of 14.2 mole %.

The mixed reactants are filtered once again and fed to the top of the reactor, from which they pass downflow through the reactor. Because of the length-to-diameter of the catalyst gauze pack, some means of encouraging even flow throughout the bed cross section is provided. The reactants pass downward through a perforated plate, the Pt/Rh/Pd catalytic gauze pack, a getter gauze pack for recovery of Pt vaporized from the catalyst as PtO_2, and a support screen or refractory packing to hold the combination and provide ΔP toward the outlet to prevent necking down of the flow profiles. Reactors may be built with conical heads to further aid in vapor dispersal over the cross section.

A waste-heat boiler is placed very closely attached to the reactor or within the reactor shell to recover a significant amount of valuable high-temperature enthalpy from the product leaving the catalyst bed. High-pressure steam is generated and, after superheating, used to drive turbines for the compressor.

The product is further cooled to a low temperature at which homogeneous oxidation to NO_2 takes place with the introduction of added air. The equilibrium (see Reaction 2) is favorable at low temperatures and also favored by high pressure because of the decline in moles (3-1/2 to 1). Finally, the NO_2 stream is absorbed in the absorber where low temperature and high pressure is also favored. The dual pressure process is designed to operate at high pressure (at 8–12 bar) in the section following the heat recovery and cooling train after the reactor.

Abatement Reactor

NO_x emissions must be kept below 200 ppm in many jurisdictions. Modern absorption units can do a good job of removing NO_x so that the tail gas contains as little as 200 ppm,[11] but lower values may require an additional absorber or a catalytic reactor in which the following reactions occur.

$$NH_3 + 1.5\ NO \rightarrow 1.25\ N_2 + 1.5\ H_2O$$

$$NH_3 + 0.75\ NO_2 \rightarrow 0.875\ N_2 + 0.5\ H_2O$$

$$NH_3 + 0.5\ NO + 0.5\ NO_2 \rightarrow N_2 + 1.5\ H_2O$$

$$NH_3 + NO + 0.25\ O_2 \rightarrow N_2 + 1.5\ H_2O$$

The tail gas at the outlet pressure of the absorption section is passed to a catalytic reactor containing a noble-metal catalyst (e.g., Pt 0.1%, Ni 3% on alumina) or a base-metal catalyst such as vanadium pentoxide, cobalt molybdenum oxide, or tungsten oxide.[5] The vandadium pentoxide is reported to be most efficient. Operating temperatures are in the range of 250–400°C, and tail gas so treated has as little as 50–100 ppm of NO_x compounds.

Catalyst Suppliers
   United Catalysts
   Engelhard

Licensed Processes
   BASF
   Gulf (now Chevron)
   Mitsubishi

Process Kinetics

The extremely complex and rapid reaction system in the partial oxidation of ammonia to nitric oxide has not yielded easily to kinetic studies. In actual plant operation, the rapid growth of thin nodular surfaces on the wire increases the surface area for reaction and causes the observed rates to be highly dependent on the gas-phase mass transfer rather than the surface kinetics. Selectivity to NO is realized by the resulting low ammonia concentration at the surface when operating in the range of 500–1000°C.

A rather thorough review of various chemical reaction rate expressions has been presented.[9] Most of the work was done at less than 600°C and subatmospheric pressure so that surface science techniques might be used. By contrast, actual plant operation is at 3–12 atm and 850–950°C. The hypothesis put forward is that selectivity to NO depends on the competition between NH_3 and O_2 for active sites. At plant temperatures and with the short contact times, the surface is rich in oxygen and, under such conditions, an Eley–Rideal equation or Langmuir–Hinshelwood noncompetitive adsorption may be useful.[8,9]

$$O(s) + NH_3(g) \rightarrow NO(g) + 3H(s)$$

where s refers to the adsorbing surface.

$$r = \frac{kP_A P_o^{1/2}}{(1 + K_A P_A)(1 + K_o^{1/2} P_o^{1/2})}$$

where subscripts A and o refer, respectively, to ammonia and oxygen.

Reasonable fit with experimental data was realized, but the authors wisely refrained from attaching profound meaning to the derived constants since they are actually lumped parameters containing undefined variables.[8]

The rate of platinum loss is a slower, more definable reaction that has significant mass-transfer characteristics. The following equation has proved useful:[17]

$$Pt(s) + O_2(g) \Leftrightarrow PtO_2(s)$$

$$r = \frac{k_b K_{eq} Po}{(1 + k_b/k_m)}$$

where k_b = rate constant for the reverse reaction that occurs when gaseous PtO_2 strikes the getter surface

K_{eq} = equilibrium constant for the reversible reaction

k_m = mass-transfer constant

15.3 HYDROGEN SULFIDE → SULFUR

Sulfur is the 13th most abundant element in the Earth's crust, occurring as elemental sulfide, organic sulfur compounds in petroleum and coal, and inorganic sulfides and sulfates.[1] The major use (86%) for elemental sulfur is in the manufacture of sulfuric acid. Other uses include rubber curing and vulcanizing agents and production of sulfurous acid for sulfide pulping ($S + O_2 \rightarrow SO_2$, $SO_2 + H_2O \rightarrow H_2SO_3$), carbon disulfide ($CH_4(g) + S(g) \rightarrow CS_2$), phosphorous pentasulfide ($2P + 5S \rightarrow P_2S_5$), sulfur concrete, detergent ingredients such as fatty-oil sulfates from fatty alcohols and SO_3 or chlorosulfonic acid, and certain dyes and pharmaceuticals using chlorosulfuric acid in the sulfating step.

Elemental sulfur is found in cap-rock deposits over salt domes and basin bedded deposits. Both are associated adjacent to anhydrite-gypsum cap rock, but the bedded deposit is not involved with a large salt dome. Sulfur deposits of both are thought to be formed by action of anaerobic bacteria on sulfates ($CaSO_4$ or anhydrite). The bacteria, over many hundreds of years, oxidized the hydrocarbons in the lower strata to CH_4 and the sulfate to $CaCO_3$.[2]

$$CaSO_4 + CH_4 \rightarrow H_2S + CaCO_3 + H_2O$$

The H_2S is subsequently oxidized to crystalline sulfur by dissolved oxygen or inorganic oxygen-containing compounds and perhaps sulfide-oxidizing bacteria.

Massive deposits of sulfur were discovered and worked over the years along the gulf coast (salt domes) and in the Delaware Basin in west Texas (bedded deposits) and also in Poland.[2] The Frasch process was used to recover liquid sulfur by injecting superheated water into the elemental sulfur deposit and pumping the sulfur to the surface. As sulfur above salt dome deposits began to be exhausted, west Texas sulfur took its place. In addition, discoveries off shore in the Gulf of Mexico began to indicate massive resources in off-shore sources. But as soon as new discoveries of oil and gas began to be dominated by crude oil with substantial organic sulfur compounds and natural gas with significant amounts of hydrogen sulfide, it became necessary to eliminate the sulfur components, particularly for environmental reasons. Use of fuels from these sources would be intolerable in a more crowded world. Environmental regulations that set the allowable emission of SO_2, for example, were soon set very low, and near-complete removal was essential. The result has been the production of large tonnages of sulfur from natural gas and refinery gases, and the natural sulfur share was dropped to around 20%. In the 1930s, sweet crude was in good supply, and natural gas H_2S content was low. In the 1940s, however, sour gas discoveries occurred, and the H_2S had to be removed to make the gas suitable for home and commercial use. Several decades later, more and more sour crude oil was entering refineries, and removal of H_2S formed during processing was in such quantities that combustion as an energy source became impossible. Economical recovery of sulfur from these sources began to show great promise and favorable economics because of the growing decline of new elemental sulfur discoveries.

The various organic sulfur compounds in liquid petroleum cuts are removed by hydrodesulfurization in which the sulfur compounds are converted to hydrocarbons and hydrogen sulfide. Natural gas sulfur is already in the form of hydrogen sulfide. In both cases, the H_2S is removed from the gas by absorption using chemical solvents that form a weak bond with H_2S, which can be broken in a stripper operating at an elevated temperature. Chemical solvents include either primary, secondary, or tertiary amines, each with special properties upon which a selection is made relative to cost, selectivity relative to other components present in the gas (e.g., CO_2), and stability. So-called *hot carbonate processes* use aqueous potassium carbonate as the chemical solvent. Other options include adsorption processes that use solid adsorbents such as molecular sieves and physical absorption processes using solvents such as dimethyl ether of polyethylene glycol, methanol, N-methylpyrrolidone, and methyl isopropylether.[2] The H_2S is absorbed under pressure in such physical absorbents and released in a series of flash drums followed by a stripper. Natural gas contains varying amounts of carbon dioxide which, as an acid gas, is removed along with the H_2S. Refinery gases can contain small amounts of hydrocarbons (up to 5%).

Refiners and gas-plant operators turned to a very old process, called the Claus process, for conversion of H_2S to elemental sulfur. In 1883, C. F. Claus obtained a British Patent (No. 5958) for the partial combustion of H_2S to sulfur.[3] The process was improved by A. M. and J. F. Chance in 1887 (British Pat. 8666). Variations on this original so-called Claus or Claus-Chance process include the several modern processes. Today, most are referred to as *Claus* or *modified Claus* processes.

Chemistry

The combined reaction system, which includes both catalytic and thermal (noncatalytic) reactions, is most complex. Although our focus is catalysis, some mention of high-temperature thermal reactions is essential to understanding both the original and the modified Claus processes. The following stoichiometric equations will provide the focus for descriptions that follow them. Note that two forms of sulfur, S_2 and S_8, are shown with S_8 at high temperatures, above 1700°F (1200 K), and S_2 at lower temperatures, below 400°F (478 K). In between, varying amounts of each sulfur species, S_2 through S_8, exist.[8] As an approximation, heats of reaction given are based on S_8 in the low-temperature region for the catalytic step and S_2 in the high-temperature thermal region that occurs in a furnace upstream of the catalytic reactor in the modified Claus process.

Most modern Claus processes consist of two reaction sections, a furnace that partially combusts the acid-gas feed containing H_2S and a catalytic reactor section of two to three stages that catalyzes the sulfur-forming reaction at the thermodynamically favorable low-temperature range. Operating inlet pressures are in the range of 5 to 12 psig (1.36–1.84 bar). Temperature range in the furnace section is 1700–2500°F (1200–1644 K). The range in the catalytic section is 450–370°F (505–461 K), with the lower temperatures occurring in the second and third beds.[5]

It is important to recognize that these equations are simply stoichiometric forms that do not define the mechanism of the complex reaction system. In particular, in the high-temperature region, the reactions no doubt involve free-radical intermediates that produce a number of complex reaction paths.

References 4, 5, and 8 are the sources of the heats of reactions listed.

1.	$H_2S + 0.5\ O_2 \rightarrow 0.5\ S_2 + H_2O$	$\Delta H = -38.0$ to -38.2 kcal @ 1700–2200°F (1200–1478 K)
1A.	$H_2S + 0.5\ O_2 \rightarrow 0.125\ S_8 + H_2O$	$\Delta H = -49.3$ kcal @ 400–600°F (476–589 K)
2.	$0.5\ S_2 + O_2 \rightarrow SO_2$	$\Delta H = -86.5$ to -86.4 kcal @ 1700–2200°F (1200–1478 K)
2A.	$0.125\ S_8 + O_2 \rightarrow SO_2$	$\Delta H = -74.5$ kcal @ 500°F (533 K)
3.	$H_2S + 1.5\ O_2 \rightarrow SO_2 + H_2O$	$\Delta H = -124.5$ to -124.6 kcal @ 1700–2200°F (1200–1478 K)
	Sum of Reactions 1 and 2 occurs in furnace.	
4.	$2H_2S + SO_2 \rightarrow 0.375\ S_8 + 2\ H_2O$	$\Delta H = -25$ to -22 kcal @ 200–700°F (367–644 K)
	Occurs in catalytic reactor.	
4A.	$2H_2S + SO_2 \rightarrow 1.5\ S_2 + 2H_2O$	$\Delta H = 10.1$ to 9.9 kcal @ 1700–2200°F (1200–1478 K)
	Occurs in furnace if sufficient H_2S is present.	

Side Reactions

Hydrocarbons present in the acid gas feed cause the production of carbon disulfide in the furnace (Reactions 5 and 6), which can react with water or SO_2 to slowly form carbonyl sulfide (Reactions 7 and 8). Another possible route to carbonyl sulfide is via hydrogenation of CO_2[11] as shown in Reactions 9–11.

5.	$CH_4 + S_2 \rightarrow CH_2 + 2H_2$
6.	$CH_4 + 2S_2 \rightarrow CS_2 + 2H_2S$
7.	$CS_2 + H_2O \rightarrow COS + H_2S$
8.	$CS_2 + SO_2 \rightarrow 2\ COS + 1.5\ S_2$
9.	$H_2S \rightarrow S + H_2$
10.	$CO_2 + H_2 \rightarrow CO + H_2O$
11.	$CO + S_2 \rightarrow COS$

Chemistry of Original Claus Process

The original Claus process was developed by Carl Friedrich Claus for the purpose of sulfur recovery from waste CaS produced in the Leblanc process for the manufacture of sodium carbonate.[8] Claus patented the process in 1883 for converting the H_2S, formed from the waste CaS stream, to elemental sulfur.[8] In 1890, the first commercial unit was brought on stream.[7]

Leblanc Process

$$2NaCl + H_2SO_4 \rightarrow Na_2SO_4 + 2HCl$$

$$Na_2SO_4 + 2C \rightarrow NaS_2 + 2CO_2$$

$$NaS + CaCO_3 \rightarrow Na_2CO_3 + CaS$$

$$CaS + H_2O + CO_2 \rightarrow CaCO_3 + H_2S$$

Claus Reaction of waste H_2S

$$H_2S + 0.5\ O_2 \rightarrow S + H_2O \tag{1}$$

$$S + O_2 \rightarrow SO_2 \tag{2}$$

This Claus reaction is Reaction 1 under the previous listing. Reaction 2, the oxidation of sulfur to SO_2, can also occur, and both reactions are essentially irreversible. Claus, however, discovered that they were kinetically controlled and that, at modest temperatures, 500°F (533 K), Reaction 1, sulfur production, was much faster than Reaction 2 when a catalyst was used.[3]

Both reactions are highly exothermic, and hence Claus designed a large kiln (reactor) with external high surface area so that reaction heat could be dissipated by radiation and the temperature kept in the region where oxidation of sulfur to sulfur dioxide was minimized.[3] As catalyst, he selected bog-iron ore and later bauxite (a high-alumina-content ore). Temperature was maintained at around 533 K (500°F) by regulating the flow of inlet air and H_2S or by recycling stack gas.[3] It was essential to maintain as low a temperature as possible without reaching the sulfur dew point, at which condition the catalyst would be deactivated by sulfur deposits. Temperature control was difficult, especially in an effort to avoid high temperatures that led to excessive SO_2 formation. Various improvements in design were made in a subsequent joint patent with J. F. Chance.

Chemistry of I. G. Farbenindustrie Process and Modern Processes

After 55 years of minor changes in the original process, I. G. Farbenindustrie developed a modified process, the major concept of which remains in use today. The kiln with catalyst was replaced by a high-temperature furnace into which was fed one-third of the H_2S feed. The stoichiometric equation No. 3, which is a combination of Reactions 1 and 2, applies and, ideally, essentially goes to completion.

Furnace

$$H_2S + 1.5\ O_2 \rightarrow SO_2 + H_2O \tag{3}$$

The exit gases from the furnace were cooled and passed to a catalytic reactor, where the remaining two-thirds of the H_2S was contacted with the SO_2 and reacted to produce sulfur at temperatures near the sulfur dew point.

Catalytic Reactor

$$2H_2S + SO_2 \rightarrow 1.5\ S_2 + 2H_2O \tag{4}$$

Combination of these two equations yields

$$3H_2S + 1.5\ O_2 \rightarrow 1.5\ S_2 + 3H_2O$$

This equation is identical to Reaction 1 when divided by 3 and thus represents the main reaction in the original Claus kiln as well as the overall reaction for the modified process. The advantage of the new process becomes apparent upon comparing the distribution of reaction heats[7] as shown in Table 15.4. The original process produced an exothermic heat, at the desired low temperature of approximately = 156 kcal, which made it difficult to maintain the favorable low temperature so important for the catalytic step. The new process produces most of the exothermic heat in the furnace, where a high temperature is desired anyway. In the catalytic step, a much smaller heat production occurs, making low-temperature maintenance much less of a problem.[7] Subsequent improvements in the I. G. Farben process consisted of several catalytic reactors in series so that intermediate cooling and sulfur removal will shift the equilibrium favorably.

TABLE 15.4
Comparison of Heat Production in Claus and Modified Claus Processes

	Furnace	Catalytic Reactor	Total
Original Claus process*	NA	−147.9 kcal	−147.9 kcal
Modified Claus†	−124.5 kcal†	− 23.0 kcal§	−147.3 kcal

Basis: 3 Moles H_2S Feed

*Reaction 1A times 3, †Reaction 3, §Reaction 4

Thermodynamic Equilibrium Calculations

Equilibrium conditions for a complex reaction system are best calculated using free-energy minimization procedures that are readily available on most process simulators and in special programs provided by licensors. All reactants and relevant products should be included. In addition, it is necessary to include all eight forms of sulfur (S_1, S_2, S_3, S_4, S_5, S_6, S_7, and S_8). Simultaneous solution of equilibrium reactions is possible; however, to gain some understanding of this complex system, the overall reaction conveniently based on one mole of hydrogen sulfide can be used.

$$H_2S + 1/2\ O_2 \rightarrow 1/x\ S_x + H_2O$$

Equilibrium calculations have been made that consider simultaneous equilibria of the various sulfur species.[5,7,8] Results are plotted in Figure 15.5 as equilibrium sulfur content versus temperature and in Figure 15.6 as equilibrium conversion of hydrogen sulfide to sulfur.

Curve 3 in Figure 15.5 is based on older thermodynamic data and the existence of S_2, S_6, and S_8, and Curve 1 is based on all sulfur entities and recent thermodynamic data, including data from both laboratory and plant operation.[8] Both show the unusual pattern of an exothermic reaction that in the lower-temperature range exhibits the usual decline in equilibrium conversion of, in this case, H_2S to sulfur with increasing temperature. But, in the higher temperature range, the equilibrium conversion increases as temperature increases above 811 K (1000°F). One must remember, however,

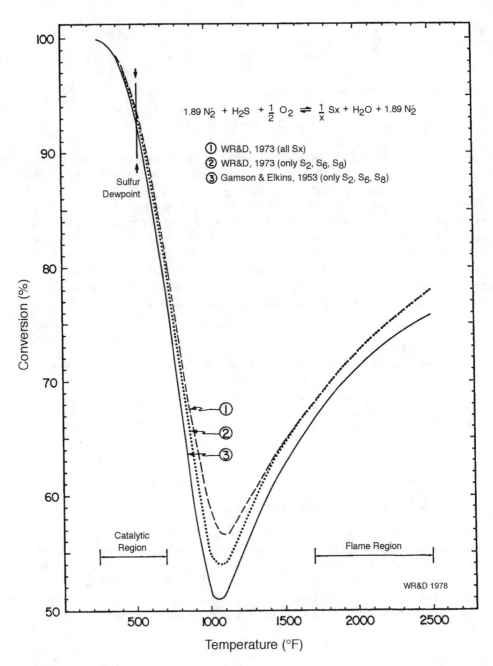

FIGURE 15.5 Equilibrium Conversions of H₂S to Sulfur (pure H₂S, 1.0 atm, no sulfur removal). Reprinted by permission: Paskall, H. G., *Capability of the Modified Claus Process*, p. 22, Sulphur Experts, Inc., Calgary, Alberta, Canada, 1979 (4th printing in 1990).

that the total reaction system includes the various species of sulfur from S_2 through S_8. The dissociation of the higher sulfur polymers such as S_8 is highly endothermic and is favored at high temperatures.

$$S_8 \rightarrow 4S_2 \qquad \Delta H = -93.7 \text{ kcal@1000 K}$$

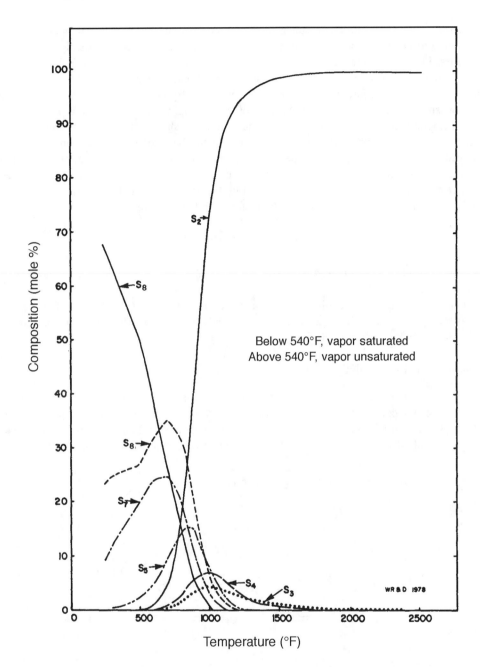

FIGURE 15.6   Equilibrium Composition of Sulfur Vapor from Reaction of H_2S with Stoichiometric Quantity of Air (total system pressure: 1.0 atm). Reprinted by permission: Paskall, H. G., *Capability of the Modified Claus Process*, p. 24, Sulphur Experts, Inc., Calgary, Alberta, Canada, 1979 (4th printing in 1990).

The net result is that the reaction system becomes less exothermic, and the sulfur yield from H_2S increases somewhat in the higher-temperature region, as shown in Figure 15.5.

Below 644 K (700°F), S_6 and S_8 predominate in the vapor phrase, while above 1200 K (1700°F), S_2 is the main sulfur form present. (See Figure 15.6.) The high exothermicity in the lower-temperature region combines the exothermicity of both the H_2S reaction and the formation of higher sulfur species (e.g., S_6–S_8). The result is vastly improved conversion at low temperatures. Ideal operation corresponds to a temperature as close to the sulfur dew point as possible.

Since the low-temperature region is dominated by sulfur species of high molecular weight (and thus present at a lower partial pressure because of fewer moles), the equilibrium conversion of H_2S tends to be shifted even farther to the formation of sulfur. By contrast, in the higher-temperature region, where one sulfur form (S_2) of lower molecular weight predominates, the partial pressure of sulfur vapor increases, causing the H_2S equilibrium to shift to the left. Unfortunately, it is somewhat futile to attempt a detailed explanation on the basis of single reactions. Free-energy minimization involving all possible species is certainly the ideal approach for such a complex system. It is most important, however, to subject such calculations, now possible by computer, to careful comparison to actual plant data before confident use of such calculations is possible.

It has been demonstrated that the catalytic step with the main reaction,

$$2H_2S + 2SO_2 \Leftrightarrow 1/xS_x + 2H_2O$$

does reach equilibrium at plant conditions. The thermal reactions in the furnace, however, are kinetically controlled.

After the start-up period, equilibrium is reached in the first 6–12 in. of the 3–4-foot bed. Over time, as the upper portions become deactivated and only a bottom section of the bed is active, the reaction becomes kinetically controlled.[8]

Mechanism for the Catalytic Step

An interesting study[9] reports the activity of various alumina preparations and pretreatments. Evidence presented suggests that both H_2S and SO_2 adsorb on basic sites such as Type A sites with the OH^- surrounded by four O^{2-} sites, which are the most basic. If the reaction occurs between adjacent sites of adsorbed H_2S and SO_2, the following mechanism may be proposed.[9]

Catalyst Types And Suppliers

Catalysts for the reaction of H_2S and SO_2 to produce elemental sulfur and water are primarily aluminas variously described as *activated alumina* or *promoted alumina*. Because of the modest unit operating pressures (1.38–1.84 bar), selection of catalyst shape and size is an important factor in maintaining low pressure drop.

Alumina catalysts: Gamma or alpha-alumina or alumina modified (promoted) by titania, zirconia, or silica.[10]

Shape: Spherical shape is the most widely used, although shaped extrudates have also been offered.

Size: 3× 6 mesh and 5 × 8 mesh spheres.

Suppliers

Acreon/Procatalyse, Alcoa, UOP LLC, and Porocel

Licensors

ABB Lummus Global, BASF, Bechtel, Fluor Daniel, Goar Allison, KTI, Linde A.G., Lurgi, Ortloff, Parsons, Pritchard, Siirtec Nigi, Stork E&C, Tecnip, TPA, UOP LLC

Catalyst Deactivation

Three major sources of catalyst deactivation require special attention and operating acumen. They are carbonaceous deposits, accumulation of sulfate deposits (called *sulfation*), and elemental sulfur condensation on the catalyst surface.

Coke Deposits

Acid-gas streams from both refinery gas and natural gas sources contain some hydrocarbons, a small portion of which is absorbed in the gas treatment system and becomes an impurity in the feed to the Claus process. Carbonaceous deposits can form on the Claus catalyst because of these hydrocarbons, particularly if they are high-molecular-weight aliphatics or aromatics often associated with refinery gas streams. As is so often true with carbonaceous deposits (coke), active sites on the catalyst are covered and catalyst activity declines. In the two major Claus process types, straight-through and split-flow (see "Process Units"), the reaction of hydrocarbons occurs as follows.[6]

Straight-Through Process (with H$_2$S content > 50%): Furnace temperature is high (1800–2500°F, 982–1371°C) and hydrocarbons are totally combusted to CO$_2$ and H$_2$O. Deposition of carbonaceous deposits on the catalyst of the Claus reactor downstream of the furnace is avoided with proper operation.

Straight-Through Process (with H$_2$S consent <40–50%): Lower furnace temperature results, and both cracking and partial oxidation occur, resulting in complexes of high carbon-to-hydrogen ratio that will deposit as coke on the catalyst in the first bed at the lower temperature.[11]

Split-Flow Process: The portion of acid-gas feed that bypasses the furnace and is then mixed with the furnace gas contains unreacted hydrocarbons, which then pass over the Claus catalyst bed and can also form carbonaceous deposits on the catalyst at a lower but undesirable rate. Deposition increases the lower the H$_2$S content of the gas and the higher the hydrocarbon content.[12]

Sulfur Deposition

Elemental sulfur will condense on the catalyst surfaces if reactor temperature reaches the dew point of sulfur. Rapid deactivation will then ensue as the liquid sulfur covers the catalyst surface. Since the exothermic reaction of H$_2$S with SO$_2$ to form sulfur is equilibrium limited, it becomes advantageous to lower the temperature in each subsequent reactor in series. This procedure is possible because sulfur is removed between stages thus lowering the dew point. Obviously, skillful operation and process control is essential. Should deposition occur, however, the sulfur can usually be revaporized[6,13] by increasing the temperature around 50°F (10°C).

Sulfation

Sulfation of alumina catalyst under Claus reactor conditions can be a major deactivating phenomenon. The formation of sulfates with the alumina catalyst deactivates the catalyst by deactivating valuable active sites. It has been suggested that sulfate formation is dependent on equilibrium between H_2S, SO_2, SO_3, O_2, and elemental sulfur. It is favored by low H_2S and high SO_2 and O_2 concentrations,[11] which is why sulfation increases in each subsequent reactor in series.

Special catalysts have been developed that exhibit resistance to sulfation.[6,11] These include promoted alumina and titanium-based catalysts.

Sulfated catalyst can be restored by reducing it with excess H_2S for 24 to 36 hr. The excess H_2S is produced by operating the plant in a deficient air mode. Such operation eliminates the sulfate by reduction and can be done after the thermal treatment to remove condensed sulfur.[8]

Ammonium Salts

Ammonia, if present in the feed to the catalytic reactor, will form ammonium salts and deposit on the catalyst, causing pore plugging in amounts exceeding 500 ppm. Ammonia can be decomposed in the furnace equipped with a special high-intensity burner.

Structural Changes

As with so many catalysts, reduced surface area and pore size of the catalyst occurs, along with some sintering and attrition. Catalyst life can be extended by careful temperature control and proper startups and shutdowns, since many of these aging problems occur during operational malfunctions. Catalyst life is reported to be 5–7 years for natural gas and several years less for refinery gas.

Regeneration

Careful regeneration of catalyst to burn off coke deposits has been practiced, but full activity does not return. After several regenerations, the catalyst may need to be replaced. As is common with all regenerations with air, avoidance of excessive temperatures is essential to prevent catastrophic changes in the catalyst crystalline structure. In this case, a temperature of 1000°F (538°C) should not be exceeded.[6]

Process Units[6,8,14]

There are two standard Claus processes in operation, straight-through and split-flow. But there are a number of other processes with special innovations for improving yields and ease of operation that are extensions of the two standard processes. Processes such as split-flow with preheating of the feed steam and a sulfur recycle process are used, respectively, for H_2S content of 10–20 mole % and less than 10 mole %. The reader should refer to rather complete coverage in publications by Sulfur Experts[8] and Kohl and Nielsen.[6]

Characteristics of Straight-Through Process (>50% H_2S in Feed)

Furnace   The entire acid-gas stream is fed to a furnace operating at 1800–2500°F (1255–1644 K). Sufficient air is mixed with the gas to combust one-third of the H_2S to SO_2 ($H_2S + 3/2 \ O_2 \rightarrow SO_2 + H_2O$). Feed must have 50% or more H_2S and usually less than 2% hydrocarbons. The remaining unreacted H_2S reacts with the SO_2 to produce sulfur noncatalytically ($2H_2S + SO_2 \rightarrow 3S + 2H_2O$). Up to 60 or 70% of the sulfur is produced in the furnace at these high temperatures.[6] If one-third of the remaining H_2S reacts with the SO_2 formed in the furnace, 50% of the total potential sulfur will have been formed.

Waste-Heat Boiler   The reaction products from the furnace are cooled by producing high-pressure steam.

Condenser   Further cooling is applied until the condensation temperature of sulfur is reached and the sulfur can be removed from the gas phase. At the lower temperature (below the dew point) required, low-pressure steam is generated.

Catalytic Reactors

The vapor from the condenser is reheated above its dew point and then enters the first adiabatic fixed-bed reactor packed with the alumina catalyst. The goal is to be above the dew point but not excessively so, since the reaction of H_2S and SO_2 is thermodynamically favored at low temperatures. ($2H_2S + SO_2 \rightarrow 3S + 2H_2O$). Although the reaction also takes place in the furnace, yields are in the range of 60 to 70%. At lower temperatures, additional sulfur is formed catalytically. By removing the sulfur formed in the furnace, the equilibrium is shifted favorably to produce additional sulfur.

As is often the case with exothermic reactions in adiabatic beds, staging proves useful, since temperature increases in the bed with conversion. Cooling the effluent from the first bed to remove sulfur followed by reheating and passing to the second bed allows additional reaction to sulfur by increasing the equilibrium conversion. In addition, since sulfur is removed prior to each stage, the dew point declines in each successive bed and permits a lower operating temperature (10–20°C above the dew point), which further improves the equilibrium condition. Plants with either two or three stages are in operation. Inlet temperatures in the several stages are in the range of 450–540°F (505–555 K) for Bed 1, 390–430°F (472–494 K) for Bed 2, and 370–410 K for a third bed.[6] A typical two-stage plant might produce 60% of the sulfur in the furnace and 25% in the first catalytic stage, followed by 7% in the second stage. A fourth stage is usually not an economical investment.

The catalytic reactors have been designed as horizontal or as vertical vessels.[6] Bed depths in the range of 3–5 ft (91–152 cm) have been reported.[6] Good distribution in such relatively shallow beds must be planned for minimum pressure drop.

Characteristics of Split Flow Process (25 to 50% H₂S in Feed)

Feeds with H_2S content of 25 to 50% contain too much inert gas to sustain a sufficient flame temperature of 1700°F (1200 K) or above, necessary for SO_2 production.[6] Operation is possible, however, if at least one-third of the acid gas is passed through the furnace to form SO_2, and the remainder is by-passed and mixed prior to entering the catalytic section. Under these conditions, only a small amount of sulfur is formed in the furnace. The catalytic reactor section is similar to that described for the straight-through process.

Since oxygen can cause sulfation of the catalyst, it is advantageous to bypass a bit less than two-thirds of the acid gases so that all the oxygen will be consumed in the furnace.

Carbon disulfide and carbonyl sulfide formed in the furnace at the necessary lower furnace temperature are passed in the combined feed to the first catalytic reactor, where a large portion can be converted to H_2S by hydrolysis, since ample water vapor is present.[6] Temperatures are in the range of 600–700°F (589–644 K), which is above the optimum for maximum sulfur production in the first bed and places a larger load on subsequent beds.

$$CS_2 + 2H_2O \Leftrightarrow CO_2 + 2H_2S \qquad (12)$$

$$COS + H_2O \Leftrightarrow CO_2 + H_2S \qquad (13)$$

Significant loss of sulfur occurs if the compounds are not converted to H_2S for processing in subsequent beds. Furthermore, CS_2 and COS, along with sulfur and sulfur compounds, cannot be released in any quantity to the atmosphere.

Yields in Claus Process

Yields of 97% H_2S converted to sulfur and recovered are generally attained in both of the standard processes, straight-through or split-flow.

Low H₂S Feeds (10–25% H₂S in Feed)

Split flow is used for this range of H_2S, but preheating of feed streams (acid gas and air) is required to assure adequate flame temperatures.[8]

Other Modified Claus Processes

A variety of proprietary processes involve clever modifications for special situations or improved sulfur recovery and economics. These processes have been categorized as oxygen enrichment, isothermal, and direct-oxidation. See Table 15.5 for short summary and Ref. 6 for a comprehensive review and description.

TABLE 15.5
Modified Claus Processes

Type	Description	Advantages	Proprietary Processes
Oxygen enrichment	Replacement of some or all of the air by pure oxygen.	Increases capacity of an existing plant but modification in the furnace are required.	(Goar Allison/Air Products) COPE SURE (BOC Gases) Oxy-Clause (Lurgi) TPA, Inc.
Isothermal Claus	Reactor designed with internal cooling coils or multitubular reactor with coolant on shell side.	Fewer reactor stages required	Clinsulf (Linde, A.G.) Catasulf (BASF), uses direct oxidation ($H_2S + O_2 \rightarrow S + H_2O$) Low H_2S, 5–15%

For more detail see Ref. 6 and additional references therein.

Claus Units for Low H₂S Acid Gas Streams

When H_2S content is less than 20 to 23%, changes in the combustion process must be made to attain adequate and stable flame temperatures. These include the following:[6]

- Preheating of the acid gas feed stream and combustion air
- Increasing oxygen content of the reaction mix by adding pure oxygen
- Addition of hydrocarbons such as methane to reaction mix

Tail Gas Treatment

Many Claus units have facilities for treating the tail gas from the plant to meet the most stringent air pollution regulations. A widely used process employs a catalytic hydrogenation and hydrolysis that converts sulfur, SO_2, COS and CS_2 to H_2S which, in turn, can be recovered by absorption and recycled. The hydrogenation and hydrolysis reactions are as follows:

$$S_x + xH_2 \rightarrow xH_2S$$

$$SO_2 + 3H_2 \rightarrow H_2S + 2H_2O$$

$$COS + H_2O \rightarrow H_2S + CO_2$$

$$CS_2 + 2H_2O \rightarrow 2H_2S + CO_2$$

Hydrogen is supplied from producers outside the battery limits. Water vapor content of Claus tail gas is adequate, since it is a product of the main Claus reactors.

Other tail gas processes include sub-dew-point operation and direct oxidation to sulfur.[6] By operating below the dew point so that equilibrium conversion increases with the lower temperature, the sulfur formed condenses on the catalyst and must periodically be removed requiring several tail-gas reactors in series and a third experiencing sulfur removal by vaporization. Processes include Sulfreen Process (Lurgi/Elf Aquitaine), CBA Process (Amoco), and MCRC Process (Delta Hudson).

Direct oxidation of tail gas ($H_2S + O_2 \rightarrow S + H_2O$) using a highly active oxidation catalyst has the advantage of an irreversible catalytic reaction with a potential 100% conversion of H_2S to sulfur. Processes include Superclaus Process (Stork), Selectox (Unocal/Parsons), MODOP Process (Mobil), Catasulf Process (BASF), and Hi-Activity (Parsons).

The details of a number of these tail gas processes are clearly described in Ref. 6.

Catalyst Type for Tail Gas

A typical tail-gas catalyst has a composition of 3.5% CoO, 14% MoO_2, and 82% alumina. Extrudates (1/3 in.) or spheres (3 × 6 or 5 × 8 mesh) are available.

Suppliers

United Catalysts, Shell, Criterion, Acreon

Licensors

Amoco, BASF, Comprimo, Exxon, Lurgi/Elf Aquitaine, Mobil, Shell/Parsons

Process Kinetics

A number of early studies on the reaction kinetics of the Claus reaction ($2H_2S + SO_2 \rightarrow 3S + 2H_2O$) have been published. But catalytic reactors in modern plants using the improved alumina catalysts have been consistently operated at thermodynamic equilibrium. Thus, reactors are designed for a generally proven space velocity of 500 to 1000 hr. Of course, additional catalyst is installed to assure economical bed life.

Suppliers of unit designs have developed proprietary programs that enable quality optimized design planned for the particular character of the acid-gas feed. Commercial programs for plant unit simulation are on the open market including Sulsim/Sulfur Experts, TSWEET/Bryan Research & Engineering, and Hysim/Hyprotech Ltd.[6]

REFERENCES, SECTION 15.1 (SULFUR DIOXIDE OXIDATION)

1. Chenier, P. J., *Survey of Industrial Chemistry*, VCH, New York, 1992.
2. Smith, G. M., *Encyclopedia of Chemical Processing and Design*, J. J. McKetta, editor, 55, p. 469, Marcel Dekker, New York, 1996.
3. Davies, P. in *Catalysis: Science and Technology*, J. R. Anderson and M. Boudart editors, Vol. 8, p. 1, Springer-Verlag, Berlin, 1987.
4. Villadsen, J. and Livbjerg, H., *Catal. Rev.-Sci. Eng.*, 17 (2) 203 (1978).
5. Urbanek, A. and Trela, M., *Catal. Rev.-Sci. Eng.*, 21 (1) 73 (1980).
6. Livbjerg, H. and Villadsen, J., *Chem. Eng. Sci.*, 27, 21 (1972).
7. Boghosian, R., Fehrmann, N. J., Boerrum, N. J. and Papatheodorou, G. N., *J. Catal.*, 119, 121 (1989).
8. Mars, P. and Maessen, J. G., *Proc. 3rd Int. Cong. Catal.*, 1, 266 (1965).
9. Glueck, A. R. and Kenney, C. N., *Chem. Eng. Sci.*, 23, 1257 (1968).
10. Boreskov, G. K. in *Catalysis: Science and Technology*, J. R. Anderson and M. Boudart editors, 3, 41, Springer-Verlag, Berlin, 1982.
11. Hansen, N. H., Fehrmann, R. and Bjerum, N. J., *Inorg. Chem.*, 21, 744 (1982).
12. Jarisen-Helm, H. and King, J. D., *Sulfur 88 Conference*, Vienna, Austria, Nov. 6–8, 1998.
13. Stiles, A. B. and Koch, T. A., *Catalyst Manufacture*, 2nd ed., Marcel Dekker, New York, 1993.
14. Donovan, J. R., Stolk, R. D. and Unland, M. L., in *Applied Industrial Catalysis*, B. E. Leach, editor, Vol. 2, p. 245, Academic Press, New York, 1983.

15. Michalek, J., *Vyt. Sk. Den. Technol.* Praze, B25, 57, (1980).
16. Albright, L. and Long, E. G., in *Encyclopedia of Chemical Processing and Design*, J. J. McKetta, editor, Vol. 55, p. 456, Marcel Dekker, New York, 1996.
17. Smith, G. M., in *Encyclopedia of Chemical Processing and Design*, J. J. McKetta, editor, Vol. 55, p. 469, Marcel Dekker, New York, 1996.
18. Schillmoller, C. M. in *Encyclopedia of Chemical Processing and Design*, J. J. McKetta, editor, Vol. 55, p. 427, Marcel Dekker, New York, 1996.
19. Boreskov, G., K., Slinko, M. G. and Beskov, V. S., *Khim, Prom.,* 44, 173 (1968).
20. Rase, H. F., *Chemical Reactor Design for Process Plants*, Vol. II, p. 86, McGraw Hill, New York, 1977.
21. Michalak, J., Simecek, A., Kadlec, B. and Bulicka, M., *Chem. Prum.,* 44, 346 (1969).
22. Livbjerg, H. and Villadsen, J., *Chem. Eng. Sci.,* 27, 21 (1972).

REFERENCES, SECTION 15.2 (AMMONIA OXIDATION)

1. Chenier, P. J., *Survey of Industrial Chemistry*, 2nd ed., VCH, New York, 1992.
2. Ohsol, E. O., in *Encyclopedia of Chemical Processing and Design*, J. J. McKetta, editor, Vol. 31, p. 150, Marcel Dekker, New York, 1990.
3. Kobe, K. A., *Inorganic Process Industries*, MacMillan Co., New York, 1946.
4. Clarke, S. I. and Mazzafro, W. J., in *Kirk-Othmer Encyclopedia of Chemical Technology*, 4th ed., Vol. 17, p. 80, Wiley, New York, 1996.
5. Thiemann, M., Scheibler, E. and Wiegand, K. W., in *Ullmann's Encyclopedia of Industrial Chemistry*, 5th ed., Vol. A17, p. 293, VCH, New York, 1991.
6. Hoftyzer, P. J. and Kwanten, F. I. G., in *Gas Purification Processes for Air Pollution Control*, G. Nonhebel, editor, Butterworths, London, 1972.
7. Gland, J. L. and Korchak, V. N., *J. Catalysis,* 53, 9 (1978).
8. Pignet, T. and Schmidt, L. D., *J. Catal.* 40, 212 (1970).
9. Stacey, M. H., in *Catalysis: Specialized Periodical Report,* Vol. 3, 98, Chem. Soc., London, 1980.
10. Loffler, D. G. and Schmidt, L. D., *AI. Ch. E. J.,* 21, 786 (1975).
11. Büchner, W., Schliebs, R., Winter, G., and Büchel, K. H., *Industrial Inorganic Chemistry*, VCH, New York, 1989.
12. Satterfield, C. N., *Heterogeneous Catalysis in Industrial Practice*, McGraw-Hill, New York, 1991.
13. Handforth, D. L. and Tilley, J. N., *Ind. Eng. Chem.,* 26, 1287 (1934).
14. Chinchem, G. in *Catalysis: Science and Technology*, Anderson, J. R. and Boudart, M. editors, Vol. 8, p. 28, Springer-Verlag, New York, 1987.
15. Neck. R. M., Bonacci, J. C., Hatfield, W. R. and Hsiung, T. H., I*nd. Eng. Chem. Process Des. Dev.,* 21, 73 (1982).
16. Gillespie, G. R. and Kenson, R. E., *Chemtech*, p. 627 (1971).
17. Nowak, E. J., *Chem. Eng. Sci.,* 24, 421 (1969).
18. Gillespie, G. R. and Goodfellow, D., *Chem. Eng. Prog.* 70, (3), 81 (1974).
19. Lee, H. C. and Farranto, R. S., *Ind. Eng. Chem. Res.,* 28, 1 (1989).

REFERENCES, SECTION 15.3 (HYDROGEN SULFIDE OXIDATION)

1. Bush, W. R. and Semrad, R. in *Encyclopedia of Chemical Processing and Design*, J. J. McKetta, editor, Vol. 55, p. 269, Marcel Dekker, New York, 1996.
2. Johnson, J. E., Taap, S. V., Kelley, R. E. and Lacako, L. P. in *Encyclopedia of Chemical Processing*, J. J. McKetta, editor, Vol. 55, p. 333, Marcel Dekker, New York, 1996.
3. Sawyer, F. G., Hader, R. N., Herndon, I. K. and Moringstar, E., *Ind. Eng. Chem.,* 42, 1938 (1950).
4. Stull, D. R., Westrum, E. F. and Sinke, G. C., *The Chemical Thermodynamics of Organic Compounds*, Wiley, New York, 1969.

5. Stull, D. R. and Siwke, G. C., *Thermodynamic Properties of the Elements*, Advances in Chemical Series 18, American Chemical Society, Washington, D.C., 1975.

6. Kohl, A. and Nielsen, R., *Gas Purification,* 5th ed., Gulf Publishing Co., Houston, 1997.

7. Gamson, B. W., and Elkins, R. H., *Chem. Eng. Prog.,* 49, 203 (1953).

8. Paskall, H. G., *Capability of the Modified-Claus Process,* Sulphur Experts, Inc. (Western Research), Calgary, Alberta, Canada, 1979. Reprinted in 1990.

9. George, Z. M. in *Sulfur Removal and Recovery from Industrial Processes,* J. B. Pfeiffer, editor, Advances in Chemistry Series, 139, p. 75, American Chemical Society, Washington, D. C., 1975.

10. Stiles, A. B. and Koch, T. A., *Catalyst Manufacture,* 2nd ed., Marcel Dekker, New York, 1995.

11. Grancher, P., *Hydrocarbon Proc.,* 57, (7), 155 (1978).

12. Grancher, P., *Hydrocarbon Proc.,* 57 (9), 257 (1978).

13. Norman, W. S., *Oil & Gas J.,* 74, p. 35 Nov. 15, 1976.

14. Gary, J. H. and Handwerk, G. E., *Petroleum Refining,* Vol. 5, p. 190, Marcel Dekker, New York, 1975.

16 Oxidation (Organics)

16.1 GENERAL

Organic oxidations must obviously be accomplished by means of selective catalysts that permit partial oxidation and discourage the favored reaction of total oxidation to carbon dioxide and water. They also require careful temperature control so as to avoid excessive temperatures that may lead to rapid thermal oxidation, catalyst damage, and runaway reaction, with a possible explosion if the reaction system reaches compositions within the flammable or explosive envelope.

Mechanisms

This subject is an elusive one for most reactions, and particularly so for organic partial oxidation reactions. It is generally agreed that the steps illustrated in Figure 16.1 occur.[1]

- Hydrocarbon chemisorbs on M_1^{n+} site and reacts with the lattice oxygen, O^{2-} associated with M_1^{n+} site producing the partially oxidized product.
- The lattice oxygen from a neighboring site, M_2^{m+}, moves to M_1^{n+} and replenishes the oxygen consumed at M_1^{n+}.
- Electrons produced at M_1^{n+} are transported to M_2^{m+}.
- Molecular oxygen is chemisorbed on M_2^{m+} and converted to lattice oxygen.

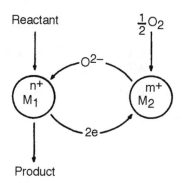

FIGURE 16.1 Selective Oxidation-Reduction Cycle. Reprinted from Dadyburjor, D. B., Jewur, S. S., and Ruckenstein, E., *Catalysis Reviews: Science & Engineering*, 19 (2), 243 (1979), by courtesy of Marcel Dekker, Inc.

Role of Promoters

In general, it may be said that promoters perform by stabilizing the favorable valance state of the catalyst, by aiding in forming the active state of the catalyst and protecting the catalyst against extensive reduction, or by increasing the electron exchange rate.[1]

Alkali and alkaline earth metals at low concentration act as electron promoters and increase the reactivity of the oxygen ion on the surface. High concentrations of these metals, however, produce inactive forms of the active species. Alkaline earth metal oxides are thought to increase the relative population of favorable O^{2-} sites.[1] Transition metal promoters act as structural promoters and tend to alter the lattice structure and produce more phases of higher activity.

Promoters that have been used commercially or in experimental work on oxidation catalysts are listed in conjunction with discussion of the several oxidation processes that follow.

16.2 ETHYLENE → ETHYLENE OXIDE

Ethylene oxide is a major building block for a number of important petrochemicals. Hydrolysis of ethylene oxide produces its most important derivative, ethylene glycol, which is consumed in copious quantities as automotive antifreeze and as the diol in the production of polyesters. Ethylene glycol reacts with either terephthalic acid or dimethyl terephthalate to produce a polymer of repeating ester units. Because of its strength and wrinkle resistance, it is used extensively in clothing, particularly in blends with cotton or wool. Other uses of polyester include bottles and film.

In producing ethylene glycol, di- and triethylene glycol are also formed, which are used in synthesis of other glycols. Triethylene glycol is used for dehydrating natural gas. Higher glycols are polymers formed by polymerization of ethylene oxide and ethylene glycol and are used as plasticizers, dispersants, and solvents.[2]

Other uses of ethylene oxide include the production of ethanolamines by reaction of ethylene oxide and ammonia. These amines (mono, di, and tri) are used in acid gas removal (H_2S, CO_2, etc.) from industrial gases and in the manufacture of surface-active agents.[1,2] Nonionic detergents, produced by reacting ethylene oxide with straight-chain alcohols (C_{10}–C_{13}) is a fast-growing portion of the detergent market, because they function in cooler wash water and require less phosphates.[1]

Chemistry

The reaction scheme on AgO catalysts, including the significant side reactions can be represented as follows:[2]

		K@500°K	ΔH@500°K
(1)	$CH_2 = CH_2 + 1/2 O_2 \longrightarrow CH_2 - CH_2$ (O) Ethylene oxide	5.75×10^6	-25.39
(2)	$CH_2 = CH_2 + 3 O_2 \longrightarrow 2 CO_2 + 2 H_2O$	4.97×10^{136}	-315.88
(3)	$CH_2 - CH_2$ (O) $+ 2 1/2 O_2 \longrightarrow 2 CO_2 + 2 H_2O$	8.63×10^{129}	-299.54
(4)	$CH_2 - CH_2$ (O) $\longrightarrow CH_3 - CHO$ Acetaldehyde	1.11×10^{13}	-27.02
(5)	$CH_3 = CHO + 2 1/2 O \longrightarrow 2 CO_2 + 2 H_2O$	7.78×10^{116}	-263.47
(6)	$[CH_2 = CH_2 + O_2 \longrightarrow 2 CH_2O]$ Formaldehyde	2.15×10^{38}	-40.64

Thermodynamics

Clearly, all reactions at the reactor conditions are exothermic and not equilibrium limited. Reactions 4, 5, and 6 are minor and less than 0.1% of the product mix. But the oxidation of ethylene and ethylene oxide is not only highly exothermic but can also occur thermally if the temperature increases excessively.

Higher temperatures not only can cause a loss in ethylene oxide yield, they increase the operating temperature so rapidly that complete oxidation occurs and a runaway reaction ensues. The development of the unique promoted AgO catalyst provides the selectivity that favors ethylene oxide production. An inhibitor, 1,2-dichloroethane, is also added in small amounts to prevent excessive total oxidation.[3,4] The dichlorethane decomposes producing Cl, which is thought to reversibly poison the more active sites that would be most active in total oxidation to CO_2 and water vapor.

Mechanism

As might be expected, various mechanistic schemes have been proposed. One of these has proved useful because of its intuitively satisfying concepts and its predictive power.[5] The following equations set out the proposed adsorption and reaction steps.

$$O_2 + 4Ag(adj) \rightarrow 2O^{2-} (ads) + 4Ag^+ (adj)^7 \qquad (7) \qquad \text{fast}$$

$$O_2 + Ag \rightarrow O_2 (ads) + Ag^{+8} \qquad (8) \qquad \text{slow}$$

$$O_2 + 4Ag (nonadj) \rightarrow 2O^{2-}(ads) + 4Ag^+ (adj)^9 \qquad (9) \qquad \text{very slow}$$

$$O_2(ads) + C_2H_4 \rightarrow C_2H_4O + O (ads)^{10} \qquad (10)$$

The term *adj* refers to adjacent site and *ads* to adsorbed condition. The oxygen atoms produced react with ethylene, producing CO_2 and water. The more reactive sites are assumed to be blocked by chlorine due to the continuous addition of dichloroethane. Hence, Step 7 may be neglected, as well as Step 9, which is extremely slow.

The oxygen atoms produced in Step 10 react with ethylene to produce CO_2 and H_2O, as can be expressed in the following step at maximum selectivity. One ethylene molecule reacts with one oxygen atom to produce one ethylene oxide, and with five oxygen atoms to produce $2CO_2$ and $2H_2O$ molecules.

$$5O(ads)$$

$$O(ads) + C_2H_4 \rightarrow OC_2H_4(ads) \rightarrow 2CO_2 + 2H_2O \qquad (11)$$

Since Steps 10 and 11 have been found to have the same order and similar activation energies, they may be combined after multiplying Step 10 by the quantity 6.

$$7C_2H_4 + 6O_2(ads) \rightarrow 6C_2H_4O + 2CO_2 + 2H_2O + 6\Delta \qquad (12)$$

where Δ represents vacant sites.

Based on Eq. 12, the maximum selectivity to ethylene oxide would be 6/7, or 85.6 percent.[5] The actual selectivity reported is 70 to 80% for the usual process with oxygen feed.[2,3,5] But the above mechanism does not account for oxidation of ethylene oxide and other side reactions that reduce the selectivity.

Catalyst Types

Suppliers and Licensors

The silver oxide catalyst is the only catalyst used. It is proprietary and offered with a licensed process by the following: CRI-Shell, Scientific Design, Nippon Shokuboi, and Union Carbide.

Catalyst Characteristics

The catalyst is indeed unique. Intuitively, one would expect that both ethylene or propylene oxides can be produced by using this catalyst. Propylene, however, simply oxidizes completely to CO_2 and H_2O, which is hardly a profitable undertaking. Other processes for direct oxidation of propylene are being developed.

The silver catalyst consists of 10–15% Ag on a refractory support such as α-alumina, silicon carbide, or silica.[2] The supports have very low surface areas (less than 1 m^2/g) and high porosity. The silver is deposited on the outer surfaces of large pores.[2] In so doing, overoxidation is discouraged. Spherical catalyst particles produce lower catalyst-bed pressure drop.

Promoters and other additives are essential for maximum selectivity. Alkali and alkaline earth salts such as those of Ba, Ce, Rb, Sn, and Sb have been used. Additives not only improve selectivity but also prevent sintering.

Much successful development work has been done on methods for depositing silver in very small particle sizes that provide higher activities.

Catalyst Deactivation

The silver catalyst has a long life, but some possible impurities in the feed may cause problems. Acetylene is a by-product of ethylene production and is converted by selective hydrogenation to ethylene in the separations section of a stream cracking olefin plant. Upsets in the hydrogenation unit could cause increased acetylene residual, and acetylene content must be monitored. Acetylene can form carbonaceous deposits in the ethylene oxide reactor, as can other hydrocarbons such as propylene and diolefins.[2] Carbon monoxide and hydrogen rapidly oxidize and cause undesired temperature increases that, at the very least, produce loss of activity due to sintering. Although chlorine is added as an inhibitor, the addition rate must be carefully controlled. Excessive amounts can cause significant catalyst deactivation. Sulfur compounds can also poison the catalyst when present in significant amounts.

The oxygen-based process has a much lower recycle and purge rate than the air-based process. For that reason, a more pronounced buildup of impurities can occur. Reports of ethane buildup indicate negative effects on both selectivity and conversion.[2]

Catalyst Recovery

Although the catalyst has a long life, sufficient activity decline ultimately occurs that warrants replacement. In such instances, the catalyst is returned to the supplier for recovery and credit for the silver. Techniques have been reported for reactivation of aged catalyst by treatment with fresh alkali salts such as Ce and Rb, but the extent of use is not available.

Process Units

There are two types of process units, air oxidation and oxygen oxidation. About 75% of ethylene oxide plants worldwide are oxygen oxidation units, which may say more about the economics than exhaustive economic studies. Special situations, however, exist where air oxidation may be preferred. It is known that some plants are designed to use either air or oxygen. The air plants lose ethylene with the necessary purging of nitrogen, and oxygen plants require a higher investment because of the need for an air separation unit.

Reactor

Both types of plants use multitubular heat-transfer reactors with 10,000 or more tubes of 25–45 mm (1–1-3/4 in.) OD and lengths of 20–30 ft.[3,6] Operating temperatures are in the range of 220–300°C and pressures of 10–20 bar. Pressure does not have much effect on conversion or selectivity, but it must be high enough to ensure adequate removal of ethylene oxide in the absorber downstream of the reactor.

Tube length and diameter must be optimized. Obviously, assuming a given catalyst quantity, lower diameters require longer lengths. They do improve heat transfer but increase pressure drop. The total amount of catalyst based on a desired space velocity must be adequate to produce the desired conversion and enable operation for a desired time between turn-around and catalyst replacement. Experience in existing plants and by licensors will be most valuable. A good operating model will allow predictions on possible excessive oxidation to CO_2 and water that might occur with larger catalyst loads. This situation could become a problem toward the end of a run, when operating temperature is being raised to compensate for declining activity.

The cooling medium in the shell side of the reactor has been varied somewhat over the years. The relatively high-temperature boiling water did not seem attractive because of the high shell-side pressure, which would increase reactor cost. Light oils such as kerosene or heat-transfer fluids were used but, in recent years, boiling water has apparently been preferred,[7] especially since lower operating temperatures are possible with improved catalysts. The high heat of vaporization of water and its nonflammable character provide a safety advantage by rapid heat removal when temperatures begin to climb. Quick acting controls that change back pressure in the steam separator drum will quickly change the temperature of the shell-side water.

To maximize selectivity, air units employ a second or purge reactor. The first reactor is operated at low per-pass conversion (less than 50%) and high selectivity. Since large amounts of nitrogen have to be purged, considerable loss of unreactant ethylene must be avoided. After ethylene oxide is absorbed from the exit stream of the reactor, a large portion is recycled to the first reactor,[2] and the remainder is sent to a purge reactor operating at higher temperature and thus high conversion but lower selectivity. The effluent from the purge reactor is then sent to a purge absorber to remove the ethylene oxide, the exit gas is partly recycled to the purge reactor, and the remainder is purged from the system.

In the oxygen process, the only significant buildup of inerts is that of by-product CO_2, which must be removed to prevent damage to the catalyst activity when present in amounts exceeding about 10%, or 20–50% when ethylene concentrations in the feed are high (above 40 percent).[2,6] In the air process, CO_2 is purged with the nitrogen after the purge reactor system. It is possible, therefore, to operate at lower conversion (~10%) in a single reactor and achieve high selectivity and yield by recycling the total overhead from the absorber to the reactor. Only a small purge is necessary to prevent accumulation of argon present in small amounts in the oxygen. Because of the small purge, very little ethylene is lost. The oxygen process produces an overall selectivity of 80–75% for an oxygen plant compared to a 68–70% for an air plant.[2,5] The lower the conversion, the better the selectivity.

A number of patents speak to the value of diluents, of which methane is of most interest. It narrows the flammability limit of the inlet gas and thereby allows higher oxygen concentrations that lead to higher selectivities.[2,6] Lower levels of CO_2 must be maintained to benefit from methane addition.

Safety

Ethylene oxide is a highly reactive compound, and the vapor decomposes explosively if heated above 560°, even without air or oxygen present. Clearly, runaway reaction is a serious problem, since it is caused by rapid lowering of selectivity so that most of existing ethylene and ethylene oxide are undergoing complete combustion to CO_2 and water. The heat of complete combustion, which is 12 times that of the reaction to ethylene oxide, rapidly heats the reactor gases to a point where free-radical oxidation also occurs. Clearly, great care and experienced licensor's procedures are most valuable. One obvious and practiced emergency procedure is a rapid shutdown of oxygen flow.

Process Kinetics

There have been a large number of kinetic studies on ethylene partial oxidation. A variety of rate forms have been proposed, tested, and declared satisfactory. It would seem prudent to avoid those with an excessive number of constants and use one that includes the effect of dichlorethane. An example that might be a useful form for adapting plant data is as follows:[8]

$$C_2H_4 + 1/2O_2 \rightarrow C_2H_4O \tag{1}$$

$$C_2H_4 + 3O_2 \rightarrow 2. \ 2CO_2 + 2H_2O \tag{2}$$

$$r_1 = \frac{k_1 P_o P_e - k_2 P_o P_e P_d^{\alpha_1}}{1 + k_5 P_o + K_6 P_e}$$

$$r_2 = \frac{k_3 P_o P_e - k_4 P_o P_e P_d^{\alpha_2}}{1 + k_5 P_o + K_6 P_e}$$

where k_1, k_2, k_3, and k_4 = rate constants

P = partial pressure (o = oxygen, e = ethylene, and d = dichloroethane)

α_1, α_2 = exponents on dichloroethane partial pressure ($\alpha_1 = 0.19$, $\alpha_2 = 0.07$)

Reference 8 lists a number of other references on kinetics of this reaction.

16.3 PROPENE → ACROLEIN

Acrolein is a highly reactive compound because of its conjugated carbonyl and unsaturated entities. Its primary commercial use is as an intermediate in the production of the essential amino acid, D,L-methionine, which is used extensively as an animal feed supplement, particularly for poultry.[1,2] It is also an intermediate in the production of acrylic acid (a very large use, for which see page 265), gutaraldehyde (used in leather tanning and as a disinfectant for hospital equipment), and certain flavors and fragrances.[2] In its pure form, acrolein is used as an effective biocide in low concentration (10 ppm) in oil field brines and irrigation systems.[2]

Chemistry

$$H_2C = CH - CH_3 + O_2 \rightarrow H_2C = CH - CHO + H_2O \qquad \Delta H = 88 \text{ kcal/mole}$$

The catalyst is bismuth molybdate which, over the years, has been improved by the addition of other metal oxides Fe and Ni and Co or K and either P, B, W or Sb to name a few.[2] Operating conditions are 300–400°C and a pressure of 1.5–2.5 bar.[1] The feed consists of propene, air, and steam in a 1:10:2 mole ratio.[3]

Thermodynamics

The main reaction is highly exothermic, and a number of side and secondary reactions produce similar or higher exothermic heats of reaction such that the typical overall reaction system generates a net 200 kcal/mole.[2] The side and secondary reactions produce acrylic acid, acetic acid, acetalde-hyde, and carbon oxides.[2]

Mechanism

A number of studies have been published on bismuth molybdate catalysts in the partial oxidation of propene (e.g., Refs. 4–7).

A general descriptive summary has been presented based on analysis of experiments with labeled carbon and oxygen.[4,9] A graphic picture of the sequence of events is given in Figure 16.2.

- The rate-determining step is oxidative dehydrogenation, which abstracts an allylic hydro-gen atom.
- The initial hydrogen abstraction involves an oxygen atom associated with bismuth.
- The second hydrogen abstraction is facilitated by the presence of molybdenum-oxygen polyhedra and the presence of a C-O bond such that the activation energy is lower than for the initial abstraction, and the rate is correspondingly higher. There is strong evidence that this second abstraction actually involves oxygen atoms solely associated with molyb-denum.
- The oxygen that appears in the acrolein product is strongly associated with bismuth in the catalyst.

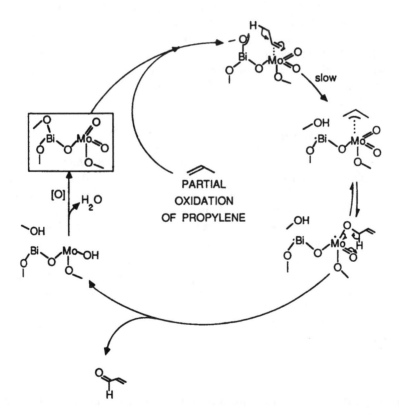

FIGURE 16.2 Mechanism for the Partial Oxidation of Propylene to Acrolein over Bismuth Molybdate Catalysts. Adapted from Grasseli.[9] Reprinted from Snyder, T. P. and Hill, G. G. Jr., *Catalysis Reviews: Science & Engineering,* 31, 43 (1989), by courtesy of Marcel Dekker, Inc.

Several investigators have noted a change in apparent orders on the propene and oxygen partial pressures in kinetic relations.[4,12,14] With laboratory-prepared catalysts without the various promoters, the reaction was noted to be first-order in propene and zero-order in oxygen at 400°C and above while, at temperatures below 400°C, the order on propene became zero and that on oxygen one-half. However, if the propene content is low relative to oxygen, the first-order on propene persists at lower temperatures. Interpretation of these findings based on the kinetics and kinetic isotopic effects have taken two different viewpoints.

1. "The partial oxidation of propene occurs via the formation of a symmetric allyl inter-mediate on the catalyst surface. Above 400°C, the abstraction of an α-hydrogen to form this intermediate is rate limiting as long as sufficient oxygen is present. At temperatures below 400°C, or if there is a lack of oxygen in the feed to the reactor, the reaction is limited by one of the steps in the catalyst reoxidation process."[4]
2. "The oxidation of propene to acrolein proceeds via the redox mechanism for which the rate of catalyst reduction (also the rate of acrolein formation) is at the steady state equal to the rate of catalyst reoxidation."[14]

Commercial catalyst developments have produced improved catalyst that include promoters that expedite movement of lattice oxygen. This fact, along with the knowledge that excess oxygen and relatively low propene concentrations have most probably lowered the temperature from 400°C to values in the lower 300°C range, at which the apparent change in kinetics occurs. See also "Process Kinetics." At conditions of very low oxygen, partial pressures oxidation of propene to CO and CO_2 occur and can dominate.[12] Consecutive oxidation of acrolein tends to occur with excess oxygen and in the higher temperature region.[12,13] But the selective catalysts in use tend to block this pathway. The real loss in selectivity at otherwise favorable conditions occurs in the homoge-neous oxidation of acrolein that can occur after the reactants and products exit the catalyst bed and produce carbon oxides and other impurities.[12] Elimination of dead space and rapid quenching can overcome this problem.

Catalyst Type and Characteristics

Commercial catalysts are usually referred to as *bismuth molybdate* (Bi_2MoO_6) but contain a number of other metal additives of Ni, Fe, Co, and K, and either P, B, VO, or Sb. If supported on a carrier, that carrier is usually silica (SiO_2). But most catalysts in use are extruded neat-metal oxides.[2] Catalyst porosity, pore-size distribution, and surface each play an important role in optimizing catalyst performance.[1] Since acrolein can be further oxidized to acrylic acid, it would seem useful to have large pores to avoid this secondary reaction.

The multicomponent catalysts in general use consist of mixed phases of oxides that can be expressed as $Me(II)_8$, $Me(III)_3$, and $BiMo_{12}O_n$ where, for example, Me(II) is elements such as Ni, Co, Mn, or Mg, and Me (III) is elements such as Fe, Cr, or Al.[10] Me(II) is present on structures isomorphous with β-$CoMoO_4$, and Me(III) is an $Fe_2(MoO_4)$-type structure.[10] Bismuth creates the high practical activity and selectivity but reduces the surface area, which suggests a covering layer of BiMo as an essential positive influence on selectivity.

Catalyst Suppliers and Licensors

Catalysts are supplied through the following process licensors in conjunction with a process license:

BP/Amoco
Knapsack

Mitsubishi Petrochemical
Nippon Shokubai
Sumitomo
Uhde GmbH

Catalyst Deactivation

The catalyst has a lifetime of 2–4 years. Since propane feed is rather pure, poisoning is usually not a problem. Deposits of coke tend to be removed by the steam fed to the reactor continuously. If high temperature excursions occur, there is always the possibility of sintering and catalyst attrition.

Over a long operating cycle, deposits and some slow attrition build to the point where pressure drop increases. Also, with aging, activity and selectivity decline to a point where additional temperature increases begin to hasten deactivation by sintering and require shutdown and catalyst replacement.

The importance of maintaining an excess of oxygen has been demonstrated by the rapid loss of activity when operating at low oxygen partial pressures.[12,13] Under these conditions of low oxygen partial pressure, the valence state of the metals composing the metal oxides will decline, and less oxygen will be available for the reaction steps.

Process Units

The high exothermicity of the oxidation reaction and high activation energy require a multitubular[1] reactor with molten salt as the heat-transfer fluid. Tubes of 2.5 cm diameter and 3–5 meters length have been used. Operating conditions vary from 1.5–2.5 bar and 300–400°C, depending on the catalyst and the stage in the operating cycle. Propene, air, and steam are fed in the ratio of 1:8–10:2–6 for the following several reasons:[1,3,8]

- Excess air is required to maintain the essential high oxidation state for high conversions and selectivities.
- Steam helps moderate temperature increases by providing a direct-contact heat sink.
- Steam reduces the oxygen partial pressure below the explosive limit.
- Steam removes coke-type deposits and helps keep the catalyst surface clean.
- There is also evidence that steam tends to displace strongly held adsorbates and thereby improves catalyst performance.
- It has also been shown that steam suppresses the formation of CO_2 by blocking the most active sites responsible for CO_2 formation. It is further suggested that steam also enhances oxidation by maintaining a high oxidation state and preventing the formation of strongly adsorbed oxygenates.[15]

After the acrolein is separated from the reactor product stream, recycle of unreacted propene can increase overall conversion and yield. Recycling is practiced only in plants producing lower conversions (e.g., 85%). Many newer plants are reporting once-through conversions of 95% and selectivities of 80 to 90% or more.[1,2,3] It may well be that continued improvement of acrolein yield may depend as significantly on reactor design as on better catalyst formulations. Some reactor designs are reported that quench the reaction products immediately at the exit of the tubes to prevent homogenous reaction of acrolein to CO and CO_2 and other products.[11]

Process Kinetics

Either mechanistic statement (see page 262) yields the same simplified kinetic expression based on a general equation[5] covering all ranges of feed composition and operating temperature.[4,14]

Statement 1 is represented by the following two equations based on (a) the abstraction of hydrogen and (b) the transfer of oxygen from a site occupied by atomic and molecular oxygen and propene and a site that is primarily occupied by atomic oxygen.

$$\frac{d(C_3H_4O)}{dt} = \frac{k_2K_1K_9P_P(K_7K_8P_O)^{1/2}}{[1 + K_1P_P + (K_7K_8P_O)^{1/2}][1 + K_9(K_7K_8P_O)^{1/2}]} \tag{a}$$

$$\frac{d(C_3H_4O)}{dt} = \frac{k_4K_1K_2K_3K_7K_8K_9P_pP_o}{[1 + (K_7K_8P_o)^{1/2} + K_7P_o + K_1P_p][1 + K_9(K_7K_8P_o)^{1/2}]} \tag{b}$$

Equation (a) can be rationalized to produce a simple equation in terms of first-order dependency for propene and zero-order for oxygen. Equation (b) rationalizes to an order of zero for propene and one-half for oxygen.[4]

Statement 2 is represented by a single equation that covers all conditions using the following arguments based on the redox mechanism.[14]

$$-\frac{dC_3H_6}{dt} = k_rP_p^x = \theta_{ox} \qquad \text{rate of catalyst reduction which is equal to rate of catalyst reoxidation}$$

$$-\frac{dC_2}{dt} = k_{ox}P_o^y = (1 - \theta_{ox}) \qquad \text{rate of catalyst reoxidation}$$

where θ_{ox} is fraction of fully oxidized sites and $1 - \theta_{ox}$ is fraction of vacant or reduced sites. P_p is propylene partial pressure and P_o is oxygen partial pressure.

At the steady state, the rates are equal, from which

$$\theta_x = \frac{k_{ox}P_o^y}{k_rP_p^x + k_{ox}P_o^y}$$

and

$$\frac{-d(C_3H_6)}{dt} = \frac{d(C_3H_4O)}{dt} = k_rP_p^x\left[\frac{k_{ox}P_o^y}{k_rP^x + k_{ox}P_o^y}\right]$$

At "high" temperatures,

(425–475°C) $$k_{ox}P_o^y > k_rP_p^x$$

and $$\frac{-dC_3H_6}{dt} = k_rP_p^x = \text{where x is found to be 1}$$

At low temperatures,

(300–350°C) $$k_rP_p^x > k_{ox}P_o^y$$

and $$\frac{dC_3H_4O}{dt} = k_{ox}P_o^y \quad \text{where y is found to be 0.5}$$

Both studies can be rationalized toward the application of a simple Power-Law equation within a narrow range of commercial operation.

$$r = kP_p^x P_o^y$$

Apparently, in the middle temperature range, more complex formulations are required. It should be noted the partial pressure range of either oxygen or propene also affects the reaction order.[14] The observed orders are also affected by the particular catalyst formulation.[13]

16.4   PROPENE → ACROLEIN → ACRYLIC ACID

Acrylic acid and its esters are used in the production of a number of important and useful polymers. The esters (methyl, ethyl, butyl, 2-ethylhexyl, and isobutyl acrylates) are produced by emulsion or solution polymerization. The polyacrylate emulsions are used in interior and exterior paints, floor polishes, and adhesives.[1] Solution-polymerized acrylates are primarily used in preparing industrial coatings. Acrylic acid itself can be polymerized to produce superabsorbing material for disposable diapers and as an essential ingredient in low-phosphorous detergents.[1] These last two uses are responsible for the increase in demand for acrylic acid.

Chemistry

Although acrylic acid was produced successfully for many years by a modified Reppe process via acetylene and carbon monoxide, essentially all acrylic acid is now produced by the catalytic partial oxidation of propene via acrolein.[2]

Acrolein is far more reactive than propene, and the second step conducted in a second reactor proceeds readily at a lower temperature than the first step. (See "Propene → Acrolein" for the essentials of the first step.)

In general, maximum operating conditions are 300–400°C for Reactor 1 and 200–300°C for Reactor 2. Side reaction products include acetic acid and some heavy boilers. The catalyst is generally described as composed of molybdenum and vanadium oxides.

Thermodynamics

As noted above, the reactions are highly exothermic with the second reaction being less so. The reactions are not equilibrium limited at the operating conditions of 200–400°C.

Mechanism

Investigation of the second step of propene partial oxidation has not been as detailed and thorough as that for the reaction to acrolein (Step 1). Although a different catalyst is used for Step 2, it is rather clear that the general redox mechanism prevails, as presented in the previous section.

Catalyst Type and Suppliers

A list of a number of patented catalysts for the second reactor is given in Table 16.1. These include both supported and unsupported types. In early developments, Mo and V were used in equal atomic amounts, but additional work demonstrated that much smaller amounts of vanadium produced a superior catalyst.[3] Elements that have been added as promoters, many of which are shown in Table 16.1, include Cu, As, V, Al, W, Ag, Mn, Ge, Au, Ba, Ca, Sr, B, Sn, Co, Fe, or Ni.[3] The modern catalyst has a high ratio of Mo to V (ranging from 2–8). These promoters improve selectivity and are thought to favorably change the relative surface concentration of molybdenum and vanadium.

In general, the catalysts have low surface areas and large pores to prevent total oxidation of product to CO_2 and H_2O and loss of yield.

TABLE 16.1
Second-Stage Catalysts Based on Patents

Company	Catalyst Composition	Reaction conditions					Results		
		ACR (%)	AIR (%)	H_2O (%)	SV (h-1)	Temp. (°C)	Conv. (%)	AA[a] select. (%)	AA[b] yield (%)
TOSOH	Mo-V/SiO_2 (Mo/V = 2-8)	8.0	44.0	48.0	3430	300	92.0	82.0	75.4
Rikagaku Res. Labs.	$Mo_{100}V_{10}Al_3Cu_{10}$/Al sponge	4.0	30.0	40.0	1000	320	98.4	97.6	96.0
Toagosei Chem. Industry	$Mo_{17.7}V_3As_{1.43}$/SiO_2	5.1	38.5	51.7	1170	320	96.5	91.0	87.8
Nipon Kayaku	$Mo_{12}V_2W_{0.5}$/SiO_2	5.9	58.8	35.3	996	220	97.8	89.0	87.0
	$Mo_{12}V_3Cu_{2.5}Fe_{1.25}Mn_{0.1}Mg0.1P_{0.1}$	2.8	24.2	73.0	2000	210	99.0	98.5	97.5
Celanese	$Mo_{12}V_3W_{1.2}Mn_3$	4.4	28.5	50.0	900	300	99.0	93.0	92.0
BASF	$Mo_{12}V_2W_2Fe_3$	3.1	27.5	43.0	1360	230	99.0	91.9	91.0
Nippon Shokubai	$Mo_{12}V_{4.8}W_{2.4}Cu_{2.2}Sr_{0.5}$/$Al_2O_3$	4.0	51.0	45.0	3000	255	100.0	97.5	97.5
		7.0	48.0	45.0	3500	255	97.0	97.0	96.1
SOHIO	$Mo_{12}V_5W_{1.2}Ce_3$/SiO_2	5.9	58.8	35.3	876	288	100.0	96.1	96.1
Sumitomo Chemical	$Mo_{12}V_3Cu_3Zn_1$/SiO_2	5.0	40.0	55.0	1000	260	98.4	96.1	94.6
Mitsubishi Petrochemical	$MO_{100}V_{20}CU_2$	4.0	50.0	46.0	850	290	99.5	95.3	94.8

[a]Acrylic acid selectivity.

[b]Acrylic acid yield.

Reprinted from Nojiri, N., Sakai, Y. and Watanabe, Y., *Catalysis Reviews: Science & Engineering*, 37(1), 145 (1995) by courtesy of Marcel Dekker, Inc.

Suppliers/Licensors

Companies listed in Table 16.1 may be assumed to be possible licensors.

Catalyst Deactivation

The lower operating temperature enables easier control of temperature rise in the reactor and reduces formation of carbonaceous deposits, which are also controlled by steam addition. Catalyst life is reported as three or four years or more. If carbonaceous deposits develop, they can be removed by air regeneration.

Process Units

As with the first stage to acrolein, multitubular fixed-bed reactors are used with tubes of 2.5 cm diameter and 3–5 m long.[1] The coolant is a molten salt. The charge to the first reactor consists of 5–7% propylene and 10–30% steam, and the rest is air plus some off gas from the absorber following the second reactor.[1] The mixture is preheated before entering Reactor 1 in the vapor state, and the product from Reactor 1 is cooled to 200°C prior to entering the second reactor. Steam plays an important role in both reactors. In each case, steam increases the rate of desorption of product from the catalyst surface and maintains a low enough oxygen concentration to assure operation below the flammability limit.[4,5]

The second reactor operates in the maximum temperature range of 200–300°C and a pressure modestly above atmospheric.[1,4,5] As the catalyst slowly deactivates over time, the operating temperature can be raised by increasing the molten salt temperature in order to maintain production and yield. A one-step process has also been described.[2]

Process Kinetics

The rate of acrolein formation in the oxidation of propene has been observed to pass through a maximum at high conversions of propene. A reaction scheme has been presented in which acrolein is formed directly from propene while acrylic acid is formed by a consecutive reaction and a side reaction directly from propene.[6]

$$C_3H_6 \xrightarrow[r_1]{1} C_3H_4O \xrightarrow[r_2]{2} C_3H_4O_2$$

with a reaction 3 (r_3) from C_3H_6 to $C_3H_4O_2$

Rate equations for the total rate of acrolein formation (r_A) and the total rate of formation of acrylic acid (r_K) have been presented.[6]

$$r_A = \frac{k_1 \cdot p_p \cdot p_A}{(1 + K_1 p_p + K_2 p_K + K_3 p_A + K_4 p_p - p_A^0)^2} - \frac{k_2 \cdot p_A^2}{1 + K_1 p_p + K_2 p_K + K_3 p_A + K_4 p_p \cdot p_A^0}$$

$$r_K = \frac{k_2 \cdot p_A^2}{1 + K_1 p_p + K_2 p_K + K_3 p_A + K_4 p_p - p_A^0} + \frac{k_3 \cdot p_p \cdot p_A}{(1 + K_1 p_P + K_2 p_K + K_3 p_A + K_4 p_p \cdot p_A^0)^3}$$

where $r_A = r_1 - r_2$, the net rate of acrolein formation

$r_K = r_2 + r_3$, the total rate of acrylic acid formation

subscripts A, K, P = to acrolein, acrylic acid, and propene, respectively

P = indicated partial pressure

p_A^0 = initial partial pressure of acrolein

K = adsorption constants

k = rate constants

These equations with multiple constants may well be greatly simplified for the narrow range of economic operation. In fact, it appears that the acrolein to acrylic acid reaction rate can be fit to an equation first-order in acrolein and zero-order in oxygen, provided adequate but not excessive oxygen is supplied.

16.5   BUTANE OR BENZENE → MALEIC ANHYDRIDE

The major use of maleic anhydride is in the production of unsaturated polyester resins in which it and phthalic anhydride are polymerized with propylene glycol to form a linear polyester with unsaturation. Dissolution with styrene monomer forms a resin of low viscosity that can be cross-linked by adding a peroxide initiator. The result is a rigid resin that can be provided with improved tensile and flexural strength upon addition of glass, Kevlar, or carbon fiber to the initial mix before adding the initiator.[1] These reinforced resins are used in producing the hulls of pleasure boats, automobile body parts, synthetic marble, and surface coatings.[2] Because of the ease of molding at close to ambient temperatures, these resins have many uses.

Maleic anhydride is also a valuable reactant in synthesizing important products such as agricultural chemicals (e.g., malathion insecticide) lubricant additives, and food acidulents (fumaric and malic acid).

Benzene was the major organic for oxidation to maleic anhydride and continues to be used extensively in Europe and the Far East, but it has been replaced by butane in the U.S.A. because butane has been a lower-cost feedstock and also avoids the toxic problems associated with benzene. New plants in Europe and the Far East are butane based except in locations where butane is not available. Both processes operate in the vapor phase, and some plants have been designed to use either process, although the catalyst must be replaced.

Chemistry (Benzene Route)

The oxidation of benzene to maleic anhydride consumes two carbon atoms, which are lost to carbon dioxide. Total oxidation of benzene consumes about 25% of the benzene. Selectivity to maleic anhydride is variously reported as 65–75%.[2,3] Operating conditions are in the range of 2–3 bar and 350–950°C.[2,3,4] The catalyst used is vanadium pentoxide promoted by molybdenum or tungsten oxides.

Thermodynamics (Benzene)

There is no thermodynamic equilibrium limitation to either the partial oxidation or the total oxidation of benzene. The process is kinetically controlled based on the selective characteristics of the catalyst. The high exothermicity indicates the need for excellent temperature control.

Mechanism (Benzene Route)

Many studies on benzene partial oxidation have been reported. An excellent summary of a rational mechanistic argument based on surface science rather than kinetics alone presents the reaction pathways as shown in Figure 16.3.[7,9]

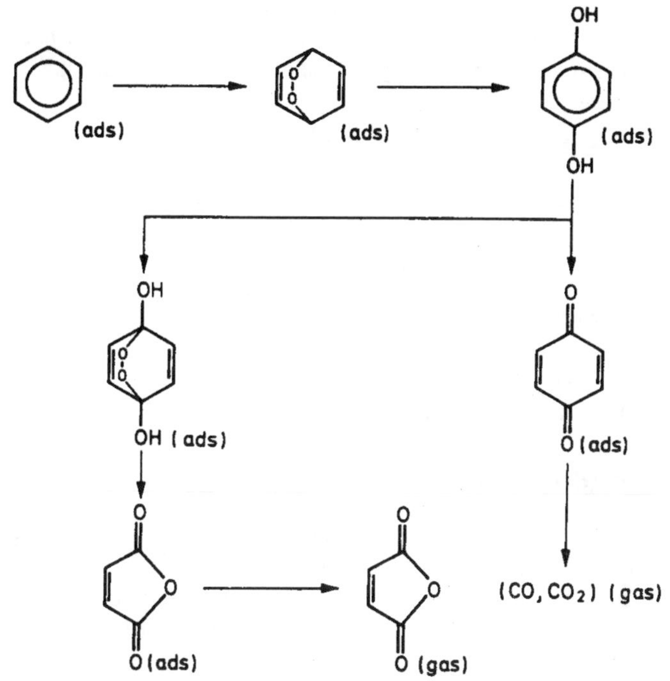

FIGURE 16.3 Reaction Pathways in the Oxidation of Benzene. Reprinted by permission: Chinchen, G., Davies, P., and Sampson, R.J., in *Catalysis: Science and Technology*, J. R. Anderson and M. Boudard, editors, Vol. 8, p. 41, Springer Verlag, Berlin, 1987.

Hydroquinone is an adsorbed intermediate that takes up oxygen and forms maleic anhydride, which then desorbs. The path to complete oxidation proceeds through the oxidation of hydroquinone to quinone and then rapid complete oxidation.

Electronic spin resonance studies suggest that molecular oxygen is chemisorbed as O_2^-. It is said to be located on V^V acceptor sites, and the organic molecule is located on V^{IV} sites. Benzene adsorbs as a stable species. The active catalyst contains two phases of mixed oxides, $Mo_6V_8^{5+}V^{4+}O_{40}$, and $Mo_4V_2^{5+}V_4^{4+}O_{25}$, and molybdenum is postulated to control and stabilize the ratio of IV and V oxidation states.[7] High benzene partial pressures will tend to reduce the catalyst and cause low selectivity, and high oxygen pressures tend to lower the V^{IV} concentration and cause poor performance. An increasing V^{IV} concentration level increases activity up to a maximum and then declines.[7] The 20–30% Mo content apparently assures the best activity/selectivity level. Aged catalyst that has both lower activity and selectivity exhibits a loss of molybdenum.[7,10]

Differences of opinion exist on the nature of the reactive oxygen. Instead of chemisorbed molecular oxygen acting as the oxidant, lattice oxygen has been proposed as the reactant. Lattice oxygen would then be the source of oxygen for the oxidation, which would then be replaced by the oxygen in the gas phase and described as a *redox mechanism*. In either case, oxygen associated with the catalyst is the reactant and is replenished from the gas phase.

Catalyst Type and Characteristics (Benzene Route)

Although there are many catalyst patents for partial oxidation of benzene, the industrial catalyst in common use is composed of V_2O_5 and MoO_3 on a low area (1–2 m²/g) refractory carrier, usually α-alumina.[5] Of the 10–20 wt% that is catalytic material, most of it is present in a mole ratio of approximately 2:1 of V_2O_5-to-MO_3. Various promoters are used, including P_2O_5 of up to 1–5 percent,[7] which is said to extend the life of the catalyst and improve selectivity. Manganese, bismuth, and tin are also claimed to extend catalyst life. Molybdenum oxide has been shown to retard secondary oxidation and stabilize the activity and selectivity of the catalyst.[7]

Catalyst Suppliers and Licensors (Benzene Route)

Supplier

Engelhard

Licensors

Alusuisse Italia (Lonza S.p.A)
Scientific Design

Catalysts are usually formed as extrudates.

Catalyst Deactivation (Benzene Route)

Sulfur compounds, alkyl aromatics, and paraffins can poison the catalyst when present above certain maxima as suggested by the licensor. Also, excessive temperature increases in a portion of the bed can volatilize molybdenum so as to move it to a cooler part of the bed and cause an overall loss of selectivity. A Danka patent[8] describes a PCl_2 or trimethyl phosphite treatment for reversing selectivity decline.[7]

Run lengths of 1–3 years are reported. Much depends on adequate temperature control so as to avoid excessive hot spots in the bed that can lower selectivity significantly. Toward the end of a run, temperature increases are required to maintain production. Not only does selectivity decline under such conditions, but tar deposits are apt to develop and hasten the time for shutdown.

Process Units (Benzene Route)

Most benzene processes use multitubular heat-transfer reactors with 13,000–22,000 one-inch OD tubes of 13–16 ft length.[2,3,4,7] Operating conditions are 350–450°C and 2–5 bar. The cooling medium is a molten salt eutectic of potassium and sodium nitrates and sodium nitrite at 350–400°C. It circulates through the shell side and thence to a waste-heat boiler that generates valuable high-pressure steam.

Benzene vapor and air are mixed so as to yield a benzene content in the combined stream of 1.4 volume percent, which is below the lower flammable limit at the preheat temperature of the inlet. The goal of reactor operation is to maintain as close to isothermal operation as possible. The higher the catalyst selectivity, the easier that task. Inevitably, at some point, the temperature will climb and create a higher temperature zone, which will move along the bed as the catalyst activity declines.

Benzene conversion is in the range of 85–98% and selectivity is usually 60–72 percent.[2,3] About one-fourth of the benzene undergoes complete oxidation, making a total exothermic heat production of 6500–7000 kcal/kg of benzene reacted.[3] This large amount of energy is, after being superheated, valuable energy for use in turbines for driving plant compressors and large pumps.

Maleic anhydride is recovered by various techniques involving azeotropic distillation and other procedures. Unreacted benzene and small amounts of MA are included in the off gases of CO and CO_2, and the mixture is incinerated.

Process Kinetics (Benzene Route)

Because of the low concentration of both benzene and product in the air mixture, assumption of first-order reactions has proved most convenient and acceptable. The large excess of oxygen suggests zero-order behavior for oxygen.[7,11] Thus, the following equations have been applied for commercial conditions. Constants should be based on the particular catalyst being used.

$$\text{Benzene} + O_2 \xrightarrow{\;1\;} MA \xrightarrow{\;2\;} CO, CO_2, H_2O$$

$$r_1 = k_1 P_B$$

$$r_2 = k_2 P_{MA}$$

$$r_3 = k_3 P_B$$

Chemistry (Butane Route)

The catalyst used is composed of vanadium/phosphorous oxides. Operating temperatures are between 400 and 480 K (higher than for benzene partial oxidation), and pressures are in the range of 2–3 bar. Major reactions are as follows:

		ΔH @ 700 K, kcal
(1)	$C_4H_{10} + 3\text{-}1/2\ O_2 \rightarrow C_4H_2O_3 + 4\ O_2$	−301.09
(2)	$C_4H_{10} + 4\text{-}1/2\ O_2 \rightarrow 4\ CO + 5H_2O$	−634.57
(3)	$C_4H_{10} + 6\text{-}1/2\ O_2 \rightarrow 4\ CO_2 + 5H_2O$	−363.53

Thermodynamics

The exothermic heat of reaction to maleic anhydride is lower for butane than for benzene, but more butane is converted to carbon oxides than for benzene (30 vs. 25%), and the selectivity is lower (53 vs. 72%). Thus, the heat generated per mole of maleic anhydride produced by butane oxidation is greater. Hence, a larger reactor or an improved catalyst is necessary for butane oxidation, not only because of the larger feed rate required to produce the same amount of maleic anhydride but also because of the higher heat removal load.

Mechanism (Butane Route)

The proposed mechanisms are certainly more complex, in keeping with the complexity of the catalyst. One such mechanism that seems to comport to the observed facts involves eight steps, as shown in Figure 16.4.[7,12,13]

All these compounds have been identified in the reaction system along with various intermediates not shown, including the inevitable formation of carbon oxides and H_2O due to combustion of maleic anhydride and unsaturated olefinic compounds. Instead of the intermediate crotonaldehyde, other workers have reported 2,3 dihydrofurane.[15]

The mechanism, as is so often the case, is still a subject of further studies. One thing is certain, with all the active species identified, it is somewhat amazing that reasonable selectivity is attainable.

Some additional insight can be realized by the work that has been done on the average oxidation state of vanadium on the catalyst surface, the V/P ratio of the activated catalyst, and the structure of the phases present.[13] There are clearly both reduced surface sites (V^{4+}) and oxidized surface sites (V^{5+}) on the activated catalyst. The dehydrogenation of butane to butene, and presumably to butadiene, takes place on the reduced sites, and the oxidation steps occur on the oxidized sites.[14] The optimal average surface valence state for vanadium is thought to be 3.9–4.5, and the optimal

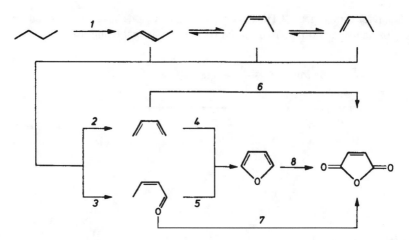

FIGURE 16.4   Reaction Pathways for Oxidation of Butane to Maleic Anhydride. Reprinted by permission: Chinchen, G., Davies, P., and Sampson, R. J., in *Catalysis: Science and Technology,* J. R. Anderson and M. Boudart, editors, Vol. 8, p. 58, Springer-Verlag, Berlin, 1987.

V/P ratio is reported to be 0.5–2, and ideally near unity.[7] These ratios are working in concert, since phosphorous, for example, stabilizes the mean valence state of the vanadium.

A high oxygen-to-butane ratio increases oxidation power and increases the V^{5+} concentration on the surface and ultimately reduces selectivity by causing excessive oxidation. Conversely, excess butane relative to oxygen will cause a higher reducing capacity (more V^{4+}), which increases reduction in the presence of inadequate oxidizing power, thereby preventing a generous yield of maleic anhydride.[13] The optimum ratio of reactants is 0.9 C_4/O_2, which amounts to about 15% butane in air, a value considerably above the lower explosion limit.

Crystalline $(VO)_2P_2O_7$ predominates in the most successful catalysts. Studies have shown that it stores oxygen by transforming into δ-$VOPO_4$.[16]

$$O_2 + 2(VO)_2P_2O_7 \leftrightarrow 4VOPO_4$$
$$V^{4+} \qquad\qquad V^{5+}$$

This reversible process provides the necessary link for the redox mechanism. Surface lattice oxygen enables oxidation of the hydrocarbon, and reoxidation of the active sites occurs by reduction of oxygen at separate sites and movement of oxide ions through the lattice.[17]

Catalyst Type and Characteristics (Butane Route)

The best catalysts are based on vanadium phosphorous oxides with crystalline $(VO)_2P_2O_7$ being the predominate form. There are hundreds of patents, and the processes and catalysts used are proprietary. In general, the catalysts vary in vanadium-to-phosphorous ratio between 0.5 and 2.0, and many are a value of 1.[7] Phosphorous is essential in stabilizing the optimum ratio of V^{5+} and V^{4+} valencies. Various alkali or alkaline earth elements are used as promoters and prolong catalyst life by slowing the loss of phosphorous.

The catalyst is prepared as extrudates or tablets and, at times, special shapes like rings are used to reduce pressure drop. Fluid catalyst has been introduced and is a spray-dried, abrasion resistant powder.

Since the conversions and yields for butane oxidation to maleic anhydride are low (50–60% yield at 90–100% conversion) compared to benzene oxidation, ways for attaining higher conversions at more favorable lower temperatures have been sought. One of the modestly successful efforts apparently in use involves producing a catalyst with higher surface area by preparing the catalyst precursor using nonaqueous solutions.[13]

Catalyst Suppliers (Butane Route)

Licensors for various processes also provide the proprietary catalyst directly or through a contracted catalyst suppliers. Engelhard is a supplier of catalyst. Licensors include the following:

Scientific Design	Huntsman
ABB Lummus Global/Lonza S.p.A	Mitsubishi
BP Chemicals	Sisas (Pantochim)

Fixed-bed catalysts are available as extrudates, and fluidized catalyst prepared to resist attrition is a powder composed of spheroidal shapes.

Catalyst Deactivation (Butane Route)

The production of the catalyst begins with heating a solution of vanadium compound such as ammonium vanadate with a reducing agent in conjunction with orthophosphoric acid. Treatment must involve procedures that will produce small particles so the finished catalyst will have medium-high surface area. The resulting product, called a *precursor*, is a complex mixture of both amorphous and crystalline phases.[2] The precursor must then be subjected to careful calcining in air to produce the desired catalyst. It is in these preparative and calcining steps where the catalyst manufacturer's expertise is essential. Careful final treatment in start-up of a new batch is essential to attain the maximum activity and selectivity for optimum performance over a period of time.

Particular care must be given not only to start-up but also in operation to avoid excessive hot spots in fixed beds that will cause movement of phosphorous from hotter to cooler regions of the bed. Such a condition, if severe, invariably reduces selectivity and shortens the life of the catalyst. The use of a volatile phosphorous compound in the feed at times to restore the catalyst to some degree by redistribution has been reported.

As with other oxidation catalysts, sulfur compounds are poisons when present in significant amounts.

Process Units (Butane Route)

Two major types of reactors have been developed: multitubular heat-transfer and fluidized-bed.

Multitubular Heat-Transfer

The reactors are similar to those described for benzene feed except they must be larger for butane feed. Selectivity is lower (50–60%), and conversion of butane is also lower (80–85%). The feed composition is 1.65% butane in air.[2] The result is a larger reactor, compressor, and condenser, which increases capital costs. Because of the higher operating temperature (420–488 K) compared to the benzene process and because of oxidic side products, alloy tubes must be used. The lower cost of butane compared to benzene, however, makes the butane process more economically attractive.

To provide flexibility of feedstocks, Scientific Design offers a process design that can be used for either benzene or butane feed. It is conceivable that the price advantage of butane could fade as other uses of benzene decline because of environmental regulations.

Fluidized Beds

Various attempts to improve maleic anhydride yields have led to the design, pilot operation, and construction of commercial fluidized-bed reactors. The fluidized-bed reactor is designed to minimize fluid backmixing and bubble growth by means of internals such as baffles,[19] but catalyst tends to be well mixed so that operating temperature throughout the bed is essentially constant. Conversions of 80% at 51% selectivity are reported.[18] As in all fluidized processes, catalyst attrition is a major concern. New processes seem to have reduced this problem considerably. The following advantages are claimed:[2,7]

- Since fluidized beds can operate isothermally because of rather complete mixing of the fluidized catalyst, optimum operating conditions are more readily attained.
- Air and butane may be introduced separately, since they are rapidly well mixed in the fluidized bed. This procedure eliminates problems in mixing the two flows in the inlet line that could produce a mixture within the flammable region. Such problems can be avoided, however, with a well-designed digital control system.
- With the butane content of the combined air-butane feed at 4% rather than 1.65% in the fixed-bed process, almost one-half as much air is required. The result is savings on compressor size and operating cost. Also, the concentration of maleic anhydride in the vapor product is about double that of the fixed-bed process, which will require smaller sizes of equipment in the recovery section.
- A relatively simple procedure is required while on stream to remove some catalyst from the reactor and replenish with fresh catalyst, thus avoiding increasing temperature over time and earlier plant shutdown.
- Excellent temperature control is attained due to backmixing of catalyst throughout the reactor and high heat transfer produced by constant impinging of catalyst particles on coiled heat-transfer surface within the bed. Reacting gases proceed through the bed without excessive backmixing.
- The fluid-bed reactor is less costly and easier to charge and discharge catalyst than the multitubular reactor.

Of course, no process is perfect, and several possible disadvantages can be put forward.

- Although the reactants are not completely mixed, some backmixing must occur that limits the maximum conversion attainable—significantly below that of the fixed-bed process. (The DuPont Company has developed and tested two-bed riser-type fluidized beds that have minimal backmixing.[17] One bed is used for oxidizing the catalyst and the other for selectively oxidizing the butane with the oxidized catalyst. High conversion and high selectivity to maleic anhydride are realized because of low gas backmixing and optimum conditions in each riser.)
- As long as the fluidized bed is operating properly, there is no problem with the ultimate feed composition being in the flammable range. If, however, a malfunction should occur in which fluidization stops or dead spots develop, a definite possibility of explosion exists unless rapid methods for detecting and eliminating the condition are included in the control system.

Purification Sections (Benzene and Butane Routes)

There are a number of patented procedures for recovering maleic anhydride from the off-gases of each of the processes previously discussed. In some cases, the major differences in licensed processes involve the recovery system, descriptions of which are beyond the scope of this handbook.

Process Kinetics

There are a number of kinetic studies on the maleic anhydride reaction system, often differing in catalyst characteristics. One of these is based on the Eley–Rideal equation, which assumes that the reactant (in this case, butane) reacts directly with oxygen from the metal oxide without a preceding adsorption step.[19] Independent observations using TPD have shown that butane is not adsorbed on the catalyst. Oxidation of maleic anhydride to CO, CO_2, and H_2O was considered redundant, leaving the main reaction and the two butane oxidation equations. The following rate equations were developed and tested by data from bench scale and pilot plant studies.[19] Reaction paths 4, 5, and 6 were shown to be redundant by regression analysis.

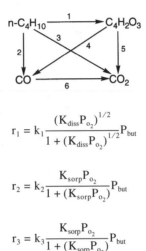

$$r_1 = k_1 \frac{(K_{diss}P_{O_2})^{1/2}}{1 + (K_{diss}P_{O_2})^{1/2}} P_{but}$$

$$r_2 = k_2 \frac{K_{sorp}P_{O_2}}{1 + (K_{sorp}P_{O_2})} P_{but}$$

$$r_3 = k_3 \frac{K_{sorp}P_{O_2}}{1 + (K_{sorp}P_{O_2})} P_{but}$$

where K_{sorp} and K_{diss} are equilibrium constants for adsorption and dissociation of O_2, respectively.

$$O_2 + s \rightarrow O_2S$$

$$O_2 + 2V^{(n+1)+} \Leftrightarrow 2OV^{n+}$$

16.6   ORTHO-XYLENE OR NAPTHALENE → PHTHALIC ANHYDRIDE

Phthalic anhydride is a reactive compound with some very important applications. These uses, in order of decreasing amounts, include production of phthalic esters, (50–60%) alkyd resins, (20%) unsaturated polyester resins (20%), and the remainder for fine chemical, dyes, pigments.[1,2] Phthalic esters are used as plasticizers. The diester, dioctyl phthalate, is 60–70% of the total phthalic esters produced, and about 80% of it is used for imparting flexibility to polyvinyl chloride.[3] Alkyd resins are used in producing tough paint coatings, and large amounts are consumed in traffic paints.[1] Unsaturated resins are made by reacting a polyol with phthalic anhydride or maleic anhydride dissolved in styrene monomer. The resulting polymer is molded with glass fibers to produce tough products like pleasure-boat hulls and other lightweight but strong molded objects.

Chemistry

Phthalic anhydride can be produced by partial oxidation of naphthalene or o-xylene. Naphthalene was the dominate feedstock up to the early 1960s, with most of it coming from coal tar. The ready availability of o-xylene from petroleum soon reduced the use of napthalene because of its inefficient consumption of two carbons in producing phthalic anhydride. Overall reactions of both feedstocks over V_2O_3 catalysts can be simplified as follows:[1,2]

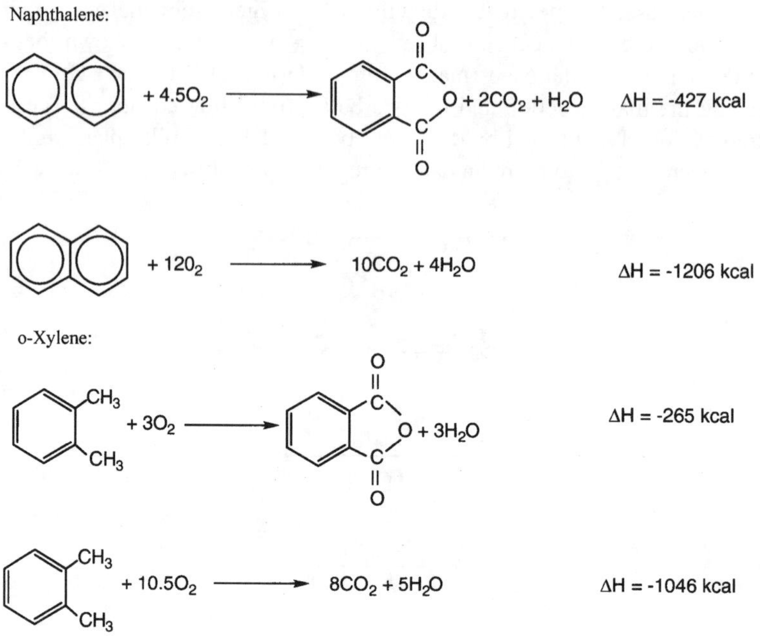

Naphthalene:

o-Xylene:

Thermodynamics

The reactions are not equilibrium limited and are highly exothermic. Since they are characterized by high activation energies, heat production increases rapidly with small temperature increases in the initial portion of a fixed-bed reactor causing significant hot spots. The total heats of reactions, based on 90% yield for naphthene feed and 80% for o-xylene feed with the remainder going to CO_2 and H_2O, are -504.7 kcal/mole and -421.2 kcal/mole, respectively.

Mechanism

Plausible reaction networks have been presented based on careful studies of laboratory-prepared catalyst. Very little has been published on commercial catalysts. Oxidation steps are generally agreed to proceed via a redox mechanism with the active vanadium site supplying the oxygen, which is then restored by oxygen from the gas phase. The reaction networks for naphthalene and o-xylene shown in Figs. 16.5 and 16.6 are based on analysis of all intermediates and products detected.[1,4,5,11]

Clearly, the o-xylene route is more complex and more susceptible to the formation of unwanted side-reaction products if significant intraparticle diffusion gradients exist. One would also expect lower selectivity for the o-xylene route, but the overall economics are clearly in its favor.

Catalyst Type and Characteristics

Although over 90% of phthalic anhydride is now produced from o-xylene, in some areas of the world, napthalene continues to be used because of its economical availability from nearby major steel mills and lack of a convenient source of petrochemical o-xylene.

The preferred catalyst for naphthalene partial oxidation was V_2O_5 promoted by K_2SO_4 on a moderately porous silica carrier and shaped as cylindrical pellets.

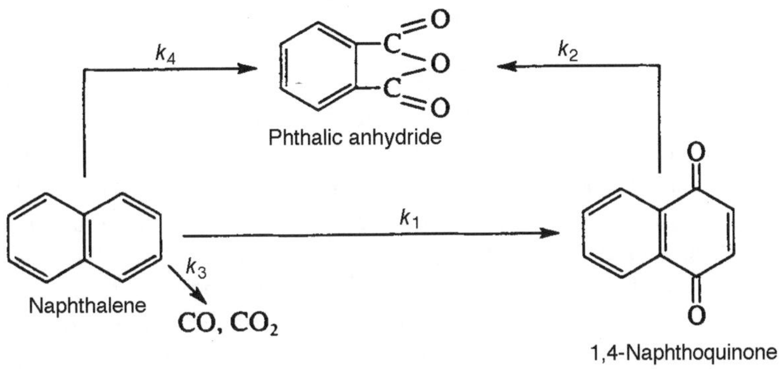

FIGURE 16.5 Naphthalene reaction network. Reprinted by permission: Bunton, R. F., Wainright, M. S., and Westerman, D. W. P., paper 14-d Chemeca 84, The Institution of Engineers, Australia, Twelfth Australian Conference, Melbourne, August 26–29, 1984.

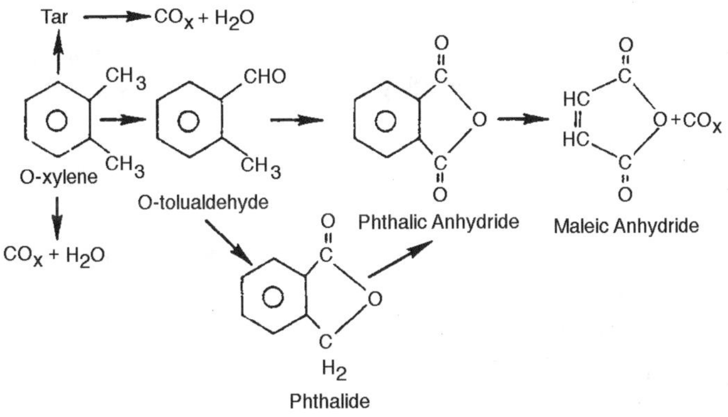

FIGURE 16.6 Ortho-xylene reaction network. Reprinted from *Applied Catalysis,* Saleh, R. Y., and Wachs, I. E., 31, 91 (1987), "Reaction Network and Kinetics of O-Xylene Oxidation to Phthalic Anhydride over V_2O_5/TiO_2 Catalysts," p. 87–98, copyright 1987, with permission from Elsevier Science.

Apparently, the napthalene route was not significantly affected by the physical nature of the carrier. But when o-xylene became an important feedstock, porous carriers were found to be detrimental to the yield. A new catalyst was developed using a low-area or a largely pore-free carrier such as silicon carbide, quartz, or porcelain.[6] The shape of the catalyst was also found to be an important factor in reducing pressure drop and increasing exterior surface area per unit mass of catalyst. A ring or half-ring shape (5–10 mm OD × 5–10 length, with hole diameter of 1.5–3.5 mm) is now favored. A mixture of V_2O_5 and TiO_2 (2–15% V_2O_5) is applied to the catalyst carrier in a very thin layer[2] of 0.02–2 mm thick.[9, 11] Numerous promoters are specified in the patent literature, including niobium, potassium, cesium, rubidium, antimony, and phosphorous. The desired promoters are added to the major ingredients, such as $VO(OC_3H_7)_3$ and $Ti(OC_3H_7)_4$, and applied as a coating to the aforementioned carrier. These new catalysts are now not only used for o-xylene feeds but also for units designed for either o-xylene or naphthalene.[11]

The complexity of the o-xylene partial oxidation network has made studies of the catalyst characteristics difficult. A study of unsupported V_2O_5/TiO_2 catalyst has yielded some interesting insights devoid of such side issues as diffusion gradients and varying carrier properties.[5]

- Vanadium occurs in two forms or active sites—a vanadium oxide surface layer and as V_2O_5 crystallites.
- The initial step of the partial oxidation of o-xylene to o-tolualdehyde occurs almost exclusively over the surface vanadium oxide monolayer.
- The o-tolualdehyde is converted to phthalic anhydride over the surface vanadium oxide and to phthalide over both the surface vanadium oxide and the V_2O_5 crystallites.
- The overoxidation of phthalic anhydride occurs on the surface vanadium oxide, as does the non-selective production of CO and CO_2 from direct oxidation of o-xylene. Almost all the CO_2 is produced directly from o-xylene. These observations, based on product distribution studies as a function of conversion, establish the fact that phthalic anhydride is produced from the intermediate, tolualdehyde, and not directly from o-xylene.
- Separate characterization studies[7,8] have confirmed the two types of vanadium oxide on the catalyst. TiO_2 in its anatase form modifies the properties of vanadium oxide, causing the formation of a vanadium oxide monolayer coordinated to the TiO_2 as surface vanadium oxide and also the formation of small crystallites of vanadium oxide. Although the crystallites are not very active and do not dramatically affect selectivity, exposed TiO_2 causes a significant drop in selectivity.
- Studies that minimize residence time produce high activation energies, which suggests the inevitable play of diffusion in exothermic fast reactions. Low-area catalyst carriers are essential in reducing or eliminating intraparticle gradients and thus undesired side reactions.

Catalyst Suppliers

Catalysts are proprietary and provided by the licensors or their designated catalyst manufacturer. Active licensors include the following:

Lonza, S.p.A.
Lurgi Öl Gas Chemie

Catalyst Deactivation

Catalyst life is quoted variously as 2–4 years, but much depends on avoiding excessive temperatures that might occur because of operating errors. High temperatures approaching and exceeding 500°C can cause rapid migration of catalyst components, which will reduce selectivity and activity significantly. Temperatures above 500°C can change the structural form of TiO_2 from anatase, the active form, to rutile, an inactive form.[1]

When impure naphthalene from coal tar is used, the various organic sulfur and nitrogen impurities reduce activity and selectivity. The low activity can be remedied by higher-temperature operation, but the selectivity remains lower, although some of the side products such as maleic anhydride are also valuable.

Some tar is produced with both feeds, but it is usually minimized by the oxidizing atmosphere. Because of the high stability of phthalic anhydride, reaction conditions are adjusted to oxidize all side reaction products, making separation of phthalic anhydride in the recovery section less complicated.

Although one might expect both organo-nitrogen and organo-sulfur compounds to be catalyst poisons, sulfur dioxide in small amounts (0.1–0.2% of hydrocarbon feed) aids in maintaining the activity of potassium promoted catalysts (V_2O_5, K_2O, SO_3). It has been suggested that the SO_3 content in the original catalyst acts as a carrier of oxygen, either to reoxidize the catalyst and/or to provide oxygen for the reaction.[11]

$$V_2O_5 + SO_3 \rightarrow V_2O_4 + SO_2 + O_2 \tag{1}$$

$$SO_2 + V_2O_5 \rightarrow V_2O_4 + SO_3 \tag{2}$$

$$2V_2O_4 + O_2 \rightarrow 2\ V_2O_5 \tag{3}$$

The oxygen liberated in Step 1 oxidizes the hydrocarbon, and the SO_2 produced is oxidized in Step 2, and the reduced vanadium ions are reoxidized by gaseous oxygen.[11,12] This phenomenon occurs in certain catalysts such as the silica-gel supported naphthene catalyst of V_2O_5 and K_2SO_4. As the SO_3 content of the catalyst is consumed, the activity declines, but a continuous small addition of SO_2 to the oxidizing atmosphere restores and maintains the activity. Curiously, the usual poison for so many catalysts, organic compounds (in this case, the napthalenes) often present in coal-tar derived naphthalene, serve as a source of sulfur oxides that aid in maintaining activity. The quantity, however, must be small, e.g., 0.1 wt% of hydrocarbon feed.[11]

Process Units

Multitubular heat-transfer reactors dominate both the naphthene and o-xylene routes. In fact, some plants have been designed for either feedstock. Such a strategy actually requires more attention in the product recovery section than the reactor. Reactors with up to 25,000 tubes of 1 in. (2.5 cm) inside diameter have been employed.

Temperature control is accomplished by a molten salt eutectic of potassium and sodium nitrites (59%/41%) with a melting point of 141°C.[2] The molten salt is circulated through a waste-heat boiler where it releases heat to boiler feedwater to produce saturated medium pressure steam (~350 psia). Operating salt temperatures are in the range of 325 to 425°C.[1] Reactor outlet temperature for fresh catalyst is 350°C for naphthalene and 375°C for o-xylene. The saturated steam can be superheated and used to drive the steam turbine for the air blower.

The temperature profile along each tube is typical of that for a highly exothermic and rapid reaction. Air and o-xylene or naphthalene are introduced at around 130°C and heated by the molten salt heat-transfer fluid circulating in the reaction shell. When the temperature of the reaction mixture reaches about 275°C with fresh catalyst, the reaction takes off, and a hot spot begins to develop with the reaction temperature rapidly reaching and then exceeding the salt temperature. Since the reaction rate increases exponentially and the heat removal increases only linearly as the reaction mixture exceeds the temperature of the molten salt, the hot spot moves to its maximum temperature. The temperature begins to decline as the reaction rate declines with increasing conversion.

Much effort has been focused on better control of the hot-spot temperature maximum. The most common procedure seems to be the use of two catalysts in the bed, including a lower-activity form in the upper section where intrinsic rates are highest because of higher concentrations of reactants. The second portion would be packed with a higher-activity form in order to increase rates when reactant concentrations are low. Alternatively, the same catalyst may be used in both sections, but the initial-section catalyst would be diluted with inert particles of the same shape as the active catalyst.

Another classic means for controlling temperature is to use a fluidized bed. Such reactors were used successfully with naphthalene feed for which the original catalyst of V_2O_5 and K_2SO_4-on-silica provided a good fluidized catalyst. Unfortunately, the silica gel support did not permit good yields of phthalic anhydride from o-xylene. Also, the newer catalyst was subject to excessive attrition. Fluidized-bed units are apparently no longer in significant use in phthalic anhydride production. One reference stated that there were only a few left, and they are located in the Far East.[9] None of these uses o-xylene as the feed.

Energy-Saving Operations

The cost of compression and preheating air is a major factor in the operating cost of a partial-oxidation reactor. Over the years, various licensors have introduced designs either involving new plants or modification of existing reactors. The focus of these designs has been to increase the o-xylene or napthalene loadings above the traditional loading of 44–46 g/m^3 STP of air to values of 60 and above. In so doing, additional heat energy will be generated and produce a significant increase in production of steam from the circulating molten salt. The additional steam can then become adequate to drive the air blower and provide preheat for reactor, without the need for electric power or general plant steam produced at the boiler house.

Since the lower explosive limit for o-xylene or naphthalene in air is in the range of 47 g/m^3 STP, the higher values are clearly in the explosive range. Studies have shown that an explosion starting in the catalyst bed would be quenched by the heat capacity of the catalyst.[10] Explosions can occur, however, in the catalyst-free space above the bed. This possibility is addressed by providing extra thickness for the upper reactor head along with an adequate combined area of rupture disks installed on the reactor bed. Since the introduction and successful implementation of this concept, the loadings have been increased in various processes to as high as 134 g/m^3 STP air with satisfactory results, including substantial energy savings.[1, 9]

Process Kinetics

A thorough review of a plethora of kinetic studies up through 1979 led to the conclusion that "most have suffered from the lack of reliable kinetic data."[11] The reaction networks are extremely complex, and oversimplification can cause real problems in obtaining useful equations. But some simplification is essential for a reliable model for industrial applications. Of course, the actual commercial catalyst of interest must be used. An appealing simplification has been used in studying the benefits of several catalysts of different characteristics.[12]

The network for partial oxidation of o-xylene can be simplified as shown by lumping the intermediates into a pseudo-compound simply called *intermediates*. Since these products are small in quantity, the simplification could be justified. The simplified rate equation for each reaction path shown is as follows:

$$r_{ij} = \frac{k_{ij}C_i}{1 + b(C_1/C_{1,0})}$$

where r_{ij} is in mol. g^{-1} s^{-1}, with C_1 and $C_{1,0}$ representing the local and initial concentration of o-xylene, respectively. The subscript i refers to the reactant, and j refers to the product. Thus, k_{12} is the rate constant for the reaction of o-xylene to intermediates.

16.7   ANTHRACENE → ANTHRAQUINONE

Anthraquinone is used to produce a variety of color-fast dyes in shades of blue and turquoise.[1] It can also be used in the production of hydrogen peroxide (H_2O_2). Anthraquinone is hydrogenated to anthrohydroquinone using Pt or Ni at 30–35°C.[3] After the catalyst is recovered and the reaction mix is exposed to air, auto-oxidation occurs, forming endoperoxide, which reacts to regenerate the anthraquinone and produce hydrogen peroxide.[3] Obviously, the major demand for anthraquinone is for dyes. The heterogeneous catalytic process for producing anthraquinone in the gaseous phase from anthracene is usually preferred when adequate supplies of anthracene are available. A liquid-phase oxidation is also practiced using a homogeneous catalyst of dichromate and sulfuric acid. Synthesis of anthraquinone from benzene and phthalic anhydride via Friedel–Crafts reaction is also practiced.

Chemistry (Gas-Phase Process)

The reaction is carried out in the temperature range of 380–420°C and a pressure of 1.5–2 bar.[4] An excess of air is used, and the reaction is not equilibrium limited at the operating conditions. Vanadium oxide is the catalyst most often used.

Mechanism

A redox mechanism as described for phthalic anhydride production seems to be accepted. A number of reaction schemes have been suggested.

An early study[5] suggested the major side reaction to be the formation of phthalic anhydride via a short-lived intermediate. The phthalic anhydride further reacts to $CO_2 + H_2O$.

A more recent detailed study has been reported of technical-grade anthraquinone from oxidation of coal-tar derived anthracene that contains many more impurities than from a typical petroleum source.[6] Twenty-nine compounds were identified, and the author was able to suggest a reaction scheme by identifying those compounds that were not initially present in the coal-tar cut. A more complex oxidation reaction system was proposed.[6] Anthracene reacting to 9, 10 anthraquinone was the main reaction, but a side reaction at a much lower rate produced 1, 4 anthraquinone that reacted in turn to 2, 3 naphthalenecarboxylic anhydride and hence to phthalic anhydride and, finally, CO_2 and H_2O. Other possible side reactions of less importance are also illustrated.

Catalyst Type and Suppliers

Vanadium oxide on Kieselguhr is used with colloidal silica added for tougher extrudates.[4,5] Promoters include alkaline metal oxides. Titanium and phosphorous compounds as additives are reported to permit operation with anthracene of below 95% purity. The impurities are apparently oxidized in the presence of these additives to carboxylic acids that can readily be separated from the product with alkaline solutions.[4]

The operating processes are proprietary. The catalyst is thought to be similar to that described for naphthalene oxidation.

Catalyst Deactivation

As do most catalysts, those used with anthracene are subject to a gradual decline in activity over time. Introduction of a small amount of ammonia injected into the air fed to the reactor has been claimed as a means for maintaining activity.[2] Excessive contamination of the product can occur if optimum temperature control is not maintained.[1]

Process Units

As for so many other partial oxidation reactions, multitubular fixed-bed reactors are used for anthracene. The anthracene is melted and then pumped to a vaporizer vessel through which steam and a portion of the air is passed to vaporize the anthracene and carry it to the top of the multitubular reactor.[4] Heat-transfer fluid circulates through the shell. Rapid quenching of the product is accomplished by introducing air after initial cooling, and energy recovery by a heat exchanger. Yields of 90–95% are reported when operating in the range of 350–430°C and 1.5–2.0 bar with excess air.

Process Kinetics

Vapor-phase kinetics over V_2O_3 catalyst has been presented based on a system of consecutive or parallel-consecutive reactions.[7, 8]

$$\text{anthracene} \xrightarrow{k_1} \text{anthraquinone} \xrightarrow{k_2} \text{phthalic anhydride} \xrightarrow{k_3} \text{maleic anhydride} \xrightarrow{k_4} CO_2$$

The reactions fit pseudo first-order equations as follows:

$-r_1 = k_1(n_a/n_t)$, rate of disappearance of anthracene

$r_2 = k_2(n_a/n_t) - k_2 (n_a/n_t)$, rate of formation of anthraquinone

$r_3 = k_2 (n_{aq}/n_t) - k_3 (n_p/n_t)$, rate of formation of phthalic anhydride

$r_4 = k_3 (n_p/n_t) - k_4 (n_m/n_t)$, rate of formation of maleic anhydride

where n_t = total moles in exit stream per mole anthracene fed

n_a = moles anthracene converted/mole anthracene fed

n_{aq} = moles anthroquinone formed/mole anthracene fed

n_p = moles phthalic anhydride formed/mole anthracene fed

n_m = moles maleic anhydride formed/mole anthracene fed

The total moles in and out are essentially constant, because the excess air used is large.

16.8   METHANOL → FORMALDEHYDE

Formaldehyde is the most active of the aldehydes and certainly one of the most reactive of all organic compounds. As such, it is used in the production of a wide variety of products including, in order of formaldehyde consumption, urea-formaldehyde (UF) resins, phenol-formaldehyde (PF) resins, 1,4-butanediol, acetal resins, pentaerythritol, hexamethylene tetramine, and melamine-formaldehyde (ME) resins. The resins are the best known, because they end up in many consumer products such as adhesives for manufacture of plywood, particle board, and fiber board. They are also used as molding compounds. Polyacetal resins produced by polymerization of formaldehyde are valuable molding plastics of high strength and durability.

Commercial formaldehyde is handled as a water solution (37–55%). Gaseous formaldehyde is not stable. Formaldehyde solutions are used as a preservative and disinfecting agent. Formaldehyde does not accumulate in the environment, for it rapidly oxidizes to carbon dioxide and water.[1,2]

Formaldehyde was suspected as a carcinogen, but this postulate has never been verified. Instead, indoor maximum allowable concentrations have been set.

Chemistry

Formaldehyde is manufactured by two major reaction processes, one using a silver catalyst and the other a metal oxide, Mo and Fe in a ratio of 1.5–2.0[1] with promoters such as V_2O_5, CuO, Cr_2O_3, CoO, and P_2O_5. The chemical reaction system differs in the two processes.

Silver Catalyst

The reaction, which takes place with an excess of methanol, has been described as oxidative-dehydrogenation. It is conducted at 600–650°C or 650–750°C, depending on the process. The following stoichiometric equations have been suggested.[1]

		ΔH kcal/mole[8]	
		900 K	**1000 K**
(1)	$CH_3OH \Leftrightarrow CH_2O + H_2$	21.99	22.27
(2)	$H_2 + 1/2O_2 \rightarrow H_2O$	−59.08	−59.24
(3)	$CH_3OH + 1/2O_2 \rightarrow CH_2O + H_2O$	−37.09	−36.97
(4)	$CH_2O \rightarrow CO + H_2$		
(5)	$CH_3OH + 3/2\ O_2 \rightarrow CO_2 + H_2O$		
(6)	$CH_2O + O_2 \rightarrow CO_2 + H_2O$		
	K @ 900 K = 1.123; K @ 1000 K= 1.658 for Reaction 1		

Reactions 1–3 are considered the primary reactions, and 4–6 are side reactions.[1] Other side reactions can produce methyl formate, methane, and formic acid.[1] Reaction 3 can be obtained by combining Reactions 1 and 2. These stoichiometric equations do not represent a mechanism, and various statements concerning the main reactions have been put forward. Since Reaction 3 is an oxidative dehydrogenation, and Reaction 1 is a dehydrogenation, it is suggested that these two reactions are the main players, with Reaction 3 producing 50 to 60% of the formaldehyde and Reaction 1 the rest.[2] The endothermicity of the dehydrogenation (Reaction 1) moderates the net exothermic heat of the reaction system. It should also be noted that Reaction 1 can be equilibrium limited, and high temperatures (700°C) are required for total conversion in this reaction.

Iron Molybdate Catalyst

In this system, the reaction is conducted with an excess of air and is described as oxidation of methanol to formaldehyde in a temperature range of 350–450°C.

$$CH_3OH + 1/2\ O_2 \rightarrow CH_2O + H_2O \qquad \Delta H = -37 \text{ kcal @ 700 K}$$

No hydrogen is produced but, at higher temperatures (>470°C), significant amounts of carbon monoxide are formed, which reduces the yield of formaldehyde.

$$CH_2O + 1/2O_2 \rightarrow CO + H_2O \qquad \Delta H = -55.99 \text{ kcal @ 700 K}$$

Other secondary reactions that have been proposed include the following[4]

$$CH_2O + 2CH_3OH \rightarrow CH_2(OCH_3)_2 + H_2O$$

$$CH_3OH + CH_3OH \rightarrow (CH_3)_2 O + H_2O \text{ dimethyl ether}$$

$$CH_2O + 1/2 \ O_2 + CH_3OH \rightarrow HCOOCH_3 + H_2O \text{ methyl formate}$$

$$CH_2O + 1/2 \ O_2 \rightarrow CHOOH \text{ formic acid}$$

The exothermic heat produced by the main reaction approaches twice that produced by the silver process.

Mechanism

Silver Catalyst:

A study[1] of the reaction of methanol on the surface of AgO suggests the following mechanism based on surface measurements involving heavy hydrogen.[3]

1. Adsorbed surface oxygen enhances both the dissociative and nondissociative adsorption of methanol.
2. The hydroxyl group of methanol interacts with surface oxygen atoms to form adsorbed CH_3O (methoxide) and H_2O.

$$CH_3OH(g) + 0(a) \rightarrow CH_3O(a) + OH(a)$$

$$CH_3OH(g) + OH(a) \rightarrow CH_3O(a) + H_2O(a)$$

$$CH_3O(a) \rightarrow H(a) + H_2CO(a)$$

$$H_2CO(a) \rightarrow H_2CO(g)$$

$$H(a) + H(a) \rightarrow H_2(g)$$

Methoxide was found to be the most abundant species and, depending on operating conditions, it will take part in various undesired side reactions. If the ratio of surface oxygen to adsorbed methoxide is high, the excess surface oxygen atoms further oxidize formaldehyde to CO, CO_2, and water. This observation corresponds to practical operation. Although oxygen is essential in the formation of formaldehyde, and methanol conversion increases with an increase in the ratio of oxygen to methanol, the formaldehyde yield will ultimately decline at higher ratios and CO and CO_2 formation will increase.

Catalyst Types and Characteristics

Silver Catalyst

The silver catalyst is employed in the form of crystals (0.5–3 mm in size), gauze, or Ag-impregnated SiC granules or spheres.[5] The gauze is made by weaving high purity silver wire (6×10^{-4} in.) into 18 mesh screen.[7] Silver is subject to poisoning by transition metals, chlorine, and sulfur compounds. Iron dust from upstream rust is particularly damaging, since it coats the surface of the catalyst. This problem can usually be avoided by using alloy steel equipment and efficient filters in upstream piping. Catalyst life is reported variously as 2–4 months and 3–8 months,[2,5] but activity can be restored after removal by an electrolytic recovery process. Because the bed is shallow, catalyst removal and replacement, although frequent by industrial standards, is relatively simple and rapid.

Fe-Mo Catalyst

The so-called iron molybdate catalyst has a composition of 18–19 wt% Fe_2O_3 and 81–82 wt% MoO_3 and contains promoters such as Cr or Co oxides.[5] The active component is iron (III) molybdate $[Fe_2(MoO_4)_3]$, but the ratio of Fe_2O_3 and MoO_3 is such that there must be excess MoO_3 in the finished catalyst. That situation has been confirmed by various spectroscopic measurements.[6] Both $Fe_2(MoO_4)_3$ and MoO_3 were identified at the surface and in the bulk.[6] Activity tests confirm that $Fe_2(MoO_4)_3$ is the active component of the catalyst. The excess MoO_3 is distributed in the bulk and the surface but appears to have no direct catalytic effect. It does, however, increase the exterior surface area per unit mass of catalyst, which results in higher conversions per mass of catalyst.[6] Another favorable effect of excess MoO_3 has been described as a source of replacement of Mo in the active $Fe_2(MoO_4)_3$, which is depleted as Mo_2O_5 over prolonged use in the hottest part of the bed.[5] This compound, often called *molybdenum blue,* is a nonstoichiometric oxide that sublimes and moves by vapor transport to cooler parts of the bed, where it condenses on the catalyst and ultimately lowers the catalyst activity in that part of the bed, and also causes increased pressure drop as sublimation increases significantly above 400°C. The excess MoO_3 in the original catalyst is thought to replace that lost from the $Fe_2(MoO_4)_3$ and maintains activity in the main part of the bed.

The catalyst is often produced as granules or spheres in 4–6 or 6–10 mesh sizes. Carriers such as an inert silica can be added to increase the crushing strength. Catalyst life is variously reported as 1 to 2 years.[2,5] As the catalyst ages, fines tend to form that can ultimately cause excessive pressure drop. The catalyst is fragile and requires careful loading procedures. It is possible, during the aging process, for excess MoO_3 to crystallize slowly and ultimately cause particle breakage, producing fines and increased pressure drop. Careful temperature control can do much to minimize such conditions. In contrast to the silver catalyst, the Fe/Mo catalyst is resistant to most catalyst poisons that might enter the reactor.

After used catalyst is removed from the reactor, it can be screened to remove inactive outside layers. The screened catalyst can then be recharged first and then each tube topped off with fresh catalyst. The screened used catalyst is reported to produce higher yields because of the elimination of excessively active sites during previous use.[7] If, however, sintering has occurred, reuse is not possible.

Licensors

Catalysts are supplied by the licensor or by an approved catalyst supplier.

Silver Catalyst
  ABB Lummus Global, Partec Resources

Fe/Mo Catalyst
  Haldor Topsoe, Petron (Haldor Topsoe is also a merchant supplier)

Process Units

There are three major methanol-to-formaldehyde processes:

- Incomplete-conversion silver process
- Complete-conversion silver process
- Fe/Mo process

Incomplete-Conversion Silver Process (Oxidative Dehydrogenation)

This is the most common silver process. The reactor consists of a shallow bed (1–5 cm thick supported on a grid, or perforated tray) in diameters of 1.5–2.0 m.[1,2,5] Operation is conducted above the explosion limit in the methanol-rich range and with less than the stoichiometric amount of oxygen. In this manner, complete conversion of oxygen is attained, but incomplete conversion of methanol results.

Since the reaction is conducted adiabatically at 590–650°C and slightly above atmospheric pressure, the net exothermicity and the reaction temperature can be controlled by the amount of air fed to the reactor. A waste-heat boiler connected close to the outlet of the catalyst bed quenches the product gases preventing, thereby, unwanted homogeneous reactions. The product gases proceed to an absorber that separates the noncondensables from the formaldehyde and unreacted methanol, both of which absorb in the absorbing liquid (water). The tail gas contains up to 20% hydrogen and can be burned to generate steam and avoid more complex and costly means of disposal.[2]

The bottoms from the absorber must be distilled in order to separate the unreacted methanol as overhead liquid distillate and a formaldehyde water solution as the bottoms product. The strength of the formaldehyde product depends on the amount of water introduced to the reactor as steam along with the recycled methanol to aid in controlling reactor temperature by providing additional heat capacity. Yields as high as 92% have been reported.[1,2]

Complete-Conversion Silver Process

In more recent years, as energy costs increased the incentive for eliminating the energy-intensive distillation (which is not required in the rival Fe/Mo process) became compelling. By operating at higher temperatures (680–720°C), it was found possible to achieve complete conversion and avoid methanol recovery and recycle along with the elimination of the distillation column.[1,2,5] The endothermic dehydrogenation rate increases rapidly with temperature increases, as does the equilibrium conversion. Yields are reported to average 90%.[1,2]

In this higher-temperature mode, the addition of water in the form of steam is very important. The steam increases the total moles present and increases the equilibrium conversion of the endothermic Reaction 1, which is favored by high temperature, low pressure, and a high value of total moles. In addition, steam prevents loss of activity due to sintering of the silver, aids in reaction temperature control by acting as a heat sink, and deters the rate of formation of carbonaceous deposits on the silver that reduce the active area.[5] There is a limit on the steam to be added, since the amount ultimately governs the strength of the formaldehyde-water solution. Additional methanol can also serve as a heat sink.

Iron Molybdate Process

This process is a direct oxidation of methanol to formaldehyde and water and is more than three times more exothermic than the silver process. Accordingly, a nonadiabatic multitubular reactor with catalyst in the tubes and a circulating heat-transfer fluid in the shell is required. The tubes are 2.5 cm OD and 1–2 m in length. Operating temperature is the range of 280–400°C at atmosphereic pressure. Valuable steam at 300 psig can readily be generated by passing the heat-transfer fluid through a waste-heat boiler.[9] High conversions (>99%) are attained, and overall plant yields reach 93%.

The feed to the reactor is on the lean methanol side of the flammable limit and operates with excess oxygen. If, however, the oxygen in the reactor is reduced to 10% by substituting tail gas for air, it has been demonstrated that the methanol content can be increased to 9–12%, thus reducing the volume of reactor gases and associated compression costs. In this operating mode, up to 80% of the tail gas can be recycled at the steady state such that only 20% must be incinerated to destroy the small amounts of side products. Since the amount of combustibles in the tail gas is very small, combustion requires addition of supplementary fuel. Alternatively, catalytic incineration is practiced, which only requires auxiliary fuel at startup.[9]

After the formaldehyde is absorbed in the water absorber, ion-exchange treatment is used to eliminate traces of formic acid that are formed by the reaction of formaldehyde with oxygen. This treatment is not required in the silver process.

Comparison of Processes

The various processes have both advantages and disadvantages that are summarized in Table 16.2. Although the economics for an 18,000 t/yr plant favors the complete-conversion silver process, the Fe/Mo process tends to be favored for smaller production units. The existing plants for each process, in descending order of numbers of plants, are incomplete-conversion silver process, Fe/Mo process, and complete-conversion silver process. Each process, in most cases, has been revamped to reduce energy costs and eliminate possible air pollution problems.

TABLE 16.2
Comparison of Formaldehyde Processes

	Silver Catalyzed Processes		Fe/Mo Catalyzed
	Incomplete Conversion	Complete Conversion*	Complete Conversion*
Operation	Methanol rich	Methanol rich	Oxygen rich
Tail gas from absorber	Burned as fuel	Burned as fuel	Not combustible; requires catalytic incineration of contaminants
Yield	92	90.5	93
Methanol recycle	Yes	None	None
Distillation required (to separate unreacted methanol)	Yes	No	No
Equipment size	Modest	Modest	Larger because of excess air
Catalyst replacement required	2–4 months	2–4 months	18–24 months
Reactor	Adiabatic	Adiabatic	Multitubular nonadiabatic
Temperature range	Lower temperature suppresses unwanted side reactions	High temperatures require careful control so as to avoid unwanted side reactions	Low temperature suppresses unwanted side reactions and favors high selectivity, although some formic acid is formed.
Economics[†] (basis: 18,000 ton/year)			
Production cost/ton	211.6	174.5	183.9
Investment cost/ton	3.7	3.3	4.0

*"Complete conversion" generally refers to conversions of methanol at 99% or more.

[†]Data from Ref. 1.

Process Kinetics

For a simple commercial model, the rate of formation of formaldehyde may be successfully correlated by a rate form first-order in the limiting reactant, which is methanol in the silver process and oxygen in the Fe/Mo process.[10,11]

Silver: $r_{HCOH} = k_m P_m$

Fe/Mo: $r_{HCOH} = k_o P_o$

where subscripts m and o refer, respectively to methanol and oxygen.

More detailed rate forms have been proposed that consider side reactions (e.g., Refs. 12 and 13).

16.9   ISOBUTYLENE OR TERT-BUTYL ALCOHOL → METHACROLEIN → METHACRYLIC ACID

Methacrylic acid is esterified with methanol to produce methyl-methacrylate, which is polymerized to form poly(methyl methacrylate). This polymer, because of its hardness and clarity, is used as cast and extruded sheet for glazing materials, biomedical appliances, surface coatings, and optical products.[1,2] It is sold in the U.S.A. under the trade names "Plexiglas" and "Lucite."

Most methacrylic acid is produced via acetone cyanohydrin by base-catalyzed reaction of acetone with hydrogen cyanide, followed by reaction with sulfuric acid to produce methacrylamide sulfate. The final stage of the process is a combined hydrolysis-esterification of the sulfate with methanol to methyl methacrylate and methacrylic acid. Methacrylic acid and methanol are recycled.[1] Recently, however, interest has developed in an isobutene or tertbutyl alcohol route similar to the successful acrylic acid process for partial oxidation of propane. Several apparently viable processes have been implemented in Japan. Since the introduction of MTBE (methyl tert-butyl ether) as a gasoline additive in the U.S.A., the demand for isobutene and tert-butyl alcohol as feedstock for the more profitable MTBE production has risen exponentially. Such is not the case in Japan, and the partial oxidation route to methacrylic acid has been successfully implemented. At this writing, however, California is restricting the use of MTBE.

The major source of C_4 fractions is from steam cracking. These fractions, all of which have close boiling points (1,3 butadiene, isobutene, 1-butene and other butenes), must be separated by means other than distillation. After removal of butadiene by solvent extraction, isobutene, with its tertiary carbocation intermediates, can be hydrated to tert-butyl alcohol and readily separated from the C_4 raffinate. The reaction is reversible, and dehydration can be implemented after separation to yield a pure stream of isobutene. Although dehydration is an important step in MTBE production, which requires an isobutene feed, tert-butyl alcohol can be used directly in the methylacryltic-acid process.

Chemistry

Isobutylene Route

Step 1

$$CH_2 = \underset{\underset{CH_3}{|}}{C} -CH_3 + O_2 \longrightarrow CH_2 = \underset{\underset{CH_3}{|}}{C} - CHO + H_2O$$

methacrolein

Step 2

$$CH_2 = \underset{\underset{CH_3}{|}}{C} -CHO + 1/2 O_2 \longrightarrow CH_2 = \underset{\underset{CH_3}{|}}{C} - COOH$$

methacrylic acid

Tert-Butyl Alcohol

Step 1A

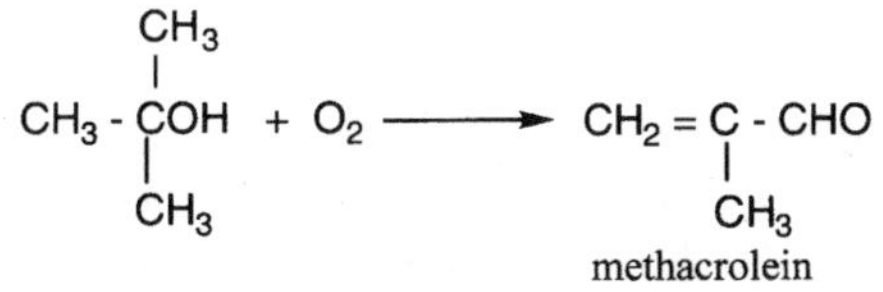

$$CH_3 - \underset{\underset{CH_3}{|}}{\overset{\overset{CH_3}{|}}{C}}OH + O_2 \longrightarrow CH_2 = \underset{\underset{CH_3}{|}}{C} - CHO$$

methacrolein

Step 2A

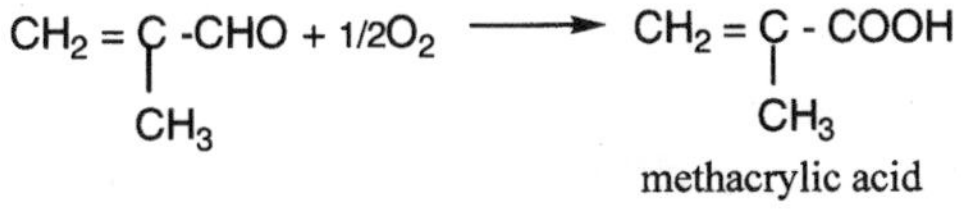

$$CH_2 = \underset{\underset{CH_3}{|}}{C} - CHO + 1/2 O_2 \longrightarrow CH_2 = \underset{\underset{CH_3}{|}}{C} - COOH$$

methacrylic acid

Step 1A is simply the sum of the dehydration of tert-butyl alcohol and Step 1 above. In either case, the methacrylic acid is removed from the aqueous phase and purified by distillation followed by esterification with methanol in the liquid phase to methyl-methacrylate.[5] Each reaction is highly exothermic and requires adequate heat removal.

Mechanism

The relatively limited use of these processes at present has not encouraged mechanistic studies. One would assume that a general redox mechanism applies for each partial oxidation step (see also sections on acrolein and acrylic acid). There are significant differences between the reactivity of isobutylene and propylene and also acrolein and methacrolein.

Although the additional methyl group on the β-carbon atom makes isobutylene more reactive than propylene, it also makes the intermediate, methacrolein, and the product methacrylic acid more reactive and susceptible to further oxidation to undesired side products.[6]

When alcohol feed is used, it is postulated that it is rapidly converted to isobutylene and water in the first stage.[6]

Catalyst Types

The route to methacrylic acid via isobutene is in a phase of rapid development in Japan as well as research and development in the U.S.A., where interest rests on possible changes in raw material costs. There are many patents and various catalyst systems proposed.

First Stage

The catalyst for oxidation of propene to acrolein has been successfully used, but improvements in preparation and composition have allowed higher yields. Various first-stage catalysts have been patented, such as the following.

- Mo/Bi/Fe/a/P/b (where a = one or more of Co, Ni, Mn, Mg, Sb, W, and b = one or more K, Cs, Tl)
- Mo-W-Te and others without Bi

Second Stage

- Phosphorous-molybdenum catalysts described as a hetero-polymolybdate salt with a composition of $[PMo_{12}O_{40}]^3$ with the corresponding acid partially exchanged with K, Cs, or Tl, e.g., $Cs_nH_{3-n}[PMo_{12}O_{42}]$.[6] Activity and selectivity apparently vary with "n" in the final catalyst. One optimum suggested is $Cs_2H[PMo_{12}O_{42}]$.

- Other dopants reported are Cu, V, As, and Sb. Replacement, for example, of a portion of the Mo by V modifies the redox properties and improves yield.[6] Catalyst stability is greatly improved by replacement of some of the hydrogens by a monovalent metal.[6]

Catalyst for both stages must be designed to discourage long residence times in the body of the catalyst so as to avoid secondary reactions that form CO, CO_2, acetaldehyde, and acetic acid and reduce selectivity and yields.[6] This goal can be realized by producing tablets no larger than 1/8 in. and with large pores.

Catalyst Deactivation

The first-stage catalyst tends, over time, to lose Mo in the inlet portion of the bed, which is the hottest section. The chemical form of the vapor is molybdic acid (H_2MoO_4), which has a significant vapor pressure and moves to the cooler portions of the bed downstream where it condenses.[6] The change in composition of the initial part of the bed causes a loss of activity. Careful temperature control can minimize these effects and prolong catalyst life.

The second-stage catalyst, over time at reaction conditions, is said to lose some of its structural characteristics related to good selectivity. These changes cause an irreversible aging phenomenon that is being addressed by additional stabilizing additives. In addition, the Mo-P salts tend to aggregate into larger crystals over time, thus reducing the active surface areas.[6] Again, careful temperature control can prolong life by avoiding accelerated migration and increased crystal size of the active component.

Licensors

Asahi Glass, Japan Methacrylic Monomer Co., and other Japanese companies with a variety of processes

Process Units

The isobutylene/tert-butyl alcohol process uses either the alcohol or the isobutylene as feed. Tert-butyl alcohol dehydrates at the inlet conditions to the first-stage reactor to form isobutylene. The isobutylene in C_4 Raffinate-I streams is easily recovered by hydrating the isobutylene (IB) to tert-butyl alcohol (IBA) followed by separation via countercurrent extraction. The TBA is then dehydrated subsequently to TB for use in producing MTBE, an octane enhancer. Alternatively, the TBA can be used directly in the partial oxidation system to produce methacrylic acid. Because of the large demand in the U.S.A. and Europe for MTBE, isobutylene, and tert-butyl alcohol, costs have increased enough to make existing acetone-cyanohydrin plants continue to be economically viable in the U.S.A. and Europe. Since Japan has not used oxygenated octane enhancers as extensively, the isobutylene/tert-butyl alcohol processes have shown favorable economics for new plants in Japan.[4]

The process consists of two packed-bed adiabatic reactors in series. Steam, air, and isobutylene or tert-butyl alcohol are fed to the first reactor in a dilute stream (5–10% TB or TBA, 10% steam, and air) as reported for a typical process.[3] The first reactor is reported to operate in the range of 300–420°C and 1–3 bar.[3,5] Yield of methacrolein is reported to be 88%,[5] but higher yields (96%) have been claimed.[3] Also, some methacrylic acid is formed in the first stage.

The second stage is variously reported to operate at 270–350°C and 2–3 bar. Obviously, if the discharge from the first reactor goes directly to the second stage, the pressure in the second stage will be less than in the first. The yield to methacrylic acid in the second stage is reported to be 86% of methacrolein fed to stage 2.

The various processes are under continued development, especially in the area of catalyst improvement, that permit valuable process changes. Initial two-stage processes practiced intermediate purification of methacrolein. New technology now eliminates this practice, and discharge from the first reactor goes directly to the second stage.[5]

Product from the second reactor is quenched to condense aqueous methacrylic acid, which is then solvent extracted and further purified and sent to the esterification reactor to produce methyl methacrylate.[3] Unreacted methacrolein is recovered from the off-gas of the quencher by absorption followed by stripping and then recycled to the second-stage reactor.

Isobutane Feed

To reduce feed costs for methacrylic acid production, Halcon SD Group proposed a process beginning with much lower cost isobutane.[4] The isobutane can be dehydrogenated in the presence of steam and the effluent fed directly to the two-stage reactor system already described. The oxygen and hydrogen remaining in the recycle gas must be removed by catalytic oxidation and the energy produced recovered. Such a process seems to be more economical for a new plant than other processes, but not so favorable as to lead to shutdown of existing processes.[4] Much depends on raw material costs that could change.

Process Kinetics

A rather thorough study of the partial oxidation of isobutene to methacrolein has been presented.[7,8] The catalyst was Bi-W/Fe-Co-Mo-K, and the study was conducted at 380–420°C and 1.3 bar. Since samples were taken along the length of the reactor operating in the gas phase, it was possible to demonstrate that methacrylic acid is generated from methocrolein and decomposition products generated from both methacrolein and methacrylic acid. Thus, it was possible to represent the reaction system by the following simplified kinetic scheme.

$$r_1 = \frac{k_1 C_b}{1 + k_b C_b} \qquad r_2 = k_2 C_m$$

$$r_3 = \frac{k_3 C_b}{1 + k_b C_b} \qquad r_4 = k_4 C_{ma}$$

where r_1, r_2, r_3, and r_4 are rates of formation of, respectively, methacrolein from isobutene, methacrylic acid from methacrolein, combustion products from isobutene, and combustion products from methacrylic acid. Subscripts b, m, and ma refer, respectively, to isobutene, methacrolein, and methacrylic acid.

Higher temperature favors methacrolein, and lower temperature favors methacrylic acid. Two stages are, therefore, advantageous, as described under "Process Units."

REFERENCE, SECTION 16.1 (GENERAL—ORGANIC OXIDATION)

1. Dadyburjor, D. B., Jewur, S. S. and Ruckenstein, E., *Catal. Rev. Sci. Eng.* 19(2), 293 (1979).

REFERENCES, SECTION 16.2 (ETHYLENE OXIDE)

1. Chenier, P. 1., *Survey of Industrial Chemistry*, 2nd ed., VCH, New York, 1992.
2. Ozero, R. J. and Landau, R. in *Encyclopedia of Chemical Processing and Design*, J. J. McKetta, editor, p. 274, Marcel Dekker, New York, 1984.

3. Welssermel, K. and Arpe, H. J. *Industrial Organic Chemistry,* VCH, New York, 1997.
4. Thomas, C. L., Catalytic *Processes and Proven Catalysts,* Academic Press, New York, 1970.
5. Kilty, P. A. and Sachtler, W. M. H., *Cat. Rev.-Sci. Eng.,* 10(1), 1 (1974).
6. Kiguchi, I., Kumazawa, T. and Nakai, T., *Hydrocarbon Proc.,* p. 69, March, 1976.
7. Zomerdljk, J. C. and Hall, M. W., *Catal. Rev.-Sci. Eng.,* 23(1) 163 (1981).
8. Petrov, E. A. and Shoper, *Applied Catal.,* 24, 145 (1986).

REFERENCES, SECTION 16.3 (ACROLEIN)

1. Ohera, T., Sato, T., Shimizu, N., Prescher, G., Schwind, H., Schwind, H. and Wesberg, 0. in *Ullmann's Encyclopedia of Industrial Chemistry,* 5th edition, Vol. Al, p. 149, VCH, New York, 1985.
2. Etzkorn, N. G., Kirkland, J. J. and Neilsen, W. D. in *Kirk-Othmer Encyclopedia of Chemical Technology,* 4th edition, Vol. 1, p. 232, Wiley, New York, 1991.
3. Weigert, W. A. and Haschke, H. in *Encyclopedia of Chemical Processing and Design,* J. J. McKetta, editor, Vol. 1, p. 382, Marcel Dekker, New York, 1976.
4. Synder, T. P. and Hill, C. C. Jr., *Catal. Rev.-Sci. Eng.,* 31, 41(1989).
5. Keulks, G. W., *J. Catal.,* 19., 232 (1970).
6. Krenzke, L. D. and Keulks, G. W., *J. Catal.,* 61, 3160 (1980).
7. Ono, T., Nakejo, T. and Hironaka, T., *Chem. Soc., Faraday Trans.* I, 79, 4–31 (1990).
8. Rylander, P. N. in *Catalysis: Science and Technology,* J. R. Anderson and M. Boudart, M., editors, Vol. 4, p. 1, Springer-Verlag, Berlin, 1983.
9. Barrington, J. D., Kartisek, C. T. and Grasselli, R. N., *J. Catal.,* 63, 235 (1980).
10. Higgins, R. and Hayden, P. in *Catalysis, Chemical Society,* Vol. 1, p. 176, The Chemical Society, London, 1977.
11. German Patent (DI 1910795, 1969), Hillenbrandt, H., Ließnz, E., Lussling, Th., Noll, E. and Simon, K. (to Degussa).
12. Keulks, G. W., Rosynek, M. P. and Daniel, C., *Ind. Eng. Chem. Prod. Res. Develop.,* 10, 139 (1971).
13. Bruckman, K. and Haber, J., *J. Catal.,* 114, 196 (1988).
14. Monnier, J. R. and Keulks, G. W., *J. Catal.,* 68, 51(1981).
15. Saleh-Alhamed, Y.A., Hudgins, R.R., and Silverston, P.L., *J. Catal.,* 161 (1), 430 (1996).

REFERENCES, SECTION 16.4 (ACRYLIC ACID)

1. Bauer, W., Jr., in *Kirk-Othmer Encyclopedia of Chemical Technology,* 4th edition, Vol. 1, p. 267, Wiley, New York, 1991.
2. Weissermel, K. and Arpe, H. J., *Industrial Organic Chemistry,* 3rd edition, p. 286, VCH, New York, 1997.
3. Ohara, T., Sato, T., Shimizu, N., Prescher, G., Schwind, H., Wesberg. O. and Marten, K. in *Ullmann's Encyclopedia of Industrial Chemistry,* 5th edition, Vol. A1, p. 161, VCH, New York, 1985.
4. Sakuyama, T., Ohara, T., Shimizu, N. and Rubota, K., *Chemtech.,* p. 250, June, 1973.
5. Naohiro, N., Sakai, Y. and Watanabe, Y., *Catal. Rev.-Sci. Eng.,* 37 (1), 145 (1995).
6. Svachula, H., Tockotein, A. and Tichy, J., *Collec. Czech. Chem. Commun.* 51, 1579 (1986).

REFERENCES, SECTION 16.5 (MALEIC ANHYDRIDE)

1. Selley, J. in *Concise Encyclopedia of Polymer Science and Engineering,* editor, Wiley, New York, 1990.
2. Cooley, S. D. and Powers, J. D. in *Encyclopedia of Chemical Processing,* J.J. McKetta, editor, Vol. 29, p.35, Marcel Dekker, New York, 1988.
3. Weissermel, K. and Arpe, H. J., *Industrial Organic Chemistry,* VCH, New York, 1993.
4. Franck, H. G. and Stadelhofer, J. W., *Industrial Aromatic Chemistry,* Springer Verlag, New York, 1988.

5. Felthouse, T. R., Barnett, J. L., Mitchell, S. F. and Mummey, M. J. in *Kirk-Othmer Encyclopedia of Chemical Technology,* 4th edition, Vol. 15, p. 893, Wiley, New York, 1995.

6. Lohbeck, K., Haferkorn, H., and Fuhrmann, W. in *Ullmann's Encyclopedia of Industrial Chemistry,* 5th edition, Vol. A16, p.53, VCH, New York, 1990.

7. Chinchen, G., Davies, P. and Sampson, R. J., in *Catalysis: Science and Technology,* J. R. Anderson and M. Boudart, editors, Vol, 8, p. 41, Springer-Verlag, Berlin, 1987.

8. US Patent 4,081,460 Danka Chemical Corp. (1975).

9. Petts, R. W. and Waugh, K. C., *J. Chem. Soc., Faraday Trans.* 78, 803 (1982).

10. Bielanskl, A., Majbar, M. S., Chrasacz, J., and Wall, W., *Stud. Surf. Sci. Catal.,* 6 127 (1980) Elsevier, New York.

11. Ramirez, J. F. and Calderbank, P.11., *Chem. Eng. Journal* 14, 19(1977).

12. Sampson, R. J. and Shooter, D. in *Oxidation and Combustion Reviews,* C.F. Tipper, editor, p.223. Elsevier, London 1965.

13. Bosch, H., Bruggink, A. A. and Ross, J. R. H., *Applied Catal* 31, 323 (1987).

14. Centi, G., Furnassari, G. and Trifiro, F., *J. Catal.,* 89, 44 (1984).

15. Trifiro, F., *Catal. Today,* 16, 91(1993).

16. Schnurman, Y. and Gleaves, J. T. in "New Developments in Selective Oxidation, II," *Studies in Surface Science and Catalysis* 82, 203 (1994), Elsevier, New York.

17. Contractor, R. M., Garnett, D. I., Horowitz, H. S., Bergna, H. E., Patience, G. S., Schwartz, J. T. and Sisler, G. M. in New Developments in Selective Oxidation II," *Studies in Surface Science and Catalysis* 82, 233 (1994), Elsevier, New York.

18. Stetani, G., Budi, F., Fumagalli, C. and Suciu, G. D., in "New Developments in Selective Oxidation," *Studies in Surface Science and Catalysis* 55, 537 (1990), Elsevier, New York.

19. Schneider, P., Emig, G. and Hofmann, H., *Ind. Eng. Chem. Res.*, 26, 2236 (1987).

REFERENCES, SECTION 16.6 (PHTHALIC ANHYDRIDE)

1. Dengler, H. P. in *Encyclopedia of Chemical Processing and Design,* J.J. McKetter, editor, Vol. 36, p. 36, Marcel Dekker, New York, 1991.

1. Franck, H. G. and Stadelhofer, J. W., *Industrial Aromatic Chemistry,* Springer-Verlag, New York, 1988.

2. Chenier, P. J., *Survey of Industrial Chemistry,* VCH, New York, 1992.

3. Buton, R. F., Wainwright, M. S. and Westerman, D. W. P., Paper No. 14d, *Chemica 84,* Twelfth Australian Conference, Melbourne, Aug. 26–29, 1984.

4. Saleh, R. Y. and Wachs, I. E., *Applied Catalysis,* 31, 87 (1987).

5. Weissermel, K. and Arpe, H. J., *Industrial Organic Chemistry,* 2nd edition, VCH, New York, 1993.

6. Bond, G. C. and Konig, P., *J. Catal.,* 77, 309 (1982).

7. Wachs, I. E., Saleh, R. Y., Chan, S. S. and Chersich, C. C., *Appl. Catal.,* 15, 20 (1985).

8. Park, C.M. and Sheehan, R.J. in *Kirk-Othmer Encyclopedia of Chemical Technology,* 4th edition, Vol. 18, p. 999, Wiley, 1996, New York.

9. Wiedemann, 0. and Gierer, W. *Chemical Engr.,* p. 62, Jan. 29, 1979.

10. Wainwright, M. S. and Foster, N. B., *Catal. Rev. Sci. Eng.,* 12(2), 211 (1979).

11. Kotter, M., Li, D.X. and Reikert, L. in "New Developments in Selective Oxidation," *Studies in Surface Science and Catalysis,* 55, 267 (1990), Elsevier, New York.

REFERENCES, SECTION 16.7 (ANTHRAQUINONE)

1. Cofranceseo, A. J. in *Kirk-Othmer Encyclopedia of Chemical Technology,* 4th edition, Vol. 2, p. 801, Wiley, New York, 1992.

2. Vogel A., in *Ullmann's Encyclopedia of Industrial Chemistry,* 5th edition, Vol. A2, p. 347, VCH, New York, 1985.

3. Weissermel, K. and Arpe, H. J., *Industrial Organic Chemistry,* 2nd edition, VCH, New York, 1993.

4. Franck, A. G., and Stadelhofer, J. W., *Industrial Aromatic Chemistry,* Springer-Verlag, New York, 1988.

5. Klopfenstein, E., thesis No. 3161, ETH, Zurich, 1967.
6. Chvatal, et al., *Collect. Czech. Chem. Commun.,* 48 (1), 112 (1983).
7. Gosh, A.K. and Nair, C.S.B., *Indian J. of Tech.,* 18, 181 (1980).
8. Gosh, A.K. and Nair, C.S.B., *Chem. Eng. Sci.,* 50 (5), 785 (1995).

REFERENCES, SECTION 16.8 (FORMALDEHYDE)

1. Reus, G., Disteldorf, W., Grundler, O. and Hilt, A., in *Ullmann's Encyclopedia of Industrial Chemistry,* 5th edition, Vol. A 11, p. 619, VCH, New York, 1988.
2. *Kirk-Othmer Encyclopedia of Chemical Technology,* 4th edition, Vol. 11, p. 929, Wiley, New York, 1994.
3. Wachs, I.E. and Macix, R. J., *Studies in Surf. Sci.* 76, 531 (1978).
4. Santacesaria, E., Morbidelli, M. and Carra, S., *Chem. Eng. Sci.* 36, 99 (1981).
5. Weissermel, K. and Arpe, 11. 1., *Industrial Organic Chemistry,* 3rd. ed., VCH, New York 1997.
6. Troung, N. V., Tittareilli, P. and Villa, P. L., *Third International Conference on The Chemistry and Uses of Molybdenum,* H. F. Barry and P. C. H. Mitchell, eds., pp. 161–165, Climax Molybdenum Company, Ann Arbor, MI, 1979.
7. Stiles, A. B. and Koch, T. A., *Catalyst Manufacture,* 2nd edition, Marcel Dekker, New York, 1995.
8. Stull, D. R., Westrum, E. F., Jr. and Sinke, G. C., *The Chemical Thermodynamics of Organic Compounds,* Wiley, New York, 1969.
9. Horner, C. W., *Chem. Eng.,* p. 108, July 4, 1997.
10. Atroshchenko, V. I. and Kusnorenko, I. P., *Int. Chem. Eng.,* 4, 581(1964).
11. Habersberger, K. and Jiru, P, *Coll. Czech. Chem. Comm.,* 37, 535 (1972).
12. Neophytides, S. and Vayenas, C. G., *J. Catal.,* 118, 147 (1989).
13. Beskov, V. S., Bibin, V., Malinovskaya, O. A. and Popov, B., *Kinet. Katal. (Eng.),* 13, 1318 (1972).

REFERENCES, SECTION 16.9 (METHACRYLIC ACID)

1. Kine, B. B. and Novak, R. W. in *Concise Encyclopedia of Polymer Science and Engineering,* p. 16, John Wiley & Sons, New York, 1990.
2. Chenier, P. J., *Survey of Industrial Chemistry,* 2nd edition, VCH, New York, 1992.
3. Gross, A. W. and Dobson, J. C., *Kirk-Othmer Encyclopedia of Chemical Technology,* 4th edition, Vol. 16, p. 474, Wiley, New York, 1995.
4. Porcelli, R. and Juran, B., Hydrocarbon Process, 65, 37, March, 1986.
5. Weissermel, K. and Arpe, H. J., *Industrial Organic Chemistry,* 3rd edition, VCH, New York, 1997.
6. Nakamura, S. and Ichihashi, H. in *Proceedings 7th International Congress on Catalysis, Tokyo* (Studies in Surface Science, Vol. 30), p. 785, Elsevier, New York, 1981.
7. Breiter, S. and Linta, H-G., *Catal. Letters,* 24, 343 (1994).
8. Breiter, S. and Linta, H-G., *Chem. Eng. Sci.* 50 (5), 785 (1995).

17 Oxychlorination

17.1 ETHYLENE → 1,2-DICHLOROETHANE → VINYL CHLORIDE

Vinyl chloride is one of the largest in monetary value of any commodity chemical, mainly because of the ever-growing use of its polymerization product, polyvinyl chloride (PVC), and vinyl copolymers. Two grades of PVC are produced: flexible and rigid. The flexible grades, which are lower-molecular-weight plasticized products, are familiar to many as wrap for raw-meat cuts in display counters, automobile fabric, electric wiring insulation, and flexible tubing. The rigid nonplasticized product is the most rapidly growing segment of the PVC market. It is predominantly used in water and sewer piping (essentially replacing copper and cast iron piping in all new installations), electrical conduit, and shoe soles. Rigid PVC is the most energy-efficient construction material other than wood.

The dominate process for vinyl chloride monomer (VCM) is the so-called *balanced* process, which produces and consumes HCl in a manner that leaves no HCl for disposal and maximizes production efficiency. The process is complex, but amenable to very large plants so as to take advantage of the economy of scale. Three separate reactions are involved, one of which, the oxychlorination of 1,2-dichloroethane, for most plants is the only step involving a heterogeneous catalyst.

Chemistry

The overall reactions for the three separate steps are[1, 2, 4]

Direct Chlorination of Ethane

$$H_2C = CH_2 + Cl_2 \rightarrow ClCH_2 - CH_2Cl \qquad \Delta H = 43 \text{ kcal}$$

1,2-dichloroethane
Liquid phase with homogeneous catalyst $FeCl_2$
40–70°C and 4–5 bar

Thermal Cracking of 1,2-Dichioroethane

$$\underset{\text{1,2-dichloroethane}}{ClCH_2 - CH_2Cl} \rightarrow \underset{\text{vinyl chloride}}{CH_2 = CHCl} + HCl \qquad \Delta H = 16.83 \text{ kcal}$$

Vapor phase at 500 → 550°C and 20–30 bar
Free-radical reaction

Oxychlorination of Ethylene

$$H_2C = CH_2 + 2HCl + 0.5\ O_2 \rightarrow ClCH_2 - CH_2Cl + H_2O$$
$$\text{1,2-dichloroethane}$$
$$\Delta H = -58.2\ \text{kcal/mole}$$

Vapor-phase at 220–240°C and 2–4 bar
Heterogeneous catalyst: $CuCl_2$-on-alumina

The net result of the first two steps is the production of HCl and vinyl chloride. The HCl can in turn be used in the third reaction, oxychlorination of ethylene, to make additional 1,2-dichloro-ethane, which can be fed to the cracking unit to produce additional vinyl chloride. By balancing the production of the three steps, it is possible to consume all the HCl. produced. This procedure is not only economical, it avoids the costly problem of disposing of the HCl from the cracking step.

Mechanism (Oxychlorination)

Oxychlorination is a complex catalyzed reaction. Although a detailed mechanism is not known, it is generally agreed that $CuCl_2$ is the chlorinating agent.[1,2,5] A typical representation is as follows:

$$C_2H_4 + 2CuCl_2 \rightarrow C_2H_4Cl_2 + 2CuCl \tag{1}$$

$$2CuCl + 2HCl + 1/2\ O_2 \rightarrow 2CuCl_2 + H_2O \tag{2}$$

Other mechanisms have been proposed, but most workers in the field find the above reaction paths helpful as a working hypothesis. As an example, cuprous chloride (Cu_2Cl_2) has been shown to reduce the induction time when added to the cupric chloride catalyst, perhaps suggesting that the first step shown above may involve a cuprous chloride–ethylene intermediate.[6,7]

Step 2 has been suggested to involve a copper oxychloride.[2]

$$1/2\ O_2 + 2CuCl \rightarrow CuOCuCl_2$$

$$2HCl + CuOCuCl_2 \rightarrow 2CuCl_2 + H_2O$$

Catalyst Type and Suppliers

There are many patents on catalyst formulation for this important oxychlorination. All employ $CuCl_2$ dispersed on an activated alumina support or other supports such as silica alumina, diato-maceous earth, activated charcoal, and pumice. Alumina is the usual choice with a surface area when calcined between 200–400 m^2/g with pores sizes distributed in the 80–600 Å range. Along with impregnation of $CuCl_2$ on the alumina, certain additives have proved most valuable. The most widely used of these is potassium chloride, which reduces by-product formations (e.g., ethyl chloride) and also reduces the volatility of copper chloride.[4] Excessive addition of potassium chloride must be avoided, since it can reduce catalyst activity. Other additives that promote the desired reaction or inhibit undesired reactions may also be added. These include other metal chlorides both alkali and rare earth metals.

Fixed-bed catalysts are formed as tablets, spheres, or extrudates in diameters of 1/8 to 1/4 in. Fluidized-bed catalysts are produced as microspheres in the range of 10–200 μm diameter.[1]

Suppliers/Licensors

There are many vinyl chloride producers who have licensed their processes and either supply a catalyst or have contracted with a catalyst manufacturer to prepare catalysts based on the licensor's patented procedure.

Some of the more commonly licensed processes include the following:

Abermale
Rhodia
PPG
Mitsui Toatsu
Geon
Toya Soda
Solutia
Shell

Catalyst Supplier
Akzo Nobel

Catalyst Deactivation

Catalyst deactivation, and thereby catalyst life, are strongly influenced by elevated temperatures, a problem common to many highly exothermic reactions. High temperatures (>300°C) invariably cause the production of side reactions including oxidation of ethylene to CO and CO_2, increased dehydrochlorination of dichloroethane to vinyl chloride (which, in turn, reacts by oxychlorination to other chlorinated hydrocarbons), and coking, followed by some breakup of catalyst in the form of powder that increases pressure drop.[2]

Catalyst in a series of fixed beds experiences the shortest life in the first bed. Discharge of catalyst in any type of reactor requires special care, since it may be pyrophoric due to coke deposits and contaminated with toxic irritants and carcinogens such as organic chlorides.[7] Disposal procedures are often contracted to a licensed disposal company.

Process Units

Because of the high exothermicity of the reaction and the need for careful temperature control, several alternative process designs are practiced.

Fixed-bed reactors using air or oxygen
Fluidized-bed reactor using air or oxygen

Fixed-Bed Reactors

Tubular reactors are used with either heat-transfer oil or boiler feedwater on the shell side. In either case, valuable exothermic heat is recovered. Saturated steam can be produced directly from the circulating boiler feedwater or indirectly by exchange of hot oil with boiler feedwater in an exchanger. This steam can then be superheated for uses such as turbine drivers or heating needs of various types. Some processes have one massive heat-transfer reactor with a shell diameter up to 5 m, tube length of 10 m, and tube diameter of 2 in.[1] Careful staging of catalyst batches is practiced to better control hot-spot temperatures. Catalyst batches are produced of different activities and then arranged with lower activities in the upper portion of the tubes and higher activities toward the exit to reach the desired reaction completion.

Alternatively, multitubular reactors (usually three) can be arranged in series with carefully controlled split additions of air or oxygen between stages.[5,1] This arrangement permits better control of temperature and more flexibility in avoiding the explosive limit. Also, formation of fewer oxidation products is claimed.[1] These smaller reactors may employ 1-in. tubes of 2 m in diameter and a shell size of 4 m.

Both alternatives use nickel alloy Type 201 for the tubes. The shell is carbon steel, but the inside of the heads and the tube sheets, which are in contact with the reactants, are clad with nickel.[1,9]

Reaction temperature is generally in the range of 220–240°C and pressure in the range of 2–4 bar with the lowest pressure occurring at the outlet.[5]

Fluidized-Bed Reactors

The other standard alternate reactor is a fluidized bed that, when operating properly, maintains close to a constant temperature. A disadvantage is the backmixing that occurs and tends to lower conversion, but this problem has been overcome by feeding an excess of air or oxygen (10–80%) and ethylene (up to 60%), depending on other operating conditions.[1]

Fluidized-bed reactors contain a number of coils for heat removal using heat-transfer oil. The hot oil is then used to generate steam or directly in process heating service. The most valuable contribution of the fluidized-bed alternative is that operation within the explosive limit is possible because of the thorough mixing of the solid catalyst particles.

Operating conditions for the fluidized bed are usually lower than the fixed-bed unit. Although the same nominal temperature range in each is sought (220–240°), hot spots in fixed beds can reach higher temperatures (e.g., 285°C). Reported temperatures for fluidized bed operation cluster around 220–225°C.[5] The lower nominal temperature in the fluidized bed serves to avoid excessive catalyst deactivation.

Pressure control is of value in each type of reactor. In the case of the fluidized bed, a higher pressure increases the reaction rate and thus conversion. Tubular fixed-bed reactor temperatures are highly sensitive to operating pressure. An increase in pressure decreases the gas flow volume causing a lower superficial velocity and a lower heat-transfer rate between the reactants and the catalyst.[5] This condition causes earlier hot spots. Lower pressure increases the superficial velocity and causes the hot spots to move down the reactor and to be lower in temperature. Thus, reactor pressure is a useful tool for reactor control. At start-up the pressure is increased for quick reaction initiation and then lowered to control the reaction temperature by affecting the heat transfer between catalyst and the flowing reactant system. Hot spots can be adjusted and mediated by this procedure. Fluidized beds do not experience hot spots, and only the reaction rate at operating temperature is affected by pressure.

Air versus Oxygen

In recent years, most new plants have been designed for use of oxygen, and many older plants have been retrofitted to enable the use of oxygen. Oxygen has the following advantages.[1,4,5,8]

- Elimination of nitrogen greatly reduces the vent stream, reducing thereby the cost of incinerating the vent gases to avoid air pollution. The oxygen-based vent stream is reported to be 20–100 times smaller than the air-based vent stream. Vinyl chloride emission into the atmosphere is strictly prohibited by environmental regulations because of its known carcinogenic characteristics. Thus, any process change that reduces the quantity of vent gases is most valuable.
- Since nitrogen as an inert helps to mediate temperature increases, and dilutes the reaction system to avoid the explosive region, its elimination required the use of excess ethylene in the oxygen-based process. Because of the higher heat capacity of ethylene, substitution of nitrogen by the same number of moles of ethylene makes for lower catalyst temperatures, which assures longer catalyst life and lower by-product production.

- Lower catalyst temperature rise at hot spots permits higher throughputs that are ultimately limited by the maximum allowable catalyst temperature.
- Recycle ethylene flow for the oxygen process can be varied to control the hot spot temperature, thus providing an additional and convenient control variable.

Of course, the oxygen process requires additional operating costs for oxygen and recycle of ethylene and additional capital costs for a compressor to recycle ethylene. Also special piping and gaskets must be used on lines carrying pure oxygen. The increased yield, lower by-product production, higher throughput, and drastically reduced incineration costs offset the added costs for new equipment. For a given production rate, a new oxygen plant actually requires lower capital costs than a new air-based plant. It is easy to understand why most plants now operating use the oxygen process.

Process Kinetics

Oxychlorination is a complex process that involves various side reactions. A preliminary kinetic study of the main reaction at one temperature (180°C) suggests a convenient Power-Law model that fits the experimental data within the standard experimental deviation.[3]

$$r_D = kP_E^l P_O^m P_{HCl}^n$$

where r_D = is the rate of dichloroethane production. Exponents 1, m, and n were found to be 0.73, 0.34, and –0.18

A Hougen-Watson surface reaction between ethylene and oxygen also fits the data. Attempts at modeling a commercial unit including side reactions would probably prove more tractable and successful if the rates of all the apparent reactions were stated in Power-Law form.

REFERENCES (DICHLOROETHANE)

1. Oreher, E. L. in *Ullmann's Encyclopedia of Industrial Chemistry,* 5th edition, Vol. A6, p. 263, 281, VCH, New York, 1986.
2. Cowfer, J. A. and Gorensek, M. B. in *Kirk-Othmer Encyclopedia of Chemical Technology,* 4th edition., Vol. 24, p. 851, Wiley, New York, 1997.
3. Carrubba, R. W. and Spencer, J. L., *Ind. Eng. Chem. Process Des. Develop. 9,* (3), 414, 1970.
4. Weissermel, K. and Arpe, H. J., *Industrial Organic Chemistry,* 3rd edition, VCH, New York, 1997.
5. Naworski, J. S. and Velez, E. S. in *Applied Industrial Catalysis*, B. E. Leach editor, Vol. 1, p. 239, Academic Press, New York, 1983.
6. Heinemann, H., *Chemtech* 3, p. 287 (1971).
7. Stiles, A. B. and Koch, T. A., *Catalyst Manufacture,* 2nd edition, Marcel Dekker, New York, 1995.
8. McPhersen, R.W.; Starks, C. M. and Fryar, G. J. *Hydrocarbon Process.,* p. 75, March, 1979.
9. Schillmoller,C.M., *Hydrocarbon Process.,* p. 89, March, 1979.

18 Petroleum Refining

The petroleum industry has had a remarkable history, beginning with the fantastic growth in oil discovery and the consistent development of ever increasingly sophisticated technology. As demand for various products for transportation and other uses grew exponentially, the technology of refining grew to meet the challenges. Crude oil, which consists of thousands of compounds, had to be separated into useful products and, ultimately, portions (cuts) in a particular boiling-point range had to be reacted to produce more valuable, lower boiling-point products. Undesirable portions had to be treated to remove or alter the offending compounds. Many of the necessary processes that were created required high temperature and pressure as well as other severe conditions. The refining industry became the pioneer in developing safe pressure vessels and many other process equipment items as well as a more scientific approach to catalyst development and application and reactor design techniques.

This chapter focuses on the major refinery processes catalyzed by heterogeneous catalysts. The chemical engineers and chemists of today have the ability to determine the actual chemical compounds in petroleum. No longer is it necessary to characterize petroleum products solely in terms of a boiling-point range. To develop predictive tools in the form of process models useful in maximizing efficient reactor operation, knowledge of the chemical nature of the many components in a particular cut of petroleum became necessary, and modern analytical techniques made it possible. Ultimately, practical models were developed in which compound types were identified, and each such group was treated as a pseudo compound. The result has been plant operation models that have greatly improved operation and profitability.

18.1 CATALYTIC REFORMING

Background

Catalytic reforming converts low-octane naphtha streams (20–50 RONC) to high-octane (90–108 RONC) gasoline blending stock using a dual-function catalyst with both metallic sites of Pt (or Pt with other metals as promoters) and acid sites supplied by the alumina support. The traditional feed is heavy naphtha (primarily paraffins and naphthenes, 200–400°F, 93–204°C BP) from atmospheric distillation of crude oil. Naphtha from delayed coking and hydrocracking and other streams high in naphthenes are also used.

The 150–200°F (66–93°C) stream from virgin naphtha, especially from paraffinic crude, will produce useful product if operation is at 200 psig or lower. Hydrocracked naphtha in this same boiling range should not be used. Also, naphtha from cracked stocks above 380°F (193°C) BP should be avoided, because it produces excessive coking.[1] The remarkable increase in octane number that reforming produces is due mainly to the conversion of essentially all the naphthenes and a sizeable portion of the paraffins to aromatics, and straight-chain paraffins to branched isomers. Aromatics can alternatively be recovered for petrochemical products, and operating conditions (low pressure and high temperature) and cut points of feedstocks can be tailored to maximize desired aromatics such as benzene, toluene, and xylenes. With the 1990 Amendment of the Clean Air Act of 1970, reformulated gasoline must not contain significant amounts of benzene. This goal can be accomplished in reforming by eliminating C_6 hydrocarbons from feed to reformers operating for gasoline blend stock production.

Chemistry

Refinery naphtha streams are composed of a large number of different molecular types in the range of C_6–C_{10} or C_7–C_{10}. A typical reformer product contains 300 different compounds. Fortunately, this complexity can be comprehended because each compound type behaves similarly, and illustration of overall reactions using compounds of a specific carbon number will be applicable to compounds with other carbon numbers.[2]

1. Dehydrogenation of cyclohexanes to aromatics (quite endothermic, very fast):

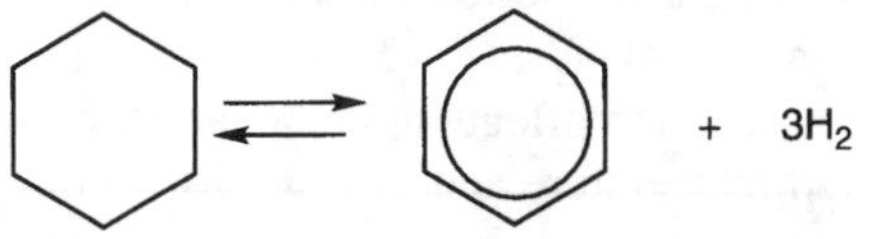

| cyclohexane | benzene |

+ $3H_2$

2. Dehydroisomerization of alkylcyclopentanes to aromatics (quite endothermic, rapid):

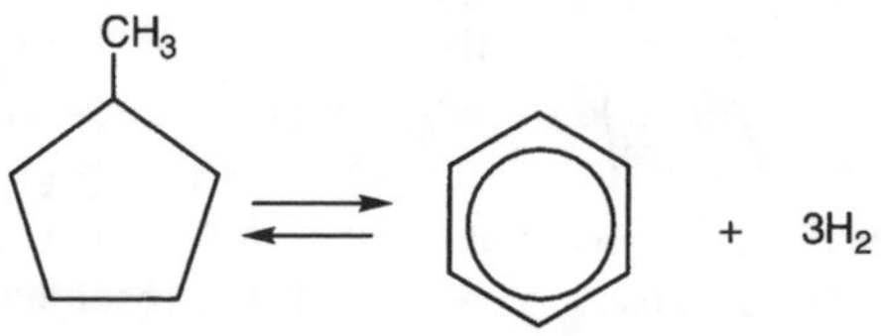

methylcyclopentane   benzene

+ $3H_2$

3. Isomerization of paraffins (mildly exothermic, rapid):

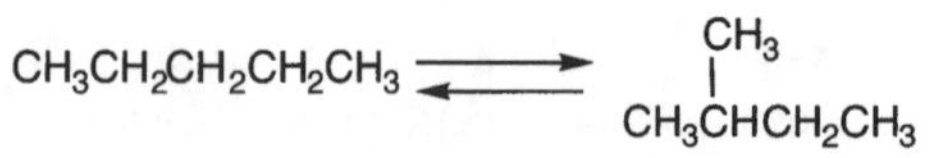

$$CH_3CH_2CH_2CH_2CH_3 \rightleftarrows CH_3\underset{\underset{CH_3}{|}}{C}HCH_2CH_3$$

n-pentane   2-methylbutane

4. Dehydrocyclization of paraffins to aromatics (endothermic, slow):

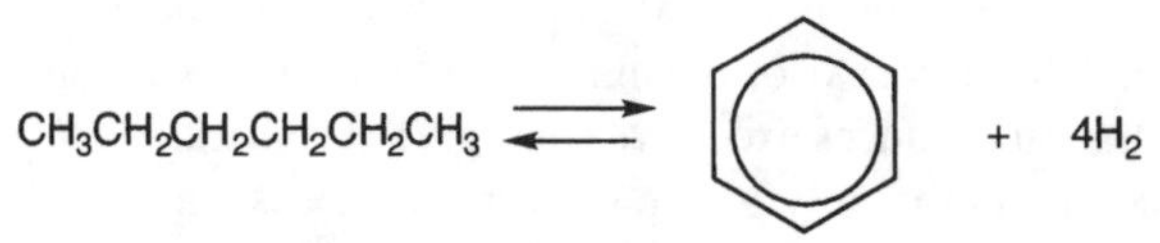

$$CH_3CH_2CH_2CH_2CH_2CH_3 \rightleftarrows$$

+ $4H_2$

n-hexane   benzene

5. Hydrocracking of hydrocarbons (quite exothermic, slowest):

$$\underset{\text{2,2,4-trimethylpentane}}{\overset{\displaystyle \underset{\overset{\displaystyle |}{CH_3}}{\underset{\overset{\displaystyle |}{CH_3}}{CH_3CCH_2CHCH_3}}}{\overset{\displaystyle CH_3 \quad CH_3}{}}} \rightleftarrows \underset{\text{isobutane}}{\overset{\overset{\displaystyle CH_3}{\displaystyle |}}{2\ CH_3CHCH_3}}$$

Since three to four moles of H_2 are formed for each benzene produced, the reforming process is a major producer of hydrogen for other refinery process units.

Each of the reactions can be thermodynamically limited under certain conditions, except hydrocracking, which is rate limited and becomes significant at higher temperatures after the endothermic reactions have subsided. Low pressure and high temperature favor Reactions 1, 2, and 4 thermodynamically. Low hydrogen concentration would favor the desired reactions, since it is a significant product, but hydrogen is essential in preventing rapid coke formation. Accordingly, a portion of the hydrogen product is recycled. Hydrogen is also a major product that is used throughout the refinery for the many hydrotreating and hydrocracking units that have become feasible because of reforming. Figure 18.1 presents simultaneous equilibrium calculations that illustrate the more favorable yields for aromatics of higher molecular weight.

Mechanism

Early in the development of Pt reforming catalysts, an overall mechanistic scheme was illustrated as shown in Figure 18.2, and it remains useful today. The dual-function nature of the reforming catalyst is based on its alumina support with its acidic centers for isomerization and cyclization and its metal centers for hydrogenation-dehydrogenation. Acid centers are also responsible for hydrocracking, but some cracking can occur on metal sites, in which case it is called *hydrogenolysis*.[3] It is also thought possible that methylcycloalkenes (e.g., methylcyclopentene) can form through the corresponding straight-chain unsaturate (e.g., hexadiene) produced on a metal site.[4]

The acid strength of the support is initially set by careful chloride addition during preparation and is maintained by close control of the chloride and water content of the feed.

Catalyst Types and Suppliers

The first catalytic reforming process, which used a molybdenum oxide-on-alumina catalyst, was jointly developed in 1939 by Standard Oil of New Jersey (Exxon), Standard of Indiana (Amoco), and M. W. Kellogg Company. In 1949, UOP introduced the first process (Platforming™) using platinum-on-alumina as a dual-function catalyst having both acidic sites and metallic sites. It soon became the catalyst of choice. Over the years, a series of proprietary processes and catalysts have been developed, largely by refiners. As primary patents have expired, a number of merchant catalysts are now available, often based on new promoters and other unique innovations. Platinum remains the major active component of all modern naphtha reforming catalysts. But, beginning in the 1970s, catalysts were introduced having one or more additional metallic components including rhenium, iridium, and tin. These bimetallic and multimetallic catalysts exhibit greatly improved stability (cycle length) and selectivity.

Platinum-rhenium catalysts are the most widely used, because they are more coke tolerant and thus provide longer cycle times (times between regeneration). In addition, it has also been possible to reduce operating pressure and take advantage of yield enhancing equilibrium conditions and energy savings while maintaining attractive cycle times. Typically, a Pt-Re catalyst will have two to four times the cycle time of a platinum-only catalyst.[5] A summary of major characteristics of catalyst types is given in Table 18.1 (see p. 306).

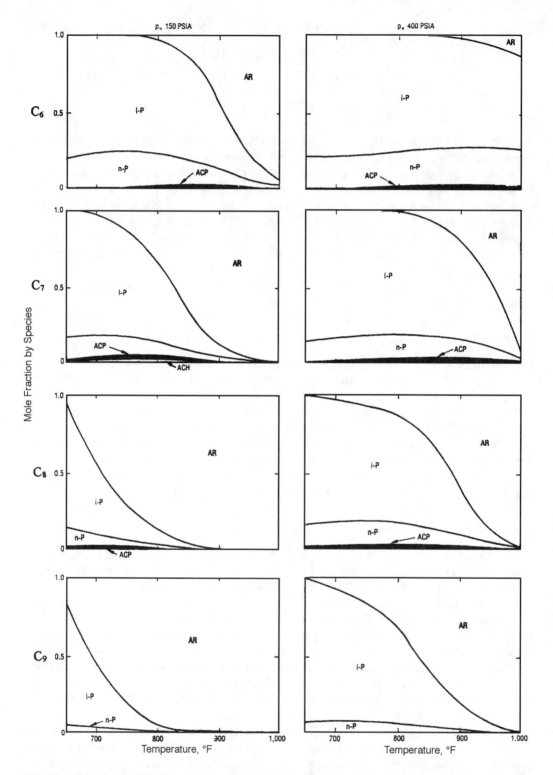

FIGURE 18.1 Equilibrium Distribution of Major Species (ACH = alkylcyclohexanes, ACP = alkylcyclo-pentanes, AR = aromatics, n-P = normal paraffins, i-P = iso-paraffins). Reprinted by permission: Kugelman, A. M., *Hydrocarbon Processing,* p. 95, January 1976.

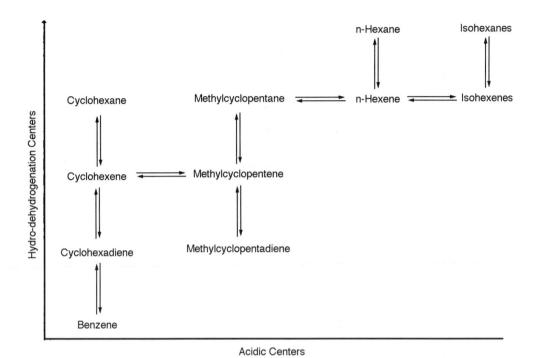

FIGURE 18.2 Dual-Function Mechanism for Catalytic Reforming. Reprinted by permission: Sinfelt, J. H., *Advances in Chemical Engineering,* 5, 37 (1964), Academic Press Inc.

Suppliers

Acreon, Criterion, Indian Petrochem., Inst. Mexicano Petrol, Kataleuna, UOP LLC, Procatalyse.

Licensors

Exxon Research and Engineering, Howe Baker, IFP, UOP LLC.

Catalyst Characteristics

Preparation

Although patents outline multiple preparative procedures that are necessary for broad legal claims, the preferred procedure is not usually apparent. However, a rather clear and valuable description of the production of a Pt/Al_2O_3 reforming catalysts has been published.[6]

The support is typically a γ-alumina in the form of $1/16 \times 0.2$ in. $(1.6 \times 5$ mm) extrudates having a pore volume of 60 cm³/100 g and a dominant pore diameter in the range of 1000 Å. The support must be as pure as practicable, with rigid standards in ppm maxima for heavy metals and sodium, which can alter the selectivity unfavorably. The extrudates are immersed in water, and chloroplatinic acid is added along with enough hydrochloric acid to stabilize the chloroplatinic acid and provide chloride ions to create sufficient acid sites for that function of the finished catalyst. The ion-exchange reactions that occur (Figure 18.3) are rapid, but the competition for sites between Cl^- and $PtCl_6^{2-}$ serves to encourage even distribution of the platinum throughout the extrudate. The high cost of platinum dictates even dispersion throughout the support for maximum utilization in the catalytic process.

After impregnation, the extrudates are dried and then calcined at 500°C (932°F) to remove residual water and increase the crushing strength of the support by controlled grain growth.

TABLE 18.1
Characteristics of Typical Reforming Catalysts

	Pt-on-Al_2O_3	Balanced Rt/Re-on-Al_2O_3			Skewed Pt/Re-on-Al_2O_3			Pt/Sn-on-Al_2O_3	Pt/Ir-on-Al_2O_3
Wt%	0.3–0.35	0.22/0.22	0.3/0.3	0.35/0.35	0.22/0.42	0.26/0.5	0.28/0.75	0.35/0.3	0.3/0.3
Maximum allowable S in feed, ppm	<5	<0.5		<1	<0.5	<0.5	<0.1	<1	<1
Protection against brief S excursions in feed	Yes	No	Modest	Modest	No	No	No	No	Yes
Comments	For feedstocks subjected to minimal sulfur removal. Higher pressure and higher H_2/oil ratio required.	Lower pressured operation possible. Lower-metals type reduces cost of catalyst, but low-sulfur feed essential. Better cycle length than Pt alone.			Increased cycle length with extra Re, but sulfur tolerance low. Requires efficient sulfur removal. Lower pressure operation possible.			Sn does not protect Pt from coking as does Re. Frequent regeneration and used, therefore, in continuous regeneration units operating at low pressures (50 psig typical). Provides increased yield of C_5+ at pressures below 200 psia.	Higher yield is possible if Pt–Re is used in first reactors and Pt–Ir is used in the last reactors where the advantages of its improved performance in dehydrocyclization can be realized.

Typical bulk densities (load densities) 39–45 lb/ft³

Notes:

1. The higher-density supports (fewer large pores with same metals loadings) are useful for existing units when it is advantageous to load a higher mass of active catalytic material in a given volume of an existing reactor.
2. Patents indicate a wide assortment of other metals that are included in smaller quantities as promoters. These include Ge; Cu, and Se, for example.
3. Catalyst sizes in inches are typically:
   1/16 extrudates 0.2 to 0.25 in length.
   1/12 extrudates are also produced for lower ΔP if feeds are clean and the bed well protected from scale and fines.
   1/20 in the shape of a clover leaf has a bed void volume 10% higher, and its shape provides shorter diffusion paths which are useful for regeneration where oxidation of the interior coke can be diffusion limited.

Source: Based on information from Refs. 18 and 19.

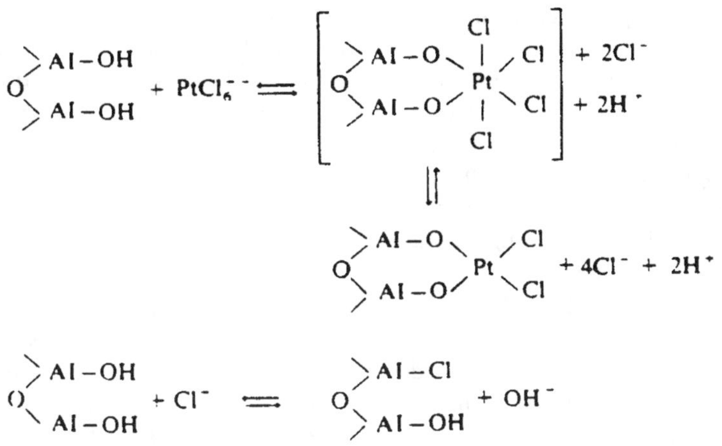

$$>Al-OH \atop O \atop >Al-OH \quad + \ PtCl_6^{--} \ \rightleftharpoons \ \left[O \begin{matrix} >Al-O \\ >Al-O \end{matrix} \begin{matrix} Cl \\ | \\ Pt \\ | \\ Cl \end{matrix} \begin{matrix} Cl \\ \\ Cl \end{matrix} \right] \begin{matrix} + \ 2Cl^- \\ \\ + \ 2H^\cdot \end{matrix}$$

$$\updownarrow$$

$$O \begin{matrix} >Al-O \\ >Al-O \end{matrix} Pt \begin{matrix} Cl \\ Cl \end{matrix} \ + \ 4Cl^- \ + \ 2H^\cdot$$

$$O \begin{matrix} >Al-OH \\ >Al-OH \end{matrix} + \ Cl^- \ \rightleftharpoons \ O \begin{matrix} >Al-Cl \\ >Al-OH \end{matrix} + \ OH^-$$

FIGURE 18.3 Exchange Reactions during Platinum Impregnation of Alumina. Reprinted by permission: LePage, J. F., et al., *Applied Heterogeneous Catalysis*, p. 486, Editions Technip, Paris, 1987.

Activation of the catalyst is done in the reactor prior to start-up by reduction with hydrogen and controlled presulfiding under optimum temperature and treatment-stream compositions that are recommended by the catalyst manufacturer. It is thought that hydrogen reduction fixes the small Pt clusters that were dispersed in a high-temperature oxygen-chloride environment by the catalyst manufacturer in the case of fresh catalyst, and by the user during regeneration. Bimetallic catalysts can be produced in a similar manner with rhenium, iridium, or tin as the other metal. In such cases, the cluster consists of the two metal atoms randomly distributed on the crystallite surface or, in the case of iridium, some selective groupings of like atoms is thought to occur.

Nature of the Catalyst Surface

Perhaps no other catalyst has been studied so thoroughly. Its astounding properties and complexity challenged many outstanding researchers to seek its mysteries. Kinetic studies on pure components were interesting but led to little concrete knowledge. Probing the nature of the catalyst surface characteristics proved most valuable and aided in tailoring catalyst recipes for improved catalysts and better operating procedures.

Extensive studies have permitted some useful generalizations:[7,8]

- Crystallite dispersion is an important prerequisite for successful catalyst use. Platinum crystallites on fresh catalyst are typically less than 35 Å. Practically speaking, the optimum preparative recipe and the redispersion procedures after regeneration may be unique for each catalyst type, and manufacturers or licensor recommendations for use should be carefully followed. Many of the desired reactions are favored on small crystallites, while undesirable reactions such as hydrogenolysis, poisoning, and coke formation, are favored by larger crystallites where high coordination-number Pt atoms and larger ensembles of Pt atoms provide the geometry for multiple metal-carbon associations. These complexes stabilize coke precursors on metal sites as well as precursors (e.g., cyclopentadiene) that migrate to the acid sites on the support and polymerize to polynuclear aromatics of low hydrogen content (0.05–0.1 H/C), i.e., coke.
- The acid function of the alumina support appears to be optimum at 0.7–0.9 wt% chlorine. At these loadings, the rate of coke formation on the alumina surface is minimized. A major coke precursor, methylcylopentane, is destroyed by hydrocracking at these higher acidity levels.

- The idealized edge of a crystallite for the several commercial catalysts may be represented as shown in Figure 18.4, with the fresh Pt/Re catalyst already sulfided which is then standard procedure for this catalyst. The Pt catalyst is shown partially coked and partially sulfided. The Pt/Sn catalyst is shown fresh. In all cases, the number of large ensembles *(strings)* of Pt atoms is reduced either by inert Sn atoms. Re atoms made inert by sulfiding, or Pt atoms made inert by partial coking. This situation favors the important dehydrogenation reactions, which are more rapid in small Pt ensembles, and discourages hydrogenolysis (which increases light ends) and coking reactions, both of which require larger assemblies of platinum.
- The iridium of the Pt-Ir catalyst is not inert, for it is an active dehydrocyclization catalyst (which is good) and an active hydrogenolysis catalyst (which is not good). But hydrogenolysis can be controlled to some degree by continuous addition of a sulfur compound in the feed (0–10 ppm). Although this action reduces the Pt ensemble size, desirable single-site hydrodecyclization is retained on both unsulfided Pt and iridium.

Deactivation

Naphtha reforming catalyst is deactivated by coke formation, poisoning, loss of chloride, fines deposition, sintering of the metallic catalyst components, sintering of the support, and heavy metals deposition.[8] The effects of all but the last two can be remedied. Since heavy metals and other poisons are restricted by feed pretreatment, coke is the main cause of short-term deactivation and determines the frequency of regeneration. Long-term effects, such as support sintering, upsets in feed purity or temperature excursions, can reduce the ultimate catalyst life.

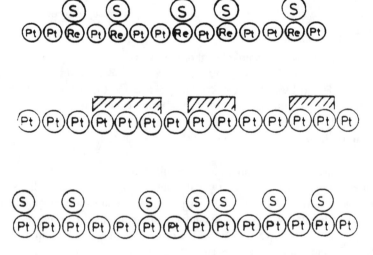

FIGURE 18.4 Idealized Edge of Various Types of Reforming Catalyst Crystallites (from top, partially sulfided Pt/Re, partially coked Pt, sulfur-poisoned Pt, and Pt/Sn with Sn being inactive). Reprinted from Biswas, J., Bickle, G. M., Gray, P. G., Do, D. D., and Barbier, J., *Catalysis Reviews: Science & Engineering,* 30, 161 (1988) by courtesy of Marcel Dekker, Inc.

Coke Formation

Coke forms rapidly at the beginning of a run but, after about 2 wt% is deposited, the coke laydown increases slowly at a steady rate.[7] As the reactor temperature is increased to compensate for loss in activity, a temperature is finally reached above which coke production accelerates, and the catalyst must be regenerated. This marked increase in coking toward the end of the run can produce plugging in a fixed-bed reactor and ultimately unmanageable pressure drop.

The early formation of coke is due to coking on the faces of clusters, which does not affect edges or corners where the major desired reactions on Pt occur. Subsequent coke formation occurs primarily on the alumina support which, because of its high surface area, can tolerate slow but constant increases in coke laydown over a longer time period.[9,10] Although metal activity does decline rapidly with coke laydown, sufficient activity remains for catalyzing all important metal site reactions. The effects of coke formation on the several reforming reactors are assembled in Table 18.2.

TABLE 18.2
Effect of Coke Deposits on Reforming Reactions

Reaction	Effect of Coke
Dehydrogenation	Occurs on metal function. Early decrease and then a slow decline, since H_2 keeps Pt partially clean.
Dehydroisomerization and isomerization	Rate limiting step occurs on acid component which is less sensitive to coke because of high area. Relative activity declines during entire run.
Dehydrocyclization	Occurs by both a monofunctional metal-site mechanism and a bifunctional (metal site and acid site) mechanism. Coking causes initial rapid decline of metal function followed by a slow decline of bifunctional function.
Hydrogenolysis	Occurs on high-coordination number face atoms, which are associated with larger ensembles of platinum. Coke will cover some of these and reduce hydrogenolysis. This effect is most noticed on monometallic platinum catalysts.
Hydrocracking	Controlled by both metal and acid function. Activity declines rapidly at first and then linearly with coke make.

Based on information from Refs. 7 and 8.

Effect of Catalyst Type on Coke Formation

Both coke formation and hydrogenolysis on Pt apparently occur on ensembles of contiguous metal atoms.[8] In the case of monometallic Pt catalysts, coke covers most of the Pt surface, leaving only islands of Pt for catalytic action. These islands can contain sufficiently large ensembles of Pt to catalyze both hydrogenolysis and coking reactions. Pt-Re, when sulfided, produces an inert ReS that separates the Pt into mostly very small ensembles that do not catalyze hydrogenolysis or coking.[8] Thus, presulfided Pt-Re/Al$_2$O$_3$ catalyst produces less coke on the aggregate of metal sites. In fact, with increased Re loadings (skewed, more Re than Pt), this effect is accentuated, and even less coke-on-Pt is produced. Since smaller Pt ensembles tend to weakly adsorb olefinic and aromatic coke precursors, these entities migrate to the acid sites of the support and form coke. Thus, it is possible to produce more coke on a bimetallic catalyst such as Pt-Re-S/Al$_2$O$_3$ without excessively deactivating the metal portion. The Pt-Re-S catalyst, however, apparently provides additional steric hindrances, perhaps partly because of its very broad distribution on the surface, and thereby makes it the most coke tolerant. It is this high tolerance to coke deposits of Pt-Re-S that has made it the preferred catalyst for most reformers. This characteristic permits operation at lower pressures and favors thereby higher aromatic yields as well as energy savings. Alternatively, lower H_2-to-oil may be used, or increased feed rate, or reduced catalyst loading.[11]

The coke make on Pt-Ir catalyst is much less than Pt, Pt-Re, and Pt-Sn.[7,12] Iridium has a higher activity for hydrogenolysis, which serves to destroy coke precursors, but this quality causes higher production of light hydrocarbons and reduces yield.[3] However, a combined reforming operation has been proposed with Pt-Re in the initial beds followed by Pt-Ir. The higher activity of Pt-Re for aromatization, which reaction occurs first, can then be used initially followed by Pt-Ir with its higher activity for dehydrocyclization in the later stages.[3]

Effects of Operating Conditions and Feed Composition on Coke Formation

The various catalysts differ in coke forming behavior, but all share similar qualitative effects of the major operating variables as presented in Table 18.3.

TABLE 18.3
Effect of Operating Conditions on Coking

Operating Variable	Result
Temperature increase	Modest increase in coke production as would be expected from a diffusion-controlled reaction with coke forming primarily after the initial line-out period on the support.
Pressure increase at constant H_2:oil ratio	Decreases coke formation by lowering the value of equilibrium coke on metal sites and reduces the production of coke precursors on the metal sites that migrate to the support and produce coke.
H_2:oil ratio increase at constant pressure	Decreases coke formation by lowering the value of equilibrium coke on metal sites and reduces the production of coke precursors on the metal sites that migrate to the support and produce coke.
Severity increase (increase in temperature, and/or decrease in H_2:oil ratio and space velocity)	Increases coke formation. Increasing severity reduces cycle time.

Based on information from Ref. 8 and other references cited therein.

Figure 18.5 summarizes general effects of feed composition on coke formation based on mean boiling point. The minimum occurs at a mean boiling point of 383 K (230°F, 190°C) for the particular catalysts used in these studies.[13] Heavier stocks produce heavy alkylaromatics that are coke precursors which increase coke production. The lighter stocks contain significant amounts of coke precursors as well.

Catalyst Poisons

The major catalyst poisons that can occur in reformer feed streams are organic components containing sulfur, nitrogen, and metals (Pb, As, P). Organic nitrogen compounds are bases and will deactivate the important acidic function of the catalyst that is necessary for isomerization and cyclization reactions. Sulfur compounds in sufficient quantity will deactivate the platinum sites, as will the metallic compounds. Hence, all streams for reformer feed containing these poisons are first passed through hydrotreating units where the compounds are converted, respectively, to NH_3, H_2S and metallic deposits on the hydrotreating catalyst. The gaseous components are removed from the product in a stripper and the product sent to the reformer. Strict limits must be placed on residual poisons in the feed sent to the reformer as given in Table 18.4.

Although sulfur compounds in the feed can seriously reduce Pt activity, controlled poisoning provides improved selectivity and activity. In particular, hydrogenolysis reactions are inhibited by the poisoning of the most active sites. Controlled addition of sulfur is practiced at start-up and strict limits on feed are imposed. However, each catalyst type is affected differently, and special procedures for each must be followed for best results. Table 18.5 summarizes some characteristics

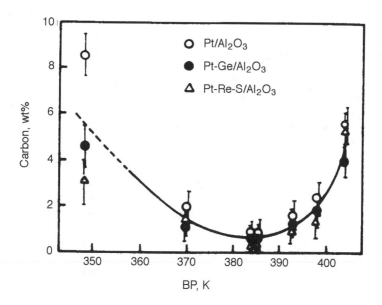

FIGURE 18.5 Carbon Deposits on Mono and Bimetallic Catalysts as a Function of Mean Boiling Point of Feed. Reprinted from *Applied Catalysis,* 32, 135 (1987), Querini, C. A., Figoli, N. S., and Parera, J. M., "Naphtha Reforming on Mono and Bimetallic Catalysts," p. 133–143, copyright 1987 with permission from Elsevier Science.

TABLE 18.4
Typical Ranges of Maximum Allowable Impurity-in-Feed Levels for Pt-Re With Equal Pt and Re Loading

Impurity	Maximum Weight Fraction
Sulfur	1 ppm (0.1–0.2 ppm for high Re catalysts)
Water	5–10 ppm
Chloride	1 ppm
Nitrogen	0.5–1.0 ppm
Lead	5–10 ppm
Arsenic	2–5 ppm

Specific suppliers recommendations for each catalyst of interest should be obtained (see also Table 18.6).

Based on manufacturers' literature.

of sulfur in the form of H_2S and its effect on the reforming catalyst. If sulfur enters as an organic sulfur compound, it is rapidly converted to hydrogen sulfide in the reactor.

Chemisorption of sulfur occurs more easily on the face atoms (high-coordination number) of metal crystallites. This adsorbed sulfur is deemed *irreversible* because it is strongly held and not easily removed by purging. By contrast, the low-coordination-number edge and corner sites adsorb sulfur less strongly, and this sulfur is called *reversible* and is controlled by the following equilibrium.[7]

$$Pt + H_2S \Leftrightarrow PtS + H_2$$

TABLE 18.5
Effect of Sulfur on Reforming Catalysts

	Pt/Al$_2$O$_3$	Pt-Re/Al$_2$O$_3$	Pt-Ir/Al$_2$O$_3$	Pt-Sn/Al$_2$O$_3$
Sulfur pretreatment (H$_2$S or organosulfide containing feed)	Reduces hydrogenolysis* by poisoning high-coordination number groups of Pt atoms	Bonds with Re and eliminates its activity for hydrogenolysis. Sulfided Re reduces formation of graphitic coke on near Pt neighbors.		
Effect of sulfur in feed	Reduces activity, but some S can be reasonably tolerated	Destroys activity because remaining metal atoms are active Pt but fewer in number than on Pt/Al$_2$O$_3$.	Iridium is an active component for hydrogenolysis and preferentially adsorbs sulfur, providing, thereby, protection of the Pt from sulfur poisoning. Low amounts of S can favor aromatization over hydrogenolysis.	Destroys activity. Tin is inert and remaining Pt atoms are easily poisoned by sulfur.

Note: Sulfur tolerance is improved at high temperatures, high H$_2$ pressures, and high severity operation (longer contact time).
*Hydrogenolysis leads to production of unwanted light ends
Based on information from Ref. 7.

High H$_2$ partial pressures cause the reaction to shift to the left, which provides a cleaning action that serves to maintain the activity of the edge and corner sites.

If feed sulfur exceeds the recommended limit, poor performance will result, including a decline in C$_5^+$ and H$_2$ and aromatics yields.[11] Ultimately, H$_2$S will appear in the recycle gas. This poisoning can be reversed by continuing operation at a lower temperature provided low-sulfur feed has been reintroduced and allowed to cause desorption of the H$_2$S. Normal operation can be realized at the desired temperature when H$_2$S no longer appears in significant amounts in the recycle gas. If the temperature is raised prior to this condition, excess coke will form, and run time is thereby reduced.[11]

Poisoning by nitrogen compounds is also reversible by procedures just described for sulfur poisoning. If it goes undetected, the ammonia produced by the reforming conditions reacts with chloride on the catalyst to produce NH$_4$Cl that deposits in downstream exchangers. The loss of chloride further depletes active acid sites.[11]

Lead and arsenic are irreversible poisons that form strong covalent bonds with the Pt sites and effectively deactivate them. Other metals that also act as poisons, although not as severe as lead and arsenic, are included in Table 18.6, along with approximate levels on the catalyst above which performance problems will ensue.

Deactivation Due to Sintering

During regeneration for coke removal, the oxidation reactions that occur produce high temperature and water vapor. Both the metal and the alumina support sinter under these conditions, with the metal sintering accelerated by the high temperature and the oxidizing atmosphere. The alumina sinters because of the high temperature and the presence of water vapor.[8] Fortunately, agglomerated platinum can be redispersed and, although sintered alumina is irreversible, catastrophic sintering of alumina normally does not occur in one cycle.

TABLE 18.6
Permanent Catalyst Poisons and Levels above which Performance Problems Expected

Poison	Level on Catalyst-wt. ppm
Arsenic	100–200
Copper	150–250
Lead	300–500
Cobalt	400–600
Molybdenum	400–600
Sodium	400–600
Phosphorus	400–600
Silicon	1000

Reprinted by permission: Pistorius, J. T., *Trouble Shooting Catalytic Reforming Unit,* Criterion Catalysts, Annual Meeting, NPRA, San Antonio, Texas 1985.

Sintering (agglomeration) of Pt is mainly influenced by high temperature. Rhenium and other metals serving as the second metal in bimetallic catalysts apparently act as barriers to the sintering of Pt by virtue of their effect on Pt interaction with the alumina. Thus, although the Pt on bimetallic catalysts does sinter, it does so more slowly or at higher temperatures. Over a number of regenerations and redispersions of the Pt, the surface area of the alumina can, because of sintering, decline to a point where the catalyst must be removed and fresh catalyst installed. Because of the high value of Pt, it is recovered by a catalyst reclaimer and reused. The customer receives credit for the Pt removed. In fact, in many cases, the customer owns the Pt inventory associated with catalyst in use or in transit.

Chloride and Water Control

The chlorine content of the alumina, for which there is an optimum value (e.g., 0.9–1.2 wt%) for each catalyst type, sets the proper acidity of the alumina that supplies the acid-catalyst portion of the catalyst.[5] If chloride content becomes too large, excess hydrocracking occurs, and more light ends are produced but less hydrogen. Water vapor is present in small amounts (1–2 ppm). If a surge in upstream units produces excess water (50 ppm), that water will strip a large portion of the chloride from the catalyst and deprive it of its important isomerization and cyclization reactions.[1,11,14] Feed water content is normally quite low (1–2 ppm). Chlorine is very slowly lost from the alumina and is replaced by a steady addition of an organic chloride or HCL (0.5–1.0 ppm of Cl in feed). Water is also added so as to produce 4–5 ppm in the feed for the purpose of distributing the chloride evenly over the catalyst bed.[14] After regeneration, the rejuvenation step (see following section) also serves to rechloride the catalyst and, again, water addition accomplishes proper distribution.

Regeneration and Catalyst Rejuvenation

As coke becomes excessive and sulfur poisoning (if it occurs) begins to deactivate the catalyst, regeneration must be initiated to burn off the coke. Air is introduced along with added nitrogen to moderate the burning rate so as to prevent excessive temperatures (>550°C, 1022°F) and resultant loss of support surface area and crushing strength.[6] The burning process requires careful control to avoid excessively low temperatures that would produce incomplete regeneration.

After regeneration, the chloride content of the catalyst has declined due to reversal of the chlorination reactions on alumina, and the metallic crystallites have undergone considerable agglomeration. A rejuvenation procedure is implemented following regeneration for the purpose of redispersing the metallic components and restoring chlorine to the alumina to maintain optimum acidity.

Rejuvenation procedures vary with the catalyst type, but some generalities based on the open literature are possible.[6,15] Metal redispersion is accomplished in an oxidizing atmosphere with chlorine as the primary reactant. Metallic chlorides (e.g., $PtCl_2$, $IrCl_3$, and $ReCl_4$) are formed from Pt, and perhaps some PtO and from IrO_2 and Re_2O_7, all of which could be present after regeneration. Evidence suggests that the chloride form for Pt might be $Pt^{IV}O_xCl_y$, an oxychloride.[8] In any event, the chloridation reactions are exothermic and reversible. Based on postulates concerning chemical vapor transport,[16] smaller crystallites, which have smaller mass per unit surface area, will reach higher temperatures than the larger crystallites when chloride formation occurs on the surface. Once these temperature differentials are established, the forward reaction ($Pt + Cl_2 \Leftrightarrow PtCl_2$) will be favored on the lower-temperature larger crystallites and the reverse reaction on the smaller crystallites. Thus, dispersion of larger crystallites occurs.

Chlorine can be added by various agents.[6] As shown in Figure 18.6, HCl is a common product when organic chlorides are employed at post-regeneration conditions. Since HCl and organic chlorides react with O_2, free chlorine is made available.

$$4HCl + O_2 \rightarrow 2Cl_2 + 2H_2O$$

$$CCl_4 + O_2 \rightarrow CO_2 + Cl_2$$

The reaction of both HCl with O_2 and HCl with alumina produces water, and control of water vapor pressure could have some effect on controlling the extent of chlorination and dispersion.

After regeneration and rejuvenation, the catalyst is reduced in hydrogen and then carefully presulfided according to manufacturer's directions before placing it back in service.

Process Units

Three types of processes are in wide use—semiregenerative, cyclical, and continuously regenerated. Each process uses hydrotreated feed, employs furnaces to reheat feed between stages, recycles a portion of the hydrogen (the rest goes to other refinery units), and separates out light ends in a stabilizer. Temperatures decline precipitously in the first stage, where the endothermic dehydrogenation of cyclohexanic naphthenes occurs. Other less rapid endothermic reactions occur in subse-

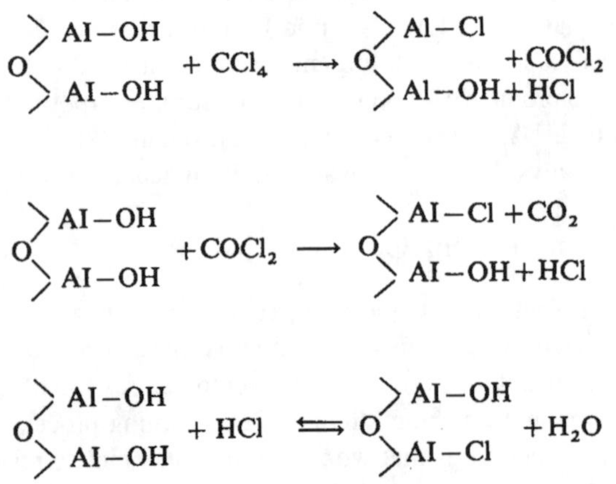

FIGURE 18.6   Reaction Steps in the Chlorination of Alumina with CCl_4. Reprinted by permission: LePage, J. F., et al., *Applied Heterogeneous Catalysis,* p. 491, Editions Technip, Paris, 1987.

quent stages, and in the final stage, exothermic hydrocracking and hydrogenolysis vie against the remaining endothermic reactions to cause very little temperature change.

Semiregenerative units have 3 or 4 fixed-bed reactors, and the unit is shutdown for regeneration periodically (every 6 to 12 months). High pressures (275–375 psig) and high hydrogen concentrations act to extend run length by reducing coke formation. But this condition causes lower aromatics and hydrogen production. Use of more coke-tolerant catalysts such as Pt/Re can ameliorate this problem. Demand for additional higher-octane product, which required higher-severity operation and thermodynamically favorable lower-pressure operation, has encouraged the use of cyclical-regenerative units and continuously regenerated units. The cyclic unit employs an additional or swing reactor along with the necessary valving and manifolding to permit removing a reactor from service on a planned cycle (one week to one month) and substituting a newly regenerated reactor for it. The continuously regenerated unit uses moving beds with catalyst passing between beds and ultimately through a regenerator and thence back to the first moving bed. Both cyclic and continuously regenerated units are operated at lower pressures (50–75 psig) which, when properly designed, offer not only higher aromatics and hydrogen production but also energy savings.

Since mass transfer within the catalyst extrudate or pellet can be a limiting factor in regeneration, catalyst sizes are usually not above 1/16 in. (0.16 cm). The small size also assures more even impregnation of the extrudate in the manufacturing process. Because of this small-size bed, depths must be limited to prevent high pressure drops through the bed. This goal can be reached in large downflow units by using spherical vessels at sufficient diameter that the modest bed depth can be accommodated in the essentially straight-sided portion of the sphere. Alternatively, a radial flow reactor can be used. In both cases, great care must be exercised in designing for efficient low pressure-drop distribution along the entire cross section of the bed. Channeling in downflow reactors is readily detected by observing temperature gradients indicated by thermocouples located radially at various depths. Unequal temperatures at all radial positions at a given bed depth would indicate channeling. Such measurements are not readily applied to radial reactors, and overall changes, such as reduction in the overall ΔT for the reactor or an extended residual burn during regeneration, must be used as indications of radial reactor channeling.[14]

Kinetic Modeling

Modeling a reaction system as complex as catalytic reforming with its plethora of compounds and reactions presents a seemingly insurmountable challenge, but major refiners developed successful models that are, needless to say, proprietary. Many man-hours were expended, but the models can be adjusted as catalysts are changed by optimizing values of constants directly from operating data.

The successful modeling[17] was made feasible by lumping hydrocarbons according to carbon numbers (Figure 18.7). Langmuir–Hinshelwood kinetic expressions were used with alterations to reaction rates applied by time-dependent activity factors on the various reactions. Model details are provided for one lump (C_6) on a catalyst that does not affect the business strategy of the developer.[17] The ultimate complete model required 30 man-years of effort and included aging kinetics as well.

18.2 HYDROPROCESSING (GENERAL)

The petroleum refining industry has a penchant for creating a number of names to describe new or evolving processes, and at times they are more confusing than helpful. Terms like *hydroprocessing, hydroconversion* and *hydrorefining* are often used in association with more specific designations such as *hydrotreating* and *hydrocracking*. It is helpful instead to classify refinery catalytic processes that use hydrogen as a means for upgrading various intermediate streams according to the catalyst type used. On this basis, *hydrotreating* and *hydrocracking* become the most useful terms.

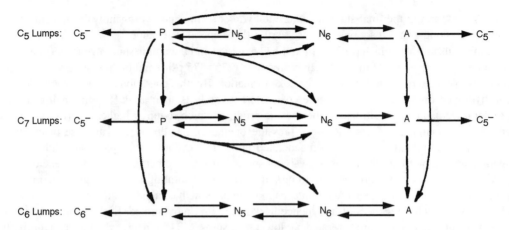

FIGURE 18.7 Reforming Lumped Reaction Network (N = cyclopentane and cyclohexane naphthenes, P = C_6^+ paraffins, A = aromatics, C_5 = pentanes and lighter). Reprinted by permission: Rampage, M. P., Graziani, K. R., Schipper, R. H., Krambeck, F. J., and Choi, B. D., *Advances in Chemical Engineering,* 13, 193 (1987), Academic Press, Inc.

Hydrotreating. Mono-functional catalysts are used that provide hydrogenation and hydrogenolysis activity and include the following reactions: hydrodesulfurization (HDS), hydrodenitrogenation (HDN), hydrodemetallization (HDM), olefin hydrogenation, and aromatics hydrogenation or hydrodearomatization (HDA).

Hydrocracking. Bi-functional catalysts are used, having both acidic and hydrogenation sites. The combination causes cracking (acid function) and hydrogenation (metallic function).

This catalyst oriented classification is quite useful, since it aids in the description as well as the understanding of the chemistry involved. There remains, however, one area for confusion. The typical hydrotreating catalysts ($Co-Mo/Al_2O_3$, $Ni-Mo/Al_2O_3$, and NiW/Al_2O_3) do exhibit some acidity associated with the sulfur-metal components. Acidities of typical hydrotreating catalysts have been measured and found to decline in the order $Ni-W/Al_2O_3 > Ni-Mo/Al_2O_3 > Co-Mo/Al_2O_3$, which order also corresponds to that for HDN activity.[44]

18.3 HYDROTREATING

Background

Prior to World War II, the German coal-tar and synthetic-fuel industry had developed catalytic high-pressure hydrogen treating processes for improving product quality. Only in the 1950s, however, were similar processes introduced in the petroleum refining industry in the U.S.A. and elsewhere. The availability of cheap hydrogen from catalytic reforming units; the growing use of crude oils containing higher organo-sulfur, nitrogen, and metallic compounds; and the stricter product and feedstock purity requirements combined to make hydrotreating processes both economically ideal and necessary. The use of the process has expanded dramatically over the years, using a variety of improved catalysts based on cobalt or nickel-molybdenum on alumina.

Today, *hydrotreating* is a term used to designate processes employing alumina catalyst supports upon which are deposited metal oxides of Co-Mo, Ni-Mo, and NiW, which are changed to the sulfides *in situ*. Various promoters are used, such as phosphorous, that can be judiciously applied to increase and optimize catalyst acidity. Such catalysts are used for hydrogen treating various feeds and products for the purpose of saturating olefins and aromatics and removing bound sulfur (as H_2S), nitrogen (as NH_3), oxygen (as H_2O), and metallic impurities (Pb, As, P, V, Ni, Cu, Si, Fe, and Na, deposited as sulfides on the catalyst). In so doing, cracking and attendant reduction in average molecular weight, if it occurs, is usually minor in extent.

Various hydrotreating applications are summarized in Table 18.7.

Chemistry

The wide assortment of functions encompassed by hydrotreating are described in terms of each accomplished chemical change: hydrogenation of aromatics and olefins, hydrodemetallization, hydrodenitrogenation and hydrodesulfurization, and hydrodeoxygenation. The dominating chemical change for a given unit has often been used to name the unit. Since hydrogenation changes unsaturated entities that cause unwanted colors and also stabilizes some feedstocks, the terms *hydrodecoloration* and *hydrostabilization* are sometimes used where applicable.[1]

The seemingly diverse character of these chemical changes surprisingly involve the same system of parallel reactions: hydrogenolysis of hetero-atomic molecules and hydrogenation of resulting unsaturated hydrocarbons[1] and inorganic entities (S, N, and O). Although the mechanisms and nature of the catalytic species are quite complex and subject to continuing investigation and debate, the overall chemical reactions can be rather simply stated as shown for the following chemical changes and reactants. All these reactions are exothermic.

Hydrodesulfurization (HDS)

The sulfur content of petroleum crude varies greatly, depending on its source, from around 0.04% in light crude to 5% in heavy crude. Crude oil may have up to 40 different organo-sulfur compounds. The disulfides that appear in various product cuts are formed during processing by the oxidation of thiols.[6]

Typical overall HDS reactions are shown in Table 18.8 (see p. 321). Mercaptans are the most easily desulfurized, while thiophenes are the most refractory. In terms of organic structure, the difficulty of sulfur removal increases in the order of paraffins, naphthenes, and aromatics. The actual reaction paths that have been postulated are complex,[12] as illustrated for dibenzothiophene in Fig. 18.8.

In addition to causing unpleasant odors in fuels, sulfur produces sulfurous and sulfuric acid upon combustion, which is linked to air and water pollution. Sulfur converted to oxides or H_2S causes major corrosion problems.[5] The original organosulfur compounds in petroleum can also cause corrosion of engines.

Hydrodeoxygenation

Organic acids are the most numerous oxygen compounds in petroleum and include a wide range of molecular weights from C_1 to C_{30}. In general, aliphatic acids occur below C_8, monocyclic are present between C_9 to C_{13}, and above C_{14}, bicycle acids dominate.[2] Other oxygen compounds include phenol and furans. The naphthenic acids are the dominant oxygen compound class in most crudes. Total oxygen content of crude oil rarely exceeds two percent.

The acids are readily decomposed into water vapor and a partially saturated organic structure by hydrotreating operations for sulfur and nitrogen removal, but the phenols and furans are more slowly decomposed. Typical reactions are shown in Table 18.9 (see p. 322).

Oxygen compounds, which produce water in the presence of hydrogen, can adversely affect the activity of sensitive catalysts such as Pt-based catalytic reforming catalysts. Oxygenated compounds in petroleum also cause fouling and product deterioration during storage and are participants in gum formation.[5] A complete review of mechanisms, kinetics, and catalyst characteristics has been presented.[1]

TABLE 18.7
Major Hydrotreating Applications

Feedstock	Remove Indicated Entities				Saturate			Purpose
	S	N	Oxygenated Group	Metals	Aromatics	Olefins	PNA	
Light straight-run naphtha	X							Protect isomerization catalyst from sulfur poisoning.
Heavy naphtha (1)	X	X	X	X		X(2)		Protect catalytic reforming catalyst from poisoning and improve reformate quality.
Kerosene (3)	X				X			Aromatics removal assures better flame without smoke. Sulfur removal reduces corrosion.
Jet fuel (3)	X	X			X			Nitrogen removal improves stability, and aromatics reduction reduces heat radiation of flame and lowers freezing point.
Residuum (5)	X	X	X	X	X	X	X	Produce high-quality vacuum gas oil for catalytic cracking, naphtha for reforming, and diesel oil from the least valuable part of the crude.
Diesel fuel	X (4)				X			Reduce sulfur to minimize pollution and reduce aromatics to improve cetane number and meet regulations for maximum aromatics allowable.
No. 2 home heating oil	X							Reduce sulfur to minimize pollution.
Fluid catalytic cracking feedstocks (heavy vacuum gas oil, heavy atmospheric gas oil from residuum hydrotreating, coke or gas oil)	X	X	X	X	X	X	X	Nitrogen compounds poison cracking catalysts. Metals alter selectivity unfavorably, and sulfur compounds cause SO_2 pollution of regenerator off-gas.

1. Light and heavy naphthas treated in combination in some refineries.
2. Olefin saturation required for cracked stock.
3. Same boiling range.
4. If diesel oil from residuum hydrotreating is too high in nitrogen, it must be hydrotreated further.
5. Operating conditions at very high pressure and temperature are in the range where thermal (non-catalytic) hydrocracking occurs. Since the catalysts used are monofunctional, catalytic hydrocracking is not significant. At thermal hydrocracking conditions 40 wt% of vacuum residuum is cracked in fixed-bed units and up to 80 wt% in ebullating beds. The conditions for atmospheric residuum are not as severe (10 wt% of the heavier compounds are cracked).

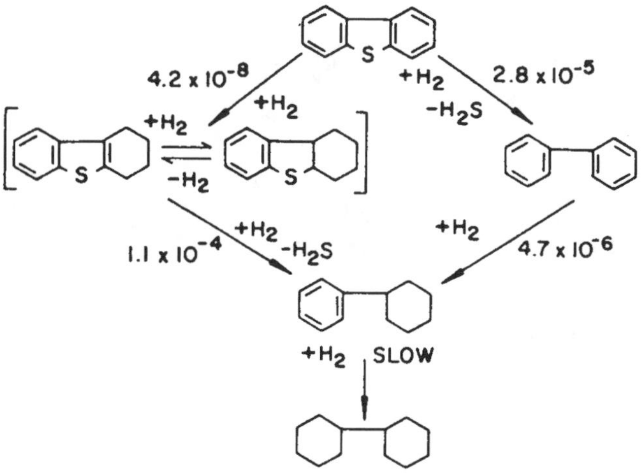

FIGURE 18.8 Proposed Reaction Network for Dibenzothiophene Hydrodesulfurization (300°C, 102 atm on sulfided Co-Mo/Al$_2$O$_3$). Values shown are pseudo first-order rate constants at modest H$_2$S concentrations. Reaction is inhibited at high H$_2$S levels. The dominant reaction is direct removal of sulfur as H$_2$S to produce biphenyl, then cyclohexylbenzene by hydrogenation and further but slow hydrogenation can lead to bicyclo-hexyl. Houalla, M., Nag, N. K., Sapre, A. V., Broderick, D. H., and Gates, B. C., *AIChE Journal* 24, 1015 (1978). Reprinted with permission of the American Institute of Chemical Engineers. Copyright © 1978 AIChE. All rights reserved.

Hydrodenitrogenation

Nitrogen occurs in crude oil within the range of 0.1–0.9%, most of which is in the higher boiling fractions that make up residues and often account for up to 95% of the nitrogen content of crude oil.[2] Nitrogen in these fractions are contained in porphyrin structures composed of four indole nuclei and a metal such as nickel or vanadium as shown in Figure 18.9.[9]

Typical overall reactions with major steps are shown in Table 18.10 (see p. 323) for some of the lower-molecular-weight nitrogen compounds. Of those shown, pyrrole and carbozole are classed as non-basic but, as can be seen, basic intermediates are produced in the reaction sequence.[1] In fact, in all cases, the hydrogenation of rings is necessary to produce aromatic or aliphatic amines before nitrogen can be removed as ammonia. Figures 18.10 and 18.11 present more detailed reaction schemes for quinoline and carbozole, respectively.[14,15,44] A reaction scheme for indole (not shown) has also been published.[16] Table 18.11 (see p. 325) illustrates a more complete group of heterocyclic nitrogen compounds, typical of light feedstocks, along with pKa values. Aliphatic amines are also present, but they rapidly react to NH$_3$ and the corresponding aliphatic. Saturated nitrogen hetero-cyclic compounds have a higher pKa value than the corresponding unsaturated ones.[7] Five-membered nitrogen heteroaromatics exhibit low pKa values. The pair of electrons associated with bound nitrogen and its basicity is involved with the π-cloud of the ring and is not available for independent base behavior.[7] In six-membered ring compounds the nitrogen electron pair is not attracted by a π-cloud and it has strong electron sharing tendencies for acid sites.

Although present in small amounts, organonitrogen compounds are potent poisons for acid sites on catalysts used for catalytic reforming, catalytic cracking, and hydrocracking. The presence of these compounds in hydrotreating feed (>25 ppm) can inhibit the other hydrotreating reactions because of the effect of these basic compounds on the mild but necessary acid characteristic of the alumina support. Fuels containing organonitrogen compounds upon combustion produce NO$_x$, which is both an atmospheric pollutant and corrosive gas in engines and furnaces.

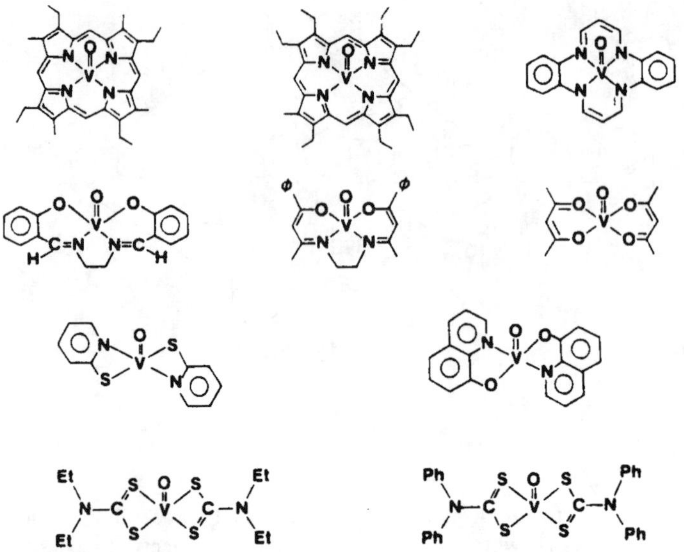

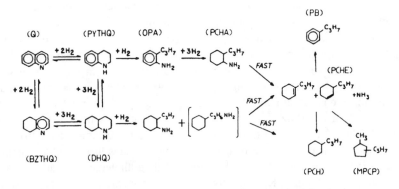

FIGURE 18.9 Vanadyl Porphyrins (first two) and Possible Nonporphyrins. Reprinted with permission from Reynolds, J. G., Riggs, W. R., and Berman, S. A. in "Metal Complexes in Fossil Fuels," R. H. Filby and F. Branthauer, eds., p. 205, *ACS Sym. Ser.,* Sep. 344, copyright 1987 American Chemical Society.

Q	- quinoline
PyTHQ	- Py (or 1,2,3,4)-tetrahydroquinoline
BzTHQ	- Bz (or 5,6,7,8)-tetrahydroquinoline
DHQ	- Decahydroquinoline (cis and trans isomers)
OPA	- o-propylaniline

PCHA	- propylcyclohexylamine
PB	- propylbenzene
PCHE	- propylcyclohexene (1,1 or 1,3)
PCH	- propylcyclohexane
MPCP	- methylpropylcyclopentane

FIGURE 18.10 A Proposed Reaction Network for Hydrodenitrogenation of a Six-Membered Ring—Quinoline on Sulfided Ni-Mo/Al$_2$O$_3$. The initial hydrogenation to 1, 2, 3, 4 tetrahydroquinoline (14 TQ) is very rapid, but since hydrogenation of 14 TQ is faster than hydrogenolysis, the preferred route is through decahydroquinoline (DHQ) and the remainder of the lower path (14, 15). Reprinted with permission from Saterfield, C. N. and Yang, S. H., *Ind. Eng. Chem. Process. Des. Dev.* 23, 11, (1984). Copyright 1984, American Chemical Society.

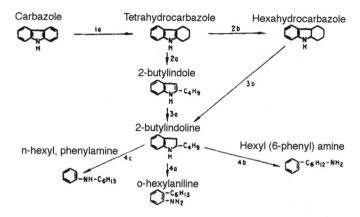

FIGURE 18.11 Reaction Scheme for Carbazone Hydrodenitrogenation. Reprinted with permission from Flinn, R. A., Larson, O. A., and Beuther, H., *Hydrocarbon Processing,* p. 129, Sept. 1963. Hydrocracking reactions convert the amines and anilines to ammonia and the aromatics to which the nitrogen group was attached.

TABLE 18.8
Example Overall HDS Reactions

1. Mercaptans

$$R-SH + H_2 \longrightarrow RH + H_2S$$

2. Disulfides

$$R-SS-R' + 3H_2 \longrightarrow RH + R'H + 2H_2S$$

3. Sulfides (aromatic, naphthenic, and alkyl)

$$R-S-R' + 2H_2 \longrightarrow RH + R'H + H_2S$$

4. Thiophenes

$$+ 4H_2 \longrightarrow CH_3CH_2CH_2CH_3 + H_2S$$

5. Benzothiophenes

$$+ 3H_2 \longrightarrow \text{(aromatic)}\;CH_2CH_3 + H_2S$$

6. Dibenzothiophenes

$$+ 2H_2 \longrightarrow + H_2S$$

Reprinted from Speight, J. G., *The Desulfurization of Heavy Oils and Residua,* 2nd edition, p. 65, Marcel Dekker, New York, 1981, by courtesy of Marcel Dekker, Inc.

TABLE 18.9
Hydrogenation of Organic Oxygen Compounds

Naphthenic acids

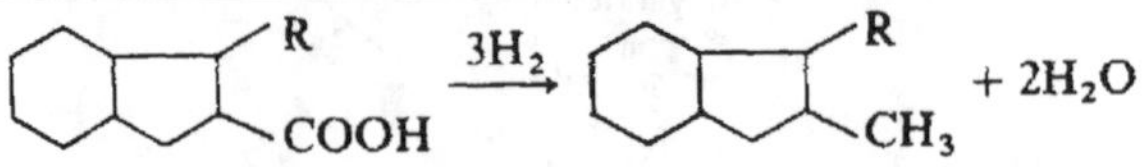

Phenols

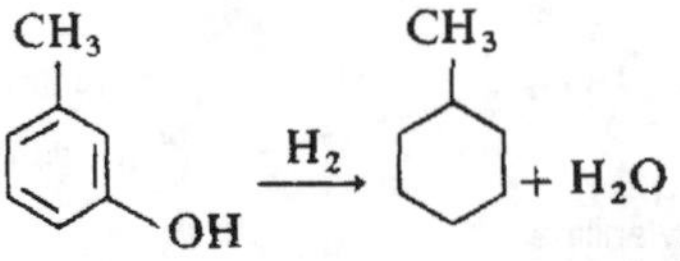

Furan

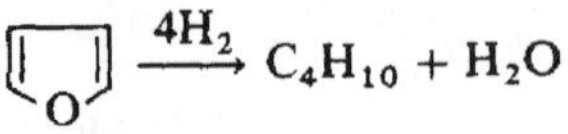

Dibenzofuran

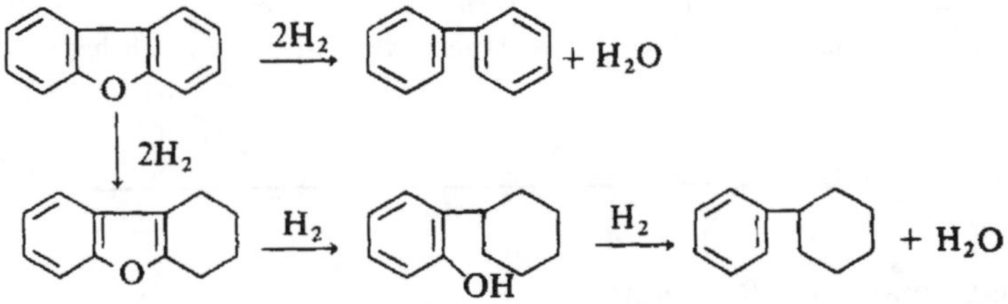

Reprinted by permission: LePage, J. F., et al., *Applied Heterogeneous Catalysis,* p. 395, Editions Technip, Paris, 1987.

Hydrogenation Of Aromatics

Of the many reactions in hydrotreating, aromatics hydrogenation (direct or as a step in hydrodenitrogenation, hydrodeoxidation, and hydrodemetallization) can be equilibrium limited. Hydrogenation is favored by low temperatures and high H_2 pressure. As in all hydrogenations, exothermic heats of reactions are high. In fact, feeds with high aromatics content must be given special attention in planning for effective temperature control, since equilibrium conversion declines dramatically with temperature increases. To that extent, aromatics hydrogenation is self-limiting, but a rapid and excessive temperature increase can sinter the catalyst and possibly exceed design conditions for the reactor and associated equipment.

Qualitatively, the equilibrium constant for aromatics hydrogenation declines in the order benzene, benzene with side chains of increasing length, polyaromatics and polyaromatics with side chains.[3,18] In the case of polyaromatics, hydrogenation occurs in steps (see Figure 18.12) with declining equilibrium constants for each successive step. If a hydrogenation step in hydrodenitrogenation becomes equilibrium limited, the desired removal of nitrogen will not occur. This situation is avoided in commercial reactors by careful temperature control. More active catalysts, of course, can alleviate such a problem by permitting lower-temperature operations, and, at the same time, allow higher throughput.[7]

Hydrogenation Of Olefins

Olefins are not present in crude oil but do occur in cracked products. They are readily hydrogenated under hydrotreating conditions, even though the reaction is inhibited by sulfur compounds and H_2S. Olefins, as always, form polymers on acid sites, but this situation is avoided by adequate hydrogen pressure. Olefins, H_2S, and sulfur compounds such as mercaptans and sulfides engage in a series of equilibrium reactions.[1] At high olefin concentration, the sulfur compounds are formed and, as olefins concentrations decline, the sulfur compounds decompose to olefins and H_2S. As catalyst

TABLE 18.10
Hydrodenitrogenation of Typical Organo Nitrogen Compounds

Amines	$R-NH_2 \xrightarrow{H_2} RH + NH_3$
Pyrrole	pyrrole $\xrightarrow{2H_2}$ pyrrolidine $\xrightarrow{H_2} C_4H_9NH_2 \xrightarrow{H_2} C_4H_{10} + NH_3$
Pyridine	pyridine $\xrightarrow{3H_2}$ piperidine $\xrightarrow{H_2} C_5H_{11}NH_2 \xrightarrow{H_2} C_5H_{12} + NH_3$
Carbazole	carbazole $\xrightarrow{H_2}$ (o-aminobiphenyl) $\xrightarrow{H_2}$ (biphenyl) $+ NH_3$
	$\xrightarrow{2H_2}$ (N-butyl indoline/tetrahydrocarbazole) C_4H_9 $\xrightarrow{2H_2}$ (cyclohexyl)$-C_6H_{13} + NH_3$
Quinoline	quinoline $\xrightarrow{2H_2}$ (1,2,3,4-tetrahydroquinoline) $\xrightarrow{H_2}$ (o-propylaniline, C_3H_7, NH_2) $\xrightarrow{H_2}$ (propylbenzene, C_3H_7) $+ NH_3$

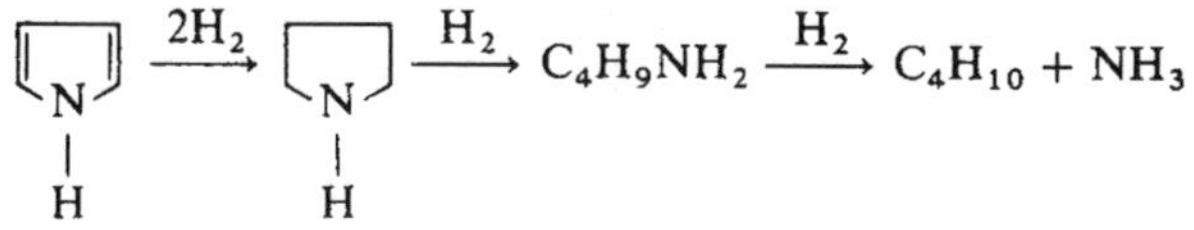

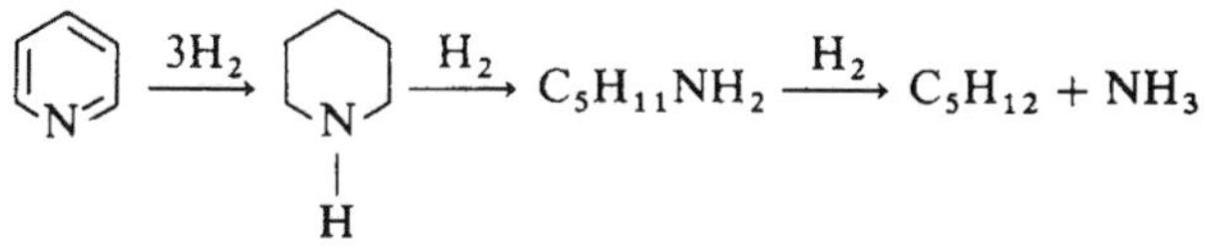

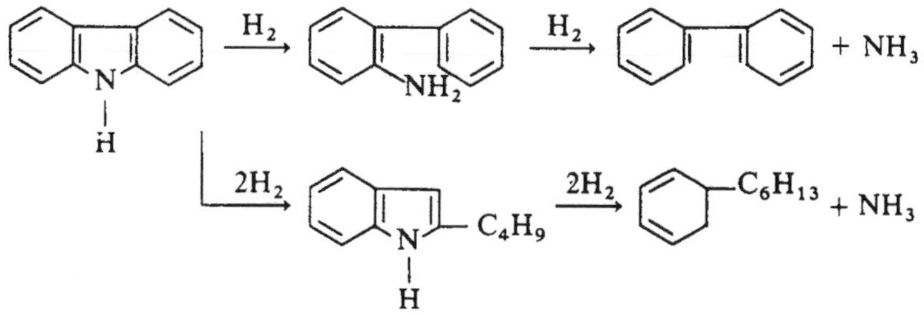

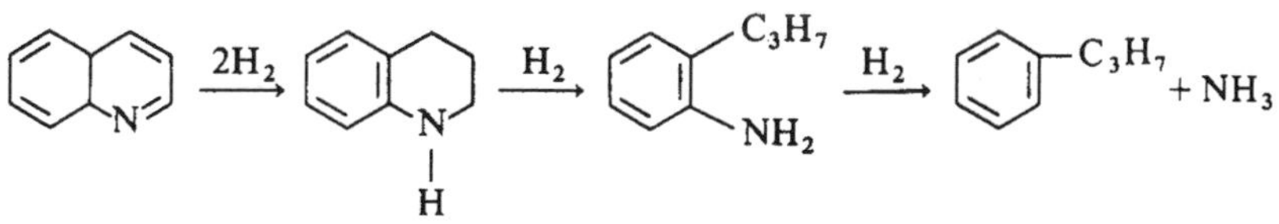

Reprinted by permission: LePage, J. F., et al., *Applied Heterogeneous Catalysis*, p. 389, Editions Technip, Paris, 1987.

activity declines toward the end of a cycle and olefins are not all being hydrogenated, removal of sulfur compounds to very low levels becomes difficult.[1]

Competitive Effects

The extreme complexity involved in even single pure component studies is exponentially increased by the competitive or interactive effects that result when almost all compound types and reactions are present in a feedstock, particularly heavy fractions at petroleum. Some generalities are summarized in Table 18.12 (see p. 326).

Hydrodemetallization

Metals occur in crude petroleum in trace quantities (ppm), as shown in Table 18.13 (see p. 326) for a number of metallic elements and several nonmetallic elements. Some, such as iron, are partially augmented by contamination due to corrosive action in storage tanks, while arsenic is thought in many cases to be due to contamination by certain components of drilling fluids, corrosion inhibitors, and bactericides. Nickel and vanadium occur in the largest amounts and predominate in the heavy ends of the crude. Table 18.13 (see p. 326) illustrates the marked differences in metals content between various crude oils.

In lighter cuts such as naphtha, the metallic or metalloidal compounds of As, Pb, P, and Cu are present in very low amounts and are easily removed in the usual hydrodesulfurization treatment.[1] These compounds are thought to be lower-molecular-weight organometallic compounds that have

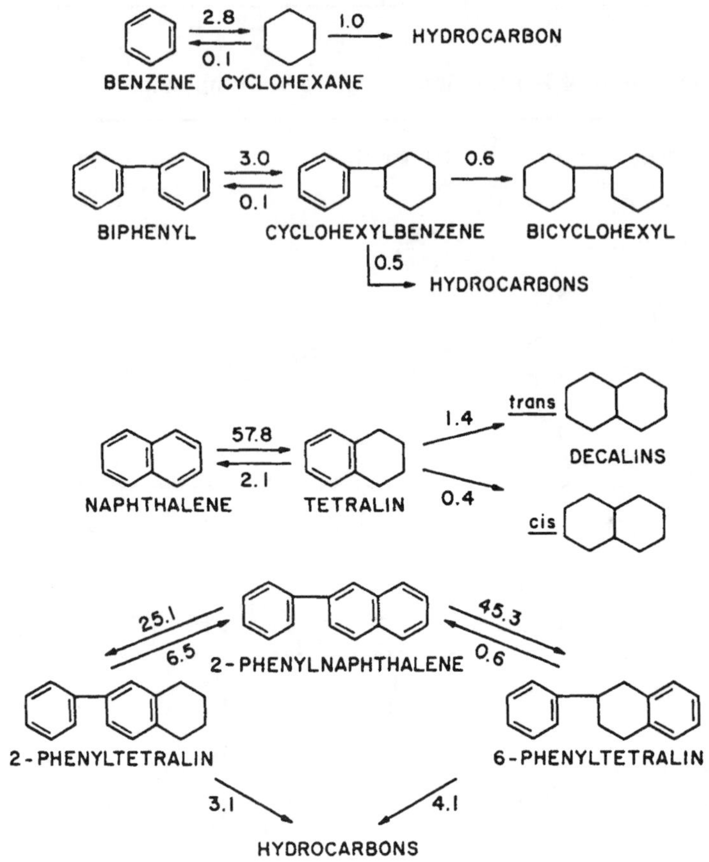

FIGURE 18.12 Proposed Network for some Hydrogenations of Aromatics on Co-Mo/Al$_2$O$_3$ at 325°C and 75 atm. Pseudo-first-order rate constants are given in cubic meters per kilogram of catalyst per second. Reprinted with permission from Sapre, A. V. and Gates, B. C., *Ind. Eng. Chem. Process. Des. Dev.* 20, 68, (1981). Copyright 1981, American Chemical Society.

been volatilized in distillation and distributed sparingly in distillates. The most troubling metallic constituents are nickel and vanadium, which are concentrated in the asphaltene and resin constituents of residuum. It is well established[8] that porphyrin complexes (chelates) of divalent nickel and vanadyl (VO) exist in petroleum. Porphyrins (Figure 18.9) have been studied since the 1930s, initially because of their geochemistry interest. Reaction networks have been proposed based on pure component studies in which a series of partial hydrogenation steps ultimately cause the release of the metal component to the catalyst surface as a sulfide (Figure 18.13). Since porphyrins are initially part of rather massive colloidal aggregates, release of the various porphyrin entities must also be described. Figure 18.13 provides a concept that is reasonable and also postulates desulfurization reactions of the resin-like portions of the aggregate, which cause their separation from the micellar clusters since the intermolecular forces have been weakened.[3] At higher temperatures (>400°C), the aggregates are further disrupted into individual components consisting of a mixture of asphaltenes, no longer held together by interlayer forces, and resins and oils.[3]

Unfortunately, not all the Ni or V content of the heavy oils can be accounted for by porphyrin structures, and an intense amount of work has been done to define non-porphyrin structures to which the substantial fraction of Ni and V might be complexed.[10] Controversy surrounds the nature of these non-porphyrin metal chelates, because convincing isolation of such compounds has proved

TABLE 18.11
Typical Heterocyclic Nitrogen Compounds

Compound	Structure	pKa
Six-membered		
Pyridine		5.2
Piperidine		11.1
Quinoline		4.9
Tetrahydroquinoline		5.0
Acridine		5.6
Five-membered		
Pyrrole		0.4
Indole		−3.6
Indoline		5.0
Carbazole		−6.0
Nonheterocyclic		
Aniline		5.0

Reprinted from Ho, J. C., *Catalysis Reviews: Science & Engineering* 30 (1) 117 (1988), by courtesy of Marcel Dekker, Inc.

elusive. Model non-porphyrin compounds that could complex with nickel and vanadyl have been suggested (Figure 18.9), and a nickel carboxylate has been isolated from a crude oil.[10] Possible carboxylic-acid ligands have been suggested (Figure 18.14). If such smaller complexes exist in the heavier cuts, they are probably part of a large colloidal aggregate.

TABLE 18.12

Interactive and Competitive Effects in Commercial Hydrotreating

	Product or Reactant Type			
Reaction Type	H_2S	NH$_3$ and/or Organo N(d)	H_2O	P$_H$(e)
HDS	inhibits (a)	inhibits		benefits
HDN	benefits (b)	inhibits	benefits	benefits greatly
HDO	benefits (c)	inhibits		benefits greatly
HDA		inhibits		benefits complex aromatics greatly

H_2S is produced by HDS, NH$_3$ by HDN, and H_2O by HDO.

Notes:

(a) Inhibiting effect reduced by increase in temperature in two-phase systems because H_2S concentration in liquid then declines.

(b) H_2S increases the rate of C-N bond cleavage possibly due to increased surface acidity.

(c) Keeps catalyst in sulfided form and prevents over reduction of the surface an essential for HDO.

(d) Strong adsorption of N compounds inhibits most of the reactions.

(e) Because HDN, HDO, and HDA are rate limited by hydrogenation, these reactions are most favorably affected by higher hydrogen partial pressures (P$_H$). Above a certain pressure typical of each feed type, other rate controlling steps take charge and further increase in P$_H$ show no advantages. The typical hierarchy of rates at high P$_H$ is HDS > HDN > HDO.

Based on Refs. 1, 3, 7, and 11.

TABLE 18.13

Ranges of Trace Elements Found in Five Crude Oils*

	Range, ppm		Range, ppm
As	0.0024–0.0655	Mn	0.01–1.2
Br	0.072–0.491	Mo	0.0–7.85
Cd	0–0.004	Na	2.92–20.3
Cl	1.47–39.3	Ni	0.609–98.4
Co	0.052–13.5	S	0.0–4694
Cr	0.0–0.64	Sb	0.0–0.303
Cu	0.0–0.93	Sc	0.0–0.004
Fe	0.7–68.9	Se	0.0052–1.10
Ga	0.0–0.3	U	0.0–0.015
Hg	0.0–23.1	V	0.682–1110
I	0.0–1.36	Zn	0.046–62.9
K	0–4.93		

Based on data from Refs. 4 and 5.

*Crude oils tested were from California, Libya, Venezuela, and Alberta.

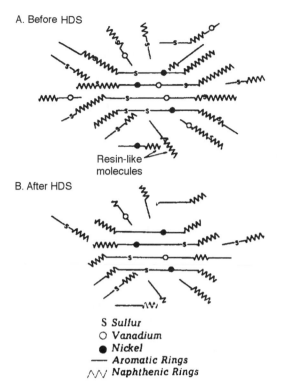

A. Before HDS

Resin-like
molecules

B. After HDS

S *Sulfur*
O *Vanadium*
● *Nickel*
—— *Aromatic Rings*
ʌʌʌ *Naphthenic Rings*

FIGURE 18.13 Qualitative Changes in Asphaltenes and Surrounding Resins during HDS Processing. Reprinted from Beuther, H. and Schmid, B. K., Section III, Paper 20-PD7, *Proceedings of the 6th World Petroleum Congress,* Section 3, p. 303. *Note:* H_2S is the volatile inorganic produced by the hydrodesulfurization.

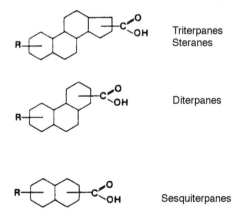

Triterpanes
Steranes

Diterpanes

Sesquiterpanes

FIGURE 18.14 Possible carboxylic acid ligands. Reprinted with permission from Reynolds, J. G. and Gallegos, E., in "Metal Complexes in Fossil Fuels," R. H. Filby and F. Branthauer, eds., p. 332, *ACS Sym. Ser.* 344, Copyright 1987 American Chemical Society.

As shown in Table 18.14, certain metals must be removed from feedstocks for subsequent catalytic processes to operate successfully and/or for safe and effective use of the resulting products.

TABLE 18.14
Negative Effects of Metallic Components in Crude Oil

Metal	Negative Effects
Ni	Deposits on catalytic cracking catalyst, causing excess coke and gas formation and causes adverse product selectivity.
V	Same as Ni. Also V is a heavy metal that is thought to cause lung disease. It also causes deterioration of fire brick and corrosion of boiler tubes and turbine blades.
Na	Sodium buildup can reduce catalyst activity. Sodium in fuels is an ash-forming constituent and corrosive.
Cu	Destabilizes various petroleum products by accelerating gum and sediment formation. Also Cu is a catalyst poison for catalytic cracking but less so than Ni or V.
Fe	An ash former and also acts as a catalyst poison-like Ni and V but to a lesser degree.
As	Permanent poison for catalytic reforming catalysts.

Unlike other hydrotreating reactions, which produce simple, easily recovered gaseous compounds of the removed element, hydrodemetallization yields metallic sulfides that deposit on the catalyst surface and in the pores. Ultimately, these deposits reach a level at which the catalysts become useless, and must be removed and replaced.

Residuum Hydrotreating

The subject of efficient use of distillation residues deserves special mention for two reasons. They contain most of the potential "troublemakers" in crude oil, both with respect to processing and to air pollution by combustion of products made therefrom. Beginning in the early 1970s, environmental regulations in both North America and Europe necessitated either removing sulfur and nitrogen from atmospheric residua to produce a usable fuel oil or installing stack gas clean-up systems at power plants for eliminating sulfur and nitrogen oxides. Hydrodesulfurization, heretofore practiced mainly on distillates, was also widely implemented in new units for atmospheric residua in the U.S.A. and Japan, and in the Middle-East refineries that exported heavy fuel mainly to Europe.[42] Oil prices rose dramatically in 1978 and declined in later years, but the supply of light and sweet crude continued a sharp decline such that refiners were facing the prospect of producing increasing amounts of heavy fuel oil for a market that was rapidly shrinking. The problem of obtaining higher-quality products (mainly transportation fuels) from this ever-increasing supply of distillation residues had to be faced and solved for vacuum residue, which is barely a liquid at room temperature and is loaded with metals, sulfur, nitrogen, and complex molecules, such as asphaltenes, resins, and polynuclear aromatics. The R&D efforts of both the oil companies and catalyst manufacturers have succeeded in developing new catalysts and reactor designs that have miraculously made catalytic treatment of vacuum residua possible and economical.

Simultaneously, a number of older processes were implemented and improved. These include removing asphaltenes and resins more or less intact as in deasphalting by solvents, or by severe thermal treatment, and total alteration of the original heavy molecules as in delayed coking.

Thermal processes, coking and visbreaking, although not requiring hydrogen, do require hydrotreating of their distillate products with the heavy gas oil fraction subsequently further processed in the fluid catalytic cracker as a portion of the total feed.[42] Direct hydrotreating of residue, by contrast, produces high-quality products that can enter the major refinery streams, such as naphtha for catalytic reforming, vacuum gas oil for catalytic cracking, and diesel oils, if the nitrogen content is low. In addition, the distillation residue can be used as blending stock for low-sulfur fuel oil or as delayed coking feed.[42]

Thermal Hydrocracking

Under residua hydrotreating conditions, catalytic hydrocracking does not occur to any significant degree because, as for all hydrotreating, monofunctional catalysts are used. The elevated temperatures required for the process, however, cause significant thermal hydrocracking, which induces free radical reactions.[39,42] The catalyst does play an important role in the thermal hydrocracking process in several crucial aspects:[42]

- The homogeneous phase, which occupies approximately two-thirds of the bed, provides the hydrogen transfer agents that saturate the cracked fragments. These agents are naphthene-aromatic compounds for which the transferred hydrogen is replenished at the hydrogenation sites on the catalyst. The transferred hydrogen reduces the free-radical reactions that initiate chain reactions, such as polymerization and polynuclear aromatics condensation reactions, that lead to coking and reduce the rate of primary and secondary cracking of carbon-carbon bonds.
- The adsorbed H_2 and H_2S on the catalyst surface also act as free-radical traps.

The moderating of free-radical reactions requires more severe conditions for cracking, but it also provides the opportunity for better control of potentially undesired reactions.

Reaction Networks

The complexity of hydrotreating reaction systems is not only compounded by the wide variety of feedstocks that must be treated but also by the fact that all the various hydrotreating reactions can occur at the same time while, in some cases, exerting competitive effects. The formidable problem of understanding both the how and why could initially be approached only by means of pure component studies. These studies often provided useful frameworks for reasoning about the chemistry when plant or pilot plant results needed such structure. In the meantime, experienced development engineers and scientists forged ahead with the creation of new and successful processes and improved catalysts, which provided those engaged in fundamental studies new challenges requiring explanations. The aspects related to chemistry essentially were involved with the sequence of chemical steps with limited knowledge of the catalyst surface and detailed characteristics. More recently, because of the advances in surface science these catalyst issues have received serious attention as detailed in the subsequent section, "Catalyst Characteristics."

Initially, hydrotreating was confined to light stocks for which hydrodesulfurization was the primary goal. This reaction, therefore, has been the one most thoroughly studied, and Co-Mo/Al$_2$O$_3$ is the most favored HDS catalyst system, but the other reactions have been given more attention in recent years.

Figures 18.8 and 18.10 provide examples of reaction networks based on kinetics determined for the indicated pure-component reactions. Typical examples of more complex molecules are shown, because they are the more refractory and dictate the severity of reactor operation. Also, the more complex molecules contain entities representative of simpler compounds (e.g., the pyridine and benzene rings of quinoline). The networks shown indicate the complexity of even the pure components, which vary in behavior with compound type and configuration. Generalizations for any given heteroatom family are precarious exercises.

Diffusion Limitations

The intricate catalytic networks just described are further complicated by diffusion limitations. Such barriers to reaction occur in two distinct ways: (1) when feedstock characteristics and operating conditions are such that certain reactions are very fast, and (2) when processing heavy feedstocks (residue) containing asphaltenes. Examples of the first situation include the initial step in hydrogenation of polynuclear aromatics when operating at high temperature and pressure, hydrogenation of olefins, and hydrodesulfurization of light gasolines, naphthas, and kerosene.[2] The second situation is caused by very large asphaltene micellar clusters. These can be of comparable size to the pores and tend to concentrate and deposit metal sulfides in the pore mouth when the intrinsic activity of HDM is greater than the diffusion rate.[20] When the pore sizes are large and the intrinsic activity is lower than the diffusion rate, more metal deposits can be accommodated all along the pore, but the hydrogenation sites essential for other hydrotreating reactions become deactivated.[20]

Catalyst Types And Suppliers

Major merchant catalyst types, primarily supported on Al_2O_3, are available from a number of catalyst manufacturers as follows:

Suppliers

　　Acreon, Akzo Nobel, BASF, Catalysts & Chemicals, Chevron Res. & Tech., Criterion, Grace Davison, Haldor Topsoe, Inst. Mexicano Petro., Katalenna, Orient, Procatalyse, United Catalysts

Licensors

　　Akzo Nobel, CD Tech, Chevron Res. & Tech., Criterion/ABB Lummus Global, Exxon Res. & Engr., Haldor Topsoe, IFP, Kellogg Brown & Root, UOP LLC

Shape Characteristics

Shape	Size (diameter, inches)	Use
Cylinder	1/8, 1/10, 1/16, 1/20	General use.
Cylinder ring	3/16, 3/32, 1/8 × 1/16	Lower pressure drop and less plugging. Used as top layer of beds or for entire bed.
Trilobe and pentalobe	1/8, 1/10, 1/16, 1/20	Overcome diffusion limitations by reducing diffusion path without increasing pressure drop. Since they pack less densely, the total mass of catalyst charged is reduced, which must be balanced by improved activity.
Extrudate	1/25, 1/32	For ebullating bed.

Although each type will catalyze all the hydrotreating reactions, catalyst selection is often based on the most difficult or limiting reaction system largely evaluated by the ability of a given catalyst to meet the most critical specification of the finished product. In general, the preferences are Co-Mo/Al_2O_3 for hydrodesulfurization, Ni-Mo/Al_2O_3 for hydrodenitrogenation and hydrodemetallization, and Ni-W/Al_2O_3 for aromatics hydrogenation. See Table 18.15 for additional details. Of course, there are many other issues impinging on final selection, such as cost, catalyst life, existing reactor design, and the optimum hydrotreating goals for all undesired components.

　　Since hydrotreating occurs at relatively high pressures and temperatures, reactors and associated equipment require significantly higher capital investments. It early became obvious that improvements in processes could most expeditiously be made by developing improved catalysts tailored to particular feedstocks and the unique and changing demands on the refining industry occasioned by new regulations affecting product characteristics. Although operating companies have continued research and development of improved engineered catalysts, the catalyst manufacturers have become major players as well. Many new formulations have been developed involving active-ingredient recipes, pore structure, and extrudate shape and size. A refiner can select from a sizeable array of alternatives, either one or several different catalysts that will enable optimum operation under newly defined product specifications with little are no new capital expenditures.

TABLE 18.15
Summary of Commercial Hydrotreating Catalyst Characteristics

Feedstock	Active Metals Content (wt%)	Main Objective	Pore Size Range, Å	Surface Area Alumina Support, m²/g
1. Naphthas through vacuum gas oil (straight run and cracked)			80–120	180–250
(a) Medium activity	3.5 CoO/12–14 MoO₃	High HDS activity		
	3.5 NiO/12–14 MoO₃*	HDS, HDN, and hydrogenation of equal importance		
(b) High activity	4–5 CoO/16–19 MoO₃	Deep HDS		
2. Residua vacuum resid				
(a) First stage: (guard bed)	2 CoO/6 MoO₃	Breakdown of asphaltenes and metals removed		
	2 NiO/6 MoO₃			
		High metals selectivity low HDS (just enough to sulfide freed metals)	180–250 20% > 1000	150 m²/g
(b) Second stage	2–3 CoO/10 MoO₃	Good HDS, fair to good HDN	More small pores, but some large pores adequate for remaining metals content	180 m²/g
	2–3 NiO/10 MoO₃			
(c) Third stage	High activity similar to 2(b)	Deep HDS		
(d) Ebullating bed	Can use a large-pore MoO₃ only catalyst followed by high activity HDS catalyst or proprietary single catalyst with both smaller pores for HDS and large pores for HDN			

*Select NiO for better hydrogenation activity (HDN, HDAr, HDCC)

Catalyst Characteristics

Preparation

Although general procedures for preparing hydrotreating catalysts have been reported (e.g., see Refs. 1 and 19), the proprietary details not available even in patents remain secret. These include details of impregnation, the production of optimum pore sizes and pore size distribution, and calcination procedures.

Cylindrical extrudates make up the major portion of hydrotreating catalysts. They are produced in two diameter ranges 1/20 and 1/16 in. (1.3 and 1.6 mm) and 1/10 and 1/8 in. (2.6 and 3.2 mm). Lengths are variable, a typical range being 5 to 10 mm. The smaller diameter is usually preferred because of intraparticle diffusion limitations that occur with fast reactions as well as with large molecules such as asphaltenes (see "Diffusion Limitations" section above). Ebullating beds use the smallest size catalyst, 1/25 in. or less. Since there is a practical lower limit to catalyst diameter (often quoted at 0.8 mm), below which pressure drop and bed plugging become excessive, shaped extrudates (trilobe, star, and fluted shapes) with multiple short paths are available in the same overall nominal diameters. Extruded rings are also produced in sizes 3/16 and 1/8 in. (4.8 and 3.2 mm). These are particularly used as top layers of a catalyst bed. They reduce pressure drop in that region where plugging often occurs and tend to distribute plugging components in the feed throughout the bed, thereby extending run time.

Two major procedures are used in producing extruded hydrotreating catalyst.[1,19] After the alumina gel is made by combining a 10% solution of $AlCl_3$ and $NH_3(OH)_2$ to a neutral pH, the resulting slurry is spray dried and then undergoes a kneading operation with peptizers, binders (e.g., nitric or hydrochloric acid), lubricants, pore-forming agents (e.g., cotton linters, powered activated carbon, polymer threads), and cobalt nitrate and ammonium molybdate. The resulting material is then extruded, dried, and calcined at 300°C to decompose the nitrate and molybdate to oxides. Alternatively, the active ingredients (Co and Mo salts) may be extruded and then impregnated sequentially with Co and Mo salts followed by a second drying and calcination.[19] This second procedure is likely to produce more Co and Mo in the outer regions of the extrudate which, in fact, can be an advantage in diffusion-controlled regimes. However, the small size of the usual extrudates serves to minimize the intensity of such gradients.

Presulfiding

Hydrotreating catalyst is delivered in the oxide form of the catalytic components and must be sulfided after placement in the reactor before actual hydrotreating of a feedstock can be initiated. Sulfiding or resulfiding must also take place after each regeneration because most of the metal sulfides are changed to oxides during regeneration. Exact procedures provided by the catalyst suppliers must be followed carefully with particular reference to avoiding excessively high temperatures. A typical procedure involves staged temperature increases to 500°F and then 550–600°F.

Sulfiding reactions, like oxidations, are highly exothermic, and good control of bed temperature is essential. In many cases, straight-run liquid feedstocks containing sulfur compounds are used, or compounds such as carbon disulfide, ethyl mercaptan, dimethyl sulfide, and butyl sulfide may be added to reduce sulfiding time. In either case, hydrodesulfurization occurs to enough degree to produce H_2S *in situ* that will serve to sulfide the Co, or Ni and the Mo sites on the catalyst. Alternatively, H_2S may be used directly with some reduction in presulfiding time, but the gas-phase treatment makes temperature control more difficult.

Care must be exercised to avoid temperatures above 400°F when H_2 is present alone. At such conditions, reduction of active metal components on the catalyst will occur to the detriment of its initial and long-term activity.

Nature of the Catalyst Surface

The most thoroughly studied catalyst is $Co-Mo/Al_2O_3$, and hydrodesulfurization (HDS) has been the reaction system of central focus, since HDS was the first hydrotreating target. Because of the similarity in chemical properties, the concepts relating to Co also apply to Ni and those for Mo apply to W reasonably well.[2] In more recent years, experts in organometallic chemistry and surface science have directed their attention to this remarkable catalyst system, some earlier concepts have been confirmed, and new more detailed understanding has also emerged. The following is a summary of what is known with some degree of confidence:[22,23,25]

- The active catalyst is a sulfided form usually created in the reactor prior to starting a run using in a mixture with hydrogen: H_2S, thiophene, or a feed containing sulfur components that react relatively easily. The result of a proper treatment is fully sulfided Co or Ni and Mo in the form, respectively of Co_9S_8, Ni_3S_2, and MoS_2.
- MoS_2 crystallites either lie with their basal planes parallel to the Al_2O_3 support surface, or they are bonded on their edges as shown in Figure 18.15.

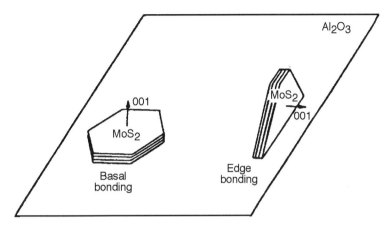

FIGURE 18.15   Orientation of Small MoS_2 Crystallites on the Surface of Al_2O_3. Reprinted from Prins, R., DeBeer, V. H. J., and Somorjai, G. A., *Catal. Rev.—Sci. Eng.*, 31 (1 and 2), 1 (1989) by courtesy of Marcel Dekker, Inc.

- Cobalt is adsorbed in ionic form around the edges of MoS_2 crystallites or as Co_9S_8 crystallites on the Al_2O_3 support surface. High temperature in the sulfidation pretreatment favors Co_9S_8. Nickel promoter is also located around the edges of the MoS_2 crystallites.
- Three major approaches to the mechanistic description for hydrodesulfurization have been used: solid state chemistry, organometallic chemistry, and surface science. Each one has weaknesses and strengths that have been competently reviewed.[22] The solid-state model is the oldest and the easiest to recall as an aid to thinking about the reaction system and problems related thereto. It is presented here for that reason, using thiophene as an example, with an asterisk to represent an active site.

$$C_4 H_4S + * \rightarrow C_4H_4S * \text{ adsorbed thiophene}$$

$$4Mo^{n+} \rightarrow 4Mo^{(n+1)+} + 4e$$

$$C_4H_4S * + 2H^+ + 4e \rightarrow C_4H_6 + S^{2-}$$

$$2H_2 + 4H^+ + 4e$$

$$S^{2-} + 2H^+ \rightarrow * + H_2S$$

A more graphic portrayal is shown in Figure 18.16, in which the thiophene is depicted as adsorbing endwise. This model has been challenged by an alternative proposal in which hydrogenation of the sulfur in thiophene occurs first followed by the hydrogenolysis step, which releases the sulfur as H_2S.[24]

- The fascinating insights provided by newer surface science studies are unfortunately burdened with the uncertainties of the characterization of surfaces, which must necessarily be conducted in high vacuum far removed from industrial conditions.
- The Co promoter ions are located in the plane of the Mo active site.[23]
- Explanations of the promoting effect of cobalt or nickel include inducing surface changes that increase the number of active Mo sites,[25] increasing electron density at Mo sites by adjacent Co or Ni ions creating thereby higher activity of the Mo,[26] and contributing to the activity itself as a sulfide.[27] Each of these proposals requires excellent dispersion of the Mo and the Co or Ni promoters.

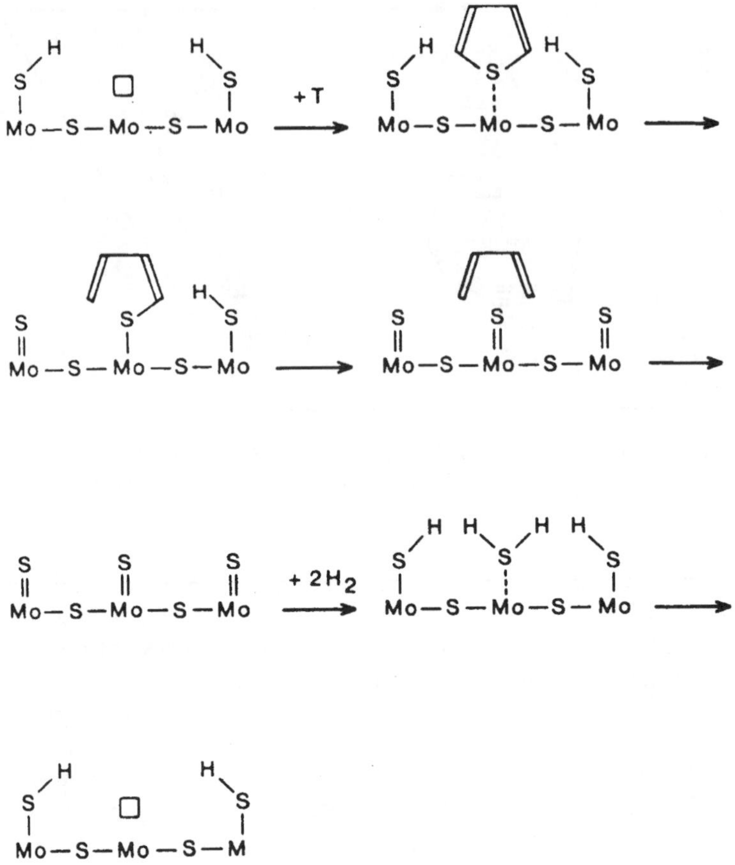

FIGURE 18.16 End-On Mechanism for the Hydrodesulfurization of Thiophene on the Edge Surface of MoS$_2$. Reprinted from Prins, R., DeBeer, V. H. J., and Somorjai, G. A., *Catal. Rev.—Sci. Eng.*, 31 (1 and 2), 1 (1989) by courtesy of Marcel Dekker, Inc.

Deactivation

It should not be surprising that a catalyst used under such severe conditions undergoes deactivation by all of the usual routes as a well as some unusual ones that have made the process a major technical challenge.

Coke Formation

Carbonaceous deposits are initially produced by strong adsorption of single-ring and multi-ring aromatics as well as polynuclear aromatics either formed by polymerization or present in the case of heavy feeds such as residuam.[1,38,34] This initial coke, which has been designated Type 1,[38] occurs rapidly in the reactor run, requiring a rapid rise in operating temperature to maintain the desired conversion. Since it is reversibly adsorbed, it reaches a steady state. Then follows a slower deactivation due to further dehydrogenation and condensation of aromatic ring clusters. Catalyst activity for HDS has been shown not to be affected by coking during this initial rapid coking period as long as the feed is asphaltene free (e.g., vacuum gas oil).[47]

In the case of heavy oils such as residua, thermal uncoupling of asphaltic clusters occurs with subsequent strong adsorption of the fragments on the catalyst surface. These deposits (Type II) also slowly dehydrogenate and condense to form large aromatic ring clusters. Most of the useful run time occurs during this period, but further dehydrogenation ultimately leaves a significant portion of large polymeric clusters of very low hydrogen content, which are converted in part to crystalline or graphitic coke (Type III), which ultimately destroys most of the remaining activity and necessitates shutdown and regeneration. Very rapid deactivation toward the end of a cycle is also facilitated by the high operating temperature, which has been increased to the point where thermal cracking dehydrogenates additional coke, and both thermal and catalytic demetallization, in the case of heavy oils, cause excessive pore plugging.

The rate of coke formation increases with molecular weight of the feed (heavier feedstocks are more aromatic), temperature increase, hydrogen partial-pressure decline (hydrogen serves to inhibit alkylation and cyclization of condensed-ring aromatic), increase in conversion level, and presence of cracked feedstocks (contain unsaturates that can rapidly contribute to condensed deposits). Coke levels prior to regeneration can vary from 3–4 wt% for straight-run naphthas to 25 wt% for residuum.[29]

Fixed-bed hydrotreating reactors exhibit a declining coke profile from inlet to outlet, and the catalyst pellets show a decline from edge to the center. In the case of residuum, the coke level declines from inlet to outlet early in the run but may produce a rising profile or an almost uniform profile late in the run, and a coke profile in the catalyst declining toward the center or a more uniform profile, depending on time-on-stream and/or catalyst pore structure.[38,42,43] Ebullating bed catalysts, because of their small size, produce uniform coke distribution.

Regeneration

Removing coke deposits by combustion, as with many catalysts, is a tedious procedure for which much previous experience is of great value. Catalyst suppliers are helpful in providing careful instructions.

The combustion process produces high heat effects due both to the combustion of coke and oxidation of sulfided catalyst components: 1.1×10^6 kcal/ton of fresh catalyst, of which 75% is due to the coke itself.[1] If proper control of temperature is not maintained, the catalyst load can be deactivated severely, or localized hot spots can develop that will destroy activity of portions of the catalyst. To avoid such difficulty, a carefully controlled burn must be accomplished by diluting air with steam or sometimes nitrogen. Typically, a mixture of about 0.5 vol% oxygen is introduced after the bed has been purged and dried with steam or nitrogen at 700°F (370°C) and a reactor pressure of 60–100 psig.[28,29*] The air-inert mixture is started and gradually increased in air content up to 1.5% O_2 while maintaining the flame front at 750°F (400°C). After the flame front has passed through the reactor, the O_2 content is increased to 2.0 vol% carefully so as to avoid temperatures above 850°F (450°C). After regeneration is completed, the reactor is purged with inert to prepare for opening or resulfiding prior to starting another cycle.

The various problems that arise due to faulty regeneration are summarized in Table 18.16. Since excessive temperatures cause the major problems in permanent deactivation during regeneration, good temperature sensing throughout the bed is essential as well as a target safe maximum temperature that will provide some security such as 850°F (450°C).

Regeneration can be done in place as just described or off-site as diagramed in Figure 18.17. Off-site treatment can also include removing Ni, V, and Fe deposits. The coke removal can be done in fluidized beds or ovens using a conveyor system so as to expose as thin a layer of catalyst as possible to the oxidizing atmosphere. Fluidization provides the best exposure of each catalyst pellet, but it must be done with expert care to avoid excessive attrition.

Catalyst Poisons

Reversible poisons that compete for active sites have already been summarized in Table 18.12 (see p. 326), where NH_3 and organic nitrogen compounds are the most troublesome in inhibiting

---

* The conditions are given as an example. Catalyst suppliers can provide more extensive details.

TABLE 18.16
Problems Resulting from Faulty Regeneration

Problem	Cause	Result
More than 1% carbon remaining on catalyst; catalyst visually still coated with black deposit	Inadequate regeneration Loss of flame front	Lower activity
Sintering of alumina support with reduction in surface area and pore size	Diffusion of Al_2O_3 from grain boundaries to the neck area between grains that constitutes pores; becomes rapid at temperatures above 1100–1200°F (594–650°C).	Reduces access to pores and therefore to the catalyst sites
Molybdenum sublimation	Above 1400°F (760°C) Mo sublimes rapidly and redeposits as large crystals at cooler spots.	Essentially destroys active Mo sites and produces a castastrophic loss in activity
Inactive complexes of Co and Ni with Al_2O_3	These complexes form significantly above 1000°F (538°C).	Significantly reduces the added activity provided by these promoter metals
Change from gamma Al_2O_3 to low-surface-area aluminas such as δ–alumina	Bed temperatures of 1500–1600°F (816–872°C) (a time-temperature phenomenon)	Massive loss in surface area and crush strength

Based on material from Refs. 1, 27, and 28.

important reactions involving the acid sites of the alumina support. Hydrogen sulfide, an almost universal bane for most catalysts, actually benefits some of the reactions and can be managed by selective operating conditions to eliminate its inhibiting effects. Permanent poisons, however, are a real problem, and most of those affecting hydrotreating do their damage largely by physical adsorption and resulting blockage of pores and catalyst surfaces rather than by irreversible adsorption on catalyst sites. The deactivation that results is permanent and cannot be recovered by regeneration. Table 18.17 summarizes these deactivating agents.

Bed Plugging

Shortened run times of hydrotreaters can occur because of the development of excessive pressure drop due to bed plugging. Because the catalyst particles are quite small and operating conditions are so severe with heavy stocks, the reactor pressure drop is very sensitive to relatively small amounts of solids buildup at any part of the bed. The primary causes are[28] inorganic particulate matter, such as iron sulfide scale from upstream units, organic particulate matter such as polymers (gums) formed in storage especially when in contact with oxygen, and catalyst fines generated by catalyst movement due to flow upsets or by fracturing because too rapid heating at start up when the initial charge might contain moisture.

Inorganic solids may be prevented from entering the reactor bed by line filters, guard beds, and metal-mesh trash baskets on top of the bed. If, for some reason, solids enter the bed in sufficient quantity to cause excessive ΔP and shutdown, the bed must be dumped, screened, and then reloaded. Catalyst fines generated in the bed can also be removed by this procedure, but great care must be exercised to avoid creating more fines. Alternatively, when most of the plugging is toward the top of the bed, the first several feet of the bed can be vacuumed off and replaced. Usually, most of this top layer consists of inert balls, but if catalyst is also removed, it must also be replaced to avoid reducing the catalyst loading in the bed.

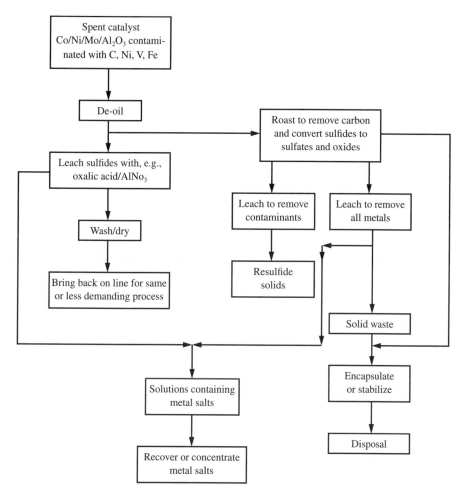

FIGURE 18.17 Steps in Processing Spent Catalyst. Redrawn from Trimm, D. L. in "Catalysis in Petroleum Refining," *Studies in Surface Science on Catalysis,* Vol. 53, p. 42, D. L. Trimm et al., editors, copyright with permission from Elsevier Science.

Organic matter can be removed during regeneration, but rapid buildup causes early shutdown for regeneration and is therefore costly. Oxygen-free storage of all feeds containing olefins is essential.

Since both pressure drop and mass transfer resistance are reduced with lower viscosity feeds, some very viscous feeds can be profitably handled by diluting with a lower viscosity stock.

Sediment Formation

As if bed plugging by coke and scale is not enough, sediment formation in hydrotreating residua can be a significant problem. The major culprit is the asphaltene fraction, which undergoes depolymerization and fragmentation to lower-molecular-weight compounds.[45] In general, this sequence of events is desired, but some of the lower-molecular-weight components having high C/H ratios do not break down further and become insoluble in the reaction medium. The insoluble carbonaceous material causes agglomeration of ebullating bed catalysts and high ΔP in fixed beds due to plugging.

Sediment formation is, of course, related to the character of the feedstock (asphaltene content, etc.). It becomes a problem at high conversions and is somewhat affected by the type of catalyst.

TABLE 18.17
Permanent Deactivating Agents

Contaminant	Effects	Typical Allowable Maxima	Source
Arsenic and lead organometallic compounds	Deposit on catalyst reduces pore volume and effective catalyst surface area. Breakthrough occurs usually before serious reduction in hydrotreating activity. Treated naphthas for reforming must contain less than 10 ppb of lead and 2 ppb of arsenic to prevent deactivation of reforming catalysts.	0.5–1.0 wt%	Present in crude oil as organometallic compounds that tend to concentrate in higher boiling fraction.
Carbon monoxide and carbon dioxide	CO and CO_2 in H_2 treat gas can inhibit HDS and HDN.	Treat gas: 2 vol% for HDS and 0.5 vol% for HDN	Synthesis gas units for H_2 production produce H_2 + CO. The CO must be removed by converting to CO_2 followed by absorption.
Nickel and vanadium organometallic compounds	Block pores and cover active surface in the form of metal sulfides produced in the hydrotreating process.	7–10% (higher for catalyst designed for metals removal)	Most concentrated in heavy stocks such as atmospheric and vacuum residue.
Silicon organometallic compounds	Silicones are cleaved at $Si-CH_3$ bonds to yield silica-gel deposits that cause pore blockage and coverage of active sites.	3–5 wt%	Silicone antifoaming agents used in delayed coking drum to prevent foaming.
Sodium	Causes pore blockage and promotes sintering of the support during regeneration.	0.2 wt%	NaCl constituent of the crude that may not be totally removed.
Sulfates	Formed during regeneration as aluminum sulfate causing lower crush strength of the catalyst support and reduced activity.	1–5 wt%	SO_2 and SO_3 produced during regeneration.

Based on information from Refs. 28 and 29.

Larger-pore catalyst are less active for asphaltene fragmentation but also produce less sediment. Catalyst suppliers can recommend a catalyst in their line that balances reasonable activity with low sediment production.

Process Units

All hydrotreating units consist of a furnace for heating the feed, a feed pump, reactor(s), exchangers and coolers (for preheating the feed and cooling the reactor output), separators, a hydrogen recycle compressor and make up compressor, and a distillation tower for separating product components. Various schemes for saving energy with emphasis on better utilization of reaction heat have been developed. Additional ancillary equipment is required for heavier feedstocks with high sulfur, nitrogen, and metals content. Table 18.18 presents some typical operating conditions for various feedstocks.

Naphtha and kerosene hydrotreaters operate totally in the vapor phase with hydrogen and vaporized feed flowing cocurrently through the fixed bed of catalyst. When distillate and heavier feeds contain substantial amounts of aromatics, intermediate cooling is required because of the

TABLE 18.18
Typical Operating Conditions and Performance Results

Conditions or Results	Naphtha, 85% SR/15% Coker	Middle Distillate, 60% SR/40% FCC LCO	VGO	Atmospheric Resid	Vacuum Resid Ebullating Bed
LHSV (hr^{-1})	4	2.5	1.0	0.25	1.0
Pressure (psig)	450	600	1000	2200	2200
Treat gas rate (SCF/bbl)	600	800	1500	3500	5000
H$_2$ purity (vol%)	75	80	85	85	85
H$_2$ consumption, as 100% H$_2$ (SCF/bbl)	75	175	300	700	1000
SOR reactor temperature (°F)					
Inlet	585	580	650	660	†
ΔT	30	50	40	70	†
Outlet	615	630	690	730	†
Sulfur removal (%)	99.96*	90	88	85	80
Nitrogen removal (%)	99.5*	50	40	40	35
Cycle life (bbl/1b)	220	150	70	10	4

*Calculated from product of 0.5 ppm each sulfur and nitrogen
†810°F WABT
Reproduced by permission: McCulloch, Donald C., *Applied Industrial Catalysis,* edited by Bruce E. Leach, Vol. 1, p. 88 (1983). Academic Press, Inc. Revised and updated by D.C. McCulloch.

high exothermic heat generated. For such cases, multiple beds in a single reactor shell are used with provision for introducing cold-hydrogen quench between beds. Radial reactors have been used for vapor-phase treating to take advantage of the lower pressure drop possible with that arrangement.

Heavy feedstocks, such as diesel oils, gas oils and residua, are liquids at operating pressures and temperatures. They are treated in down-flow reactors with hydrogen and the liquid phase proceeding cocurrently through the bed. The gas is the continuous phase, thus the term *trickle bed* with reference to the liquid flow. Great care must be exercised in ensuring good distribution throughout the bed to avoid bypassing and hot spots and ultimately high pressure drop and poor treat performance. Means for accomplishing good distribution while avoiding excess energy-consuming pressure drop include the following:

1. Effective distribution of gas and liquid
   - Require a minimum of 0.3 psi/ft pressure drop (70 mbar/meter) to ensure adequate distribution of flow.[1]
   - Provide a well designed distributor tray (see Appendix) that is checked for alignment at turn-around so as to prevent channeling.
   - Use inert balls larger than catalyst on top of the bed to minimize ΔP and also to be able to filter out solids that may have reached the reactor. In some cases, these inerts are placed in metal baskets so that they can easily be removed and replaced. Other inert shapes are also offered including cylinder rings and star-shaped pieces, both of which provide lower pressure drop and more surface. To eliminate highly exothermic aromatics hydrogenation later in the bed, inert packing at the bed inlet may be replaced by packing with a low loading of Ni-Mo sulfide.
   - Most trickle beds operate at a surficial mass velocity in the range of 1500–5000 lb/hr/ft^2. Any attempt to operate below 1000 lb/hr/ft^2 will cause maldistribution.[29]

2. Minimization of pressure drop
   • Use as low a superficial velocity as possible.
   • Reduce feed viscosity for very heavy oils by diluting with a less viscous feed.
   • Reduce pressure-drop increase during run time by careful control of temperature and maintaining adequate hydrogen pressure in order to reduce coke buildup in the bed voids.

Although middle distillates usually operate in the liquid phase, it is possible to have partial vaporization at the more severe conditions that are required for some stocks or that occur toward the end of the run. Vapor-phase systems (such as kerosene) that may undergo very mild treatment could operate partially in the liquid phase. Such operation can cause operating problems due to channeling of the small quantities of liquid produced that is not adequately distributed.

Residua often contain some solids, and adequate filter elements need to be installed to avoid plugging early in the run. Furthermore, a guard bed is often used for fixed-bed units to remove metals such as vanadium and nickel that deposit on the hydrotreating catalyst and rapidly cover the active sites, a problem that is a major issue in hydrotreating vacuum residue. The guard reactor is often designed for use of an inexpensive *throw-away* catalyst, such as Mo/Al_2O_3, with larger pores that provide easy access for asphaltenes to enter, break down (disaggregate), and give up their metal constituents on to the pore surfaces. These deposits are then sulfided by the H_2S present in the reaction system. Some desulfurization and denitrogenation occurs, but most of these reactions take place in the two or three main reactor(s) that follow, where metal deposits are minimal. Such an arrangement may require two guard beds where metal content exceeds 400 ppm[42] to permit switching guard beds when the first one becomes saturated with metals, thus continuing the run without interruption.

As in the case of kerosene hydrotreating, when saturation of high concentrations of aromatics in heavy stocks produce large exotherms, intermediate cooling is necessary between multiple beds in a single reactor or between multiple reactors. This cooling can be accomplished by using recycled gas, makeup H_2, and recycled liquid. Since the last bed in a trickle bed system will exhibit the lowest H_2 pressure, there is merit in flowing makeup hydrogen countercurrent to the trickle-bed flow so as to maximize hydrogen partial pressure in the region where the saturation reactions are approaching equilibrium.[23]

Alternatively, several processes have been developed to handle the problem of high metal content typical of vacuum residua. These processes use ebullated beds that allow removal of catalyst and addition of a corresponding amount of fresh catalyst during continuous operation.[31,32] Because of the complete mixing in the reactor, temperature is constant throughout the reactor, but lower conversions due to the lowered rate of reaction necessitates the use of several reactors in series.

When using standard hydrotreating catalysts with a low-acidity alumina support, it is possible at elevated temperatures (750–850°F) to produce substantial thermal hydrocracking, which hydrogenates multi-ring structures in the case of vacuum residuum and then fragments them by a free-radical mechanism. Because the reaction is not catalytic, it is not diminished by increasing hydrogen partial pressure.[3]

Advantages of the ebullated bed include isothermal operation and reduction of coke formation due to hot spots as in fixed beds, elimination of mass-transfer gradients between liquid and catalyst, elimination of plugging as in fixed beds, and removing and replacing a portion of the catalyst periodically without shutdown. Disadvantages include a need for sophisticated temperature control because of possible instabilities in such back-mixed systems, catalyst attrition and agglomeration, and need for special equipment to separate catalyst fines from the product.[39] By contrast, the more frequently used fixed-bed units afford better temperature control, a wider range of space velocities, minimal catalyst attrition and agglomeration, and lower capital costs. But disadvantages include mass transport limitations, the need for quenching to control temperature, bed plugging, and the necessity of shutting down to replace all or part of the catalyst.

When processing residuum in ebullating beds, a portion of the catalyst must be discarded periodically and replaced because of pore plugging by Ni and V sulfides. A less expensive catalyst must be used either in the early beds such as described for fixed-bed guard reactors, Mo/alumina catalyst, or a Ni-Mo catalyst with low loading can be used, both with large pores. These will be excellent for demetallization but only moderately good for desulfurization and hydrogenation. Such catalysts could be used in the first several reactors followed by a more effective desulfurization and hydrogenation catalyst. A proprietary catalyst has been developed that is superior for both requirements.[32]

A process has been developed that uses ebullated beds or so called *expanded-bed, back-mixed reactors* in series that simultaneously removes metals, sulfur, and nitrogen and operates at a high enough temperature (~810°F) to cause substantial thermal hydrocracking.[32] The various lighter distillate products are valuable for refinery feedstocks, while the much lower quantity of residue product can be sent to the coker.

Process Kinetics

Hundreds of studies on pure components contained in various hydrotreating feedstocks have been made that have resulted in useful understandings of the various reactions. A thorough review has been published.[46] Studies on actual commercial feedstocks are relatively rare in the open literature, and any useful attempt for model building must be simplified if successful implementation is to occur in any reasonable time period. What the process engineer needs is rate expressions for the disappearance of S, Ni, O, metals, and unsaturation, along with deactivation rate expressions especially for heavy-oil hydrotreating.

Since each feedstock consists of a number of the target compounds within the boiling range, it is logical to expect each compound to exhibit a different rate of reaction determined by its molecular structure and complexity. Three techniques for developing useful kinetic models have emerged to tackle this extreme complexity. They can be listed in an increasing order of the development work required and model reliability.

1. Base the rate equation on the most refractory compound in the feedstock. In the case of sulfur, dibenzothiophene is difficult to react and therefore, has the slowest rate of reaction.

    This approach makes possible the sizing of reactors for a given residual S, N, O, metals, or unsaturation. It does not, however, provide a means for following the reactor profiles, since more easily reacted components will be reacted earlier in the bed and set the temperature profile initially.

2. Lump all compounds of a given heteroatom (e.g., S) for a particular feedstock into one pseudocompound defined by an appropriate order that results in a reasonable fit of the observed data.

    This procedure is based on the fact that narrow cuts in various regions of the boiling range of a specific feedstock disappear at a rate proportional to the concentration of undesired component (e.g., S) remaining to the first power. Thus, for a number of such reactions occurring simultaneously at different rates (larger and more complex members of the group reacting more slowly), the apparent rate will be correlated by an order between 1 and 2. The most narrow boiling stocks will show orders near one, and the wider boiling will exhibit second-order behavior.

3. Divide the feedstock into many more lumps representative of both compound types and sizes.

    Each of these lumped sets of compounds will show first-order behavior in sulfur or other heteroatoms (N and O), and the many rates so described can be used in the differential mole balances and heat balances to follow the reaction system through the bed when adequate representation of catalyst deactivation is included. Simultaneous solution of such a mass of equations is, of course, possible. The major barrier to implementation is adequate definition of components of feedstocks and the gathering of reliable data for a good fit of the kinetic model. Major refiners have developed such programs and use them successfully. Progress with this technique has been based largely on the spectacular advances in analytic methods.[35,36] Initially, because of analytical limitations, only a few lumps could be selected and defined with any measure of confidence, although many useful models were developed and provided valuable concepts for future work[37] and enabled rather detailed description of molecular types in the feedstock. In particular, liquid chromatography coupled with field-ionization mass spectroscopy combined with gas chromatography and mass spectroscopy can be used to determine the carbon number distribution of various homologous series with identical structural groups. For purposes of the model, each set of structural isomers, defined as molecules having the same set of structural groups but arranged in different ways, are lumped, and one isomer is selected to represent the set. Since chemical and physical properties of such a set for isomers are similar, such lumping is reasonable.

In addition, the best known understanding of apparent reaction mechanisms for various structural types based on pure-component studies can be encoded as reaction rules, and rules for catalyst deactivation rates can be obtained from pilot-plant studies that are catalyst and feedstock specific.

Although all of this effort represents a major R&D expenditure, once developed, such a model can produce significant returns by aiding in setting optimum operating conditions for hydrotreating units of all types. Finally, practical physical and performance characteristics of products were developed from structural increments using a group contribution method based on structural increments. The resulting model can allow operations people to determine the best operating conditions for maximizing desired products and product qualities, even when feedstock characteristics vary during an operating cycle.

Examples of Simple Equations

Since hydrotreating reactions are inhibited, Langmuir–Hinshelwood equations are often conveniently used. As an example, the simplest rate equation for an industrial feedstock for any hydrotreating reaction is often written as follows:

$$r = k[W(1 - x)]^{\alpha} P_H$$

where W = weight fraction of component (e.g., sulfur) in feed
         x = conversion
         P_H = hydrogen partial pressure
         α = apparent order

Hydrogen sulfide inhibits HDS and some other reactions while it accelerates HDM. In the case where it inhibits, the following simple Langmuir–Hinshelwood form may be used.[3]

$$\frac{k}{k_o} = \frac{1}{1 + K_S P_S}$$

where k = rate constant with H_2S present
         k_o = rate constant with H_2S not present
         K_S = equilibrium adsorption constant
         P_S = partial pressure of H_2S in reactor

For the case where the most refractory compound is used, the order is one. For a lump of all components of a given type, each will exhibit an order somewhere between 1.5 and 1.7 with respect to the weight fraction of the heteroatom type being considered.[42]

In hydrodesulfurization of residuum, a complex array of organosulfur compounds must be considered as well as the inhibiting effects of H_2S and asphaltenes. The simplest approach involves an equation of the following form written for each major sulfur compound type, in the feedstock.[3]

$$-\frac{dS}{dt} = \left[\frac{P_H^n}{(1 + K_A W_A + K_S P_S)^m}\right] \sum_{i=1}^{j} k_i S_i$$

where k_i = rate constant for compound, i

K_A = chemisorption constant for asphaltenes

K_S = chemisorption constant H_2S

P_H = partial pressure of H_2

P_S = partial pressure of H_2S in reactor

S = weight fraction of sulfur in the liquid phase

S_i = weight fraction of sulfur associated with component, i

t = residence time

W_A = weight fraction of asphaltenes in liquid phase

If the hydrogen partial pressure does not change significantly, that term can be considered a constant and become a multiplier of the rate constant. In a given feedstock, hydrogen partial pressure will affect rate up to a certain pressure and then have no further effect. However, higher pressure may continue to have an inhibiting effect on coke formation.

Reactivity Comparisons

It is most useful to compare reactivities of various compounds involved in hydrotreating. Such comparisons are conveniently done by studying pure compounds at the same reaction conditions and reporting relative rates as pseudo-first-order constants.[33] These studies are valuable in providing a good idea of the most refractory compound types.

Modeling Catalyst Deactivation

The kinetics of deactivation is particularly essential for modeling heavy-oil hydrotreating, and many rate forms have been proposed that include the major deactivating phenomena of coking and metal deposition. Relatively reliable deactivation models are known to be in use commercially. They prove most valuable in predicting performance of a particular feedstock, including catalyst life and, in turn, the probable unit shutdown date.

Some insight in the crucial modeling parameters can be gained from the open literature. Two examples, both for residuum with one for deactivation due to demetallization and the other due to coking, are instructive.

Demetallization of Asphaltenes[48]

Molecular size distributions of vanadium compounds determined from size exclusion chromatography are used in a reactor model that includes reaction and diffusion in catalyst extrudates. The model is then used to calculate activity and product vanadium size distributions. Conservation equation in each zone is

$$C_o = C_I + \exp(-\eta k/SV/I) \tag{1}$$

where C_I and C_o = concentration of V or S in and out, respectively, in each zone

η = effectiveness factor

k = first-order overall rate constant based on total catalyst surface in a zone

SV = space velocity

I = total number of zones set for the model

The effectiveness factor in this situation of pore-size reduction accounts for both diffusion resistances and catalyst deactivation. It is obtained by determining the concentration profile in the catalyst from the equation for diffusion into a cylinder using first-order kinetics.

$$\frac{d}{rdr}\left(rD_e\frac{dC}{dr}\right) = k_s(r_p/r_{po})C \tag{2}$$

where k_s = intrinsic rate constant per unit surface area

D_e = effective diffusivity

r = catalyst particle radius

r_p/r_{po} = ratio of pore radius, after metals deposition to that before

C = metal concentration

The effective diffusivity, D_e, is determined as a function of initial void fraction, tortuosity, and the ratio of molecular diameter to pore diameter.

The rate of deposition of metal per unit length is given by

$$dm/dt = \alpha r_p k_s MC \tag{3}$$

where   a = number of metal sulfide molecules per molecule of organometallic reactants

M = molecular weight of metal sulfide

k_s = rate constant per unit surface area modified by deactivation

m = mass of metal deposited

$$k_s = k_{so}(1 - m/m_s) + k_{sc}(m/m_s) \tag{4}$$

where k_{so} = rate constant for fresh catalyst

k_{sc} = rate constant for a covered surface

m_s = value of m for complete active-site poisoning

From the mass of metals deposited, the volume remaining in the pores and effective diffusivity can be calculated throughout the particle, and thus the effectiveness factor, which is then available for use in Eq. 1, the conservation equation for a bed zone (or increment).

Clearly, the constants obtained from bench scale or pilot-plant will vary greatly with feedstock and catalyst type.

Proprietary Models

Much more sophisticated and more accurate models have been developed using the lumping technique backed by many years of careful study of the reactions and plant data. The reader is referred to an early article describing the lumping procedure that provides a foundation for molecular-based modeling.[35]

18.4   HYDROCRACKING

18.4.1   BACKGROUND

Catalytic hydrocracking was originally developed in Germany in the late 1920s for hydrogenating coal slurries at very high pressures to produce useful liquid fuels. It did not come into great favor in petroleum refining in the U.S.A. until the 1960s, when large quantities of hydrogen were being produced by catalytic-reforming units.

As the name implies, hydrocracking combines catalytic cracking and hydrogenation by means of a bifunctional catalyst to accomplish a number of favorable transformations of particular value for certain feedstocks (see "Chemistry"). Although it is a versatile process for converting a variety of feedstocks ranging from naphthas through heavy gas oils to useful products, its most unique characteristic involves the hydrogenation and breakup of polynuclear aromatics. Fluid catalytic cracking encourages dehydrogenation and condensation of these massive structures, which end up as coke. Of course, some of the large aromatic complexes, once partially hydrogenated in hydro-cracking, can proceed to dehydrogenate and form coke on the hydrocracking catalyst. But significant portions of lesser-size complexes do crack and are hydrogenated into useful product constituents.

Typical feedstocks include vacuum gas oil, fluid-cracked cycle oil, atmospheric gas oils, heavy cracked gas oil, and coker and thermally cracked light gas oils. Gas oils usually dominate the feed mix, which may include smaller amounts of the other stocks listed and even some residuum. Chemically, the typical feed consists of a mixture of alkylated polyaromatic/naphthenic and branched/straight-chain paraffinic/olefinic molecules.[2]

In many refineries, the hydrocracker serves as the major supplier of jet and diesel fuel components (middle distillates). By choosing a catalyst with high selectivity for middle distillates, a feed of vacuum gas oil, and favorable operating conditions, it is possible to maximize the production of diesel and jet fuel products.[1,2,3] Of course, other products are produced as well, such as gasoline, fuel gas, LPG, and FCC feed components. Any one of these instead can be maximized by a similar approach of feed and catalyst selection and optimization of operating conditions (temperature, hydrogen pressure, and feedstock cutpoints[4]). If one is free to select a particular feedstock, naphtha favors LPG; atmospheric gas oil, gasoline and diesel; and vacuum gas oil, diesel.[3]

Because of the high pressures required and hydrogen consumption, hydrocrackers are both more costly to build and to operate than fluid catalytic crackers. But they are ideal for middle distillate production using various gas-oil stocks and are also being used more extensively for cracking residuum to useful products and feedstocks for other units. Because of the high capital costs, most new developments are made possible by new catalysts that can be adapted to existing units.

18.4.2 MILD HYDROCRACKING

Mild hydrocracking refers to a process for treating vacuum gas oil in which 25–40% of the gas oil is converted to low-sulfur distillates and fluid catalytic-cracking feed.[2] Conventional hydrotreat-ing catalysts (Ni-Mo/Al$_2$O$_3$) are used at higher severity so that thermal hydrocracking will occur (see "Hydrotreating"). In fact, it has been possible and expeditious to convert existing hydrotreating units designed for low pressure (700–1000 psig hydrogen pressure) to mild hydrocrackers with minimal capital cost.

In recent years, some mild hydrocracking has been done using bifunctional hydrocracking catalyst in a stacked-bed arrangement. The first bed consists of a Ni-Mo/alumina hydrotreating catalyst for converting organo-nitrogen and organo-sulfur compounds to NH$_3$ and H$_2$S. The second bed in the stacked arrangement employs a bifunctional Ni-Mo/zeolite catalyst, which provides significantly higher activity for hydrocracking such that conversions to lighter components reach 65 to 75%. Because of the higher activity of the zeolite, it is possible to operate at lower temperatures even toward the end of the run so that equilibrium-limited hydrogenation reactions are favored. Ammonia is a reversible poison for acidic catalysts, but its effect can be reduced by raising operating temperature, which in the case of zeolite is a favorable option, but is not so practical with less-active amorphous silica-alumina.[9] Obviously, feedstocks with high organo-nitrogen content cannot be effectively processed without intermediate removal of NH$_3$ and H$_2$S between stages.

18.4.3 COMPLETE HYDROCRACKING

The standard hydrocracking unit is designed and operated to crack the entire feedstock to lighter more valuable products (100% conversion of original boiling range). In all cases, the hydrocracker must be preceded by a first-stage consisting of a hydrotreating reactor to accomplish HDS and HDN so as to protect the hydrocracking catalyst. In addition, the pretreatment serves to improve the hydrocracking reactivity of the feed by partially hydrogenating a portion of the aromatics content the feed.[2] This dual goal is reached using an active catalyst that can operate effectively below the equilibrium hydrogenation temperature (see also "Process Units").

Chemistry

As is the case in other catalytic processes in petroleum refining, many hundreds of reactions occur in hydrocracking, but the general types can be neatly summarized. Cracking (cission) and isomerization occur on the acid sites of the catalyst, and saturation (hydrogenation) occurs on the hydrogenation sites. Although cracking is endothermic, the exothermicity of the various hydrogenations is large, making the net heat of the reaction system highly exothermic.

The major reactions with reference to compound type are[4,12]

- Hydrogenation of aromatics and olefins
- Hydrogenation of polyaromatics
- Hydrodealkylation
- Hydrodecyclization of polyaromatic/naphthenic complexes
- Isomerization of paraffins and naphthenes
- Cracking of iso-paraffins, alkylnaphthenes, and alkylaromatics (hydrocracking)
- Coking (multi-ring aromatics, resins, and asphaltenes alkylating with olefins and cyclizing with loss of hydrogen to ultimately yield coke deposits on the catalyst)

For cracking of polynuclear aromatics to occur, partial hydrogenation must take place first (see Figure 18.18). Side chains are easily removed by cracking and subsequently hydrogenated. Since mono-aromatics are valuable, it is important to avoid excessive hydrogenation of these. If hydrogenated, they must be dehydrogenated later in a catalytic reformer. These aromatic hydrogenation reactions are equilibrium limited and, since exothermic, become unfavorable at elevated temperatures. The problem is demonstrated in Figure 18.19, showing equilibrium aromatics percentage for the hydrogenation of one ring of benzene and multi-ring compounds. The reaction temperature must be controlled by interstage cooling to attain the desired aromatics reduction of polynuclear

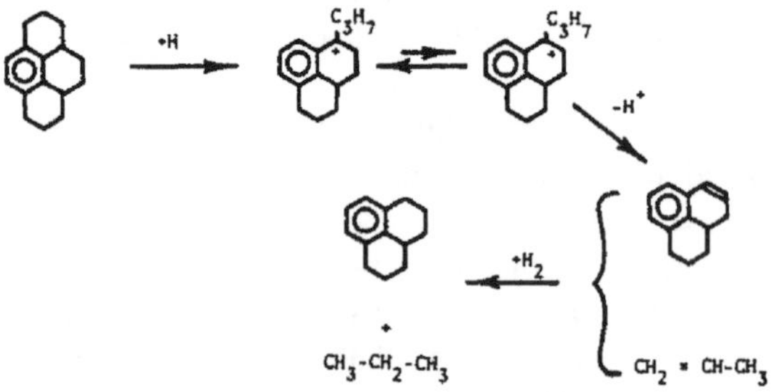

FIGURE 18.18 Hydrocracking of Partially Hydrogenated Pyrene (decahydropyrene). In this study, pyrene was hydrogenated under mild conditions so that hydrocracking could be observed separately. Reprinted with permission from Haynes, H. W., Parcher, J. F., and Helmer, N. E., *Ind. Eng. Chem. Process. Des. Dev.*, 22, 401 (1983). Copyright 1983, American Chemical Society.

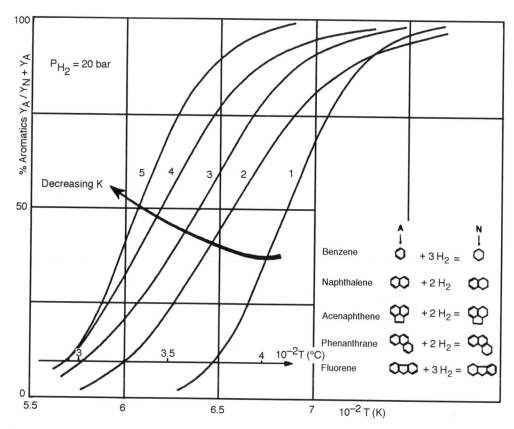

FIGURE 18.19   Partial Hydrogenation of Polycyclic Aromatics Compared with Benzene Hydrogenation. Reprinted by permission: LePage, J. F., et al., *Applied Heterogeneous Catalysis*, p. 368, Editions Technip, Paris, 1987. Originally from a thesis by B. Samanos.

aromatic rings. To avoid excessive hydrogenation of benzene and other single-ring aromatics, an excessively low temperature must be avoided so that the equilibrium will favor the single-ring aromatic. The delicate balancing required is accomplished by a first stage hydrotreater in which partial hydrogenation of the polynuclear aromatics occurs, making them vulnerable to hydrocracking in subsequent stages where conditions for minimizing hydrogenated single-ring aromatics are achieved along with the cracking reactions.

Catalyst Types And Suppliers[4–6,12]

A variety of catalysts are offered, based on variations in the relative strengths of acidity (cracking functions) and hydrogenating activity. In general, for example, strong cracking functions (strong acidity) combined with weak to moderate hydrogenation functions produce low-boiling products such as gasoline and lighter. Catalysts with weaker acidic functions but stronger hydrogenation functions, by contrast, produce middle distillates more effectively. These generalizations must be supplemented by other characteristics that the catalyst manufacturer can use to produce efficient catalyst for particular operations. See Table 18.19.

Hydrogenation and acidic functions, in the order of increasing activity for several hydrocracking catalysts under low-sulfur conditions, have been reported[12] as follows: hydrogenation (Ni-Mo, Ni-W, Pt-Pd), acidity (Al_2O_3, Al_2O_3/halogen, SiO_2/Al_2O_3, zeolites). Although Co-Mo can also be used, it is more costly. The non-noble metals are converted to sulfides and are not, therefore, sensitive

TABLE 18.19
Effects of Acidity and Hydrogenation Characteristics on Hydrocracking Products

Desired Reaction	Acidity	Catalyst Characteristics		
		Hydrogenation Activity	Surface Area	Porosity
Hydrocracking Conversion	Strong	Moderate	High	Low to moderate
A. Naphthas to LPG. Gas oils to gasoline				
B. Gas oils to jet and middle distillate	Moderate	Strong	High	Moderate to high
Gas oils to high V.I.				
lubricating oils				
Solvent deasphalted oils and residua to				
lighter products				
Hydroconversion of non-hydrocarbon	Weak	Strong	Moderate	High
constituents sulfur and nitrogen in gas oils				

Reprinted from Ward, J.W., in *Preparation of Catalysts III: Studies in Surface Science and Catalysis*, G. Poncelet and P. Grange, eds., 16, p. 587. Copyright 1983, with permission from Elsevier Science.

to organo-sulfur compounds present in modest amounts in the feed. Noble-metal catalysts, when used without the presence of sulfur compounds, remain in the metallic form and will hydrogenate all aromatics. If a sulfided form is used (usually maintained by H_2S in the feed), single aromatic rings are not hydrogenated, which can be an advantage for producing high-octane product or aromatics for petrochemicals.

Catalyst stability is greatly improved by incorporating even small amounts of Pt on zeolite because of its superior ability for hydrogenating coke precursors.[10] Both Pd and Pt zeolite hydrocracking catalysts require feeds with low concentrations of organo-nitrogen and organo-sulfur compounds. Both compound types affect the noble metals, and organo-nitrogen is particularly poisonous to the zeolite. The H_2S in the hydrotreated feed must be removed if complete hydrogenation of single-ring aromatics is desired. In the case of non-noble metals, some H_2S (<30 ppm) must be maintained to keep the metals in their active form. The noble-metal catalysts are finding increasing use for final reactors in a series when hydrogenation of aromatics is desired.

Acid functionality and support structure is most frequently supplied by a crystalline zeolite or a mixture of zeolite and an amorphous support such as silica-alumina or alumina.[12] Amorphous-silica-alumina is now mainly used in two-stage units in which the first stage involves conventional hydrotreating to remove organo-nitrogen as NH_3 and organo-sulfur S as H_2S. Separation between stages eliminates a major portion of these gases and protects the silica-alumina from poisoning by ammonia and organo-nitrogen compounds, since it is very sensitive to these poisons. In such an arrangement, the acid function provided by the silica-alumina produces a higher yield of middle distillate production than zeolite-based catalysts.

Most other hydrocracking processes now use zeolite-based catalysts. Such catalysts are more resistant to coke formation, less sensitive to NH_3 and H_2S, operate at a lower temperature because of higher activity, and exhibit longer run lengths between regenerations and longer life.

New and improved zeolites for use in hydrocracking catalysts continue to be developed. It was discovered early on that a high-area and stable catalyst (up to 1000°C) could be produced from sodium Y-zeolite by exchange (see "Catalyst Preparation"). The resulting low-sodium catalyst exhibited a reduction in the unit-cell parameter.[7] Later, a process was developed for producing low-sodium Y-zeolite by steaming.[8] The resulting zeolites have reduced unit-cell dimensions, the size of which can be controlled by the length of the steaming treatment. Superior properties compared to ammonium-Y and dry-air calcined zeolites include[4]

- Improved thermal and hydrothermal stabilities
- Higher activity (operate at lower temperature)
- Fewer acid sites (lower gas production)
- Larger pores (better access of larger molecules)
- Improved selectivity (increasing with lower unit-cell size)

Steam treatment removes a portion of the aluminum from the zeolite crystalline structure, creating thereby a higher-silica zeolite. The aluminum remains in the structure as alumina but is no longer part of the crystal lattice. Various chemical treatments have been developed for removing alumina from the structure and inserting silicon in place of the alumina, thereby obtaining a high-silica zeolite. A variety of procedures for producing such zeolites have been described that were initially developed for fluid catalytic cracking but have been applied to hydrocracking zeolites as well (see "Fluid Catalytic Cracking, Catalyst Types and Suppliers").

In addition to the already mentioned variables, catalyst manufacturers have the zeolite content of hydrocracking catalyst as another variable useful in optimizing catalyst performance for specific product states (see Table 18.20).

TABLE 18.20
Zeolite Content Related to Desired Primary Product

Desired Primary Product	Approximate Typical Support Composition*	Characteristics
Gasoline	80% zeolite, 20% binder	High activity which allow lower temperature operation and thus reduced cracking to light gases (C_1–C_4)
Middle distillate	10% zeolite, 70% alumina, and 20% binder[†]	Lower activity but higher selectivity for middle distillate
Mixed products or changing product goals	Intermediate amounts of zeolite	By changing severity can change primary product, but activity and selectivity may be less than for a product specific catalyst.

*Metal-free basis, weight percent
[†]Partial substitution of alumina with silica-alumina increases selectivity to middle distillate, and as silica-alumina content is further increased activity increases; but selectivity does not, beyond the initial increase.

Catalyst Suppliers and Licensors

Suppliers

Acreon, Akzo Nobel, Catalyst & Chemicals, Chevron Research & Tech., Criterion, Grace Davison, Haldor Topsoe, Kataleuna, Orient, Procatalyse, UOP LLC, Zeolyst (marketed by Criterion)

Licensors

ABB Lummus Global, Chevron Research & Tech., IFP, Kellogg Brown & Root, Shell Global Solutions, Veba Oel Tech., UOP LLC

Catalyst Size and Shape

1/16- and 1/8-in. extrudates as cylinders or shaped cylinders (multilobe) and spheres

Pretreatment

Pretreatment of catalyst on line involves activation of the metallic components. The non-noble metals are activated by presulfiding similar to procedures used for hydrotreating catalysts. Dilute

H$_2$S in H$_2$ is used or a mercaptan in kerosene. Temperature is raised to 700°F (371°C) after the major sulfiding exotherm subsides, thereby ensuring the desired complete sulfiding.

Noble metals must be reduced in hydrogen as the temperature is raised to 700°F. This careful procedure avoids aggolmeration of the Pt or Pd by removing water initially adsorbed on the zeolite which, when present, encourages aggolmeration. Both platinum and palladium catalysts are highly sensitive to the activation procedure and time and temperature play a crucial role.

Catalyst Preparation[4,5,11,12]

Zeolite-based hydrocracking catalysts have become the preferred catalysts for hydrocracking. Amorphous silica-alumina, although used primarily for middle distillate production, is also used in combination with gamma alumina and zeolite to create catalysts for special purposes. Metals are added in various ways in the final production of the catalyst.

Non-noble Metals[12]

The most frequently used metal combinations are Ni-Mo and Ni-W with weight percentages of 2–8% for Ni and 12–30% for the Mo and W components. If Co is used the same quantity as that for Ni is required. Various methods for effective deposition of the desired metal components are used as summarized in Table 18.21 for non-noble metals. If the acidic support is silica-alumina, the ion exchange option is not possible.

TABLE 18.21
Techniques for Deposition of Non-noble Metals

Method	Description	Special Characteristics
Impregnation	A mixture of zeolite, alumina, and binder (e.g., 20% zeolite, 60% alumina, and 20% binder) is pelletized or extruded and then impregnated by pore saturation, either sequentially with each metal component or simultaneously using a solution of metal compounds.* The catalyst is then dried and calcined at 900–1000°F.	Uneven metal distribution is possible.
Comulling	Components of zeolite, alumina, binder, and metallic compounds are mulled together with sufficient water to make a paste which is then extruded and cut into.	If done with great care, even distribution of metals is possible.
Exchange and impregnation	One component is added to the zeolite by ion exchange during the production of the desired zeolite form. At pellet or extrudate stage, the other component can be added by impregnation.	Special characteristics are made possible by dispersion of the exchanged component suitable for low loadings.

*Compounds used include oxides, nitrates, and carbonates of nickel; ammonium heptamolybdate; molybdenum oxide; and ammonium dimolybdate. Solubilizing agents may also be used.

Information source: Refs. 4 and 12.

Noble Metals[12]

Noble metals can be incorporated in the support by the methods given for non-noble metals, but ion exchange is a preferred method. The exchange capacity of zeolite (~2%) is more than needed for the highly active Pt or Pd components. Tetramine salts such as Pd(NH$_3$)$_4$ (NO$_3$)$_2$ are produced in the process from the metal chloride and concentrated ammonium hydroxide. The catalyst is then pelletized or extruded with an alumina binder and dried and calcined at >800°F (423°C) in air.

Zeolite Preparation

Refer to "Fluid Catalytic Cracking" for more detail on zeolites. Although a wide variety of zeolite preparations are used in fluid catalytic cracking, two types are common in hydrocracking. Because fixed-bed reactors are used, extended run times are essential. Therefore, zeolites for hydrocracking must have high activity and stability.

The most widely used zeolite for hydrocracking begins with a sodium Y-zeolite that is exchanged with ammonium ions, displacing thereby a goodly portion of the sodium. This treatment is followed by heating in live steam at 350–600°C for several hours followed by further ammonium-ion exchange to yield a low (>0.2%) sodium zeolite with good stability and with a reduced unit cell. Aluminum atoms are removed from the crystalline structure producing a higher SiO_2/Al_2O_3 ratio, which has been found to have higher thermal and hydrothermal stability.[12]

Deactivation

Hydrocracking catalysts, because of the dual functionality, are subject to deactivation by coking and by a decline in either the hydrogenation function or the acidic function, or both. Coke formation is a major but partially controllable deactivating factor. Unlike fluid catalytic-cracking catalyst, hydrocracking catalyst has a hydrogenation function that can prevent excessive coke production by hydrogenating coke precursors. It is therefore possible to continue operation from 1 to 5 years between regeneration, depending on the feedstock.[12] After initial rapid laydown, coke formation occurs gradually and deactivation caused thereby is overcome by gradually raising operating temperature. Ultimately, the operating temperature limits of the reactor and associated equipment cause shutdown for regeneration or replacement of the catalyst. Regeneration may be done in place or off site (see also "Hydrotreating").

Hydrogen partial pressure plays an important role in providing sufficient hydrogen chemisorption on the catalyst and dissolved hydrogen in the liquid that can act in hydrogen transfer. Also, it is advantageous to have strong hydrogenation activity which is highest for Pt and Pd followed by Ni-W and Ni-Mo catalysts.

Regeneration

Regeneration of hydrocracking catalysts by careful burning of the coke as described for hydrotreating usually restores a substantial portion of the original activity of non-noble metal catalysts such as Ni-Mo and Ni-W. Platinum and palladium, however, are usually deposited on zeolites by ion exchange and are finely distributed on the zeolite surfaces. The driving force for agglomeration is thereby enhanced especially during long exposure to ammonia, water vapor, and hydrogen sulfide at high temperature, which can be exacerbated during regeneration.

One method for redispersion of the noble metal reported involves treatment with aqueous ammonia solution. It is hypothesized that the metal dissolves in the ammonia and reacts to the Pt or Pd tetramine complex and is then distributed by ion exchange in the same manner as in preparing the original catalyst.[12] Removal of excess ammonia followed by calcination restores the metal to close to its original activity.

Catalyst Poisons

Organo-nitrogen compounds are strongly adsorbed poisons for the acidic sites of both zeolite and amorphous silica alumina. Zeolites of high acidity are more sensitive to organo-nitrogen compounds than amorphous silica-alumina but, in both cases, these compounds must be decomposed prior to entering the hydrocracking beds. Ammonia produced by this decomposition in the hydrotreater can be passed on in the feed to the hydrocracker if zeolite-based catalysts are used, because their higher acidity makes them more tolerant to ammonia, which is also a potential poison. Ammonia in sufficient quantity, or more accurately, at higher partial pressure, can affect activity. An expression written in terms of required operating temperature (ROT) to attain a particular conversion has been suggested.[13]

$$ROT = C + K \ln(P_A)$$

where   C and K = empirical constants related to catalyst type and feedstock
        P_A = the partial pressure of ammonia

Thus, at higher partial pressures of ammonia caused by higher ammonia content or an increase in operating pressure, activity declines. Also, light naphtha produced in conventional hydrocracking has an octane number five points higher when produced under low rather than under high ammonia partial pressures.[13] This advantage is thought to be due to a higher isomerization ratio and higher benzene production due to the greater catalyst acidity for the low ammonia operation. There are many other observations possible from various combinations of operating conditions and catalyst recipes that produce different ratios of hydrogenation-to-acid activities. Clearly, the practice of passing the hydrotreated feed to the hydrocracking section without intermediate removal of ammonia is workable in most cases. With highly active zeolites, it is possible to raise reaction temperature to overcome activity losses due to ammonia. Obviously, for very high loadings of ammonia, even this technique may prove impractical due to equipment temperature limitations.

Hydrogen sulfide in modest amounts is needed to retain proper sulfiding of the non-noble metals. However, when sulfided, the noble metals, although still active, can no longer hydrogenate single-ring aromatics, which may be desired for gasoline or petrochemical-aromatics production. If total hydrogenation is sought, as well as higher hydrogenation activity, the treat gas products must be removed prior to the noble-metal bed.

Process Units

Because of the substantial heat released by hydrocracking and the equilibrium limitations of various essential hydrogenation steps at high temperatures, it is necessary to design multi-bed reactors with interstage cooling. In most cases, recycle gas is used for this purpose. Reactors are operated as trickle beds with cocurrent flow of H_2 and liquid feed.

The various process configurations and applications are summarized in Table 18.22.[14] Process and catalyst innovations have made possible the production of a wide variety of products via hydrocracking of all types of gas oils, cycle oils from fluid catalytic-cracking, and naphthas.[14] The resulting products are used in motor, diesel, and aviation fuels, lube oils, and petrochemical feedstocks, to name a few. Unfortunately, the term *stage* used in the table is defined in different ways by different companies. It would be better just to describe each type in terms of the number of reactors and separation units.

Operating conditions vary for different feeds and reactor designs. Typical ranges are 1500–2500 psig, 650–740°F (343–393) for hydrotreating reactor and 600–700°F (316–371°C) for hydrocracker with contact times of 0.5–1.5 hr for each reactor.[15]

Process Kinetics

As is the case with other complex petroleum refining systems, lumped pseudocomponents constitute the only rational approach to successful modeling. Hydrocracking is such a versatile process involving many feed alternates, operating variables, and catalyst choices that some detailed prediction of reactor performance is needed to determine optimum operating conditions and to monitor catalyst activity. Operating companies have developed models based on rational lumping, but these are proprietary. One such model, however, has been described and is offered on the open market.[16]

TABLE 18.22
Hydrocracking Process Units

Type	Description	Comments
Single reactor	Single multi-bed reactor followed by gas-liquid separation with substantial removal of H_2S and NH_3 from recycle gas by the water wash and fractionation, with provision for recycle of fractionator bottoms.	Common in mild hydrocracking units for middle distillate production from vacuum gas oil and some atmospheric gas oil. Fractionator bottoms not recycled for mild hydrocracking.
Single catalyst	Conventional hydrotreating catalyst used at higher severity to produce thermal hydrocracking as well as hydrotreating reactions.	
Two catalysts	Conventional hydrotreating catalysts (Ni-Mo/Al_2O_3) in initial bed(s) followed by Ni-Mo/zeolite for significant catalytic hydrocracking.	NH_3 and H_2S created in initial bed passes on to zeolite hydrocracking catalyst which can tolerate NH_3, if not excessive.
Two-reactors* series flow (integral single stage)§	Hydrotreater reactor feeds directly into hydrocracker with zeolite-based catalyst followed by gas-liquid separation and substantial removal of H_2S and NH_3 from recycle gas by the water wash. Fractionation produces a bottoms liquid product for recycle to extinction. A small fractional purge of this liquid prevents excessive build-up of polycyclic aromatics (coke precursors).	Lower capital and operating cost. Not good for feeds with very high nitrogen and sulfur content. Favored for vacuum gas oils and heavy cracked gas oils for maximum jet fuel and diesel production.†
Semi two-stage	Output from hydrotreater is fed to an interstage separation unit where NH_3 and other light compounds are flashed from the liquids. The liquid is then fed directly to the hydrocracking reactor which is followed by a fractionator.	For high S and N feedstocks.
Separate hydrotreater/ hydrocracker stages	Flashed liquid from hydrotreater is fed to a fractionator where lighter products are removed and bottom product sent to the hydrocracker.	Common arrangement for gasoline production by hydrocracking of atmospheric gas oils and cycle oils. Coker and thermal cracked gas oils also added to feed. Non-noble metal catalysts used.
Three reactors (two-stage)§	Instead of fractionating after the hydrotreater, fractionation occurs after a first hydrocracker. Then bottoms from fractionators goes to second hydrocracker.	This arrangement permits complete control of temporary poisons such as H_2S and NH_3 to the last reactor noble metal catalysts so that strong hydrogenating capacity is available. Content of H_2S totally eliminated so that all aromatics will be hydrogenated. Used to maximize high octane gasoline production from atmospheric gas oil and coker gas oil.

*Recycle of fractionator bottoms permits complete conversion. When recycle is not used, the process is called once-through hydrocracking; and, with modern catalysts, conversions can be as high as 80 percent.

†Also used to produce naphtha cracker feedstock from various gas oils by reducing polyaromatic content; and, by once-through operation to produce catalytic cracker feed by removing organo-sulfur and nitrogen and hydrogenating naphthenes and aromatics.

§Stage in this context refers to each hydrocracking reactor.

Information Sources: 2 and 14.

Since the hydrocracking unit invariably includes a hydrotreater as a pretreater, the model includes its reaction system as well, along with separators, makeup gas system, purge systems, and recycle/quench-gas system.[15] The feed is divided into carbon-number pseudocomponents (lumps), each of which is further classified into aromatic carbon, naphthenic carbon, and paraffinic carbon fractions. Sulfur, nitrogen, and olefinic groups are assigned to each pseudocomponent. Kinetics for the reaction rate of each reaction is set as first-order for the hydrocarbon and selected orders for hydrogen. The resulting differential continuity equations along with differential heat balances are solved simultaneously for the plug-flow cocurrent reactor. Seven reactions are considered: desulfurization, denitrogenation, olefin saturation, naphthenic ring opening, aromatics separation, demetallization asphaltene reduction, and cracking. An activity factor is included as a function of time and coking characteristics of the feed/catalyst combination. Results with the model on commercial units have been most encouraging, although consistency checks with equilibrium constraints is computationally complex.[17] No doubt this model, like other such models, is undergoing continued refinement and becoming even more useful. The reader is referred to Ref. 18 for background on more rigorous models.

18.5   ISOMERIZATION

Background

Isomerization of butane and of pentane and hexane mixtures constitute another group of processes designed to increase the octane of components in the gasoline boiling range. Butane is isomerized separately to isobutane for use in the alkylation unit in which it reacts with olefins such as butenes and propene from catalytic cracking units. The products include isooctane and other high-octane products that became more essential blending components as lead additives were phased out of the gasoline pool. As new oxygenate additives such as MTBE (methyl tertiary-butyl ether) were tested and used, isomerization of n-butane again became more important. The isobutane thus produced was dehydrogenated to the iso-olefin and then reacted with ether to produce MTBE (see MTBE). There are also processes for isomerizing C_4 and C_5 olefins directly. Hydrocracking and catalytic cracking units also produce isobutane, and many refineries so equipped do not require an isomerization unit for n-butane.

In the case of pentane/hexane mixtures, often called *light straight-run naphtha (LSR)*, a separate isomerization operation is carried out to produce primarily isopentane and the hexane isomers (2 and 3 methylpentanes and 2,2 and 2,3 dimethylpentanes). By recycling unreacted pentane and hexane, octane numbers of the final product reach values of 87–89 RON.[1] Thus, a valuable addition to the gasoline pool is created that not only contributes to the octane rating but also supplies the necessary volatility needed to meet vapor-pressure specifications for gasoline.

Isomerization of the light straight-run fraction of the gasoline pool has increased in importance as tetra-ethyl lead was removed from gasoline and later as restrictions on aromatics, particularly benzene (RON = 100), were imposed. Since the C_5/C_6 paraffinic fraction of gasoline is helped the most by octane additives, it became an important fraction for upgrading.[3] As benzene was removed, the C_5/C_6 fraction became an even more important candidate for upgrading in octane number.

The first butane commercial isomerization unit was placed on stream in late 1941 and, by the end of World War II, there were 38 plants operating in the U.S.A., all based on Friedel–Crafts chemistry but of varying proprietary designs.[2] Each of five successful processes employed aluminum chloride as the catalyst promoted by hydrochloric acid. These plants were crucial elements in supplying isobutane for alkylation to isooctane and other high-octane components to be blended into aviation gasoline essential for the war effort. Although the process did the job, the units were difficult to operate and maintain.[2] Corrosion and catalyst bed plugging were recurring problems that caused high operating and maintenance costs.

With the advent of dual-function catalytic reforming catalysts and the success of the reforming process, researchers realized the effective isomerization activity of these catalysts and began to study how a catalyst of this dual-function behavior could be adapted to isomerization of C_4s and C_5/C_6 mixtures. The catalysts offered today for this purpose are the latest versions of this now established line of light hydrocarbon isomerization catalysts. They have been designed to have both isomerization activity and hydrogenation activity that serves to prevent undesired side reactions and is also essential in the main reaction. Two major types are in general use, noble metal (e.g., Pt) on chlorided alumina and noble metal on zeolite, both of which provide an acid function that simultaneously serves as a support. The noble metal provides the hydrogenation function essential for preventing excessive coke buildup. The reaction is carried out in the vapor phase. In the temperature range of 250–500°F (120–260°C) and at pressures of 250–400 psig (18–28 atm) depending on the catalyst.[1] The hydrogen-to-hydrocarbon mole ratio is maintained between 1:1 and 4:1, depending on catalyst type and operating conditions.[4] Very little makeup hydrogen is required. Because of the presence and action of hydrogen, the dual-function processes are often referred to as *hydroisomerization*.

Chemistry

Isomerization is mildly exothermic, and equilibrium is favored by lower temperatures. In isomerizing pentane, the equilibrium isopentane content is 64 mole percent at 500°F (260°C) and 82 mole percent at 248°F (120°C). The hexane equilibria as shown in Figure 18.20 is more complex. Equilibrium values of several of the isomers are not much affected by temperature, but 2,2 dimethylbutane is produced preferentially from the remaining hexane as the temperature is lowered, thus replacing the low-octane hexane 24.8 by the isomer with the second highest octane, 91.8.[2]

Clearly, the ideal catalyst would provide sufficient activity to operate at low, equilibrium-favorable temperatures. That goal has been the thrust of catalyst development since the first introduction of dual-function hydroisomerization catalysts. Initially, however, these catalysts operated in the range of 500–850°F, 260–454°C.[3] Equilibrium was not reached, and significant quantities of normal pentanes and hexanes were separated from the product and recycled.[4]

The next generation of catalysts, both Pt-on-chlorinated alumina and Pt-on-zeolite, have higher activities and operate in a lower temperature range where isomerization equilibria are favorable.[4,5,7,9] The following reaction scheme has been suggested. Butane is used as an example to avoid the

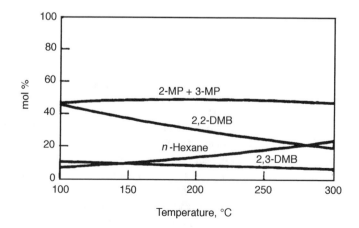

FIGURE 18.20 C_6 Isomer Equilibria. Reprinted by permission: N. A. Cusher in *Handbook of Petroleum Refining Processes,* 2nd edition, R. A. Meyers, ed., p. 9.22, copyright McGraw-Hill Companies, Inc., New York, 1996.

complexity of portraying a number of isomers.[7] Skeletal isomerization of the carbonium ion suggested to occur via a cycloalkyl intermediate.

$$CH_3{-}CH_2{-}CH_2{-}CH_3 \xleftrightarrow{Pt} CH_3{-}CH_2{-}CH{=}CH_2 + H_2$$

$$CH_3{-}CH_2{-}CH{=}CH_2 + [H^+][A^-] \longrightarrow CH_3{-}CH_2{-}\overset{+}{C}H{=}CH_3 + A^-$$

$$CH_3{-}CH_2{-}\overset{+}{C}H{-}CH_3 \longrightarrow \underset{H^+}{\overset{C\ \ CH_3}{\bigtriangleup}} C{-}C \longrightarrow CH_3{-}\underset{+}{\overset{CH_3}{\underset{|}{C}}}{-}CH_3$$

$$CH_3{-}\underset{+}{\overset{CH_3}{\underset{|}{C}}}{-}CH_3 + A^- \longrightarrow CH_3{=}\underset{+}{\overset{CH_3}{\underset{|}{C}}}{-}CH_2 + [H^+][A^-]$$

$$CH_3{-}\underset{+}{\overset{CH_3}{\underset{|}{C}}}{=}CH_2 + H_2 \longrightarrow CH_3{-}\overset{CH_3}{\underset{|}{C}}H{-}CH_3$$

Alternatively, it has been suggested that, at the low temperatures of operation, it is unlikely that dehydrogenation will occur on platinum at a sufficient rate. Instead, it has been postulated that hydrogen is dissociated on the noble metal and spills over on to the zeolite.[6,8]

Catalyst Suppliers and Licensors

Suppliers
Acreon, Akzo Nobel, Engelhard, Procatalyse, United Catalysts, Zeolyst

Licensors
ABB Lummus, Global, CD Tech/Lyondell, Engelhard, IFP/HRF, Kellogg Brown & Root, Phillips Petroleum, UOP LLC

Catalyst Details
Pt or Chlorided Pt-on-alumina, Pt-on-zeolite (0.3 wt% Pt typical)

Form
Extrudates, spheres, cylinders

Catalyst Characteristics

The alumina-based catalysts are descendents of catalytic reforming catalysts upon which extensive investigations have been pursued. The material covered in the section on "Catalytic Reforming" will be helpful in understanding the nature and roles of both the acid sites and the Pt metal sites. The acid activity is extremely crucial in a viable isomerization process, which requires substantial activity at low temperatures where the equilibrium is favorable. This goal is achieved by higher levels of chloride on the alumina and continuous addition of organic chloride (CCl_4, $CHCl_3$, CH_2Cl_2, etc.). Since HCl is produced in excess to the chloride that attaches to the catalyst, the system must be dry to prevent corrosion.

The catalysts using zeolite as the source or reservoir of acid sites employ a low-sodium zeolite, perhaps HY or HZSM-5 with high acidity. The zeolite will catalyze the reaction without Pt present on the surfaces, but that activity declines rapidly.[6] Apparently, Pt stabilizes the activity by splitting H_2 into hydride and proton ions that provide a continuous supply of these ions on the zeolite surface. Pore size and pore-size distribution must be such that reacting hydrocarbons can gain access to the active sites with ease.

A base-metal catalyst on alumina has also been introduced with an operating range between chlorided Pt-on-alumina and Pt-on-zeolite. The potential user should consider each catalyst available, since each has significant advantages that apply to different process situations.

Although zeolite-based catalysts have been steadily improved in isomerization activity, there remains a need to operate at temperatures higher than the chlorided alumina-based catalyst (400–500°F, 200–260°C compared with 250–300°F, 120–150°C). Equilibrium is closely approached in both instances, but that at the lower temperature is more favorable for maximum production of isomer and octane improvement. Butane isomerization requires lower temperatures than C_5/C_6 isomerization, and chlorided Pt-alumina catalysts have been favored for the production of isobutane by isomerization of normal butane.

Makeup hydrogen is governed mainly by the presence of benzene in the C_5/C_6 feed. Benzene is rapidly hydrogenated, consuming six moles of hydrogen for each benzene mole. Hydrocracking of C_7 paraffins that may be present in some feeds also uses hydrogen. Because of the higher operating conditions for the zeolite catalysts, hydrocracking is more apt to occur with these catalysts when heavy impurities are present.

Catalyst Deactivation

The chlorided Pt-alumina catalysts are deactivated by organic sulfur and nitrogen compounds. Water is a particular problem, since it strips chloride from the catalyst and causes loss of acid activity. Water must be removed from the feed and the hydrogen makeup by efficient molecular-sieve driers. In addition, the HCl present in the reaction system from the organic chloride additions becomes highly corrosive in the presence of liquid water that can accumulate in lower temperature regions of the unit. Sulfur compounds act as a temporary poison but, if present in the feed above one ppm, can seriously affect the platinum activity. Nitrogen compounds neutralize active sites, and ammonia reacts with HCl to form NH_4Cl, which can plug exchangers. It is imperative that hydrotreated feed be used from which sulfur and nitrogen compounds have been removed.

The Pt-zeolite catalysts are more rugged and tolerant of water and sulfur compounds. Sulfur in amounts up to 150 ppm do not affect activity and, since organic chloride promoter is not used, water in amounts typical of light hydrocarbon streams and hydrogen refinery streams is not a problem. Hence, driers on feed and hydrogen makeup streams are not required.

Catalyst Regeneration

Because of the low-temperature operation associated with Pt-alumina catalyst, coking is negligible, and regeneration is not necessary. Catalyst life of up to 10 years has been reported if water and other contaminants such as sulfur are excluded from the system. Since units using Pt-zeolite operate at higher temperatures, coke laydown is possible, particularly during some reactor upset. In such situations when coke accumulates and regeneration becomes necessary, the process is straightforward. By contrast, regeneration of the chlorided Pt-alumina system is impossible, because the acid sites are destroyed by direct regeneration. One commercial catalyst has been described[9] that has up to 14% chlorine bound to the surface and does not require continuous addition of organic chloride (CCl_4). It has been reported as regenerable by a regeneration-rechlorination method perhaps similar to that used for catalytic reforming catalysts.

Process Units

For the chlorinated Pt-alumina catalyst, a feedstock C_3/C_6 that is rich in n-butane is used.[5,7] The feed is dried in a dryer containing molecular sieve, if necessary. Hydrogen makeup, after passing through a similar dryer, joins the feed into which is injected an organic chloride in small amounts. The combined flow then proceeds through two adiabatic reactors in series with intermediate cooling. This arrangement allows higher-temperature operation in the first reactor and thus higher reaction rates. The second reactor, as is common for exothermic adiabatic reactor systems, is then operated at a lower inlet temperature so that the remaining conversion can occur at the most favorable equilibrium condition. After cooling and flash separation of the product, the liquid product is stabilized. Unreacted n-butane is recycled via the deisobutanizer.[7] The two-reactor system permits removing one reactor per catalyst replacement without shutting down. Reactor operating conditions are reported to be in the range of 120–150°C (248–302°F) and 15–30 bar.

The zeolite-based system is not burdened with water removal. Because of the higher-temperature operation, however, sufficient n-paraffins remain in the product that recycle of these paraffins is indicated. Product issuing from the isomerization reactor is flash-separated with the liquid passing through a vaporizing exchanger to the adsorption bed and the flash overhead passing through a compressor and heater, followed by the desorption bed. These beds are cycled between adsorption and desorption duties. The molecular sieve is designed with a pore size that accepts the n-paraffins but not the isoparaffins. The product isomers pass out to a stabilizer where light components are removed, such as some hydrogen that was dissolved in the flash liquid and small amounts of light hydrocarbons (C_3 and C_4) from minor cracking reactions and from the original makeup hydrogen.[7] The desorption gases consist of hydrogen and n-paraffins, which are recycled to the reactor.

These units also have two reactors in series. Reactor operating conditions are reported to be in the range of 200–260°C (392–500°F) and 26 bar. Since some hydrocracking can occur, and coking can develop over an extended period, regeneration after some years may be necessary if upsets occur. This possibility creates no serious problem, since chlorine is not present, and careful and successful regeneration is possible when indicated.

The Pt-zeolite process is the favored process for upgrading the octane number of light straight-run naphtha (C_5-C_6). Referring to Figure 18.20 for equilibrium compositions of C_6 isomers, low temperature favors 2,2 dimethylbutane and has little affect on 2- or 3-methylpentane or 2,3 dimethylbutane. Since 2,2 dimethylpentane has research octane number (RON) of 93, it is separated, and the other components are recycled to form additional 2,2 DMP and eliminate more lower-octane components. The C_5 components are also improved by recycle. The resulting RON can be as high as 92 for the product mixture.[7]

New catalysts continue to be developed, and users should stay in contact with licensors and catalyst suppliers for information on improved catalysts.

18.6   OLIGOMERIZATION (POLYMER GASOLINE PRODUCTION)

Background

The term *polymerization* has been used in refinery parlance for many years when referring to processes producing lower-molecular-weight polymers from C_3 and C_4 olefins as high-octane components for gasoline. Such relatively short-chain polymers are properly called *oligomers* and the process termed *oligomerization,* which terms are now becoming more frequently used.

Actually, this process is one of the first catalytic processes used in refineries. It has been making a modest comeback to provide additional high-octane blending stock to offset both the lead and benzene reductions mandated for motor gasoline. Feed consists of propene and/or butenes plus some propane and butane, which serve as diluents.

Chemistry

As would be expected, these oligomerization reactions are the reverse of some cracking reactions. It is not surprising, therefore, that they are catalyzed by acid catalysts and are exothermic. The following exothermic heats of reaction in the liquid phase are averages for all isomers in each reaction:[1]

	Average ΔH
$2C_3H_6 \rightarrow C_6H_{12}$	−366.6 kcal/kg C_6
$C_3H_6 + C_4H_8 \rightarrow C_7H_4$	−311.2 kcal/kg C_7
$2C_4H_8 \rightarrow C_8H_{16}$	−274.9 kcal/kg C_8

A portion of the C_6, C_7, and C_8 olefins produced react further with propene and butenes.[1] Examples are

$$C_3H_6 + C_6H_{12} \rightarrow C_9H_{18}$$

$$C_3H_6 + C_7H_{14} \rightarrow C_{10}H_{20}$$

$$C_4H_8 + C_6H_{12} \rightarrow C_{10}H_{20}$$

Table 18.23 provides a typical product distribution for a proprietary process.

TABLE 18.23
Product Oligomer Distribution (UOP Catalytic Condensation Process)

Feed source: FCC C_3–C_4 fraction

Operation: once through

Product	Distribution, wt%
Unreacted C_3–C_4	5.0
C_6 oligomer	4.4
C_7 oligomer	41.3
C_8 oligomer	24.0
C_9 oligomer	15.8
C_{10} + oligomer	9.5

Reprinted by permission: Tajbl, D.G. in *Handbook of Petroleum Refining Processes*, R.A. Meyers, ed., p. 143, copyright McGraw-Hill Companies, Inc., New York, 1986.

The reactions are acid catalyzed and involve carbenium ion mechanisms. The original catalyst, phosphoric acid deposited on Kieselguhr (an inert porous diatomaceous earth), continues to be the most widely used.[2] Amorphous silica-alumina catalysts are also used. Propene oligomerization proceeds through a secondary carbenium ion.

In the case of C_4 olefins, tertiary carbenium ions are formed, and the reaction is much faster in accordance with established carbenium-ion rules.[3] The sequence of events is much more complex because of the many isomers possible. The steps in dimer formation from isobutylene are illustrated in Figure 18.21, which would be followed by production of trimers or higher, depending on the character of reactor control.

Catalyst Types And Suppliers

Two major types of catalysts are offered commercially. The most widely used, phosphoric acid on Kieselguhr, was developed by V. N. Ipatieff at UOP.[2] A mixture of phosphoric acid with Kieselguhr

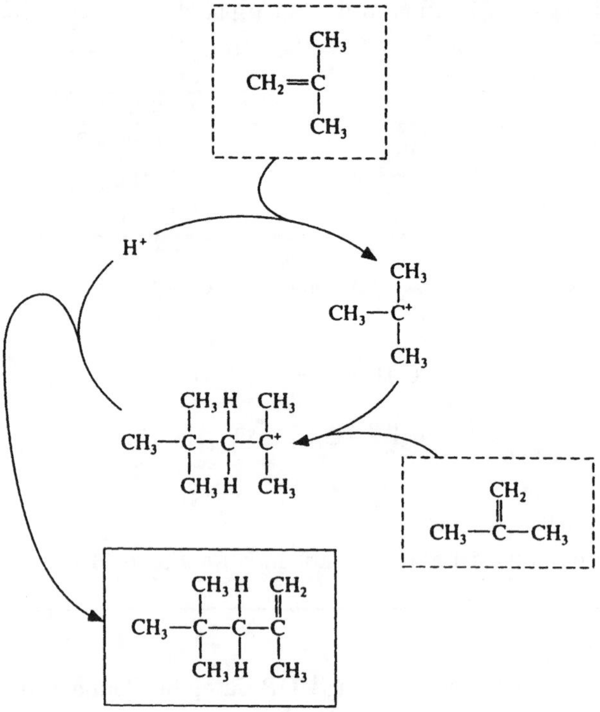

FIGURE 18.21   Catalytic Cycle for the Dimerization of Isobutylene. Reprinted by permission of John Wiley & Sons, Inc.: Gates, B. C., *Catalytic Chemistry,* p. 54, Wiley, New York, copyright © 1992.

solidified when heated and could then be extruded into cylindrical shapes or formed into spheres (actually, spheroids).

Spherical-shaped beads of amorphous silica alumina similar to moving-bed cracking catalyst are also available and offer high mechanical and thermal stability together with necessary acidity. Pore structure favors dimerization.

Operation at high pressure with the phosphoric-acid solid catalyst keeps the catalyst washed clean of higher-molecular-weight polymers, which would otherwise deposit over time on the catalyst.[1] Table 18.24 lists the merchant suppliers and licensors of proprietary catalysts and processes.

Catalyst Manufacture

One procedure for producing the phosphoric acid catalyst involves the following steps:[4]

- Spray the acid in a coating device on to the Kieselguhr powder or extrudates, allowing sufficient time for the liquid to fully soak into the supports.
- If in powder form, extrudates are subsequently produced.
- Extrudates are dried, which converts a substantial portion of the acid into solid P_2O_5 well distributed on the support.

Beaded silica-alumina is produced by methods used in manufacturing silica-alumina catalytic cracking catalysts. Aqueous solutions of sodium aluminate and sodium silicate are mixed and polycondensed by acid-base neutralization to form a hydrated gel. The gel is dispersed in hot oil dropwise, which coagulates the drops into spheroids. The spheroids are then dried.

TABLE 18.24
Oligomerization/Catalyst Suppliers and Licensors

Catalyst Type	Supplier	Description
	Suppliers	
Phosphoric acid on Kieselguhr	United Catalysts	70% H_3PO_4
	UOP LLC	Spheroids: 1/4 5/16 in., 5/16 × 1/4 in.
		Extrudates: 1/4 and 3/16 in.
Amorphous Silica-alumina	Solvay Catalysts	Beaded, 2–3 mm spheres
	*Licensors**	
Phosphoric acid on Kieselguhr	UOP LLC	Extrudates: 1/4 and 3/16 in.
Phosphoric acid on Kieselguhr	Instituto Mexicano del Petroleo	Cylindrical pellets
ZSM-5 zeolite	Raytheon Engineers and Constructors/Mobil	Used in fluidized bed system with reactor and regenerator

*There are also liquid-phase catalyst processes which are not in the heterogeneous category covered in this book.

Catalyst Deactivation

Being an acid catalyst, basic organonitrogen compounds are reversible poisons that must be removed from the feed, usually by a water wash. Sulfur compounds, although they do not affect the catalyst, are removed upstream in an amine/caustic wash system, since they must not be present in the product. Some water content must be maintained in the feed (350–400 ppm) to provide sufficient ionization to produce proton sites. If excess water is used, the catalyst is deactivated because of softening and compacting of the acid deposit that not only destroys the initial desirable distribution of acid sites but also tends to increase pressure drop as agglomerates develop[6] and ultimately plug the bed. Large excesses of water can also produce damaging corrosion in the reactor. Too little water tends to permit excessive water loss from orthophosphoric acid (H_3PO_4), which causes conversion to metaphosphoric acid, which is catalytically inactive.[6]

The catalyst cannot be regenerated, but it has a reasonable life expressed as 100–200 pounds of polymer per pound of catalyst.[2]

The operating pressure (400–1200 psi, 27.6–82.7 bar) was found to be important in preventing deactivating deposits of higher molecular weight from forming on the catalyst. Depending on the feed composition and temperature, pressure can be set to assure at least a mixed-phase fluid, which washes away large polymer molecules deposited on the catalyst surface.[1] Pressure also drives the reaction to the product, and thus lower temperatures can be used with higher pressures. Oxygen in the feed has also been reported to promote deposits (tarry substances) on the catalyst that reduce activity.[7]

Process Units

Because of the high heat of reaction, good control of reactor temperature is necessary to favor the desired short-chain polymerization. Operating conditions are in the range of 100–200°C (212–392°F), and pressure as high as 100 atm has been reported.[5] Temperatures above 220°C (428°F) cause the onset of hydrogen-transfer reactions, which lead to diolefin production and subsequent massive deposits that deactivate the catalyst. As is conventional in such situations, either a multistage adiabatic reactor with cold-shot cooling between stages or heat-transfer reactor with multiple tubes can be used. In the adiabatic case, a single downflow reactor is used with multiple stages (beds) of various depths (small at the top and larger at the bottom, where temperature increase is lower). The feed is diluted with LPG, and the required amount of water is added. It then passes through four or five beds with direct contact cooling between the intermediate beds by cold LPG quench. The LPG is recovered in the separation unit and recycled.

The multitubular reactor uses tubes packed with catalyst and a cooling medium of boiler feedwater. Jacket temperature is controlled by the back pressure on the steam drum. Generated steam is used for heating the feed. As is usually the case for exothermic reactor systems, the adiabatic reactor is less costly but has higher utility costs than the tubular reactor.[1]

Operating conditions are 300–425°F (175–235°C) and 400 to 1200 psi (27.6–82.7 bar). The higher pressure and lower temperature is favored for the tubular reactor.

Silica-alumina beads can be applied in similar reactor types. Detailed information on licensed processes listed in Table 18.24 are, of course, not in the public domain. The use of shape-selective catalysts (e.g., ZSM-5) has been implemented in a licensed process listed, and the process itself uses a fluidized bed and a small regenerator.[8]

18.7 FLUID CATALYTIC CRACKING

18.7.1 HISTORICAL BACKGROUND

Ever since Henry Ford created the Model T in 1908, American's love affair with the automobile has continued unabated, as has the demand for gasoline. The portion of crude oil containing components boiling in the gasoline range soon became inadequate, and large-scale thermal cracking of heavier portions of the crude to produce more gasoline product was introduced. The first successful commercial process was invented by W. M. Burton in 1912. It was a batch process that was carried out using shell stills under modest pressure (73–95 psig). Then followed (1914–1922) the introduction of various continuous flow processes.[1] As higher-compression engines were developed, a gasoline that did not self-ignite in the cylinder during the compression stroke and cause an explosion, called *knocking*,[2] was needed. Initially, this problem was solved by operating thermal crackers at higher temperatures in a mixed vapor-liquid phase process, which produced more branched-chain olefins in the gasoline boiling range.[1,3] These branched-chain components provided a fuel not as subject to self-ignition. After many years of research at General Motors, Thomas Midgley and associates prepared a successful anti-knock compound, tetraethyl lead, in 1921. Commercial production began in significant amounts in 1924 by the Ethyl Corporation, which was founded for that purpose as a joint venture by General Motors and Standard Oil of New Jersey. (Standard, now Exxon, had patents on processes for making tetraethyl lead and other related patents.) Tetraethyl lead acts as a free-radical trap and prevents rapid chain reactions from taking place at the high temperature of the cylinder. Since not much TEL is needed (~2 ml/gallon) to accomplish this feat, the product proved economical, even though its cost per unit volume was high.

The need for the best possible octane rating of the base stock persisted so that TEL consumption could be minimized. Improvements in cracking processes made a giant leap when Eugene Houdry in 1936 ushered in a major shift to catalytic processes in refining by inventing and commercializing the first successful catalytic cracking unit with his partners Socony–Vacuum (now Mobil) and Sun Oil Company. An acid-treated clay (natural silica-alumina) was used, which he found caused cracking at temperatures less than required for thermal cracking and produced a gasoline product with higher octane due to higher amounts of branched-chain olefins and paraffins and benzene. Two or three catalyst beds were used. As coke built up on the catalyst, deactivation occurred, and the deactivated bed was taken off stream for purging with steam and regeneration by air. The feed was switched to another reactor, which had been previously regenerated and stored much heat that was then used by the endothermic cracking that took place as the feed passed through it. The cycles were about 30 minutes in length and, like any semicontinuous process, produced product characteristics that varied with the time on stream. To avoid excessive pressure drop, catalyst particles in the range of two-thirds of an inch were used. This size proved most workable, but it also presented diffusion resistances for some of the reactants and for the regeneration step as well.

The inconvenience of cyclical operation was early recognized, and Houdry and his partners developed means for moving catalyst from reactor to regenerator and back again using bucket elevators (1941) and air lifts (1952). The same size catalyst was used, but special hard beads were developed that were hardy but continued to offer undesirable diffusion resistances.

A group of companies[*] led by Standard of New Jersey and the M. W. Kellogg engineering firm (now Kellogg Brown & Root), not wishing to pay high royalties for the Houdry process and also wanting to develop a continuous process, pooled their resources in a group designated the Catalytic Research Association. It was a massive effort in catalyst and process development.[4] There were a total of 1,000 researchers and staff of which 400 were Standard of New Jersey employees. Since Standard (now Exxon) had already begun studies on ways to move powdered catalyst between the reactor and regenerator, they continued this work on an expanded scale. Based on some detailed experimentation by Warren K. Lewis and Edwin Gilliland, sponsored by Jersey at MIT, some basic concepts of catalyst flow in a gaseous fluid were developed. They found that gas and catalyst in proper proportions would flow under differential pressure just like a liquid. Using these concepts and other observations, they proposed a continuous process involving transfer of the catalyst in the fluidized state. Jersey researchers developed this technique using two columns, one for upward flow of catalyst in vaporized feed to the reactor and the other for downward flow in a denser catalyst-vapor mixture from the spent catalyst hopper for which the difference in densities in the two columns provided the driving force.[4] The same concept was used for both the reactor and regenerator (see Figure 18.22).

The first commercial unit was built at the Standard of New Jersey (now Exxon) refinery in Baton Rouge, Louisiana, and went on stream May 25, 1942. It was a stunning success that was followed rapidly by units at Jersey's Bayway, New Jersey, and Baytown, Texas, plants in 1943. Although Jersey held the state of the art patents, other members of the consortium made significant contributions. M. W. Kellogg (now Kellogg Brown & Root), in particular, was involved with the mechanical aspects of the design of the first pilot plant and first commercial unit. Success came at a fortuitous time when U.S.A. plunged into World War II, and demand for high-octane aviation gasoline soared.

As soon as the war started, Jersey made its patents and know-how available to all domestic refiners, and other members did the same. In short order, two more units (Model II) were added at Baton Rouge that remain in operation today, at five to six times the original capacity.[4] Other companies built fluid units and, by the end of the war, 32 units were operating in the U.S.A.

The Model II unit was introduced at Baton Rouge by Exxon one year after Model I became operational. Several months later, an identical Model II, located adjacent to the first Model II, was added. Obviously, both were planned almost simultaneously with Model I because of the war-time need for high-octane aviation gasoline. To reduce the height of the original Model I and reduce pressure drop and cyclone loading, downflow of catalyst in both reactor and regenerator was provided. By the 1950s, the typical unit was less than half the height of the original Model II units. Many improvements in proprietary internals have been introduced such as high efficient cyclones, improved feed and steam injection and dispersion, power recovery from regenerator flue gas, and countless other innovations.[39] The vastly improved highly active catalyst made possible a change from a large back-mixed reactor with inevitable bypassing to nearly plug flow in a riser reactor involving cocurrent flow of catalyst and feed. It is interesting that upflow of catalyst and reactants was part of the original Model I design, but for another reason.

Innovations continue to improve the process and equipment. The interested reader will find the Half-Century Special *Oil & Gas Journal* series beginning with Ref. 39 and also Ref. 5 (Parts 1

---

[*] Standard Oil of New Jersey (Exxon), M.W. Kellogg, Standard Oil of Indiana (AMOCO), Anglo Iranian Oil Co., British Petroleum, Royal Dutch Shell, Texaco, and Universal Oil Products (UOP).

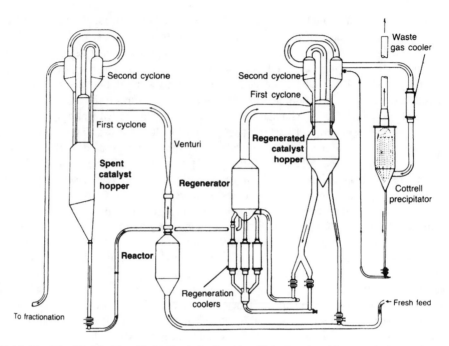

FIGURE 18.22   The World's First Fluid Catalytic Cracking Unit. Reprinted by permission: Reichle, A. D., *Oil & Gas Journal,* p. 41, May 18, 1992. Upward flow of catalyst and vapor occurred in both the reactor and regenerator. Separation of catalyst and vapor took place in each case via a cyclone separator. A pressure differential of 27 psi was developed in the spent catalyst standpipe due to the static head of the catalyst in the hopper and the height of the standpipe of 45 ft. Similarly, a pressure of 31 psi was developed in the 113 ft of regenerated-catalyst standpipe because of the higher density of the flowing catalyst (less aeration) and the 113 ft of standpipe height.[39]

and 2) of great value in providing more details up to the present. There are now more than 360 fluid cracking units operating throughout the world, and the U.S.A. has the largest number.[4] Catalyst improvement has also continued unabated. Beginning with the first synthetic catalyst produced for commercial use by Davison Chemical Company (now Grace Davison) in 1942, the science and technology of fluid catalyst has been steadily advanced by Grace Davison and other manufacturers. More than 230 types of fluidized catalysts are marketed,[4] and it is now possible for a refiner to select a catalyst essentially tailor made for a particular cracking unit and feedstock mix. Obviously, the technical service of suppliers is essential.

In the U.S.A., fluid units produce 38% of the gasoline pool and feed for another 12% provided by alkylation.[3] The feed to these units consists (primarily vacuum gas oil) on the average of 73.2% virgin gas oil, 9.4% cracked gas oil (coker, and visbreaker gas oil) and 8.9% atmospheric residuum.[3] The remainder at 2–3% includes deasphalted oil and vacuum residuum. Refiners will at times charge poor quality feeds such as slops residua and extracts for upgrading. Negative effects on the fluid-catalytic reactor can be minimized by keeping the fraction charged at low values.[3]

Although fluid catalytic crackers can be operated successfully with almost any mix of feedstocks, excessive amounts of polynuclear aromatics are not handled well, because they are not cracked. Instead, they lose hydrogen and condense into larger agglomerates that ultimately become coke and reduce catalytic activity. Regeneration simply consumes the deposit by combustion. By contrast, hydrocracking hydrogenates polynuclear aromatics stage-wise, which then crack to useful products. Therefore, hydrocracking, although requiring higher investment and operating costs, is preferred for heavy ends that are high in polynuclear aromatics.

18.7.2 PROCESS DETAILS

Chemistry

Catalytic cracking of feedstock such as gas oil involves hundreds of reactions and thousands of components. Obviously, the fine detail of all this chemistry is not completely understood, but general descriptions of major reaction types are possible.

A useful summary of the main reactions in fluid catalytic cracking is presented in Figure 18.23 using a simplified approach. The actual reaction steps can be quite involved, as shown in Figure 18.24 as a scheme for gas oil cracking. Note the various contributions to the refinery product mix and to coke production on the catalyst. The heavy hydrocarbon mixture (gas oil) illustrated tends to produce more coke than smaller molecules because of the wide variety of reactive unsaturated species that are formed. Highly reactive diolefins undergo oligomerization and Diels–Alder cyclo-addition with olefins to produce ring compounds that undergo dehydrogenation to aromatics with side chains that, in turn, form polynuclear aromatics via dealkylation and condensation. These condensed masses grow into large deposits of either coke or tar.[6]

Over many years, the reactions of organic compounds under the influence of acidic catalysts were described in terms of positively charged intermediate species then called *carbonium ions*. More recent developments have occasioned the adoption of new IUPAC nomenclature as follows:

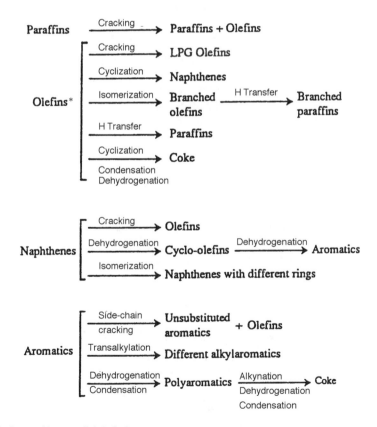

*Mainly from cracking, very little in feed.

FIGURE 18.23   Main Reactions in Fluid Catalytic Cracking. Reprinted from Scherzer, J., *Catalysis Reviews: Science & Engineering,* 31 (3), 215 (1989) by courtesy of Marcel Dekker, Inc.

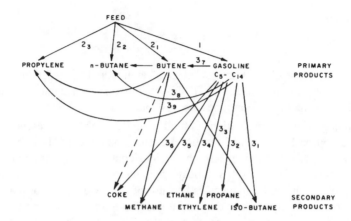

FIGURE 18.24 Mechanistic Scheme for Cracking Gas-Oil. Reprinted by permission: John, T. M. and Wojciechowski, B. W., *Journal of Catalysis,* 37, 240 (1975), Academic Press, Inc.

- *Carbocation.*   General term for positively charged species.
- *Carbenium ions.*   Dicoordinated (e.g., $C_6H_5^+$, $R_1 - CH = C^+ - R_2$) and tricoordinated (e.g., CH_3+, $R_1 - CH_2 - CH^+ - R_2$). To put it simply, a carbenium ion is a cation in which the charged atom is carbon.[34]
- *Carbonium ions.*   Four or five coordinated (e.g., $R_1 - CH_2 = CH^+ - R_2$, $R_1 - CH_3^+ - CH_2 - R_2$, CH_5^+), where R represents a hydrogen atom or an alkyl group.

This change essentially eliminates the term *carbonium ion* from discussions on fluid catalytic cracking, since most cracking reactions involve carbenium ions at commercial operating conditions. Some studies, however, suggest that carbonium ions are involved in several steps in cracking over ZSM-5 cracking catalyst.[7]

Reactant Types

Most studies on reaction pathways and mechanisms have involved pure components. Even when the reaction system is restricted to a pure-component feed, the combined primary and secondary reactions that occur can become quite complex. It is useful, however, to consider the major reactions involved under fluid cracking conditions for the various compound types (paraffins, olefins, naphthenes, and aromatics). In each case to follow, the acid sites on the catalyst are shown as both Brønsted and Lewis sites.

Paraffin Cracking

The reaction scheme for paraffin cracking is shown in Figure 18.25. It is initiated by a small amount of olefin produced by thermal cracking of a paraffin.

Significant characteristics of paraffins are as follows:[5]

- Cracking rate increases with chain length up to nC_{16} and then declines.
- Isoparaffins crack faster than normal paraffins.
- Paraffins crack faster than aromatics.
- Paraffins tend to produce products of three or more carbon atoms. Noncondensible gases such as methane, ethane, and ethylene are present in the product in very small amounts.[6]

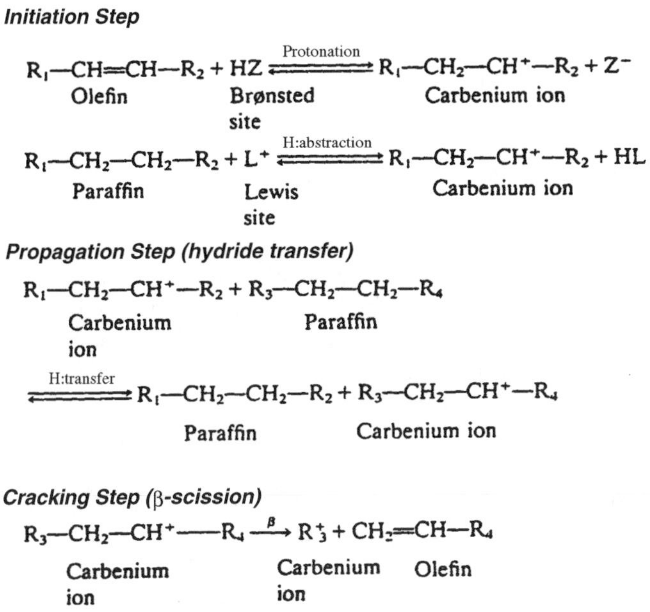

Initiation Step

$$R_1—CH\!\!=\!\!CH—R_2 + HZ \xrightleftharpoons{\text{Protonation}} R_1—CH_2—CH^+—R_2 + Z^-$$

  Olefin   Brønsted   Carbenium ion
                      site

$$R_1—CH_2—CH_2—R_2 + L^+ \xrightleftharpoons{\text{H:abstraction}} R_1—CH_2—CH^+—R_2 + HL$$

  Paraffin   Lewis   Carbenium ion
                      site

Propagation Step (hydride transfer)

$$R_1—CH_2—CH^+—R_2 + R_3—CH_2—CH_2—R_4$$

    Carbenium   Paraffin
    ion

$$\xrightleftharpoons{\text{H:transfer}} R_1—CH_2—CH_2—R_2 + R_3—CH_2—CH^+—R_4$$

           Paraffin   Carbenium ion

Cracking Step (β-scission)

$$R_3—CH_2—CH^+—\!\!—R_4 \xrightarrow{\beta} R_3^+ + CH_2\!\!=\!\!CH—R_4$$

    Carbenium   Carbenium   Olefin
    ion   ion

FIGURE 18.25 Reaction Scheme for Catalytic Cracking of Paraffins. (Note that β-scission in the cracking step refers to cleavage in the position beta to the positive charge.) Reprinted from Scherzer, J., *Catalysis Reviews: Science & Engineering,* 31, 295 (1989) by courtesy of Marcel Dekker, Inc.

Olefin Cracking

The main olefin cracking reaction may be illustrated as follows:[8]

$$R - CH_2 - CH = CH - CH_2 - CH_2 - R^1 + H^\oplus \rightarrow$$

$$R - CH_2 - {}^\oplus CH - CH_2 - CH_2 - CH_2 - R^1 \rightarrow$$

$$R - CH_2 - CH = CH_2 + R^1 - CH_2 - {}^\oplus CH_2 \rightarrow$$

$$R^1 - CH = CH_2 + H^\oplus$$

The additional olefins produced are valuable but, because of their high reactivity in hydrogen transfer, they form diolefins, which are significant coke precursors (see discussion of coke formation in a following section).

Significant characteristics of olefins are as follows:

- Olefins crack faster than any other compound type in the feed.
- As with paraffins, longer chains crack more easily than shorter chains, and iso-olefins more readily than normal olefins.

Cycloparaffins (Naphthenes) Cracking

Cycloparaffins crack, via the formation of a cyclic carbocation, to the olefinic-chain isomer. Naphthenes with side chains experience both cleavage of the side chain and of the rings.[8,9] These compounds are excellent hydrogen donors, and the dehydrogenation of the ring leads to aromatics formation. (See section "Other Reactions.") The hydrogen thus made available can have a useful effect in suppressing coke formation to some degree.[8]

Significant characteristics of cycloparaffins are as follows:

- They crack at rates similar to isoparaffins.
- The ease of cracking of side chains declines in the order of tertiary, secondary, and primary.
- Side chains crack more easily than the ring.

Aromatics Cracking

Cracking of aromatics without side chains is negligible. Alkyl side chains are cleaved at the bond attached to the ring, yielding an olefin and the intact aromatic, benzene or a polynuclear aromatic. Cracking occurs between the aromatic carbon and the adjacent carbon of the side chain. The rate of this reaction increases in the order of primary, secondary, and tertiary alkyl chains. Cracking rate also increases with the length of the chain within each alkyl-type attachment. For example, tertiary butylbenzene cracks faster than butylbenzene, and hexylbenzene cracks faster than butyl-benzene.

As is true in so many homologous series, the first member behaves quite differently. Toluene does not crack to methane and benzene but undergoes disproportionation to benzene and xylene.

$$2\ C_6H_5CH_3 \rightarrow C_6H_6 + C_6H_4(CH_3)_2$$

toluene   benzene   xylene

Xylene isomerizes and disproportionates as follows:[8]

$$o\text{-xylene} \Leftrightarrow m\text{-xylene} \Leftrightarrow p\text{-xylene}$$
$$\downarrow \qquad\qquad \downarrow \qquad\qquad \downarrow$$
toluene + trimethylbenzene

Secondary Reactions

Although the major reactions are cracking reactions, many secondary reactions occur that affect product quality either positively or negatively, depending on the particular reaction, operating conditions, and type of catalysts. Modern management of these variables is crucial to maximizing refinery profit and ability.

Many of these secondary reactions also involve acid catalyst sites of varying strengths as shown in Figure 18.26.[8,35]

Isomerization

Double-bond shift, skeletal, and cis-trans isomerization occur via carbocation intermediates involving hydrogen transfer (proton and hydride transfer steps).

- Sketal Isomerization

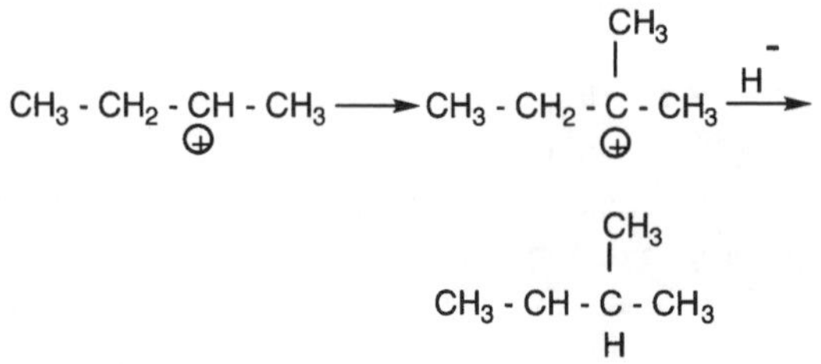

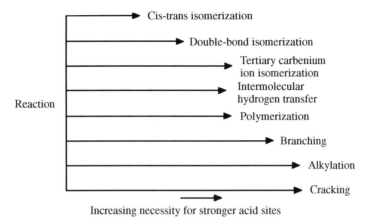

FIGURE 18.26 Required Relative Strength of Acid Sites in Indicated Carbenium-Ion Reactions. Reprinted from Wojciechowski, B. W. and Corma, A., *Catalytic Cracking,* p. 34, Marcel Dekker, New York, 1986 by courtesy of Marcel Dekker, Inc.

Branched-chain hydrocarbons formed by isomerization are subject to rapid cracking, and thus isomerization increases indirectly the yield of cracked products.

- Double-Bond Shift and Cis-Trans Isomerization
  These two reactions are thought to occur independently.[10] A concerted reaction involving a switch mechanism produces only double-bond shift (Scheme 2, Figure 18.27). The catalyst (Cat-H+) lies below the plane of the three central atoms, and the rate of reaction is governed by the capability of the active catalyst site to simultaneously donate and accept a proton.[10]

  The reaction leading to cis-trans isomerization involves complete transfer of a proton to form a carbenium ion that in turn produces both cis and trans isomers. Some minor amount of double-bond shift can occur as in Figure 18.27, Scheme 1, depending on the original configuration of the reacting molecule.

Cyclization and Aromatization

Cyclization is another example of intramolecular hydrogen transfer in which an olefin, carbenium ion forms a cyclic carbenium ion that can proceed to alkyl benzenes and stable paraffins from olefins by intermolecular hydrogen transfer.[5]

$$\overset{\oplus}{CH_3 - CH - (CH_2)_3} = CH = CH_2 \longrightarrow$$

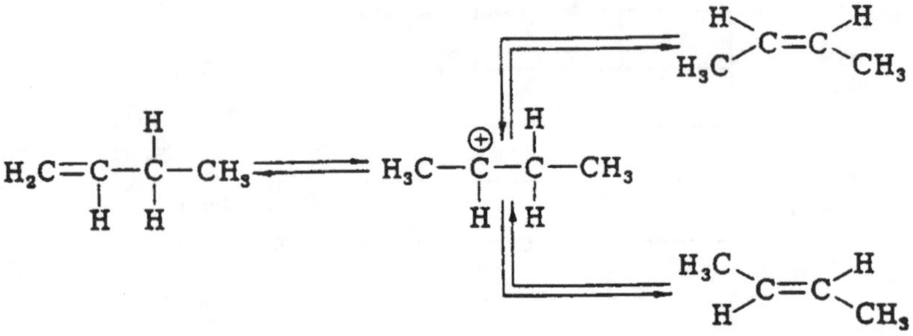

Scheme 1

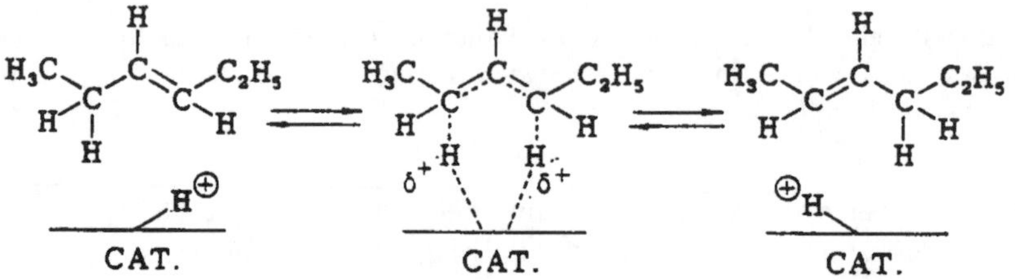

Scheme 2

FIGURE 18.27   Reaction Schemes for Isomerization. Scheme 1, cis-trans isomerization. Scheme 2, double-bond shift. Reprinted by permission: Brouwer, D. M., *Journal of Catalysis,* 1, 22 (1962). Academic Press, Inc.

The result of these hydrogen transfers is the conversion of highly reactive olefins into more stable entities, paraffins and aromatics. It is thought that cyclization reactions may depend on the geometry of surface sites, which suggests that different types of zeolites may show disparate behavior.[8]

Dehydrogenation

What is referred to as *dehydrogenation* in catalytic cracking is largely the result of hydrogen (hydride and proton) transfers by resonance stabilized carbenium ions such as shown above. In fact, the general series of reactions in which highly reactive olefins are converted to more stable paraffins and aromatics is uniquely characteristic of zeolite cracking catalyst, and this has assured its wide use by producing higher yields of gasoline components and less coke precursors.[5,11]

$$\text{olefin} + \text{naphthene} \rightarrow \text{paraffin} + \text{aromatic}$$

$$3C_nH_{2n} + C_mH_{2m} \rightarrow 3C_nH_{2n+2} + C_mH_{2m-6}$$

Coke Formation

The term *coke* is a general term that refers to, in a practical sense, the material that remains on the catalyst after purging with nitrogen or steam at a defined temperature and time period.[8] The ratio of hydrogen to carbon in catalytic cracking coke is in the range of 0.3 to 1.[12] Coke rapidly deactivates the catalyst as it blocks pores, building up largely on the exterior surfaces and regions between particles. The concept of a moving catalyst is based on the need to move a continuous stream of catalyst from the reactor to the regenerator where air is introduced to burn-off the coke. The exothermic heat of burn-off is used in providing the heat for the endothermic cracking reactions.

Many studies on pure compounds and simple mixtures have led to some useful insights but, just as in the case of the cracking reactions, the interactions between the various constituents are most difficult to quantify. In general, however, coke is deposited due to a series of reactions involving olefin oligomerization and then cyclization and dehydrogenation (aromatization) followed by dealkylation and condensation.[5] Surface species formed transfer hydrogen to gas-phase olefins, which results in lower hydrogen content of the coke that ultimately remains on the catalyst.[8] This coke consists of hydrogen-deficient polyaromatic hydrocarbons (see Figure 18.28), but the composition is complex and can include complexes of higher hydrogen content. The complexes shown were extracted from catalyst used in vacuum gas-oil cracking.[36] Since it has been observed that similar structures are found in cracking light alkanes, coke characteristics may depend on zeolite properties more than feedstock.[37]

The acidity of the zeolite used and its pore structure influence the amount and character of the coke formed. Highly acidic Y-zeolites with a large-pore network produce more coke than zeolites with low acidity and narrow pores.[5]

Not all the coke is produced via the catalytic processes just described, which are directly related to the catalyst acid function. Four categories of coke have been defined in the fluid catalytic cracking process.[5,13]

- *Catalytic coke.* Produced via acid-site reactions on the cracking catalyst (amount reduced by using zeolites with smaller unit cell sizes and lower acidity).
- *Cat-to-oil coke.* Results from hydrocarbons remaining on the catalyst after the catalyst stripper (amount reduced by increasing the average pore diameter of the catalyst matrix and decreasing the matrix surface area).
- *Carbon residue coke.* Results from high-boiling refractory components in the feed as indicated by high Conradson carbon residue (CCR) in the feed (amount reduced by combining with lower CCR feed to successfully operate unit).

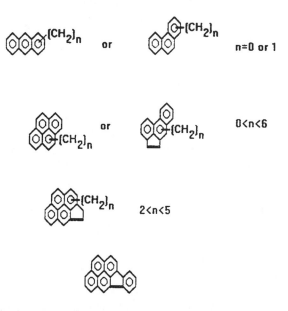

FIGURE 18.28 Compounds Found in Soluble Coke. Reprinted with permission from Turlier, P., Forissier, M., Rivault, P., Pitault, L., and Bernard, J. R., in *Fluid Catalytic Cracking,* M. L. Occelli and P. O'Connor, eds., ACS Symposium Series, 571, p. 98, copyright 1994, American Chemical Society. *Note:* $[CH_2]_n$ represents various aliphatic chains such as $n[-CH_3]$, $-[CH_2]_{n-1}$, and $-CH_3$.

- *Contaminant coke.* Results from metallo-organics in feed (Ni, V, Fe, Cu), which are ultimately deposited as oxides in the regenerator and are reduced to metallic or suboxide forms that act, especially in the case of nickel, as a dehydrogenation catalyst, which leads to increased hydrogen and coke formation. Vanadium also can partially destroy the zeolite structure in the presence of steam due to the formation of vanadic acid, which not only attacks the catalytic components but migrates freely throughout the particle. Contaminant coke can be reduced by metals passivation higher catalyst addition rate and improved catalyst selection.

In the usual gas-oil cracking operation, catalytic coke accounts for 50–65 wt% of the total coke produced.[5,13] Carbon residue coke and contaminant coke on catalyst in low-metals gas-oil service is small. As feeds with significant metals content and high CRC levels are used, the coke make can be dominated by CRC and contaminant coke.

Catalyst Types And Components

When crystalline zeolites were first developed, their activity was too high to use directly in the existing fluid catalytic cracking units. They were instead incorporated as small (1 µ) particles into a silica-alumina or clay matrix, which, upon spray drying, produced microspheroids of 60 µ average particle size. The first such commercial catalyst was introduced by the Davison Chemical Company (now Grace Davison) in 1964. Although perhaps not realized at the time, this combination of components provided a pallet of variables that has enabled continuous innovation to meet the dramatically changing needs of the petroleum refining industry. Furthermore, the reactor/regeneration system proved amenable to drastic changes in operating conditions, often requiring retrofitting but rarely abandonment. In fact, the world's oldest fluid catalytic cracking units constructed at Exxon's Baton Rouge refinery in 1942 continue to operate profitably because of upgrading of equipment and, perhaps more importantly, greatly improved catalysts.

Because refinery requirements can vary at different sites, and optimization of operations is so critical in the commodity business, it is not surprising that catalyst manufacturers have produced more than 250 different types of FCC catalysts. In fact, it is possible to tailor make a catalyst for the special needs of a particular refiner. FCC catalyst manufacturers are involved in continuous research and development and provide expert technical service to the refining industry.

It is not useful to attempt to describe each type of catalyst recipe, but much can be comprehended by focusing on the characteristics and purposes of the various major components of the finished catalysts in the order zeolite, matrix, and additives. Since many variations of each of these categories exist, the number of possible catalyst recipes is quite large indeed.

Zeolites

The most common crystalline form of silica-alumina used as a cracking catalyst ingredient is the Y-type zeolite (Figure 18.29), which is a member of the faujasite family. It is produced in its sodium form (NaY), which is inactive because it lacks strong acid sites. Active catalysts are produced by ion exchange of the sodium with rare earth elements or with NH_4^+ ions. In the case of NH_4^+ exchange, the catalyst is calcined at 572–982°F (300–500°C) to drive off NH_3 and leave protons at the former Na sites,[8] which bond with oxygen atoms in the lattice (see Figure 18.30). A mixture of rare earth chlorides is used to exchange sodium with multivalent rare-earth cations that hydrolyze to create acid sties.[38] For example,

$$La^{3+} + H_2O \rightarrow La(OH)^{2+} + H^+$$

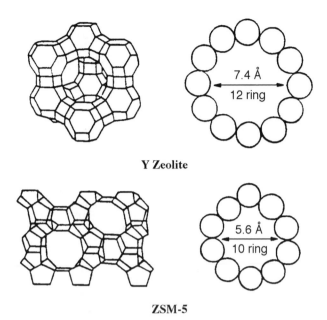

Y Zeolite

ZSM-5

FIGURE 18.29 Framework Structures and Projections of Y and ZSM-5 Zeolites. Chen, Y. and Deguan, T. F., *Chemical Engineering Progress,* p. 32, Feb. 1998. Reproduced with permission of the American Institute of Chemical Engineers. Copyright © 1988 AIChE. All rights reserved.

Both approaches lead to zeolites with high cracking activity. The rare-earth forms, however, are the more stable and resistant to thermal or hydrothermal deterioration, which can destroy the crystal structure and thus the favorable three-dimensional large-pore structure.

Rare-earth ions not only stabilize the crystalline structure up to 1600°F (871°C) but also increase the catalytic activity. The sites left vacant so as to preserve neutrality when Na^+ or H^+ are exchanged produce large dipole moments between the rare earth and the nearest neighbor Al^- ions. These dipoles produce strong electric fields, causing a charged electron distribution that increases the acidity of the active sites. Although rare-earth exchanged zeolites produce higher gasoline yields, the gasoline cut is lower in octane. By combining the higher octane producing qualities of ultra-stabilized zeolite (USY) with the higher gasoline yield of rare-earth exchanged zeolite, an economically optimum octane/gasoline yield can be realized. This goal is accomplished by partial exchange of USY catalysts with rare earths.

Much greater stability can be induced by removal of framework aluminum so as to increase the SiO_2/Al_2O_3 ratio in the zeolite with reduction in unit cell size. This removal is accomplished by steaming at temperatures above 760°C.

Removal of some of the aluminium from the zeolite framework decreases the acid- site density, since these sites are associated with the framework aluminum. Low sodium levels also enhance the vanadium tolerance of the partially dealuminated zeolite.[15] Reduced acid-site density due to dealumination favors less hydrogen transfer reactions, and thus lower gasoline selectivity but higher octane. Hydrogen transfer reactions are bimolecular and can be reasoned to involve components on adjacent sites. Thus, if site density is reduced, the hydrogen-transfer is inhibited, and more olefin production with higher octane properties along with less coke results.[15] The reduced aluminum zeolites are called *ultra-stabilized Y-zeolites*.The dealumination process also involves removal of sodium to low levels. Sodium ions destabilizes the zeolite and also cause lower octane product because of partial neutralization of strong acid sites.[5,14]

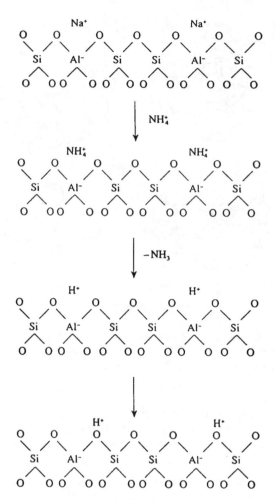

FIGURE 18.30   Reaction Sequence in the Formation of Brønsted Acid Sites. Reprinted from Humphries, A., Harris, D. H., and O'Connor, P. in *Fluid Catalytic Cracking Science and Technology, Studies in Surface Science and Catalysis,* J. S. Magee and M. M. Mitchell, Jr., eds., Vol. 76, p. 41, copyright 1993 with permission of Elsevier Science.

Four major modified Y-type zeolites are widely used. They are designated as REY, REHY, USY, and REUSY as described in Table 18.25. Various proprietary methods of ion exchange and structural modification can be used to tailor each one of these types to specific needs. In so doing, levels of exchange, residual sodium level, and SiO_2/Al_2O_3 ratio are significant variables. Table 18.26 provides a detailed performance comparison of the four zeolites along with the effect of a typical active matrix.

Matrices

A number of proprietary matrices are used by catalyst manufacturers, which provide another series of adjustable variables that can be combined in various ways with specific zeolites to permit a variety of options for specific refinery needs.

The matrix consists of synthetic amorphous silica, alumina, or silica-alumina along with a natural clay such as kaolinite, halloysite, or montmorillonite.[5] The synthetic components have modest catalytic effects and also serve, along with the clay, to bind the active zeolite particles,

TABLE 18.25
Major Catalytically Active and Stable Forms of Y-Zeolites

Type	Preparation	Characteristics
Rare-earth exchanged Y-zeolite (REY)	NaY + RE³⁺ → REY	High concentration of active sites. The higher the greater yield of gasoline and light cycle oil and the lower C_3–C_4 olefins, coke production, and RON octane.
Rare-earth exchanged, hydrogen Y-zeolite (REHY)	NaY + RE³⁺→ RENaY→ NH_4^+ calcination → RE, NH₄Y calcination REHY	Lower rare earth content than REY but higher than REUSY. Lower octane than for REUSY but high gasoline selectivity.
Ultra stable Y-zeolite (USY)	NaY + NH₄⁺ → Na, HY → NH_4^+ calcination (in steam) → USY + Na solution	Lower activity requiring larger amounts of zeolite and/or more severe operating conditions. Has better hydrothermal stability. Provides high LPG selectivity, high olefins yield, low coke make, higher octane.
Rare-earth exchanged ultra-stable hydrogen Y-zeolite (REUSY)	USY + Re³⁺ → REUSY	Higher gasoline selectivity and lower octane than USY due to increase in unit cell over that of USY.

Notes:

1. RE = rare earth, usually in chloride form (cerium, lanthanum, neodyminum)
2. Most ion exchange involves two steps with intermediate heating. Low sodium content is assured by such a procedure which includes filtration and washing after each step to remove the displaced sodium.
3. Addition of rare earth increases equilibrium unit cell size.

producing thereby (through spray drying) stable and tough microspheroids that can resist rapid attrition in the reaction and regenerator. In addition to the binding action, the matrix provides a stable pore structure for diffusion of molecules to the active zeolite grains. Fashioning an appropriate pore size and structure can add some selectivity control through the physical process of excluding certain large molecules. Additional matrix functions include a moderating effect on zeolite activity by its diluting action, a heat-transfer effect that protects the zeolite, and a sodium trapping effect that increases zeolite stability by capturing residual sodium that migrates from the zeolite during use.

Originally, the matrix was designed to have as little activity as possible but, in later years, it was found that a controlled activity of the matrix could be useful in certain situations as summarized in Table 18.27 (see p. 377).

Since the zeolite and the matrix activities and selectivities are quite different, it follows that the ratio of zeolite-to-matrix (often reported as a surface area ratio, Z/M) would substantially affect performance as the following examples indicate.[5,16]

- As Z/M increases, the selectivity pattern approaches that of pure zeolite. At low values of Z/M, the selectivity and yield approach that of the matrix.
- At constant zeolite, increasing matrix activity causes increased total catalyst activity.
- Dry gas and C_3/C_4 yields decline considerably as Z/M increases, and gasoline selectivity improves.

TABLE 18.26*
Summary of Cracking Selectivities

	Zeolite Type					Matrix
	USY	REUSY		REHY	REY	
Dry gas yield	←		Low		→	High
C₃/C₄ yield	High	←	Moderate	→	Low	High
C₃/C₄ olefins	High	←	Moderate	→	Low	High
Coke/conversion	Very Low	→		Low	→	Very High
Gasoline selectivity	Moderate	←	High		→	Low
Octane potential	High	Moderate		Low	→	NA
Cracking activity						
650–900°F material	←		High		→	Moderate
900+ °F material	Moderate	→		Low	→	High
Steam/V/Na stability	Moderate	→		Low	→	High
Ni dehydrogenation activity	←		Low		→	High

C_3/C_4 yield, C_3/C_4 olefins

Reprinted by permission: Wear, C. C. and Mott, R. W., Grace Davison, Paper AM-88-73, National Petroleum Refiners Association Annual Meeting, March 20–22, 1998, San Antonio, Texas.

- At constant conversion, a low Z/M catalyst causes an increase in light-cycle oil, coke, and dry gas yields but produces significant cracking of bottoms into more useful components.
- Olefin/saturate ratio in the LPG cut increases with lower Z/M.
- High-Z/M catalysts produce less coke and gas but more bottoms when compared to low-Z/M catalyst at the same conversion. Because of this fact, it is possible, provided unit and mechanical constraints permit, to operate at a higher severity with the high Z/M catalyst and attain a comparable conversion of bottoms as that realized with low Z/M catalyst.

Clearly, zeolite-to-matrix composition must be tailored for the particular unit and feedstocks. Since pure matrix activity is never ideal, catalyst manufacturers take great care to see that the equilibrated catalyst (average characteristics of the catalyst in the operating reactor) is never dominated by the character of the matrix.[16] Furthermore, the two components are not totally independent variables. Certain proprietary matrices have been observed to act synergistically and effect conversion patterns that tend toward simultaneously eliciting the best characteristics of both the zeolite and the matrix while moderating the undesired effects of each.

In addition to the variables of zeolite and matrix activity, unit limitations are important factors that must be considered in catalyst selection. Table 18.28 (see p. 378) is exemplary. The column designated "Limit" refers to the limiting factor such as mechanical and downstream capacity or to a process requirement such as a desired octane. The column "Zeolite Unite Cell" refers to the unit cell size, which, it will be recalled, correlates inversely with desired performance. Thus, the lower unit cell size produces increasing octane potential, higher olefinicity, and better coke selectivity.[16]

Additives

Other components are added to cracking catalysts for special desired effects. These include ZSM-5 zeolite, CO oxidation promoter, SO_x emissions reducer, metal passivator, or trap.

TABLE 18.27
Active Matrix Components

Component	Activity Level	Characteristics	Functions	Additional Functions
Amorphous silica	Low	Has low activity because of lack of acidic OH groups. When combined with clay, high surface areas and high pore volumes can be developed for improved diffusion of large molecules.	Can be advantageous when cracking heavy feeds. Produces less coke and gas, permitting thereby operation at higher severity.	
Amorphous or alumina	Medium	Exhibits increased activity, stability and attrition resistances. Acidic OH groups (Brønsted sites) as well as Lewis sites are present. Activity can be altered by coprecipitation with metal oxides, which reduce strong acid sites and increase medium and weak acid sites.	Reduces sulfur oxides emissions from the regenerator when processing sulfur-containing feeds. (Also provides same functions as silica-alumina listed below commensurate with its lower activity).	
Amorphous silica-alumina	High	Various preparations are possible: low alumina (10–13%) and high alumina (25%) with acidity proportional to alumina content. Surface area can be varied from 100–600 m^2/g based on the preparative procedure. Pore size distribution and pore volume also a function of preparation.	Cracks large molecules into fragments that can diffuse freely to the zeolite components.	
Amorphous silica-alumina				Preferentially cracks heavy aliphatics into components in the light cycle oil range. Cracks heavy components containing metals (Ni, V, etc.) and bind them to matrix, thereby protecting zeolite from deactivation and, in the case of V, structural collapse. High acidity strong adsorption of nitrogen compounds which protects zeolite acid sites. Both olefin/paraffin ratio and gasoline octane are improved because of lower hydrogen-transfer rates due to scattering of active sites.

Based on information from Refs. 5, 8, 16, and 17.

TABLE 18.28
Recommended Catalyst Properties for Specific Process Limitations

		Recommended Catalyst Properties	
Goal	Limit	Zeolite Unit Cell	Matrix Activity
Gasoline yield and octane	Blower	Low	Low
Gasoline yield and octane	Compressor	Moderate	Low
Gasoline yield and octane	C_3/C_4 capacity	Moderate	Moderate to high
Balanced yield: gasoline and distillate (LCO)	Octane	Low	Moderate to high

Reprinted by permission: Wear, C.C. and Mott, R.W., Grace Davison, Paper Am-87-51, National Petroleum Refiners Association Annual Meeting, March, 1988, San Antonio, Texas.

Note: Air blower limitation occurs at high coke levels, and compressor limitation occurs at high dry-gas production.

ZSM-5

ZSM-5 zeolite, which was patented in 1972, is a unique zeolite, the structure of which provides shape selective properties that enhance the octane of the gasoline range of the feed by systematically removing low-octane components. ZSM-5 has an unusual pore structure, with pore size smaller than that of the faujasite-based zeolites. Figures 18.29 and 18.31 depict the framework of five-membered rings that, when joined, form two types of channels with ten-membered ring openings, one elliptical and one circular. The major selectivity that serves to improve octane is depicted in Figure 18.32, where the exclusion of higher-octane branched paraffins, olefins, and aromatics occurs, while large straight-chain paraffins, which are low in octane, can enter and be removed from the gasoline boiling range by cracking to small olefins and paraffins. Thus, the octane of the product is improved. Some isomerization also occurs so that the light hydrocarbons (C_3–C_5) that are produced are branched-chain olefins.[5]

CO Oxidation Promoter

When coke is burned off in the generator, it is advantageous to achieve full combustion to carbon dioxide. In so doing, the total heat of combustion produced per unit mass is increased, and the amount of coke on the catalyst required to provide heat for the cracking reaction is minimized.

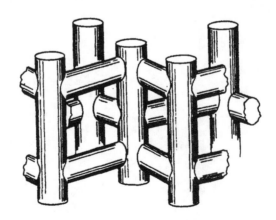

FIGURE 18.31 Channel Structure of ZSM-5. The vertical channels are elliptical (5.7×5.1 Å), and the horizontal channels are nearly circular (5.4–5.6 Å) and sinusoidal. Reprinted by permission: Kokotailo, C. T., Lawton, S. L., Olson, D. H., and Meir, W. M., *Nature,* p. 437, March 30, 1978. Copyright 1978, Macmillan Magazines, Ltd.

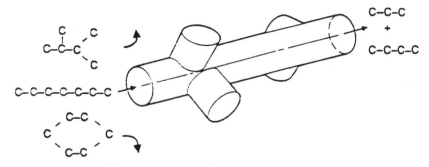

FIGURE 18.32 Illustration of Shape-Selective Cracking with ZSM-5 zeolite. Reprinted from Scherzer, J., *Catalysis Reviews: Science & Engineering,* 31 (3), 215 (1989) by courtesy of Marcel Dekker, Inc.

Furthermore, excessive CO in the regenerator gases can cause afterburning in the dilute phase of the regenerator, the heat from which might do mechanical damage to that region of the regenerator. By contrast, the dense-phase catalyst absorbs the additional heat of combustion and minimizes the temperature rise. Obviously, however, a more heat-stable cracking catalyst is required when CO combustion promoters are used.[8]

The oxidation catalysts of choice are Pt or Pd supported on alumina or silica-alumina and added with the cracking catalyst or added directly to the FCC matrix during manufacture in trace quantities.

SO$_x$ Reduction

Heavy feedstocks often contain organosulfur compounds, a portion of which becomes part of the coke deposit. In the regenerator, about 90% of sulfur on the coke is converted to SO$_2$ and 10% to SO$_3$, which prove to be costly and difficult-to-remove pollutants in the regenerator off-gas. Sulfur dioxide, when released to the atmosphere, is oxidized to SO$_3$. Then, along with SO$_3$ present in the off-gas, it combines with atmospheric moisture to produce the highly destructive acid rain. This problem can be eliminated by hydrotreatment of the feedstock or, when sulfur content is not excessively high, by catalyzing the conversion of SO$_2$ and SO$_3$ to H$_2$S, which is readily removed by amine absorption. Catalyst used are metal oxides such as Al$_2$O, MgO, and rare-earth oxides-on-alumina. These may be added to the FCC catalyst during manufacture or physically mixed with FCC catalyst. A proposed general reaction sequence involving reactions on the additive surface is as follows (M = metal):[5]

Regenerator

$$SO_2 + 1/2\ O_2 \rightarrow SO_3$$

$$MO + SO_3 \rightarrow MSO_4$$

Reactor

$$MSO_4 + 4H_2 \rightarrow MS + 4H_2O$$

$$MSO_4 + 4H_2 \rightarrow MO + H_2S + H_2O$$

Stripper (steam)

$$MS + H_2O \rightarrow MO + H_2S$$

These sulfur-reducing additives tend to be deactivated by silica when contained in the matrix as silica or in conjunction with high-silica, silica-alumina. Steam reacts with silica to produce silicic acid vapors that deposit on the metal oxide additive, and over time the additive is deactivated.[19]

Metals Passivations and Traps

Heavy feedstocks, in particular, contain Ni and V organic compounds that ultimately are deposited on the catalyst as metal oxides that negatively affect the yield of desired products. These deposits, fortunately, can be deactivated. Deposited nickel can be rendered inactive by adding an inorganic metallic compound of antimony or bismuth, which reacts with nickel to form an alloy (e.g., Sb-Ni) that changes the electronic properties of Ni so as to deactivate it.[20] Vanadium can be passivated by a tin compound. In both cases, the passivating agent can be impregnated on the catalyst or added separately to the reactor. The vanadium oxide (V_2O_5), which is in a mobile form, combines with the tin to form a non-mobile mixed oxide of vanadium and tin. Other metals reported for passivating vanadium include zinc, manganese, and barium.

A number of specially prepared large-pore metal oxides and natural clays have been patented as traps for Ni and V oxides. These include alumina or silica, zirconia, lanthanum oxide, barium titanite, etc.[5] Particles of one of these are added to the catalyst matrix during the manufacturing process.

Catalyst Suppliers and Licensors

Suppliers

Akzo Nobel, Engelhard, Grace Davison, Inst. Mexicano de Petro., Intercat (FCC additives), PQ Corp. (zeolites).

Licensors

ABB Lummus Global, Engelhard, Exxon Research & Engineering, Kellogg Brown & Root, Shell International, Stone & Webster/IFP, UOP LLC.

Forms

Microspheres 58, 64, 72 microns (typical grades)

Catalyst Characteristics for Heavy Oil Cracking

The term *residium* or *resid* is used to describe heavy oils such as tower bottoms from atmospheric and vacuum crude distillation as well as heavy gas oils and vacuum gas oil. These products contain significant amounts of polynuclear compounds, resins, and asphaltenes, which are complex compounds with a strong tendency to form coke. They are high in molecular weight and include metal-porphyrin structures that decompose under reactor conditions to leave metal deposits (e.g., Ni and V) that have deleterious effect on catalyst activity and selectivity.

The decline in the heavy fuel oil market and increasing demand for gasoline imposes a need to crack these heavy compounds to more useful sizes. Hydrotreating with hydrocracking has become a major tool for this purpose (see "Hydrocracking"), but some refiners have developed processes for fluid catalytic cracking of resid feedstock. Such processes require extensive revamping of a FCC unit, and many others use hydrotreating and hydrocracking to produce a quality feedstock for an existing FCC unit. Alternatively, a number of plants are able to use an existing FCC unit when only about 20% of the feed is resid and the rest is light cycle oil.

Catalysts for either total resid feed or 20% resid feed must have special characteristics to successfully and economically handle the difficult-to-process compounds. In general, higher rates of coke formation occur because of the polynuclear compounds that are coke precursors. Conversion is lower, as is gasoline yield. Gasoline octane is higher due to greater aromatics content. Catalyst activity declines more rapidly than with the usual gas-oil cracking.

Many of these disadvantages can be reduced by operating at higher temperature, increasing the fresh catalyst addition rate, and by using catalysts especially designed for this difficult operational challenge. Major characteristics of such catalysts are summarized in Table 18.29.

TABLE 18.29
Special Characteristics of Resid FCC Catalysts

Characteristic	Necessary Catalyst Property
Low coke and dry gas production	Y zeolite with small unit-cell size.
	Increased zeolite/matrix activity ratio but with sufficient matrix activity to crack heavy-oil components.
	Increased pore size in matrix to help in improving stripping efficiency so that hydrocarbons held in the pores are readily accessible to steam purging.
High thermal and hydrothermal stability	Ultrastable zeolite with very low Na content.
	Large pore matrix is less likely to sinter in the regenerator.
Abrasion resistance	Rapid deactivation requires high catalyst circulation rate, which requires a tough catalyst to avoid excessive abrasion.
	Large pore matrix produces a larger catalyst particle.
Metal resistance	Additives must be able to render metals such as Ni or V harmless or nearly so, by passivation or cracking (see section on "Metals Passivation").
	Lower matrix surface area reduces metal dispersion.
Resistant to nitrogen poisoning	High activity zeolite, an active matrix, and a broad pore size distribution.
	Many organonitrogen compounds in petroleum are strong bases and are temporary poisons that adsorb strongly on acid sites and reduce activity and selectivity. High activity-zeolite and matrix provide enough active sites to continue operation even after some are poisoned by nitrogen compounds.
Effective reduction of sulfur	Adequate reducing additive to handle sulfur oxides produced in the regenerator.
Good cracking ability of large molecules (bottoms cracking)	Large-pore, high-pore volume active matrix so that large molecules can be readily cracked by the matrix, the smaller parts of which can then penetrate the matrix and react to desirable products on the zeolite particle.
	Smaller size zeolite particles are stabilized by high silica-alumina content to provide more surface for cracking large molecules.

Based on discussions in Refs. 5 and 6.

Catalyst Manufacture[5,21]

Production details on the many types of fluid cracking catalysts are covered by numerous patents and proprietary techniques. The process is most complex, and the major purpose of presenting a brief description based on literature is to illustrate the complexity of the process. The chances of success for a novice are slim.

The first step involves the production of NaY crystallites. Initiator (seeds) is prepared by mixing at high-agitator rpm a slowly added solution of sodium aluminate into a solution of caustic and sodium silicate.[5] The mixture is aged unagitated for 12–16 hr below 120°F. This seed material is then added, along with sodium aluminate, to a mixture of sodium silicate, water, and recycled mother liquor from the crystallization step. Since the silicate in the mother liquor is in the form of a disilicate, aluminum sulfate (alum) is added in appropriate amounts to neutralize and change it to form a silica-alumina gel.

Crystallization temperature must not exceed 218°F (103°C) so as to prevent formation of unwanted forms of crystalline zeolite. Furthermore, once crystallization is complete, the mixture must be quenched to 150°F (66°C) to avoid transformation of Y-type to an unwanted P-type.[21]

Crystallites are separated from the mother liquor and are water washed using rotary filters or horizontal belt filters.[5] The resulting NaY will have a silica-to-alumina ratio of 5 or higher on a surface of 800 m^2/g and a particle size of 1–5 μm.

Ultra-stabilized zeolite Y (USY) is prepared by ammonium exchanging the NaY crystals to remove sodium to about 3 wt% followed by calcination at 1400°F (760°C) in the presence of steam in rotary calciners. This step increases thermal and hydrothermal stabilities of the zeolite by moving some of the aluminum from framework to nonframework positions, which reduces the unit cell size.

The matrix components (e.g., silica, alumina, or silica-alumina) are then mixed as slurries with the clay and zeolite, and additives. Mixers must be high-shear type because of the thixotropic nature of the slurries. Some means of ensuring adequate up-and-down motion, such as circulating recycle pumps, is needed to produce uniformity.[21] The mixed slurry is then pumped to a spray dryer where it is atomized through a spray nozzle on a rotating fluted wheel (+6000 rpm) into the heated chamber constituting the dryer shell. The flow of hot gases can either be cocurrent or mixed cocurrent-countercurrent. Spray dryers in catalyst plants are very large pieces of equipment that produce near-spherical particles with a mean size of 70 μ and within a carefully controlled distribution. The viscosity and other physical properties of the slurry, along with atomizing procedures, are important variables in controlling the mean size and size distribution.

The dried newly formed catalyst is ion-exchanged with $(NH_4)_2SO_4$ to remove residual sodium. The exchange is accomplished on a belt filter by spraying the exchange solution over the catalyst on the moving belt upon which filtration occurs simultaneously. A final water wash on the traveling belt is followed by flash drying in a rotary kiln at 300–400°F (149–204°C). The product then goes to storage into which additives not made as part of the catalyst may be added.

One should understand that the sequence of events just described is one of many. Other sequences must be used, depending on the type of matrix and many other desired catalyst characteristics. The utility of the description primarily lies in the demonstration of the complexities and multiple variables that must be controlled in FCC catalyst manufacture. Catalyst manufacturers must maintain the highest standards of quality control to supply constant characteristics for each catalyst type and batch. In so doing, a competent technical service and research staff is essential.

Catalyst Deactivation

The genius of fluid catalytic cracking technology is that it was developed to handle the most rapidly deactivating catalyst known and became the most versatile and profitable unit in the refining industry.

Coking

Even though modern riser cracking units (see "Process Units") operate with contact times of only a few seconds, after which catalyst is rapidly separated from the reactants, coke on the catalyst can reach levels of 5 to 6% by weight of the feed. Coke not only covers active sites but also retards access by partially blocking pores. Such retardation in the matrix limits access to the zeolite particles that supply a major portion of the desired activity. A large portion of the coke is deposited on the zeolite components.[8] Since they make up a small fraction of the total catalyst particle, they are deactivated at a faster rate than the matrix. Coke selective catalysts have been developed that reduce coke make, but complete elimination of coke is not sought. The heat required for the endothermic cracking reaction is largely supplied by the exothermic heat of regeneration produced in burning the coke on the catalyst after it has been stripped of oil clinging to the catalyst, leaving the catalyst separator. Adjustment of the feed preheater outlet temperature and catalyst circulation rate provides the means for adding any additional heat content required for the endothermic reaction with the control variable being the desired reactor temperature.

In addition to catalyst characteristics required for reducing coke production given in Table 18.29, which are most important for resid cracking, certain operating conditions and design aspects can also be useful. These include shorter contact time in the reactor section between catalyst and feed, rapid separation of catalyst and oil vapors at the riser discharge, and efficient steam stripping of the spent catalyst by minimizing channeling.

Coke formed on the catalyst can be attributed to four major sources.[22] Catalytic coke is that produced by a series of reactions catalyzed by the active acid sites on the zeolite and the matrix. Cat/oil coke is produced from hydrocarbons occluded in the pores of the circulating catalyst leaving the stripper and is a function of the catalyst circulation rate, catalyst pore volume, and stripping efficiency. Contaminant coke is that catalyzed by metal poisons deposited on the cracking catalyst (Ni, V, Fe, Cu). Feed/nondistillable or Conradson carbon coke is formed from high-boiling refractory feed components that deposit on the catalyst surface as a residue. Conradson carbon is determined by an ASTM proscribed procedure that involves destructive distillation of a feedstock in the absence of air and at one atmosphere. The nonvolatile compounds produce a carbonaceous deposit, which is reported as a weight percent and correlates empirically with the coke-forming tendencies of these heavy high-carbon content compounds in the feed.

Regeneration

Stripped catalyst enters the regenerator along with combustion air. The exothermic reaction that ensues raises the catalyst temperature some 180–360°F (100–200°C) above the reaction temperature. Because of the higher stability of zeolite catalysts, regeneration temperatures as high as 1350°F (732°C) or more are possible. Actually, certain ultrastable catalysts can tolerate even higher temperatures, but regenerator metallurgy may not permit temperatures above 1400°F (760°C) unless special internals are used.

Coke Yield

Coke yield of the FCC unit, defined as the weight ratio of coke combusted to the oil fed to the reactor, can be determined by an overall heat balance as depicted in Figure 18.33 and summarized as follows:[23]

$$\text{Coke yield, wt\%} = \frac{[\Delta H_R + \Delta H_{RX} + \Delta H_D + \Delta H_{RC}\ ]100}{[\Delta H_C + \Delta H_A + \Delta H_L\ ]}$$

where ΔH_R = enthalpy change of feed at reactor inlet conditions to reaction products at reactor exit conditions, Btu/lb of feed

ΔH_{RX} = heat of cracking, Btu/lb of feed

ΔH_D = enthalpy change of any diluents (i.e., steam or lift gas) from reactor inlet to reactor exit, Btu/lb of feed

ΔH_{RC} = enthalpy change of any recycle stream from reactor inlet to exit, Btu/lb of feed

ΔH_C = heat of combustion for burning coke to CO, CO_2 and H_2O, Btu/lb of coke

ΔH_A = enthalpy change for air between inlet and outlet of the regenerator, Btu/lb of coke

ΔH_L = heat losses from both reactor and regenerator, Btu/lb of coke

The heat of combustion for the coke is some average value depending on the degree of oxidation of the carbon and the relative amount of hydrogen present. Heats of combustion of carbon to CO and CO_2 and of hydrogen to water do not vary much with temperature in the range of interest.

	kcal/g mole[24]	BTU/lb	Basis
$C + 1/2\ O_2 \rightarrow CO$	26.77	4015	C
$C + O_2 \rightarrow CO_2$	94.32	14,160	C
$H_2 + 1/2\ O_2 \rightarrow H_2O$	59.24	106,632	H_2

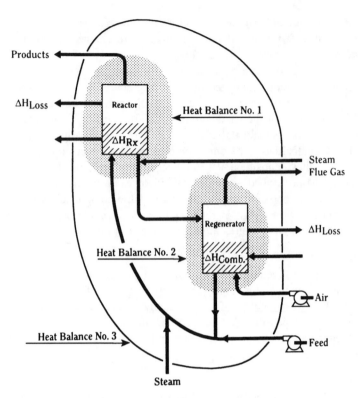

FIGURE 18.33   Heat Balance for Fluid Cracking Unit. Reprinted from Upson, L. L.; Hemler, D. and Lomas, D. A., in "Fluid Catalytic Cracking Science and Technology," *Studies in Surface Science and Catalysis,* J. S. Magee and M. M. Mitchell, Jr., Editors, Vol. 76, p. 397, copyright 1993, with permission from Elsevier Science.

The hydrogen content of coke depends on the efficiency of stripping of heavy hydrocarbons from the catalyst prior to discharging it into the regenerator. The hydrogen portion of these large molecules has a high heat of combustion per unit mass of hydrogen. Thus, if these large molecules are stripped, the average heat of combustion of the remaining material on the catalyst actually declines which, according to the heat balance, increases the *coke yield,* the mass of coke removed/mass of feed.[23]

The effect of the extent of coke burning to CO or CO_2 is more intuitively obvious. Burning a portion of the coke to CO causes a lower average heat of combustion and increases coke yield. The reader should recall that thermodynamics is independent of path. The heat balance does not tell us how the particular steady state occurs. It merely shows what conditions must apply on stream entering and leaving the envelope.

The term *delta coke* is defined as the difference between the coke on the catalyst entering the regenerator and that leaving the regenerator, both expressed as mass of coke per unit mass of catalyst. Delta coke can be related to coke yield by the following expression.[23,25]

$$\Delta C = \frac{\text{coke yield}}{\text{cat/oil}}$$

where cat/oil = the mass ratio of catalyst circulation rate to the oil feed rate.

Delta coke is a path-dependent property, since the cat/oil ratio depends on feedstock properties, processing conditions, and catalyst characteristics.[23] These relationships are summarized in a following section on process units.

By a heat balance around the regenerator, the regenerator temperature can be related to the delta coke, which can be approximated for efficient regenerators as the coke content of the regenerated catalyst.[23]

$$T_{RG} = T_{RO} + \frac{\Delta C}{c_p}[\Delta H_A - \Delta H_A - \Delta H_{LRG}]$$

where T_{RG} = Regenerator dense phase temperature

T_{RO} = Reactor outlet temperature

ΔH_{LRG} = Heat loss from regenerator/lb of coke

$\Delta C \approx C_{spent}$ for many modern-day operations

Poisons

FCC catalyst is poisoned by metallic decomposition products from organometallic compounds, organonitrogen compounds, polynuclear aromatics, and residual coke. The effects on catalyst performance are summarized in Table 18.30 along with methods for ameliorating their actions.

Process Units

Modern fluid cracking units are designed to achieve short contact time, plug flow in the reactor, rapid separation of catalyst and reactants, efficient steam stripping, and effective temperature control in the regenerator. These goals are accomplished by combining a so-called *riser* or *transfer-line reactor* with a specially designed catalyst separator vessel and stripper, and also a regenerator designed for highly efficient combustion.

Prior to the 1960s, a fluid catalytic cracking unit consisted of a large dense-bed reactor that was fluidized at such a gas velocity that mainly all the catalyst but fines created by attrition remained in the reactor. Since the amorphous silica-alumina catalyst used at the time had a low activity, contact time in the reactor dense phase was high, and a large bed was required, which produced a backmixed fluidized dense bed. To reduce coke formation and other undesirable side reactions, conversion was modest, and heavy gas oil was recycled to extinction.[3]

The invention of zeolite-containing catalyst changed all these operating conventions. Activity of the new catalyst was orders of magnitude greater than that of the amorphous silica-aluminas. Because of the high activity of the crystalline zeolites, it became apparent that the advantages of higher-temperature operation and shorter contact times were most beneficial. Under such conditions, desired reactions would be favored, and undesired secondary reactions such as coking would not. The pure zeolite, being excessively active, was mixed with a support of modest activity so as to obtain a controllable catalyst. Short contact time was realized by using the rising portion of the transfer line from the regenerator as a moving plug-flow reactor and converting the large dense-bed reactor as a separator for catalyst and reacted gases. The plug-flow type operation in the riser further favored primary reactions over undesired second reactions. Refiners and engineering firms developed clever designs for modifying existing units to riser design, thus avoiding large new capital outlays for entirely new units. Inevitably, however, expansion and new refineries required new designs totally based on optimum riser and separator design combined with the latest regenerator technology.

Reaction Section

Figure 18.34 depicts one such unit, Exxon's Flexicracking unit (see Appendix for several other designs). The flow can be followed beginning at the oil feed point in the riser where the feed meets the regenerated catalyst. Multiple nozzles are used, arranged around the circumference of the riser,

TABLE 18.30
Catalyst Poisons

Poison	Mode of Action	Effect on Catalyst Performance	Methods for Overcoming Deleterious Effects
Organo-metallics	• Contained in heavy fractions of feed. • During C–C bond cleavage, metals are deposited and undergo oxidations and reductions as catalysts circulate from regenerator to reactor.	• Permanent poisons. • Heavy metals promote uncontrolled afterburning (CO) in the regenerator. Heavy metal deposits require higher catalyst circulation rate and higher catalyst replacement rate.	• Eliminate by hydrotreating prior to cracking. • Use catalysts with large pores and low surface-area active matrices and with high zeolite loading.
Ni	• Nickel complexes are reduced in the reactor and oxidized in the regenerator. This action generates active dehydrogenation sites on both matrix and zeolite. • Deposited Ni tends to deactivate with age.	• Causes dehydrogenation, which increases coke and dry gas production, including hydrogen which limits compressor throughput.	• Passivate nickel by passivating agent (complex antimony, bismuth, or cerium compounds) added to the feed or the catalyst form alloys (e.g., Ni-Nb) that render the nickel inactive. • Trap nickel using certain inorganic oxides such as low-area alumina incorporated in the matrix that causes Ni to agglomerate.
V	• Vanadium deposits on the catalyst are changed to the V^{5+} oxidation state in the regenerator and then vary between 5^+ and 4^+ as the catalyst traverses between regenerator and reactor. • Acts as dehydrogenation catalyst but to a lesser degree than nickel. • The most damaging effects occur as vanadium migrates from the matrix surface to the zeolite particles. In the presence of steam in the regenerator vanadic acid is formed $V_2O_5(s) + H_2O(v) \rightarrow 2H_3VO_4(v)$. • This volatile species spreads throughout the catalyst and catalysis the hydrolysis the SiO_2/Al_2O_3 framework of the zeolite, destroying its crystallinity.	• Has same but lesser effect as Ni due to dehydrogenation (often quoted at 1/4 the dehydrogenating activity). • Major loss in activity occurs due to loss of surface area particularly of the zeolite where most of the highly active surfaces resides. • Up to 300 fold drop can occur at high V loading (~10,000 ppm).	• Passivate with organo-tin or organo-tin and Sb compounds. These are thought to complex with the vanadium and both reduce the dehydrogenation activity and render the V less mobile. • Use metal traps incorporated in the matrix such as alkali and alkaline-earth oxides which complex with the vanadium under regenerator conditions.
Fe	Small amount of iron porphyrins in crude oil. Iron, if present in large amounts, occurs as tramp iron such as upstream corrosion products. As with Ni, iron is an active dehydrogenation catalyst.	Same effect as Ni but about one-third more active in degrading surface area.	Often not major problem if proper removal of tramp iron occurs.
Na	Neutralizes active sites and destroys zeolite crystallinity similar to vanadium.	Same effect as vanadium but no dehydrogenation activity.	Often not a problem unless desalter is malfunctioning.

Based on information from Refs. 5 and 23.

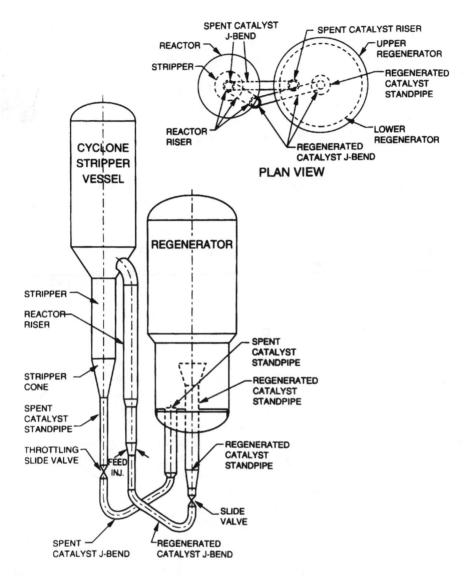

FIGURE 18.34 Exxon Flexicracking III R Fluid Catalytic Cracking Unit. Detailed proprietary internals are not shown. It is in these areas that many improvements are introduced. Some are adaptable to existing units. Reprinted by permission: Ludwig, P. K., in *Handbook of Petroleum Refining Processes,* 2nd edition, R. A. Meyers, ed., p. 36, copyright McGraw-Hill Companies, Inc., New York, 1996.

and are designed to produce small droplets so as to realize more rapid evaporation of the droplets and rapid contact and heat transfer with the catalyst particles. Steam or water is also introduced to aid in dispersion of the feed. Steam partial pressure in the riser reduces that of the reactants and reduces thereby secondary reactions to coke and light gases. Apparently, steam also reduces the rate of dehydrogenation of metal oxides on the catalyst and thus reduces the undesirable side reactions produced by reduced Ni and V.[3]

In a few seconds (2–3), the reacting vapors and catalysts emerge from the riser to the cyclone stripper vessel, where rapid separation of catalyst and reactants occurs. Products leave through a cyclone separator, and catalyst proceeds downward through a stripper section with steam flowing countercurrent at a typical rate of 2 lb/1000 lb of catalysts.[3] Baffles are arranged so as to facilitate good mixing of catalyst and steam to assure maximum removable of hydrocarbons. The length of this stripping section must be long enough to desorb hydrocarbons from the catalyst exterior surface and the pores.

Inertial separation is the usual technique for rapid separation of catalyst and product gases at the terminus of the riser. Flow from the riser is forced to make a 180° turn so that the catalyst flows down to the stripper, while the product gases turn 180° and out through the cyclone. Other reactor designs have used direct discharge of the riser into cyclones.

Regeneration Section

Spent catalyst from the stripper flows down the catalyst standpipe into the regenerator in which air is distributed through a plate, dome, air ring, pipe grid or a Christmas-tree distributor with nozzles. The pressure in the regenerator is often set higher than in the reactor, because higher pressure in the regenerator favors higher combustion rates while low pressure in the reactor promotes higher yields and selectivities. Obviously, the pressures and pressure differentials must favor economical design and reasonable height differentials between the reactor and regenerator to provide adequate static head to move the spent catalyst to the higher-pressure regenerator. Typical reported reactor and regenerator pressures are 15 and 25 psig, respectively.

The air passes up through the regenerator and causes combustion of the coke. Flow is most often in the turbulent-bed region with superficial gas velocities of 1.0–3.5 ft/s. This turbulent condition produces better contacting than lower velocity bubble flow and provides improved contacting efficiency resulting in higher coke-burning capacity.[23] These higher flow rates require improved cyclone design and better design of diplegs that return catalyst to the dense phase from the cyclone separator.[23] The fluidized-catalyst bed in the regenerator continues to be operated in a back-mixed mode. One licensor (UOP) uses a so-called *fast-fluidized bed* with 3–10 ft/s superficial velocity. In order to initiate combustion, some of the hot regenerated catalyst (typically 1350° F) is recirculated to the bottom of the regenerator combustor vessel where it is mixed with the cooler spent catalyst (spent catalyst is typically around 970° F). The regeneration rate is greatly increased by this procedure, since the flow tends to be more like plug flow rather than backmixed. The catalyst recirculation can also help in controlling afterburn.

Higher temperatures improve regeneration efficiency, and most regenerators now operate between 1350 and 1400°F (732 and 760°C), which condition requires stainless steel internals.[3,23] When operating with resid or high-resid containing feeds, more coke is produced, and excessive temperatures may occur in the regenerator. These temperature increases can be avoided or overcome by one or more of the methods summarized in Table 18.31. Special catalysts that have lower coke-forming tendencies can also prove helpful, and the initial response to a sharp temperature rise, at times, can be handled by simply lowering the preheat temperature.

Energy Recovery

The hot gases leaving the regenerator are capable of being converted to higher level work by discharging the gases through an expander, which in turn drives the regeneration-air blower. Connection in tandem with a motor or steam turbine makes it possible to make up for shortfalls in required energy generated by the expander. An efficient fines removal system must precede the expander.[3]

Catalyst Flow Control

Two slide valves, one on the spent catalyst line from the reactor and the other on the regenerated catalyst line to the riser, along with a pressure control valve on the regenerator flue-gas line, control the flow of catalyst to the riser and from the regenerator. A level controller maintains a constant catalyst level, and thus static head in the cyclone stripper vessel, by controlling the flow of catalyst

TABLE 18.31
**Methods for Avoiding Excessive Regenerator Temperatures Due to Increased Coke
Production**

Method	Description	Advantages and Disadvantages
Partial burning to CO	Maintain constant air rate but increase catalyst (increased cat/oil ratio).	• A complex control problem. • No added duty for air blowers.
Cooling coils in regenerator bed	Boiler feedwater is circulated through coils and out to steam separator drum.	• Requires increased catalyst circulation rate. • Since coil is inside regenerator, water must always flow through tubes to prevent coil damage by high temperature. Operating flexibility is reduced. • Good heat transfer from fluidized bed. • Requires increase combustion air.
External catalyst cooler	External vertical exchangers that circulate catalyst from top of bed to the bottom of the bed. Water passes through the tubes, forming steam.	• Good control flexibility by varying catalyst flow. • Catalyst can cause erosion. • Requires increased catalyst circulation and increased combustion air. • Excellent heat transfer from fluidized solids.
Water injection	Water is injected directly in the reactor feed.	• Requires increased catalyst circulation. • Requires increased air rate • Steam produced has other advantages, since it reduces activity of metal contaminants and reduces secondary reaction by lowering hydrocarbon partial pressure.
Reduce preheat		• Requires increased catalyst circulation.

Based on information from Refs. 3 and 23.

through the slide valve to the regenerator. The slide valve on the regenerated catalyst standpipe controls the flow of regenerated catalyst from the regenerator to the riser.[3] It is actuated by a temperature-recorder controller that senses the riser temperature and actuates the slide valve so as to provide a flow of hot catalyst needed to maintain the desired temperature. The reaction section cyclone separator is kept at constant pressure by control of the wet product-gas compressor, which discharges to the vapor recovery unit. Thus, the flue-gas slide valve in the regenerator acts as the primary control of differential pressure. When a higher reactor-regenerator ΔP is required, the control valve closes to a smaller opening, causing an increase in regenerator pressure relative to the air blower discharge pressure.[3]

Process Variables

Fluid catalytic cracking units are the most complex process units in existence and as such require careful study and analysis when making decisions about operations and catalyst selection. Operators have at their disposal a number of operating variables, summarized in Table 18.32. Changes in these variables can cause dependent variables to change automatically, since the control system is necessarily designed to reach a new steady-state energy balance rapidly. The variable most frequently affected is the catalyst circulation, which increases or decreases to balance the energy requirements between the reactor and regenerator.[3]

TABLE 18.32
Effect of Major Operating Variables on FCCU Performance[3,23]

Variable	Controlled by	Effect
Reaction (riser) temperature	• Catalyst circulation rate • Increase in cat/oil ratio increases riser temperature	Increase in temperature: • Increases conversion due to higher temperature and higher catalyst loading in riser • Increases gasoline yield up to a point but at excessive temperature overcracking occurs and gasoline decreases and light ends increase • Increases regenerator temperature • Increases gasoline octane • Increases olefins at constant conversion especially in heavy gasoline fraction • Increases aromatics in heavy ends (Effects are complex because of wide variation in activation energies for various compound types as well as the differences in compositions of different feedstocks.)
Reactor pressure	• Suction conditions at recovery section compressor	Not much latitude for changing Increase causes: • Increases conversion and gasoline yield • Modestly decreases the LPG • Coke increases slightly
Hydrocarbon partial pressure	• Steam flow to riser	Increased HC pressure due to lower steam rate: • Increases coking • Decreases olefins and octane • Effects RON more than MON
Feed preheater temperature	• Firing rate of feed furnace	Increased feed preheat temperature at constant reactor temperature (valve-controlled unit): • Decreases cat/oil ratio • Decreases coke yield • Decreases conversion • Increases regenerator temperature
CO combustion efficiency	• Catalyst combustion additives or high temperature and blower air rate	• Additives/or increased air rates reduces afterburing and resulting excessively high temperatures in dilute phase. • Increased dense-phase temperature results if operating with partial CO combustion and combustion promoter is increased. Dense bed temperature increases due to CO → CO_2 reaction, causing more coke to be removed.
Regenerator temperature (see Table 18.31)		
Catalyst addition rate and equilibrium catalyst removal	• Slide valves	• Maintains desired catalyst activity in the reactor section including controlling equilibrium catalyst metals content. • Existing catalyst can be replaced by a catalyst selected for a new feed or product distribution by daily additions over a 3–6 month period.

PROCESS MODELS

The development of useful models for fluid catalytic cracking has always faced enormous complexity involving thousands of reactions and compounds plus reactor flow dynamics that are most difficult to define. In the days of the old amorphous catalyst, the reactors involved backmixing of the catalyst and considerable bypassing of the reacting vapors occurred. The modern riser-type reactors have simplified the flow dynamics somewhat. Plug flow of gas and solid in combination with a slip velocity between solid and gas can be used after the mixing portion of the riser, where the condition is in between backmixed and plug flow. Other uncertainties in the main part of the riser, of course, remain. These include possible temperature differences between the catalyst and gas phase and radial catalyst density variations due to higher densities near the wall.

Even if somewhat idealized fluid dynamics are used, execution of the model can yield indications of nonideal regions, such as the feed-catalyst mixing section, that needed design attention and innovation. For example, the mixing section can be modelled as a well mixed region of the reactor and then compared with actual performance to determine the impact of imperfect mixing.[27]

The development of useful models for the chemical kinetics of fluid catalytic cracking have been major strides in parallel with the increase in computer speed and capacity and with the vast improvement of analytical techniques. Such procedures as liquid chromatography, field ionization mass spectroscopy, and gas chromatography/mass spectroscopy have made feasible the characterization of a feedstock in terms of thousands of molecular types.

Table 18.33 (see p. 393) summarizes FCCU kinetic models, providing references, description, and unique characteristics. Early on, it was realized that, as computers made complex calculations feasible, complex mixtures remained too vast in many instances to consider every compound separately. For example, there are five isomers for five carbon-atom olefins and the number increases exponentially for each additional carbon reaching 536,113,477 isomers for 25 carbon-atom olefins.[28] Thus, the concept of lumping large numbers of compounds into a smaller number of pseudocomponents began to evolve. Initially, however, the lumps were few and were defined by the total feed and products such as gasoline rather than chemical types.

Although pure-component rate studies indicated first-order behavior for the disappearance of individual reactants, when a lump covers a wide range of molecular weights, the actual behavior of the lump approximates second-order kinetics. This situation occurs because some isomer components crack readily while others do so slowly. Thus, the rate appears to slow down with conversion, which is typical of second-order rate expressions.[29]

The table shows a progression in sophistication, reliability, and predictive capabilities. Obviously, details of the models in use today by refiners are not available in the open literature. But one must assume that structurally oriented models of various types are in use. FCCU feedstocks can be analyzed and organized into structural lumps. The results can be entered into such models so that operations can quickly predict the best operating conditions for maximum profitability. Because of the direct tie to chemistry of the compounds in each of many lumps at the level of molecular structure, the model is more solidly based on catalytic chemistry.[33] It is certainly capable of providing sufficient detail to predict subtle but significant changes in product properties.

18.8 OXYGENATES

ISOBUTYLENE + METHANOL $\rightarrow$ METHYL TERT-BUTYL ETHER

The major use of methyl tert-butyl ether (MTBE) is as an additive to gasoline for enhancing the octane number and to serve as a source of oxygen for assuring complete combustion of gasoline components that usually are only partially oxidized. These include aromatics and various other hydrocarbons as well as partially burned products of combustion such as carbon monoxide.

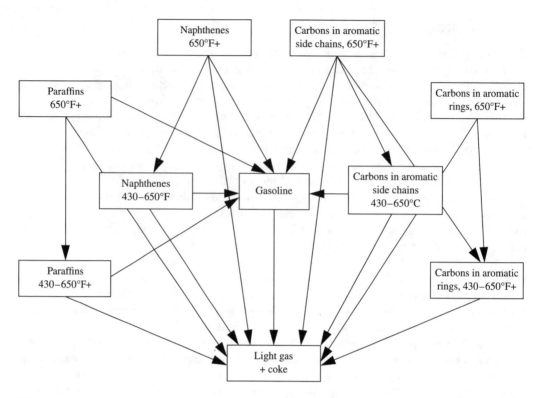

FIGURE 18.35   Ten-Lump Model for 1970s. Reprinted from Krambeck, F. J. in "Kinetic and Thermodynamic Lumping of Multicomponent Mixtures," ACS Symposium, G. Astarita and S. J. Sandler, eds., copyright 1991, p. 123, with permission from Elsevier Science.

Although MTBE had been used by several refiners for blending with unleaded gasoline as early as 1979, increased use began in the 1980s, when tetra-ethyl lead antiknock fluid was being removed from the gasoline mixture in accordance with federal phase-out orders. The major surge in demand, however, occurred beginning in 1990 when the Clean Air Act of that year required at least 2% oxygenate in ozone non-attainment areas. Since the number of these areas increased, the demand for MTBE skyrocketed. Although other oxygenates were available, including ethanol, MTBE was the most attractive economically and technically.

At this writing, California has ordered the phase-out of MTBE as a gasoline oxygenate by the end of 2002.[1] This action was taken when MTBE was found in polluted ground water and drinking water.[2] Some sort of waiver, however, from the 1990 Clean Air Act must be obtained to eliminate MTBE from the gasoline. Since ethanol could conceivably be used, it has two strikes against it: high vapor pressure and not enough available. Although MTBE has growing uses, such as an extraction agent, and as a solvent, there would surely be excess capacity if it were phased out as a fuel additive.

MTBE has been used as a means for removing isobutylene from C_4 mixtures. The isobutylene in the feed can be converted to MTBE, which can be separated by distillation and then readily cracked to isobutylene and methanol.[3] If idle capacity develops in the refineries, such a route for isobutylene separation could prove attractive. Since butylenes form azeotropes they are difficult to separate, and sulfuric-acid extraction, which creates waste disposal problems, is frequently used. The conversion of MTBE to isobutylene, which is used by several refineries, may become more widely adopted.

TABLE 18.33
Fluid Catalytic-Cracking Kinetic Models

Model Type	Circa	Description	Characteristics
3-lump	Late 1960s Refs. 29 and 30	Complex reaction system divided into 3 lumps—gas oil feed, gasoline, and dry gas plus coke.	• Lumps are defined in terms of boiling range or compound class. • Second-order kinetics. • Catalyst decay expression based on Voorhies Law.[31] $$k_{actual} = e^{-at}k_{intrinsic}$$ where a = catalyst decay coefficient t = catalyst residence times • Analysis of feedstocks are limited to bulk properties, and thus the feedstocks and products were defined mainly in terms of boiling points, thereby limiting the number and detail of lump definitions. • Model was useful in correlating experimental studies and in providing guidance for commercial-unit operation. • Main disadvantage: not able to readily determine the effect of feed composition on the rate constants for different feedstocks.
10-lump	Mid 1970s Refs. 29 and 32	Lumps added were more closely related to chemical type with subcategory of boiling range (see Figure 18.35).	• Gave improved accuracy for a variety of feedstocks. • Improved accuracy realized largely because of separation of aromatic rings and aromatic rings with side chains which react quite differently. • Both this model and the 3-lump model assume fixed cut points between the several cracked products, but in reality they vary depending on changing refinery objectives. • Supplemental correlations were supplied to adjust yields and product properties accordingly. • Rate equations used an adsorption term for the heavy aromatic rings. For the lump: $$r_j = \frac{k_j a_j \rho}{1 + K_h C_{ah}}$$ where k_j = rate constant ρ = density of gas a_j = moles j/g of gas K_h = heavy aromatic ring adsorption constant C_{ah} = wt% aromatic rings in heavy fuel oil
Continuous mixture	Late 1980s Refs. 27 and 28	A continuous property index of commercial interest, boiling point, is used.	• Second-order principle earlier established for individual lumps is extended to cover both the unconverted material above the gasoline end point, but any cut point. Thus the entire lump of gasoline plus unconverted heavy oil and distillate is represented as a second-order disappearance with the rate constant expressed as a continuous function of the cut point. • Although successful in predicting gasoline selectivity versus conversion, the method provides no information on chemical composition of the products and multiple product properties. • Empirical correlations are required to estimate the rate constant vs. boiling-point relation.
Structure-oriented lumping	Early 1990s Ref. 33	Although details of its use for FCCU have not been published, the technique is summarized in this section.	Utilizes the power of modern analytical techniques to provide a definitive link between compound types and appropriate lumping with copious rules on behavior of various chemical types and their properties so that model predictions can be readily transferred to industrial product properties upon which refinery operations are based.

Chemistry

The reaction is a straightforward etherfication, which is an exothermic reaction catalyzed amacro-porous acidic ion-exchange resin.

$$CH_2 = C - CH_3 + CH_3OH \rightleftharpoons CH_3 - O - \underset{\underset{CH_3}{|}}{\overset{\overset{CH_3}{|}}{C}} - CH_3$$

isobutylene   methanol   methyl-tert-butylether

Heats of reaction for the all liquid system have been calculated.[4]

$$\Delta H @ 323 \ K = -9.43 \ kcal$$

$$\Delta H @ 363 \ K = -10.41 \ kcal$$

Operating conditions vary with the process used.

Temperature: <100°C (variously reported as 60–90°C)
Pressure: 10–20 bar
Catalyst: acidic ion-exchange resin
Phase: liquid with some units mixed phase

The only significant by-product is diisobutene. However, if butadiene is present, oligomerization of isobutylene and butadiene can occur.

Thermodynamics

This exothermic reaction is favored by low temperature and an excess of one reactant. An excess of methanol is used in practice, because it also prevents significant dimerization of isobutylene.[5] Figure 18.36 is a plot of temperature vs. isobutylene equilibrium conversion at various metha-nol/isobutylene ratios, based on a C_4 feed containing 45% isobutylene.[4]

Mechanism

A detailed mechanism was presented based on surface science together with reaction studies. It has been postulated that the reaction proceeds in the gel phase as the rate controlling step with concurrent sorption equilibrium between components in the macropore liquid and the gel phase.[4]

Catalyst Type and Suppliers

The catalyst is a macroporous, sulfonic-acid, ion-exchange resin. An example is sulfonated poly-styrene crosslinked with divinylbenzene.[7] In recent years, suppliers are producing uniformly sized particles (with a uniformity coefficient of 1.1) by jetting monomers through a grid of equal-size holes into a suspension fluid.[6] Such uniformly sized beads have higher exchange capacity and lower pressure drop. Information on the use of uniformly sized beads for MTBE is not available.

Suppliers

Bayer AG, Dow Chemical, Mitsubishi Kasei, Rohm and Haas

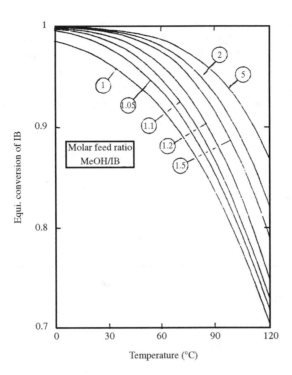

FIGURE 18.36   Isobutylene Conversion vs. Temperature at Various Methanol/Isobutylene Ratios. Basis: feed containing 45% isobutylene. Reprinted from *Chemical Engineering Science,* 45 (6), 1605 (1990), Rehfinger, A. and Hoffman, U., "Kinetics of Methyl Tertiary Butyl Ether Liquid Phase Synthesis Catalyzed by Ion Exchange Resin-1. Intrinsic Rate Expression in Liquid Phase Activities," p. 1605–1617, copyright © 1990, with permission from Elsevier Science.

Licensors[8]

Arco Chemical, CDTECH, Edeleanu, IFP, Neste Oy. Engineering, Phillips Petroleum, Snaam-progetti, SpA, Sumitomo Chem., UOP/Hülls, UOP LLC

Catalyst Deactivation

Long catalyst life of over ten years has been reported.[5] Excessive temperatures can harm the polymeric catalyst, and suppliers can provide safe operating temperatures for their products. Temperatures around 130°C are reported to destabilize the ion exchange resin. Also, temperatures above 100°C can cause fouling of the resin by polymer formation in the reaction mixture.[9] Some feedstocks, such as cuts from FCC units, may contain basic nitrogen compounds, which obviously destroy the effectiveness of active acid sites because of strong adsorption. A water wash of the feed will remove these offending compounds.[9] If a feed contains diolefins and sulfur and nitrogen compounds in unacceptable amounts, selective hydrogenation is required.

Process Units

The most common reactor system consists of two adiabatic fixed-bed reactors in series with intermediate cooling and operating in the liquid phase. Although the reaction is mildly exothermic, a significant concentration of isobutylene in the feed can cause an excessive temperature increase, which at the very least reduces the equilibrium conversion but can also injure the catalyst. Some designs employ both intermediate cooling in a heat exchanger and direct contact cooling in the first reactor by circulating a portion of cooled Reactor 1 effluent.

Several processes use a catalytic distillation unit as the second reactor (CDTECH, UOP, and IFP). This unit consists of distillation trays and an intermediate section packed with proprietary catalytic section containing additional ion-exchange resin. As the distillation action separates methanol from the product MTBE a more favorable equilibrium condition is created that drives the reaction more toward completion. Catalytic distillation creates a mixed-phase reaction system.

TAME (tertiary amyl methyl ether), ETBE (ethyl tertiary butyl ether), and DIPE (dipropyl ether) can be produced in similar units as described for MTBE.

Process Kinetics

A rate equation has been proposed for the all-liquid reaction system using activities from the UNIQUAC method rather than concentrations.[4]

$$r = \left(\frac{a_I}{a_M} - \frac{a_{MT}}{K_2 a_M^2} \right)$$

where a_I, a_M, and a_{MT} are, respectively, activities of isobutylene, methanol, and MTBE. This study presents a very thorough and rigorous procedure along with details on activity and equilibrium calculations.

MTBE → Isobutylene + Methanol

As described in the previous section, the production of MTBE from mixed C_4s and subsequent decomposing of the MTBE to isobutylene is a valuable alternate route for the separation of isobutylene from certain refinery streams.

Chemistry

Since the desired reaction is the reverse of MTBE formation reaction and is endothermic, it is favored by higher temperatures and lower pressure (e.g., 170–220°C and 7 bar). At these conditions, the reaction system was in vapor form.

Catalyst

An alumina catalyst modified by surface silica has been reported[3] that was selective and avoided equilibrium-favored side reactions to a great extent. Selectivities to isobutylene were greater than 98 percent.

Process Units

Several refiners have been using the production of MTBE followed by cracking of the MTBE to isobutylene as a means for removing isobutylene from C_4 streams. Details have not been published.

REFERENCES, SECTION 18.1 (CATALYTIC REFORMING)

1. McCoy, C. S., Edgar, D. and Gardner, A.S., *Today's Refinery,* January, 1993.
2. Ciapetta, F. G., and Wallace, O. N., *Cat. Rev. Sci. Eng.* 5, 67 (1971).
3. Sinfelt, J. H., *Bimetallic Catalysts*, Wiley, New York, 1983.
4. Sinfelt, J. H., *Advan. Chem. Eng.*, 5, 37 (1964).
5. Little, D. L., *Catalytic Reforming*, Pennwalt Books, Tulsa, 1985.
6. LePage, J. F., et al. *Applied Heterogeneous Catalysis*, Editions Technip, Paris and Gulf Pub., Houston, 1987.

7. Biswas, J., Bickle, G. M., Gray, P. G., Do, D. D. and Barbier, J., *Catal. Rev.-Sci., Eng.*, 30(2), 161 (1988).

8. Parera, J. M. and Figoli, N. S., *Catalysis: Specialist Periodical Report*, Vol. 9, p. 65, Royal Soc. Chem., London, 1992.

9. Barbier, J., Corro, J., Zhang, Y., Bournville, J. P., and Frank, J. P., *Appl. Catal.*, 13, 245 (1985).

10. Barbier, J., Corro, J., Zhang, Y., Bournville, J. P., and Frank, J. P., *Appl. Catal.*, 16, 169 (1985).

11. Edgar, M. D., "Catalytic Reforming of Naphtha in Petroleum Refineries," in *Applied Industrial Catalysis*, Vol. 1, Bruce E. Leach, ed., Academic Press, New York, 1983.

12. Ludum, K. H. and Eischens, R. P., "Symp. Catal. on Metals, Div. Petro. Chem.," Am. Chem. Soc., New York, April 4–7, 1976.

13. Querini, C. A., Figoli, N. S., and Parera, J. M., *Appl. Catal.*, 32, 133 (1987).

14. Pistorius, J. T., "Trouble Shooting Catalytic Reforming Units," Annual Meeting NPRA, San Antonio, Texas (1985). Also available from Criterion Catalysts.

15. Gates, Bruce C., *Catalytic Chemistry,* Wiley, New York, 1992.

16. Schafer, H., *Chemical Transport Reactions,* Academic Press, New York, 1964.

17. Ramage, M. P., Graziani, K. R., Schipper, P. H., Krambeck, F. J., and Choi, B. C., *Adv. Chem. Eng.*, 13, 193 (1987).

REFERENCES, SECTION 18.2 (HYDROPROCESSING)

See Ref. 44 in Sec. 18.3.

REFERENCES, SECTION 18.3 (HYDROTREATING)

1. LePage, J. F., et al. *Applied Heterogeneous Catalysis,* Technip. Paris, 1987. (English translation distributed by Gulf Publishing Co., Houston).

2. Kinghorn, R. R. F., *An Introduction to the Physics and Chemistry of Petroleum,* John Wiley & Sons, Ltd. London, 1983.

3. Speight, J. G., *The Desulfurization of Heavy Oils and Residua,* Marcel Dekker, New York, 1981.

4. Yen, T. F., *The Role of Trace Elements,* Ann Arbor Science Publishers, Ann Arbor, Mich., 1973.

5. Valkovic, V., *Trace Elements in Petroleum,* Petroleum Pub. Co., Tulsa, 1978.

6. Speight, J. G., *The Chemistry and Technology of Petroleum*, 2nd edition, Marcel Dekker, New York, 1991.

7. Ho, J. C., *Catal. Rev-Sci. Eng.* 30 (1), 117 (1988).

8. Branthaver, J. F. in "Metal Complexes in Fossil Fuels," R. H. Filby and F. Branthauer, eds., p. 188–204, *ACS Sym. Ser.* 344, American Chemical Society, Washington, D.C., 1987.

9. Reynolds, J. G., Riggs, W. R. and Berman, S. A., in "Metal Complexes in Fossil Fuels," R. H. Filby and F. Branthauer, eds., p. 205, *ACS Sym. Ser.* 344, American Chemical Society, Washington, D.C., 1987.

10. Fish, R. N., Reynolds, J. G. and Gallegos, E., in "Metal Complexes in Fossil Fuels," R. H. Filby and F. Branthauer, eds., p. 332, *ACS Sym. Ser.* 344, American Chemical Society, Washington, D.C., 1987.

11. Furimsky, H., *Catal. Rev.-Sci. Eng.* 35 (3) 421 (l983).

12. Mitchell, P. C. H., *Catalysis,* eds., C. Remball and D. A. Dowden, Vol. 4, p. 125, The Chem. Soc. London, 1981.

13. Gates, B. C., *Catalytic Chemistry,* Wiley, New York, 1992.

14. Saterfield, C. N. and Cocchetto, J. F., *Ind. Eng. Chem. Process Des. Dev.* 20, 53 (1981).

15. Saterfield, C. N. and Yang, S. H., *Ind. Eng. Chem. Process Des. Dev.* 23, 11 (1984).

16. Olive, J. L., Biyoko, S., Moulinas, C. and Geneste P., *Appl. Catal.*, 19, 165 (1985).

17. Krishnamurthy, S., Panvelker, S. and Shah, Y. T., *AIChEJ,* 27 (6), 994 (1981).

18. Sapre, A. V. and Gates, B. C., *Ind. Eng. Chem., Proc. Des. Dev.*, 20, 68 (1981).

19. Stiles, A. B., *Catalyst Manufacture: Laboratory and Commercial Preparations,* Marcel Dekker, Inc., New York, 1983.

20. Adams, C. T., Del Paggio, A. A., Schaper, H., Stork, W. H. J. and Shiflett, W. K., *Hydrocarbon Proc.*, Sept. 1989.

21. Wade, R. A. and Wei, J., *J. Catl.*, 93, 122 (1985).

22. Prins, R., de Beer, V. H. J. and Somorjai, G. A., *Catal. Rev.-Sci. Eng.* 31 (1 & 2), 1 (l989).

23. Topsoe, N. Y. and Topsoe, H., *J. Catal.*, 84, 386 (1983).

24. Bussell, M. E. and Somorjai, G.A., *J. Catal.*, 106, 93 (1987).

25. Farragher, R. L. and Cossee, P., in *Proc. 5th Intl. Congress on Catalysis,* 1972, North–Holland, Amsterdam, 1973, p. 1301.

26. Harris, S. and Chianelli, R. R., *J. Catal.*, 98, 17 (1986).

27. deBeer, V. H. J., Duchet, J. L. and Prins, R., *J. Catal.*, 72, 369 (1981).

28. McCulloch, D. C., in *Applied Industrial Catalysts,* Vol. 1, ed. B. E. Leach, Academic Press, 1983.

29. Edgar, M. D., Johnson, A. D., Pistorius, J. T. and Varadi, T. 1984 NPRA Annual Meeting March 25–27, 1984. (Available from Criterion Catalysts.)

30. Keilberg, L., Zeuthen, P. and Jakobsen, H. I., *J. Catal.*, 143 (1), 45 (1993).

31. Keating, C. J. and MacArthur, J. B., *Hydrocarbon Process.*, 59, 12, 101 (1980).

32. Beaton, W. I. and Bertolacini, R. J., *Catal. Rev.-Sci. Eng.*, 33, 281 (1991).

33. Suchanek, A. L. and Hamilton, G. L., Nat'l Pet. Ret. Assoc. Annual Meeting, AM-91-35, San Antonio, TX, 1991.

34. Lewis, J. M., Kydid, R. A., and Boorman, M., in *Progress in Catalysis: Studies in Surface Science and Catalysis,* Vol. 73, p. 451, Elsevier, Amsterdam, 1992.

35. Quann, R. J. and Jaffe, S. B., *Ind. Eng. Chem. Res.,* 31, 2483 (1992).

36. Sullivan, R. F., Bodnszynaki, M. M. and Fetzer, J. C., *Energy Fuels,* 3, 603 (1989).

37. Weekman, V. W., *Chem. Eng. Prog. Monogr. Ser.,* 11 (1979).

38. Beuther, H., Larson, O. A. and Perrota, A. J., in *Catalyst Deactivation,* B. Delmon and G. Froment, eds., 6, 271, Elsevier, Amsterdam, 1980.

39. Bartholomew, C. H. in "Catalytic Hydroprocessing of Petroleum and Distillates," based on *Proceedings of the AIChE Spring National Meeting,* Houston, Texas, 1993, ed. M. C. Oballs and S. S. Shih, p. 1, Marcel Dekker, New York, 1994.

40. Tamm, P. W., Hernsberger, H. F. and Bridge, A. G., *Ind. Eng. Chem. Process Des. Dev.,* 20, 263 (1981).

41. Beaton, W. L. and Bertalacini, R. J., *Catal. Rev.-Sci. Eng.* 33 (3 and 4), 281 (1991).

42. LePage, J. F., Chatila, S. G. and Davidson, M., "Resid and Heavy Oil Processing," Editions Technip, Paris, 1992.

43. Thomas, M., *Appl. Cat.,* 15, 197 (1985).

44. Aboul–Ghett and Summan, A. M., *Adv. in Hydrotreating Catalysts,* M. L. Occelli and R. G. Anthony eds., Elsevier, Amsterdam, 1989.

45. Stanislaus, A., Absi–Halabi, M., and Khan, Z., in "Catalytic Hydroprocessing of Petroleum and Distillates," based on *Proceedings of the AIChE Spring National Meeting,* Houston, Texas, 1993, ed. M. C. Oballs and S. S. Shih, p. 159, Marcel Dekker, New York, 1994.

46. Girgis, M. J. and Gates, B. C., *Ind. Eng. Chem. Res.* 30, 2021 (1991).

47. Wiwel, P., Zeuthen, P. and Jacobsen, A. C., in "Catalyst Deactivation 1991, *Studies in Surface Science and Catalysis,* eds., H. Bartholomew, and J. B. Butt, Vol. 68, 257, Elsevier, Amsterdam, 1991.

48. Sughrue, E. L., Adame, R., Johnson, M. M., Lord, C. J. and Phillips, M. D., in "Catalyst Deactivation 1991," *Studies in Surface Science and Catalysis,* eds., H. Bartholomew, and J. B. Butt, Vol. 68, 281, Elsevier, Amsterdam, 1991.

49. Trimm, D. L., Catalysis in Petroleum Refining: *Studies in Surface Science and Catalysis,* eds., D. L. Trimms, S. A. Kashah, M. Ahsi-Halabi and A. Bishara, Vol. 51, p. 41, Elsevier, New York, 1990.

50. Flinn, R. A., Larson, O. A. and Beuther, H., *Hydrocarbon Proc.,* p. 129, Sept. 1963.

REFERENCES, SECTION 18.4 (HYDROCRACKING)

1. Sullivan, R. F. and Meyer, J. A., ACS Symposium Series, No. 20, *Hydrocracking and Hydrotreating,* Chapter 2, p. 28, (1975).

2. George, S. E. and Foley, R. M., "Hydrocracking Catalyst Applications—Criterion/Zeolyst Approach," NPRA Annual Meeting, March 17, 1991, San Antonio, Texas.

3. Speight, James G., *The Desulfurization of Heavy Oils and Residua,* Marcel Dekker, New York, 1981.

4. Scherzer, J. and Gruia, A. J., *Hydrocracking Science and Technology,* Marcel Dekker, New York, 1996.

5. Ward, J. W., *Fuel Proc. Tech.*, 35, 55 (1993).

6. Scott, J. W. and Bridge, A. G., *Adv. Chem. Ser.* 103, 113–129 (1971).

7. McDaniel, C. V. and Maher, P. K. *Soc. Chem. Ind.*, London, 186 (1968).

8. Ward, J. W., U.S. Patent 3,929,672 (1975).

9. Dufresne, F., Quesada, A., and Mignard, S., in *Catalysis in Petroleum Refining: Studies in Surface Science and Catalysis*, D. L. Trimm, S. Akashah, M. Abasi-Halabi, and A. Bisharo, eds., Vol. 53, p. 301, Elsevier, New York, 1990.

10. Sachtler, W. M. H. and Zhang, Z., *Adv. in Catal.*, 39, 129 (1993).

11. Saterfield, C. N., *Heterogeneous Catalysts in Industrial Practice,* 2nd edition, McGraw-Hill, New York, 1991.

12. Ward, J. W., in *Preparation of Catalysts, III: Studies in Surface Science and Catalysis,* G. Poncelet and P. Grange, eds., Vol. 16, p. 587, Elsevier, Amsterdam, 1983.

13. Nat., P. J., Schoonhoven, J. W. F. M., and Plantenga, F. L., in *Catalysis in Petroleum Refining: Studies in Surface Science and Catalysis*, D. L. Trimm, S. Akasha, M. Absi–Halabi, and A. Bisharo, eds., vol. 53, p. 399, Elsevier, New York, 1990.

14. Ward, J. W., in *Preparation of Catalysts, III: Studies in Surface Science and Catalysis,* G. Poncelet and P. Grange, eds., Vol. 16, p. 417, Elsevier, Amsterdam, 1983.

15. Speight, J. G. "The Chemistry and Technology of Petroleum," 2nd edition, Marcel Dekker, Inc., New York, 1991.

16. Powell, R. T., *Oil and Gas J.*, p. 61, Jan. 9, 1989.

17. Krambeck, F. J. in "Kinetic and Thermodynamic Lumping of Multicomponent Mixtures," ed. G. Astarita and S. I. Sandler, Elsevier, New York, 1991.

18. Quan, R. J. and Jaffe, S. B., *Ind. Eng. Chem. Res.*, 31, 24–83 (1992).

REFERENCES, SECTION 18.5 (ISOMERIZATION)

1. Gary, J. H. and Handwerk, G. E., "Petroleum Refining: Technology and Economics," 3rd ed., Marcel Dekker, New York, 1994.

2. Rosati, D. in *Handbook of Petroleum Refining Processes,* edited by R. A. Meyers, p. 5–39, McGraw-Hill, New York, 1986.

3. Asselin, G. F., Bloch, H. S., Donaldson, G. R., Haensel, V. and Pollitzer, E. L., Am. Chem. Soc., Div. of Pet. Chem., Prepr. 17 (3), B4 (1972).

4. Satterfield, C. N., *Catalysis in Industrial Practice,* 2nd. ed., McGraw-Hill, New York (1991).

5. Weiszmann, J. A., *In Handbook of Petroleum Refining Processes,* edited by R. A. Meyers, p. 5.47, McGraw-Hill, New York, 1986.

6. Kouwenhoren, H. W. and van Zijll Langhout, W. C., *Chem. Eng. Prog.* 22, (4), 65 (1971).

7. Cusher, N. A. in *Handbook of Petroleum Refining Processes,* edited by R. A. Meyers, p. 9.7 and 9.15, McGraw-Hill, New York, 1997.

8. Fujimoto, K., Maeda, R., Almoto, K., *Preprints Div. of Pet. Chem., Amer. Chem. Soc.* 37 (3), 768 (1992).

9. Greenough, P. and Rolfe, J. R. K., in *Handbook of Petroleum Refining Processes,* edited by R. A. Meyers, p. 5.25, McGraw-Hill, New York, 1986.

REFERENCES, SECTION 18.6 (OLIGOMERIZATION)

1. Tajbl, G. D. in *Handbook of Petroleum Refining Processes*, R. A. Meyers, ed., p. 143, McGraw-Hill, New York, 1986.

2. Gary, J. H. and Handwerk, G. E., *Petroleum Refining: Technology and Economics*, 3rd edition., Marcel Dekker, New York, 1994.

3. Gates, Bruce C., *Catalytic Chemistry,* Wiley, New York, 1992.

4. Stiles, A. B., *Catalyst Manufacture,* Marcel Dekker, New York, 1983.

5. Thomas, C. L., *Catalytic Processes and Proven Catalysts*, Academic Press, New York, 1970.
6. Nelson, W. L., *Petroleum Refinery Engineering*, 4th edition, McGraw-Hill, New York, 1958.
7. Jones, E. K., *Advances in Catalysis*, Vol. VIII, p. 219, Academic Press, New York, 1956.
8. Refining Processes, '94, *Hydrocarbon Processing*, p. 142, Nov., 1994.

REFERENCES, SECTION 18.7 (FLUID CATALYTIC CRACKING)

1. Williamson, H. F., Andreano, R. L., Daum, A. R. and Klose, G. C., *The American Petroleum Industry* (1899–1959), Northwestern University Press, Evanston, IL, 1963.
2. Shell Oil Staff, *The Petroleum Handbook,* Elsevier, New York (1983).
3. Montgomery, J. A. and Staff, *Guide to Fluid Catalytic Cracking*, Part One, Grace Davison, Baltimore, ML, 1993.
4. Veocci, A., Jr., *The Lamp,* 13 (1992).
5. Scherzer, J., *Catal. Rev. Sci. & Engr.,* 31 (3), 215 (1989).
6. Decroocq, D., *Catalytic Cracking of Heavy Petroleum Fractions*, Editions Technip, Paris, 1984.
7. Haag, W. and Dessau, M. R., *Proc. 8th Intern. Congr. Catal.*, Vol. 2, Berlin, 1984.
8. Wojciechowski, B. W. and Corma, A., *Catalytic Cracking,* Marcel Dekker, New York, 1986.
9. Speight, J. G., *The Chemistry and Technology of Petroleum,* Marcel Dekker, New York, 1991.
10. Brouwer, D. M., *J. Catal.,* 1, 22 (1962).
11. Gates, B. G., *Catalytic Chemistry,* Wiley, New York, 1992.
12. Naccache, C., in *Deactivation and Poisoning of Catalysts,* J. Ondar and H. Wise, eds., Marcel Dekker, New York, 1985.
13. Cimbalo, R. N., Foster, R. L. and Wachtel, S. J., *Oil & Gas J.,* p. 112, May 15, 1972.
14. Pine, L. A., Maher, P. J. and Wachter, W. A., *J. Catal.* 85, 466 (1984).
15. Szostak, R. in *Introduction to Zeolite Science and Practice: Studies in Surface Science and Catalysis,* H. van Bekum, E. M. Flanigen and J. C. Jansen, eds., Vol. 58, p. 153, Elsevier, New York, 1991.
16. Wear, C. C. and Mott, R. W., Paper AM-88-73, NPRA Annual Mtg., March, 1988, San Antonio, TX.
17. Ward, J. W., in *Applied Industrial Catalysis*, Vol. 3, B. C. Leach, ed., Academic Press, New York, 1984, p. 271.
18. Csicsery, S. M., *Zeolites,* 4, 203 (1984).
19. Rheaume, L. and Ritter, R. E., *ACS Sym. Ser.,* 375, 146 (1988).
20. Gall, J. W., Nielsen, R. H., McKay, D. L. and Mitchell, N. W., *NPRA Annual Mtg.,* San Antonio, TX, March, 1982, Paper No. AM-82-50.
21. Woltermann, G. M., Magee, J. S. and Griffith, S. D. in *Fluid Catalytic Cracking: Science and Technology: Studies in Surface Science and Catalysis,* J. S. Magee and M. M. Mitchell, Jr., eds., Vol. 76, p. 105, Elsevier, New York, 1993.
22. Cimbalo, R. N., Foster, R. L. and Wachtel, S. J., *Oil & Gas J.* 70 (20) 122 (1972).
23. Upson, L. L., Hemler, C. L. and Lomas, D. A., in *Fluid Catalytic Cracking: Science and Technology: Studies in Surface Science and Catalysis,* J. S. Magee and M. M. Mitchell, Jr., eds., Vol. 76, p. 385, Elsevier, New York, 1993.
24. Stull, D. R., Westrum, E. F., Jr. and Sinke, G. C., *The Chemical Thermodynamics of Organic Compounds,* Wiley, N.Y., 1968.
25. Wear, C. C. *Catalagram,* No. 75, 4 (1987), Grace Davison, Baltimore, MD.
26. Corella, J. and Frances, E., *ACS Sym. Ser.,* 452, 165, (1991).
27. Krambeck, F. J. in *Chemical Reactions of Complex Mixtures,* F. J. Krambeck and A. V. Sapre, eds., p. 42, Van Nostrand Reinhold, New York, 1991.
28. Krambeck, F. J. in *Kinetics and Thermodynamic Lumping of Multicomponent Mixtures*, G. Astarita and S. I. Sandler, eds., p. 111, Elsevier, New York, 1991.
29. Weekman, V. W., Jr., *AIChE Monograph,* 75, No. 11, 1979.
30. Weekman, V. W., Jr. and Nace, D. M., *AIChEJ,* 16, 397 (1970).
31. Voorhies, A., Jr. *Ind. Eng. Chem.,* 37, 318 (1945).
32. Jacob, S. M. B., Gross, S. E. and Weekman, V. W., Jr. *AIChEJ,* 22, 701 (1976).
33. Quann, R. J. and Jaffe, S. B., *Ind.-Eng. Chem. Res.,* 31, 2483 (1992).

34. Saterfield, C. H., *Heterogeneous Catalysis in Industrial Practice,* 2nd edition, McGraw-Hill, New York, 1991.

35. Scherzer, J., *Octane-Enhancing Zeolitic FCC Catalysts,* Marcel Dekker, New York, 1990.

36. Turlier, P., Forissier, M., Rivaut, P., Pitault, L. and Bernard, J. R., in *Fluid Catalytic Cracking*, M. L. Occelli and P. O'Connor, eds., ACS Symposium Series 571, p. 98, American Chemical Society, Washington, D. C., 1994.

37. Bernard, P., Rivault, D., Nevicato, D., Piteut, I., Forissier, M. and Collet, S. in *Fluid Cracking Catalysts,* M. L. Occelli and P. O'Connor, eds., p. 143, Marcel Dekker, New York, 1998.

38. Humphries, A., Harris, D. H. and O'Connor, P., in *Fluid Catalytic Cracking Science and Technology, Studies in Surface Science and Catalysts,* J. S. Magee and M. M. Mitchell, Jr., Vol. 76, p. 41, Elsevier, New York, 1993.

39. Reichle, A. D., *Oil & Gas J.,* p. 41, May 18, 1992.

REFERENCES, SECTION 18.8 (OXYGENATES)

1. *Chemical Engineering,* p. 29, April, 1999.

2. *Chemical Engineering,* p. 56, November, 1998.

3. Fattore, V., Mauri, M. M., Oriani, G. and Paret, G., *Hydrocarbon Proc.,* p. 101, August, 1981.

4. Rehfinger, A. and Hoffman, U., *Chem. Eng. Sci.,* 45 (6), 1605 (1990).

5. Scholz, B., Butzert, H., Neumeister, J. and Nierlich, F., in *Ullmann's Encyclopedia of Industrial Chemistry,* 5th edition, Vol. 16, p. 543, VCH, New York, 1990.

6. *Chemical Engineering,* p. 63, September, 1992.

7. Saterfield, C. N., *Heterogeneous Catalysis in Industrial Practice,* 2nd edition, McGraw-Hill, New York, 1991.

8. *Petrochemical Processes*, '99, *Hydrocarbon Proc.,* p. 156, March, 1999.

9. Davis, S. in *Handbook of Petroleum Refining Processes,* 2nd edition, R. A. Meyers, ed., McGraw-Hill, New York, 1996.

19 Synthesis Gas and Its Products

19.1 HISTORICAL BACKGROUND

The words *synthesis* and *synthetics* are widely used today by both the scientific and lay communities. But in the early 1900s, they were first ushered into the general vocabulary by the discovery of a process for fixing atmospheric nitrogen by synthesizing ammonia from its elements, nitrogen and hydrogen. Fritz Haber, a chemistry professor with significant earlier industrial experience, and Carl Bosch, a practical industrial chemist for BASF, are credited with the development of the ammonia synthesis process. Hydrogen was obtained initially from water gas but was soon more efficiently produced via the catalyzed water-gas shift reaction.[1] Nitrogen was first obtained from distillation of liquefied air, but a more economical source, producer gas, was rapidly developed.

It is useful at this point to define some terms that originated in the early days of the synthesis gas, town gas, and steel industries, since they continue in use today.

Water Gas or Blue Gas[2]

This was originally obtained by passing steam through a bed of incandescent coke. The homogeneous reaction system produced a product of CO and H_2 which was called *water gas,* and that term continues in use today.

$$C + H_2O \rightarrow CO + H_2$$

$$K_P @ 600 \text{ K} = 5.05 \times 10^{-5}$$

$$K_P @ 1000 \text{ K} = 2.61$$

This water gas reaction becomes favorable at higher temperatures, and the rate becomes significant at 1000°C and above.

Another reaction that also took place at high temperatures was called the *water-gas shift reaction.* At the high temperatures, it reached equilibrium toward CO and water and produced a gas low in CO_2, so-called *town* or *city* gas.

$$CO_2 + H_2 \Leftrightarrow CO + H_2O$$

Producer Gas or Blow Gas[2]

Producer gas is formed by the partial combustion of carbonaceous fuel (coal, coke, straw, etc.).

$$2C + O_2 \rightarrow 2CO$$

$$C + O_2 \rightarrow CO_2$$

Steam is introduced in a bed of solid carbonaceous material of at least 50% carbon content, and the following reactions occur along with the water-gas reaction between water and carbon monoxide.

$$C + 2H_2O \rightarrow CO_2 + 2H_2 \text{ @ 1000 to 1900°F (538–1038°C)}$$

$$C + H_2O \rightarrow CO + H_2 \text{ @ 2200°F and above (1204°C)}$$

The second reaction is favored at the higher operating temperatures that were normally used. The process was introduced in 1861 and applied to the production of heat for open-hearth furnaces in the steel industry.

The gas produced from coke was about 60% nitrogen and 40% carbon oxide. It was used in the steel industry for fuel and became an early source of nitrogen for ammonia synthesis. The water-gas shift reaction was used to react the producer gas with water to yield the desired mixture of H_2 and N_2 with the product CO_2 being removed by water scrubbing.

19.2   MODERN SYNTHESIS GAS PRODUCTION

In modern times, the reactions in these two processes have been adapted to the production of synthesis gas with the use of catalysts that improve yields and permit the adjustment of the H_2 to CO ratio of the product as required for different downstream processes. Synthesis of methanol and ammonia consumes the majority of synthesis gas output. Natural gas has continued to be the major carbon source (CH_4), but naphtha is used particularly in areas where natural gas is not available. The molar ratio of hydrogen to carbon monoxide in the reformer output can vary between 6 for methane (natural gas) reforming to 2 for naphtha reforming.[3]

During the so-called oil and gas shortage, which was more a geopolitical crisis than a resource crisis, major efforts were implemented in Europe and the United States to develop improved processes for the manufacture of town gas and synthetic natural gas from other sources such as coal, lignite, and (later) naphtha. In addition, much research on producing chemicals from synthesis gas (C_1 chemistry) was pursued. But once incentive systems were adjusted, the world was soon awash with oil and gas. Many European countries that were totally dependent on imported oil soon had oil and natural gas in abundance from the North Sea. Steam reforming with natural gas became the preferred feedstock for synthesis gas, and naphtha tended to be used only when low-cost natural gas was not available. Naphtha had two strikes against it. Capital costs were higher, and other uses for naphtha (such as steam cracking to ethylene, propylene, and other valuable products) proved more profitable. Accordingly, steam reforming will be discussed mainly in terms of methane, although a section on naphtha steam reforming will follow.

The uses of synthesis gases can be conveniently divided into products requiring hydrogen-carbon oxide mixtures, pure hydrogen, and pure carbon monoxide. Table 19.1 summarizes the major types of gas produced by various steam-reforming plant schemes. As shown in this table, each of these processes requires a steam reforming step followed by additional operations including, in several cases, additional reactors or special separations.

19.2.1   USES

The largest user of synthesis gas (H_2 and CO) is for methanol synthesis, which requires a stoichiometric ratio of 2:1 hydrogen-to-carbon monoxide. Another major user of synthesis gas is for ammonia production, which requires a stoichiometric ratio of 3:1 hydrogen-to-nitrogen and requires maximizing hydrogen production in the reformer and the introduction of air in a subsequent reactor, which introduces the needed nitrogen and maximizes synthesis gas production by converting

TABLE 19.1

Production of Various Types of Gas by Tubular Reforming of Gaseous Hydrocarbons

Type of Synthesis Gas	Typical Process Stages	Typical H_2O/C ratio, mol/mol	Typical Pressure at Reformer Tube Outlet, MPa	Typical Reformed Gas Temperature (primary reformer), °C
CO	Desulfurization, tubular reforming with recycled CO_2 scrubbing. CO separation (e.g., low-temperature separation, COSorb, PSA, or membranes).	2–3	1–3	850–900
H_2	Desulfurization, tubular reforming, HT and LT shift conversion, CO_2 scrubbing, methanation (standard arrangement).	4–5	1.5–3	800–900
	Desulfurization, tubular reforming, HT shift conversion, PSA, or membranes.	3	2–2.5	800–900
Oxosynthesis gas	Desulfurization, tubular reforming, CO_2 scrubbing partial separation of hydrogen by PSA, addition of imported CO_2 upstream of reforming, if necessary.	2.5	1–2	850–950
Methanol synthesis gas	Desulfurization, tubular reforming, or additional secondary reformer.	2.5–2.8	2–3	850–900 700–800
NH_3 synthesis gas	Desulfurization, tubular reforming (primary reformer), secondary reforming, HT and LT shift conversion, CO_2 scrubbing, methanation.	3.5	3–4	780–830
Town gas	Desulfurization, tubular reforming, HT shift conversion, CO_2 scrubbing.	3.0	1–2.5	670–750
Reduction gas	Desulfurization, tubular reforming, or top gas recycling upstream of tubular reforming.	1.25–1.5 0.65	0.2–0.3 0.15–0.2	850–1000 900–1000

PSA = pressure-swing adsorption, HT = high temperature, LT = low temperature.

Reprinted by permission: Renner, H.J. and Marschner, F. in *Ullmann's Encyclopedia of Industrial Chemistry*, 5th edition, Vol. A12, p. 195, Wiley/VCH, Weinheim, Germany, 1989.

methane to carbon monoxide and hydrogen. Subsequent reactors remove all traces of carbon monoxide. Another major user of synthesis gas is oxosynthesis of olefins to aldehydes by homogeneous catalysis with H_2 and CO in a stoichiometric ratio of one-to-one. The aldehydes are then hydrogenated to valuable oxoalcohols.

Several carbonylation reactions requiring pure CO are major routes to important organic chemicals. Hydrogen, of course, is necessary for the voluminous numbers of hydrogenations conducted daily ranging from high tonnage commodities to specialty chemicals.

19.2.2 METHANE STEAM REFORMING (METHANE + WATER → CARBON MONOXIDE + HYDROGEN)

Chemistry

The reactions are catalyzed by nickel deposited on a sturdy carrier such as a refractory alumina or a ceramic material. Operating conditions are in the range of 800–1000°C and 8 to 35 bars.[3]

1. Steam reforming reaction, $CH_4 + H_2O \Leftrightarrow CO + 3H_2$

Temp., K	ΔH, kcal	K_p
800	+53.21	0.031
1000	+53.87	26.12
1123	+54.33	530.26
1273	+54.36	9301.00

2. Shift reaction, $CO + H_2O \Leftrightarrow CO_2 + H_2$

Temp., K	ΔH, kcal	K_p
500	−9.5	130.0
700	−9.05	9.0
1000	−8.31	1.4
1273	−7.67	0.6

Since the methane reforming reaction is highly endothermic, it requires temperatures in the range of 1000 K or more to realize high conversions to carbon monoxide and hydrogen. The catalyst promotes rapid reactions such that equilibrium is essentially reached at the discharge end of the reformer. By contrast, the shift reaction, although rapid and reaching equilibrium, requires lower temperatures to take place under favorable conditions for maximum conversion of CO to CO_2 when hydrogen production is the goal. Accordingly, separate shift-converters operating at lower temperatures are employed in such instances.

Operating pressure is usually set in a range that minimizes capital costs for both the reformer system and downstream units utilizing the reformer product. With pressure set, the steam-to-carbon ratio in the feed to the unit and the outlet temperature of the raw product at the reformer outlet are the operating variables that control the composition of the outlet product.[4] Product gas compositions are successfully predicted by assuming equilibrium or an approach to equilibrium by means of a ΔT approach. When used, the ΔT approach is usually based on operating experience with a given unit.

Table 19.2 provides an interesting series of product compositions based on such equilibrium calculations for various steam reformers producing gas for the ultimate uses indicated.

Steam-to-carbon ratio and reformer outlet temperature are varied to produce H_2/CO values ranging from 1 to 2.8 in the reformer outlet product. Such products can be used after purification either directly in a downstream process or, as in the case of NH_3, subsequent shift reactors are used to reach the desired composition. Of course, rather different compositions are possible when the feedstock is changed as shown with naphtha feed. Considerable production of methane, for example, is then possible.

Side Reactions

In a process conducted at elevated temperatures, various side reactions, both noncatalytic and catalytic, can occur. The most significant of these is carbon formation, which can create operating problems if not minimized. The following reactions summarize the typical carbon forming pathways that are generally considered to be possible.

TABLE 19.2
Calculated Product Gas Compositions[a]

Final Product	1	2	3	4	5	6	7	8	9
						Town Gas	SNG[b]	Oxo-alcohols	Reducing Gas
	NH_3	NH_3	H_2	H_2	MeOH				
Feedstock	CH_4	CH_4	naphtha	CH_4	CH_4	naphtha	naphtha	CH_4	CH_4
p_{exit}/MPa	3.3	3.3	2.7	2.7	1.7	2.4	3.5	1.7	0.5
T_{exit}/K	1073	1103	1073	1123	1123	948	788	1138	1223
H_2O/C_nH_m mol/C atom	3.7	2.5	4.5	2.5	3.0	2.4	1.6	1.8[b]	1.15
H_2O/vol%	44.30	34.17	49.62	31.09	32.20	45.58	47.70	25.21	4.28
H_2	39.12	44.74	34.60	48.59	50.28	25.62	8.25	28.05	70.92
CO	5.04	7.59	5.33	9.22	9.53	3.25	0.43	25.91	22.44
CO_2	6.00	5.49	8.03	5.24	5.42	10.14	11.18	19.71	0.90
CH_4	5.54	8.01	2.42	5.86	2.57	15.41	32.44	1.12	1.46

[a]$\Delta T_R = -10$ K, $\Delta T_s = 0$ K
[b]$\Delta T_R = 0$ K
1 MPa = 10 atm. abs.

Reprinted by permission: Rostrup-Nielson, J.R. in *Catalysis: Science and Technology,* J.R. Anderson and M. Boudart, eds., Vol. 5, p. 1, Springer-Verlag, New York, 1984.

Note: ΔT_R = temperature approach to reformer equilibrium, ΔT_S = temperature approach to shift equilibrium.

3. $CH_4 \Leftrightarrow C + 2H_2$

	ΔH, kcal	K_p
700 K	20.40	0.111
800 K	20.82	0.701
1000 K	21.40	10.02

4. $2CO \Leftrightarrow C + CO_2$

Boudouard Reaction	ΔH, kcal	K_p
800 K	−41.19	91.2
900 K	−40.99	5.176
1000 K	−40.78	0.525

5. $CO + H_2 \Leftrightarrow C + H_2O$

	ΔH, kcal	K_p
800 K	−32.39	22.75
900 K	32.44	2.36
1000 K	−32.47	0.384

Carbon basis: graphite

Reaction 5 is a combination of reactions 2 and 4. Reactions 3, 4, and 5 have been used for analyzing carbon-forming tendencies.[12,25] Combinations of these, such as 3 and 5, 4 and 5, and 3 and 4, yield, respectively, the steam-methane equilibrium, the shift equilibrium, and the carbon dioxide-methane equilibrium.

Mechanism

Actual mechanistic descriptions, as opposed to empirical kinetics, have been difficult to formulate because of the complexity of the reaction systems and the variety of catalysts and catalyst preparation procedures. Patents indicate two major general preparative procedures: *coprecipitation* and *impregnation*. In coprecipitation, a solution of nickel nitrate and very small particles of alumina hydrate are combined in a basic solution to form a precipitate of basic nickel carbonate distributed throughout the alumina hydrate.[14] Various typical steps consist of washing, drying, calcining, and pelletizing or extrusion. Coprecipitation of nickel nitrate and aluminum nitrate at a pH of 7 or slightly above has also been described. Coprecipitation of nickel nitrate and magnesium nitrate and sodium aluminate has also been described[13] which, upon further processing and calcination, yields a magnesium aluminate distributed throughout the pellet, along with NiO or nickel aluminate.

Alternatively, similar catalysts can be produced by impregnating cylinders, rings, or spheres of the carrier followed by calcination.[14]

Either procedure leads to some admixture of carrier and nickel oxide which, upon calcination, produces a mixture of nickel-rich (NiO) crystallites with a small amount of Al^{3+} ions and others rich in alumina with a small amount of Ni^{2+} ions.[15,16] As shown in Figure 19.1, the nickel-rich crystallites can be considered essentially NiO, and the low nickel crystallites have been claimed to be nickel aluminate. Upon reduction, the nickel-rich crystallites becomes elemental nickel buried within the mesopores of an amorphous alumina. These crystallites are the active phase of the catalyst.[16] The proposed nickel aluminate phase is difficult to reduce. The carrier and various additives (promoters) stabilize the crystallite, preventing thereby rapid agglomeration and commensurate activity decline.

Reaction on the active crystallites has not been fathomed in satisfactory detail, but the steps shown in Figure 19.2 convey a reaction scheme relative to the catalyst surface.[13] This surface is composed of nickel crystallites and the alumina carrier, or nickel crystallites and free magnesium oxide and the magnesium aluminate carrier. In both cases, water is considered to be dissociatively adsorbed on the alumina or magnesium oxide as hydroxyl groups. The character of the carrier, therefore, must play some role in the overall reaction.

Consistent with the concept of Figure 19.2, the water-gas shift reaction is depicted in Figure 19.3 with step 5 being the controlling step.[13] Commercial operation suggests that the shift reaction

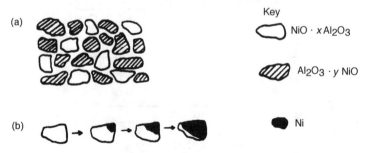

FIGURE 19.1 Model for (a) Calcined Catalyst and (b) States in the Reduction Process. Reprinted by permission: Alzamora, L. E., Ross, J. R. H., Krulssink, E. C., and Van Reijen, *Journal Chemical Society, Faraday Transactions, II* 77, 665 (1981). The Royal Society of Chemistry, Cambridge, England.

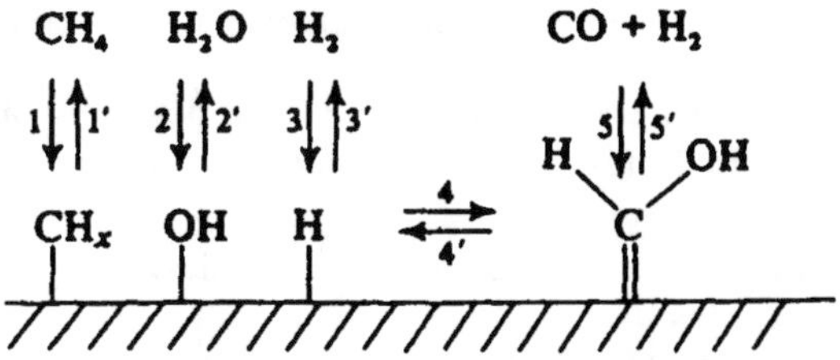

FIGURE 19.2 Reforming Reaction Pathway. Other intermediate species, e.g. $(CH_2OH)_{ads}$ and $(CHO)_{ads}$, may be involved.[13] Reprinted by permission, Ross, J. R. H. in *Surface and Defect Properties of Solids: Specialist Periodical Reports,* Vol. 4, p. 34, The Royal Society of Chemistry, Cambridge, England.

Catalyst

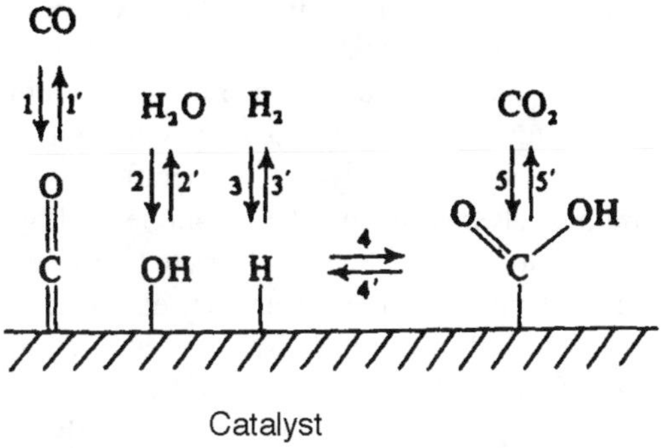

FIGURE 19.3 Shift Reaction Pathway. Reprinted by permission, Ross, J. R. H. in *Surface and Defect Properties of Solids: Specialist Periodical Reports,* Vol. 4, p. 34, The Royal Society of Chemistry, Cambridge, England.

quickly reaches equilibrium. But in regions where the reforming reaction is limited by pore diffusion of methane, the effective shift rate is also limited.[4] See also *Kinetics* in this section.

It should be emphasized that other concepts of mechanisms are also in vogue. For example, a proposed path involving carbide formation has been suggested for the water gas synthesis reaction.[17]

$$CH_4 \rightarrow C_{ads} + 2H_2$$

$$C_{ads} + H_2O \rightarrow CO + H_2$$

The interested reader can study the several reviews listed in the references.[4,5,12,13,15]

Catalyst Types

Catalyst manufacturers offer a wide variety of catalysts and are experienced in selecting the best catalyst type based on feedstock characteristics and furnace design. The reforming reaction is very rapid and occurs at high temperatures in multiple catalyst-packed tubes installed in a direct-fired furnace. Such conditions require a rugged, high-crush-strength carrier in a form that will permit maximum entry to the active surface. In regions of maximum temperature, the reaction becomes diffusion controlled, and much attention must be given to catalyst shape, pore size, and overall dimensions. Options such as very small catalyst size and high internal surface area from very small pores are not feasible. Small catalyst size causes unacceptable pressure drop. Very small pores, although adding significantly to the measured surface area, are diffusion limited for the fast reforming reactions and are essentially not usable.

These impediments are overcome by providing catalysts with the following characteristics:[9]

1. Catalyst shapes that maximize outside surface area per unit volume while providing sufficient size to minimize pressure drops, including:
   - Raschig-ring shape
   - Cylinders with multiple holes (e.g., seven axial holes)
   - Ring with seven spokes
2. Sturdy carriers resistant to thermal shock and with high crush strength, of which a major portion is retained even after reduction of the NiO prior to startup.
   - Refractory alumina
   - Ceramic (e.g., magnesium aluminate)
   - Calcium aluminate, titanate
3. High-activity catalysts, when placed near the inlet of the tubes, can prolong tube life.
   - High-activity catalysts, such as smaller sizes and/or catalysts with higher exterior surface area, permit an earlier close approach to equilibrium in the inlet region and a lower rate of temperature rise. A corresponding reduction in the maximum tube skin temperature thus occurs which, over years of operation, assures longer tube life.
4. Combination of certain catalyst types may be recommended by the catalyst supplier to optimize performance for a particular feed and reactor furnace type.
   - As the reaction rate declines in the lower part of the tubes because of the close approach to equilibrium, larger catalyst sizes may be used to reduce pressure drop.
   - Since the number of moles flowing increases by a factor of two, pressure drop reduction is a significant goal in the last half of the catalyst bed.

Catalyst Carriers

A variety of catalyst carriers are available with characteristics applicable to specific reformer design and operation. Both refractory alumina and ceramic magnesium aluminate carriers offer high crush strength and stability including maintenance of crush strength over long periods on stream. Calcium aluminate, a cement, is very hard and has high initial crush strength but loses a significant amount of strength during use, especially in higher-pressure operation. High partial pressure of carbon oxides at high temperature apparently react with the calcium and break bonds that originally contributed to the crush strength. For this reason, calcium aluminate carriers tend to be used for low-pressure operation. A calcium aluminate titanate with titanium oxide substituted for part of the alumina has been developed that exhibits good retention of crush strength because of the ceramic-type bonds developed. This catalyst can be applied for more severe conditions.

Typical chemical analyses for the several catalysts may be summarized as follows:

Carrier	Nickel as NiO, wt%	SiO_2, wt%
Refractory alumina	12–20	<0.05
Magnesium aluminate	16–18	<0.2
Calcium aluminate	16–25	<0.2
Calcium aluminate titanate	19–25	<0.2

Silica must be minimized to avoid significant volatilization of the silica on downstream apparatus.

Sizes

Catalyst sizes are mostly 5/8 in. in diameter and from 1/4–3/4 in. in height. Diameters of 3/4 in. are also produced. Surface areas are in the range of 3.5–5 m²/g. Forms include cylinder with holes, rings, and rings with spokes.

Additional catalysts are discussed under "Steam Reforming of Higher Hydrocarbons and Naphtha."

Catalyst Suppliers and Licensors

Suppliers

   BASF, Dycat International, Haldor Topsoe, Synetix, United Catalysts

Licensors

   ABB Lummus Global, Davy Process Tech., Foster Wheeler, Haldor Topsoe, Jacobs Engineering, Kellogg Brown & Root, Technip, Selas, Krupp-Uhde, Synetix

Catalyst Deactivation

The severe operating conditions for steam reforming can potentially exacerbate catalyst deactivating processes such as carbon formation, sintering, and poisoning. Great care must be exercised to avoid such problems, otherwise catalyst life can be severely curtailed and reactor tubes damaged beyond repair.

Carbon Formation

Carbon formation in methane steam reforming occurs by methane cracking (Reaction 3).

$$3. \quad CH_4 \Leftrightarrow C + 2H_2$$

$$K_p @ 700 \text{ K} = 0.11$$

$$K_p @ 800 \text{ K} = 0.701$$

$$K_p @ 1000 \text{ K} = 10.02$$

This endothermic reaction becomes more favorable at higher temperatures (800 K and above). Since the main reactions (water gas and water-gas shift) are rapid, and can be made even more so by a highly active catalyst, equilibrium is rapidly reached in the first $1-4$ m of a reactor tube.[4,6] This conclusion is especially true in the catalyst pores of the low-effectiveness-factor catalyst. An approximation of the possibility for carbon formation can be obtained by calculating the minimum steam ratio below which carbon formation can occur at a given temperature, assuming simultaneous equilibrium of both the main reactions and the methane cracking reaction.[4,5,6] The main synthesis gas forming reactions are competing for methane with the carbon forming reaction. With adequate steam in the reaction mix, the synthesis gas reaction is able to consume the methene preferentially and prevent carbon formation.

$$CH_4 \underset{\longleftarrow}{\overset{H_2O}{\rightleftharpoons}} \text{ synthesis gas}$$
$$\updownarrow$$
$$\text{carbon}$$

The steam ratio is expressed as moles water per mole of carbon in the hydrocarbon. Such calculations show that, at usual methane steam reforming conditions, no carbon will form. However, this condition assumes equilibrium, which may not exist in the open system inside the tube where radial temperature gradients occur.[4,5,6]

   This radial gradient may cause a deviation from the reforming equilibrium calculated at the mean catalyst temperature. Figure 19.4 depicts such condition by plotting temperature vs. tube length. Three temperatures are shown: the mean catalyst temperature (T_{CM}), the carbon-forming

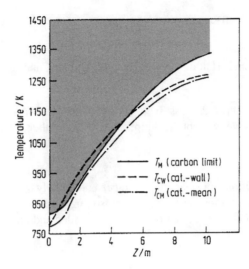

FIGURE 19.4 Carbon-Forming Limit vs. Axial Distance in Reformer Tube (H_2O/CH_4 pressure ratio = 1.3, discharge pressure = 4 bar, average heat flux = 72 kw/m²). Reprinted from Rostrup-Neilsen, J. R. and Christiansen, L. J., in *Proc. 6th Simposio Ibero-Americano de Catalise*, Rio de Janeiro, p. 1615, 1978.

limit temperature (T_M) above which carbon production is predicted, and the catalyst temperature at the tube wall (T_{CW}). The cross-hatched region is the region of carbon formation.[4,6] Note that the catalyst at the wall exceeded the limit temperature, T_M, in the early portion of the tube (between 1 and 4 m). The resulting over temperature causes *hot bands*.[6] Such hot bands may be the result of high heat flux and low catalyst activity.[6] Lower activity reduces the rate of these highly endothermic reactions and thus the rate of heat absorption. Higher mass velocities can alleviate such conditions.

Calculated values of the minimum H_2O/C ratio for various temperatures are usually conservative estimates. Most thermodynamic tables use graphite as the basis for carbon. It has been demonstrated that the carbon formed is actually a slightly different form referred to in the literature as Dent carbon.[5] It permits lower steam ratios (see Figure 19.5). Another view explains the difference as due to the higher surface energy of the very small-diameter whisker, which is indeed graphitic.[4] However, conservative operation would suggest a higher ratio based on operating experience, furnace design, feedstock, and product CO/H_2 ratio desired (see Table 19.1).

The worst possible outcome is extensive carbon formation. Unlike the coke so common in many hydrocarbon reaction systems, the carbon formed by methane cracking forms as rather sturdy needles or whiskers. The carbon grows as a filamentous structure with the nickel particle at the top. As it grows out from the catalyst interior, it can cause permanent structural damage to the catalyst and require shutdown and replacement.

If significant amounts of higher hydrocarbon impurities are present in the methane feed, pyrolysis carbon can also be formed. See the section on "Steam Reforming of Higher Hydrocarbons."

Sintering

Although the catalyst carrier is rugged and somewhat temperature resistant, small nickel crystallites essential for high activity can grow in size via surface diffusion at temperatures above 700–800°C.[6] It is known that stable micropores of the carrier can impede crystal growth if the nickel crystallite is of the same order of magnitude as the pore diameter.[6] Catalyst suppliers have various proprietary approaches to this problem.

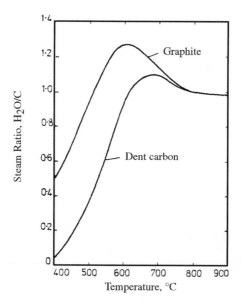

FIGURE 19.5 Minimum Steam/Carbon Ratio below Which Graphite and Dent Carbon Will Form. Reprinted by permission: Bridger, G. W., in *Catalysis: Specialists Periodical Reports,* Vol. 3, p. 64, the Royal Society of Chemistry, Cambridge, England, 1980.

Poisons

Sulfur compounds are major poisons of nickel catalysts, since they are strongly chemisorbed on the active nickel surface and cause rapid loss of catalytic activity beginning at the tube-inlet region. As activity declines in a region, the heat consumption required by the endothermic synthesis-gas reaction declines. Higher tube-wall temperatures result that can shorten tube life. Unreacted feed gas will react at higher temperatures in downstream sections making carbon formation more likely.[6] Although desorption of sulfur compounds such as H_2S is possible, it is a very slow and impractical procedure. Removing chemisorbed sulfur with steam involves the following suggested reactions.[4,6]

$$NiS + H_2O \rightarrow NiO + H_2S$$

$$H_2S + 2H_2O \rightarrow SO_2 + 3H_2$$

High-temperature steaming is required to remove the SO_2 from the carrier. The second reaction is not highly favored ($K_p = 3.5 \times 10^{-7}$ @ 700°C) and is therefore inhibited by even a small amount of hydrogen.[6]

Sulfur poisoning can occur due to an operating upset with resultant loss of activity. In such cases, it is possible to remove the sulfur as H_2S by reducing the feed flow by 90% and the steam rate by 75% while maintaining the highest allowable tube-wall temperature for 12–24 hr.[9] This procedure maintains the catalyst in the reduced conditions so that normal operation can be rapidly restored. Prolonged sulfur poisoning that leads to carbon formation requires a general regeneration aimed at both carbon and sulfur removal (see "Regeneration"). In summary, a thorough effort to avoid sulfur contamination of the catalyst is the solution. If the feed gas contains modest amounts of H_2S (or COS, RSH, CS_2), adsorption in a bed of ZnO is most effective and can produce a treated feed with less than 10 ppb of sulfur. Large amounts of H_2S can be initially removed by absorption in an amine solvent followed by ZnO adsorption.[4]

$$ZnO + H_2S \Leftrightarrow ZnS + H_2O \qquad Kp \text{ @ 573 K} = 5.9 \times 10^6$$

Other known poisons include arsenic, chlorine, and alkali compounds. It is thought that arsenic alloyed with nickel and As_2O_3 particularly reduce catalytic activity of alkali promoted catalysts.[6,7,8] Small amounts of lead and silica in the feed or stream can, over time, cause pore blockage and often foul downstream heat exchangers. Silica, which is volatilized in high-temperature regions, is kept below 0.3 wt% to avoid such problems. Chlorine can also cause deactivation at amounts in the feed greater than 1000 ppm.

Regeneration

Whisker carbon, over a period of time, can break down into a dense material that is difficult to remove under reducing conditions. Regeneration can be accomplished and operations continued successfully if the catalyst carrier has not been structurally damaged. All feed flow is terminated, and steaming at 700°C or above is continued until CO_2 in the outlet gas reaches a very low value. Alternatively, a lower temperature (450°C) can be used if a controlled addition of a small amount of air with the steam is used. The air oxidizes both the carbon and the nickel,[6] creating the possibility of hot spots if the process is not carefully monitored.

If the catalyst has been poisoned by sulfur, it can be removed along with the carbon by the steaming process. Catalyst suppliers need to be consulted for exact procedures for the particular carrier being used. Sulfur dioxide can react with some carriers, forming sulfates that are not removable.

Relatively fresh whisker carbon is easily removed by steam via the water-gas reaction.

$$C + H_2O \Leftrightarrow CO + 2H_2$$

This regeneration only requires an increase in the steam-to-methane ratio to a value of about ten,[6] followed by normal operation.

Process Units

Catalytic steam reforming of methane was invented by BASF in the late 1920s, but the first commercial application was designed and operated in 1931 by Exxon (then Standard Oil of New Jersey) at its Bayway and Baton Rouge Refineries to produce hydrogen for refinery use. Improvements have continued throughout the subsequent decades.

Natural gas feed has the advantage of usually being free of significant amounts of heavy hydrocarbons and complex sulfur compounds. In addition, it is delivered by pipeline at up to 70 bar (1000 psig) so that feed compressors are not required for reforming at 35 bar in the reformer. Methane content is in the range of 85 volume percent, with ethane and propane composing most of the remaining hydrocarbon content. At the lower inlet temperatures of the reformer tube, methane formation is briefly favored[5] but, as the temperature rises rapidly, direct conversion to CO and H_2 is favored. (See "Heavy Natural Gas and Naphtha Steam Reforming.")

Desulfurization

In most cases, sulfur content in pipeline natural gas is low and consists almost exclusively of H_2S, which can be removed by a ZnO adsorbent bed. Should excessive amounts be present that would cause rapid saturation of the ZnO unit, pretreatment with an amine solution in a separate column followed by clean-up in a ZnO bed is practiced. Natural gas supplied for domestic purposes contains purposely added mercaptans which, in small amounts, can be removed above 350°C by the ZnO bed.

Reactor Furnace

The treated feed is preheated in steam-heated exchangers and then in the convection section of the furnace after being joined by preheated steam. It is then distributed to multiple rows of identical tubes via headers. The tube ensemble is supported by a fixed tube support at the top or bottom but located outside the firebox.[10] By this means, the tubes can freely expand longitudinally from the support. Counterweights or spring hangers are used to relieve the entire tube assembly of a significant portion of its dead weight.[10] Pigtails or hairpin connections are made between headers and each tube in a row to allow for expansion. A similar arrangement at the outlet of the tubes is used.

Two types of firing arrangements are in common use: top-fired and side-fired (see Appendix). The tubes in a side-fired furnace are heated by the radiant side walls, while top-fired tubes are heated mainly by radiating combustion products from the large gas burners.

Two types of side-wall furnaces are in use. The vertical-wall type employs an arrangement that deflects the flame to the refractory wall rather than allowing it to impinge directly on the tube surfaces. The terrace-wall type uses burners located at the bottom of each section, which impinge at the slanted refractory wall. Since gas must be supplied to burners that are accessible to an outer wall, the side-fired furnaces require a separate radiant box for each row of tubes. Both types have been successfully used. Top-fired furnaces have fewer but larger burners and are more compact and thus amenable for large units.[11]

Tubes are generally in the range of 75–125 mm OD (3–5 in.), 10–20 mm wall thickness (0.39–0.79 in.), and 9–15 m (29.5–49 ft) in length, depending on the reformer type.[10]

Although low pressure is favorable to the desired equilibrium conversion, higher pressure provides more compact and energy-efficient design, since the high-pressure natural gas requires no costly compression, and downstream uses of the synthesis gas usually involve significant operating pressure. To operate effectively at higher pressures, the operating temperature must be increased. Newer tube alloys have made such operations possible. There are a number of such alloys, but the most frequently used high-temperature alloy is a 25 Cr 35 Ni, Nb. The older alloy, often referred to as HK 40, which is 25 Cr, 20 Ni, continues to be the preferred selection for low-pressure units.

Calculation of the outlet gas composition is readily and frequently done by simultaneous equilibrium calculations using actual outlet reaction gas temperature or an experience-based approach. A kinetic model is needed, however, to not only determine the required tube length but also the heat and mass-transfer processes both in and outside the tube so as to yield complete profiles of tube and reactant temperatures along with pressure drop in the tube and heat flux. This is illustrated in Figure 19.6.

The entire hot piping and multitube system must be mechanically designed as a single unit and based on detailed stress analysis of the hanging unit with calculation of hanging, expansion, and rupture stress.[11]

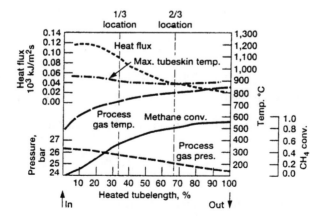

FIGURE 19.6 Top-Fired Reformer Profiles Based on Simultaneous Heat-Transfer and Reaction-Kinetics Calculations. Reprinted by permission: Johansen, T., Raghuraman, K. S., and Hackett, L. A., *Hydrocarbon Processing,* p. 120, August 1992.

Catalyst Loading

As is the case for any multitubular reactor, careful loading of each tube is essential for even distribution of reactants to each tube so as to avoid side reactions such as carbon formation in tubes with lower mass velocities. Even loading is accomplished by lowering and emptying a plastic or canvass sock containing catalyst at within one to two feet of the bed support and continuing at this same distance from the top of the bed as it is being formed. Each tube is loaded to the same height, and the pressure drop through the tube is tested. Any tube deviating more than 5–10% from the average ΔP for all tubes should be emptied and reloaded.

Operating Conditions

Operating temperatures and pressures vary according to the ultimate synthesis gas application. See Table 19.1. Typical ranges vary between 800–1000°C and 8–35 bars (100–500 psig). The higher pressure and temperature range is used in most modern steam reformers.

Subsequent Treatment of Steam Reforming Product

Methane steam reforming yields a product with a ratio as high as 6.5 hydrogen to carbon monoxide. This ratio can be varied somewhat by altering operating temperature and steam/methane ratio. The important processes that use synthesis gas produce H_2/N_2 ratios of one, two, three, or pure hydrogen. These values are accomplished by subsequent operations following the steam reformer.

The most frequently used procedure for increasing the CO content is the reverse shift reaction, which is favored at high temperatures and in the presence of excess carbon dioxide. Reference to Table 19.1 will show that four out of the eight processes listed use shift conversion. The shift reactor and process is discussed beginning on p. 418, followed by a description of each of the major synthesis processes along with the use of shift conversion where applicable.

Process Kinetics

Kinetics of the steam-reforming reaction (Reaction 1) have been studied extensively since 1933 with little overall agreement. Extensive reviews have been published,[5,12,13] and it is known that useful design models have been developed by catalyst suppliers and process licensors. The reaction system is really quite complex, involving not only the main reaction but also the shift reaction and carbon formation. A plant design model must be accurate enough to enable calculation of heat fluxes in each portion of the tube length and radial gradients in areas of possible high tube skin temperature. The simplest possible rate equation is helpful for the modeling problem that must be capable of accurate predictions of tube length, temperature distributions, and heat fluxes. One simple form that has been reported as used in plant design is based on the distance from equilibrium.[5]

$$r = k(P_M - P_M^*)$$

where P_M = the actual value of methane partial pressure at any position in the tube

P_M^* = the corresponding equilibrium value of methane partial pressure

This equation can also be written in terms of partial pressures of the reversible reaction.

$$r = k\left(P_M P_W - \frac{P_{CO} P_H}{K}\right)$$

where K = the equilibrium constant for the reforming reaction

As written, the previous equation does not take account of the effectiveness factor, which is less than 0.1 in most commercial operations. Accordingly, more complete rate forms have been proposed.[4]

$$r = \eta k_{iv}(1 - \epsilon)(C_M - C_M^*)f(P_W, P_H)$$

But $\eta = \dfrac{3}{\phi}$ for this low effectiveness factor and ϕ for a first-order reversible reaction is

$$\phi = \frac{D_p}{2}\left[\frac{k_{iv}(K + 1)}{KD_{eff}}\right]^{0.5}$$

$$\therefore r = \frac{(C_M - C_M^*)(1 - \epsilon)^6}{D_p}\{D_{eff}k_{iv}[K/(K + 1)]f(P_W, P_H)\}^{0.5}$$

where C_M = concentration of CH_4

C_M^* = concentration of CH_4 @ equilibrium

D_{eff} = effective diffusivity

D_p = equivalent particle diameter

$k_{i,v}$ = intrinsic rate constant based on catalyst volume

K = equilibrium constant

P_W = partial pressure of H_2O

P_H = partial pressure of H_2

ϵ = catalyst-bed void fraction

η = effectiveness factor

The equation can also be written in terms of partial pressures of methane and also on a unit mass of catalyst. The term involving the function of partial pressures of H_2O and H_2 is the main source of variation in rate equations. The aforementioned reviews will provide the background for the interested professional. One should be aware, however, that the best kinetic data and design and simulation models in this case are proprietary programs developed by catalyst manufacturers and licensors. Customers can receive valuable assistance from these sources.

The shift reaction under reforming conditions is known to be very fast and is often been assumed to be at equilibrium so that composition of the reaction mix could be followed by the reforming equation and simultaneous equilibrium calculations for the shift reaction. Extensive reviews of both laboratory and plant data, however, suggest that the shift reaction deviates significantly from equilibrium at low methane conversions.[12] At higher conversions of methane (~80%), equilibrium is achieved rapidly. Two explanations of this problem have been proposed.[4] One is a suggested kinetics based on two paths involving the reforming reaction.[12,18]

The CH_2O represents the surface residue from dissociative chemisorption of methane and steam, which decomposes to H_2 and CO or reacts with steam to give H_2 and CO_2.[12] The balance between these two routes represents the approach to equilibrium of the overall shift reaction. Increases in steam increase CO_2 production and the approach to equilibrium.

A second explanation suggests that, because the shift reaction is inherently very fast, it can be limited by the rate of diffusion of methane into the pore structure of the catalyst.[4] At lower conversions, the rate of methane reaction is high, and pore diffusion could become limiting.

Often, Power-Law rate equations have provided satisfactory expressions for following the shift reaction in the steam reformer. One such equation for the forward rate is[19]

$$r = kP_{co}^* P_W^{(1-x)/2}$$

The order on the water pressure must be in the range such that

$$0 > \frac{1-x}{2} < 1$$

Additional kinetics can be found in Refs. 23 and 24.

19.2.3 OTHER STEAM REFORMER TYPES AND PROCESSES

Over the years many improvements in steam reformer design have occurred along with the development of special reformer units that can be added to existing processes to increase capacity and/or reduce energy consumption. The characteristics of such units are summarized in Table 19.3.

19.3   HIGH AND LOW TEMPERATURE SHIFT CONVERSION (CO + $H_2O \rightarrow H_2 + CO_2$)

Although shift conversion occurs in the steam reforming reactor, it is more favored thermodynamically at lower temperatures. If production of hydrogen with high purity or a product of high hydrogen content is the operating goal, shift conversion is used to convert CO to CO_2. The CO_2 is then readily separated by absorption via any of the various acid-gas removal processes using alkanolamines or alkaline salt solutions followed by CO_2 recovery by means of a stripper. The lean solvent is then recycled to the absorber.

Lower-temperature operation requires different catalysts to assure favorably high reaction rates, but a simpler reactor design is possible because of the less severe operating conditions.

Chemistry

The reaction can be conducted at two different temperature levels and with two different catalysts in series. The so-called *high-temperature shift reactor* converts a major portion of the CO content to CO_2, and the low-temperature shift reactor removes most of the remaining CO using a more active catalyst with good activity in the favorable lower-temperature range. The stoichiometric reaction is the same in each.

$$CO + H_2O \Leftrightarrow H_2 + CO_2 \qquad (1)$$

High-Temperature Shift

Catalyst: iron oxide-chromium oxide		
Typical temperature range: 350–400°C		
Temp., K	ΔH, kcal	K_p
600	−9.29	26.65
700	−9.05	9.0

TABLE 19.3
Description and Characteristics of Various Reforming Processes

Type	Description
Methane or natural gas reforming	See section on "Methane steam reforming."
Naphtha steam reforming	See section on "Naphtha steam reforming."
Secondary reforming	An adiabatic fixed-bed reactor following the primary reformer is used to reduce residual methane in the gas from the primary reformer and add the necessary nitrogen to attain the 3:1 hydrogen-to-nitrogen ratio required for ammonia synthesis. See "Ammonia Synthesis."
Prereforming	An adiabatic fixed-bed reactor contains a higher-activity catalyst that enables some reforming to take place at lower temperatures and effect the removal of higher hydrocarbons prior to the regular steam reformer. The catalyst is similar to the regular steam reforming catalyst but higher in nickel content. By heating the mixed feed in the convection section of the steam reformer to 500°C, the fired-duty (radiant section of the reformer) is reduced, saving energy or, alternatively, permitting increased primary reformer capacity. Excess steam production can also be eliminated. In addition, the catalyst adsorbs the common poisons such as sulfur more strongly at the lower operating temperature and protects the more difficult to replace catalyst in the main reformer. Since some sulfur compounds will leak through the sulfur removal system, the preformer will act as an effective guard bed that will prolong catalyst life on the primary reformer. Higher hydrocarbons such as present in naphtha and, to a lesser extent, in natural gas cause carbon lay down in the primary reformer. The prereformer prevents this from happening in the primary reformer by converting C_2 and higher hydrocarbons to methane.
	The overall reaction is endothermic for a methane feed, and the temperature decline toward the end of the bed favors the exothermic shift reaction and methanation reaction. Since both of these reactions consume carbon monoxide, a typical prereformer will produce an effluent containing mainly hydrogen and methane. Thus, prereforming can be attractive for ammonia or hydrogen plants. Changes to steam-to-carbon ratio and temperature can be used to alter the outcome.
	Primary-reformer tube life may be increased when fed from a prereformer, since the tube-wall temperature may be lower because of the reduced severity required in the primary reformer.
	Prereforming advantages must be considered along with added costs for installation of convection section heating for the prereformer feed. Reports suggest that prereforming is best applied to existing units.
Post reforming	This process employs a multitubular reactor with catalyst inside the tubes. A portion of the feed is introduced at the bottom of the tubes and flows upward through the catalyst bed and out the top of the tubes, which are open to the reactor shell where it is joined by the effluent from the primary reformer. The combined gases provide heat for the catalyst bed and reacting gases. The technique can provide increased capacity and save energy for an existing unit and is especially useful for plants designed for production of rather pure hydrogen.
Gas-heated reformer	Another concept for recovering both high- and medium-temperature heat within a refractory lined multitube heat exchanger designed to exchange heat from the primary reformer effluent with a by-passed primary-reformer feed (CH_4 + steam). This permits increased capacity for the primary reformer and energy savings. It is reported to be useful for hydrogen and ammonia plants.

TABLE 19.3
Description and Characteristics of Various Reforming Processes (continued)

Type	Description

Autothermal reforming	This combines steam reforming with partial oxidation. In so doing, endothermic heat for steam reforming is supplied by the exothermic partial oxidation. The large reactor vessel consists of a refractory lined unit, the top portion of which contains the oxygen burner. Burners of various designs have been used, each of which provides thorough mixing of the feed and oxygen as well as means for cooling the burner length. Oxygen or air and steam flow through the burner inner tube, which is jacketed for cooling. Methane or primary reformer effluent (depending on the particular application) flows in an annular space around the burner. Combustion occurs between a portion of the synthesis gas or methane, and the very hot mixture flows to the larger diameter portion of the bed containing the catalyst bed. The catalyst is protected by layers of ceramic balls. The catalyst is produced as very strong refractory Raschig rings, similar in active ingredients to regular reforming catalyst but larger in size.
	Operating conditions:
	   1200–1250°C inlet to bed
	   20–40 bar or higher
	   Lower pressure can cause soot formation.
	   Higher allowable temperature provides greater operating flexibility.
	Uses: It has been used in ammonia and methanol plants and also for producing hydrogen or carbon monoxide. Uses have included standard or secondary reforming.
	Economics: It offers lower capital costs, but operating costs hinge on the charge for pure oxygen, which can vary more than that of compressed air. Oxygen is often supplied by an outside source but can be provided by a supplier adjacent to the plant and via a long-term contract.

Oxygen secondary reforming	See section on "Ammonia Synthesis."

Based on Refs. 3, 10, 11, 20, 21, and 22.

Low-Temperature Shift

Catalyst: copper oxide-zinc oxide		
Typical temperature range: 200–250°C		
Temp., K	**ΔH, kcal**	**K_p**
473	−9.56	230
523	−9.45	87.4

Mechanism

There appears to be no general agreement on a mechanism for either the high-temperature or low-temperature shift catalyst. Early proposals based on an oxidation-reduction mechanism for the iron-based, high-temperature catalyst postulate alternate oxidation and reduction of the partially reduced surface of the catalyst oxide.[1,2]

$$H_2O + * = H_2 + O* \tag{2}$$

$$CO + O* \rightarrow CO_2 + * \tag{3}$$

Although a stoichiometric-number technique has been shown to indicate that the proposed reaction sequence of reactions 2 and 3 is not valid,[2,3] rate equations developed therefrom have shown good agreement with various experimental studies. One mechanism that does pass the criteria of the stoichiometric-number technique is as follows:[3]

$$CO \rightarrow CO*$$

$$H_2O \rightarrow 2H* + O*$$

$$CO* + O* \rightarrow CO_2*$$

$$CO_2* \rightarrow CO_2$$

$$2H* \rightarrow H_2$$

The third and fourth steps are said to be rate controlling.[3]

High-Temperature Shift Catalyst (HTS) Types and Suppliers

An iron oxide-chromium oxide catalyst without additional support material is used. It is prepared either by (a) kneading fine iron oxide and chromium oxide (CrO_3) powder together followed by adding of water, calcining, and pilling or by (b) coprecipitation of ferrous sulfate and chromic acid at an adjusted pH followed by filtration and calcining.[4] The coprecipitated catalyst is more active.

Typical Catalyst Properties

Analysis: 89% Fe_2O_3, 9% Cr_2O_3
Some varieties are promoted by 2% CuO and are reported to be more active and capable of operating at lower H_2O/C ratio.
Tablets: 9.5×4.5 mm, 6×3 mm, 9.5×9.5 mm, 6×6 mm
Operating Temperature: 350–400°C

Upon reduction, the iron oxide is converted to Fe_3O_4, and the final reduced state consists of a solid solution of Fe_3O_4 and Cr_2O_3 of the spinal type.[2] Chromium oxide is considered to be a stabilizer of the active reduced iron oxide, thereby hindering sintering at higher temperatures.[2] It has also been suggested that Cr_2O_3 also increases the active area of the catalyst. Cr_2O_3 content over 14% is not useful because the overage forms a separate phase.

Catalyst Suppliers
BASF, Dycat International, Haldor Topsoe, Synetix, United Catalyst

Licensor
Kellogg Brown & Root, Haldor Topsoe

Deactivation of HTS Catalysts

Poisoning
Actually, the iron-chrome HTS catalyst is fairly resistant to poisoning. It is resistant to sulfur compounds such that it only loses activity modestly rather than completely, as in the case of nickel reforming or low-temperature shift catalyst. In the range of 70–1000 ppm of H_2S, for example, activity declines proportional to $(P_{H_2S})^{-0.55}$ have been reported.[5,6] Further significant sulfur poisoning

is unlikely, because the upstream gas feed to the reformer is necessarily desulfurized to avoid catastrophic poisoning of the sulfur sensitive reforming catalyst. HTS activity is not affected until the sulfur content of the feed exceeds 200 ppm.[8] Sulfur contamination is the result of incomplete sulfur removal from the methane feed due to malfunction of the adsorbing unit or from a fresh reforming catalyst.

Chlorine compounds in amounts above 1 ppm in the feed to the shift converter will cause deactivation.[5] Lower quantities are not a problem, but they become one when a second stage LTS catalyst follows the HTS reactor.

Pore Blocking by Solids

Solids that can enter the system via steam, especially during supply upsets, include silica and phosphorous compounds. These impurities, along with dust and soot from upstream sources, can block the pores of the catalyst.[6] The result is significant activity decline because of the loss of available surface area. Unsaturated hydrocarbon impurities will polymerize on the catalyst and, over time, cover active surface with a polymer barrier. Apparently, small amounts of nitric oxide in the feed promote this polymer formation.[2] Use of inert balls on top of catalyst beds is very effective in capturing solid deposits. If sized properly, they will not plug rapidly.

Sintering

The active iron-chromium spinal is subject to sintering at temperatures significantly above 500°C and can occur during operating upsets such as an upstream cooling train malfunction. The reduced area resulting from sintering will ultimately require higher operating temperature. The result can be a shorter catalyst life.

Active Catalyst Deterioration at Low Steam Rates

As synthesis plant operations began to seek significant savings in energy costs, it was noted that, at H_2O/C (or H_2O/methane for methane feed) below 3.5, significant amounts of methane, along with other hydrocarbons and oxygenates, were produced. This situation not only reduces hydrogen production but also feeds compounds to an LTS unit that will produce noticeable deactivation of the LTS catalyst.[7]

The phenomenon is caused by reaction of CO with some portion of the shift Fe_3O_4 active catalyst to form Fe_5C_2, an active carbide phase. This Fe_5C_2 is a Fischer–Tropsch catalyst type that is used to produce various hydrocarbons from CO and hydrogen.

$$5Fe_3O_4 + 32CO \Leftrightarrow 3Fe_5C_2 + 26CO_2$$

Soon after plant operation became limited by the degree of reduction in H_2O/C ratio, catalyst suppliers developed a new catalyst containing CuO, which acts to prevent the formation down to reported rather low steam-to-carbon values as low as 2.[7]

The copper content is subject to sulfur poisoning and, where necessary, a ZnO layer can be installed on top of the HTS catalyst bed.

Shutdown and Reuse

The HTS catalyst can be easily damaged even during careful oxidation at shutdown. When not necessary to remove the catalyst, a continuous flow of inert gas is recommended during shutdown so that the catalyst remains safely in its oxidized form.

Low-Temperature Shift Catalyst (LTS) Types and Suppliers

The low-temperature shift catalyst is a mixture of copper oxide and ZnO in a ratio of 1:2, or some other ratio approaching 1:1, with alumina added in place of some of the ZnO. In addition, promoters such as Cr_2O_3, MnO, or other metal oxides have been used. Chromium oxide has been used in place of alumina. Preparative procedures are much more critical for the LTS catalyst than for the HTS catalyst. Coprecipitation of the metals as metal nitrates by pH adjustment with ammonium bicarbonate is reported.[4] The oxides thus formed are intimately intermixed by this procedure, which is essential for high activity and stability. High dispersion appears to be essential for high activity, which in turn means close contact between small crystals of copper and zinc oxide.[6] In the reduced active form, metallic copper and ZnO would be in close contact, and it has been suggested that such a metal-semiconductor junction would serve to stabilize unusual but essential oxidation states. ZnO in excess also serves to protect the copper from inadvertent sulfur poisoning (see section below on poisoning).

Aluminum oxide also serves as a stabilizer for copper which, by itself, would be easily sintered. There is no doubt that these LTS catalysts require great skill and proven proprietary techniques in their manufacture.

Typical Catalyst Properties

	Analysis	
CuO	ZnO	Al_2O_3
33	65	<2
42	47	10
33	39	33

Tablets: 6 × 3 mm, 4.5 × 2.5 mm, 4.5 × 3.4 mm
Operating temperature: 200–250°C

Suppliers

Haldor Topsoe, Synetix, United Catalysts

Licensor

Kellogg Brown & Root, Haldor Topsoe

Deactivation of LTS Catalyst

Poisons

The common poisons are sulfur and chlorine compounds and silica. Normally, sulfur compounds such as H_2S or COS are removed from the methane reformer feed in the ZnO adsorber beds. However, during times of upset such as even short-time periods of high-sulfur feed, break-through sulfur will occur and pass to the HTS reactor. In such instances, the HTS catalyst can safely adsorb the H_2S, for example, and protect the LTS bed. But the HTS catalyst has a limited capacity such that H_2S will at some point break through to the LTS bed. It is for this reason that LTS catalysts containing excess ZnO are used such that the upper portion of the bed can serve as a sulfur guard.

Sulfur poisoning of LTS catalyst is rather complex. In the early part of the bed, zinc sulfide forms, but farther down the bed, sulfur, as H_2S, is chemisorbed, the extent of which depends on operating conditions.[6] This action will, of course, cause deactivation of active sites. An interesting calculation has been presented based on an assumed ammonia plant of 1000 MTPD capacity with all the sulfur being retained by reaction of ZnO and chemisorption and reaching full poisoning of 2000 ppm after 2 years. The amount of sulfur deposited per day was calculated to be 50 parts per billion by weight.[6] Clearly, the best efforts to remove sulfur from all streams that ultimately flow through the LTS reactor are essential. These include hydrocarbon feed, steam, quench, lubricating oils (in compressors, pumps, and other machinery), and air introduced in a secondary reformer.

Chloride compounds, although not a serious problem for steam reforming catalysts and HTS catalysts, are a major and permanent poison for LTS catalysts. It is thought that CuCl and $ZnCl_2$ are formed, which, by mass transport, cause copper crystal growth (sintering) and significant loss of catalyst activity.[6] In addition to installing a bed of chlorine adsorbent (e.g., CaO/ZnO or alkalized alumina) upstream of the ZnO adsorbent bed prior to the reformer, another bed is placed above the LTS catalyst, composed of a proprietary adsorbent.

Several somewhat surprising situations can occur if both sulfur and chlorine compounds enter an LTS bed.[6]

- If sulfur was adsorbed first, chlorine that follows will lower surface area and cause sulfur to desorb from the inlet portion of the bed and deactivate the lower section.
- If chlorine enters the bed first, the loss in surface area will reduce the sulfur capacity and thus resistance to sulfur poisoning.

Chloride compounds can enter via the same sources listed above for sulfur compounds. In addition, air introduced to the upstream secondary reformer may contain chlorine or chlorides from air contamination by chlorine or salt water spray.[8]

Silica enters feed streams from steam supplies, quench water, catalyst bed-support extrudates, brick linings, and even possibly from reforming catalysts (small amounts). It deactivates by covering active sites and blocking pores. Inert balls or rings of proper sizes installed on top of the bed can filter out such solids as silica and dust or soot without excessive increase in pressure drop.

Sintering

In addition to the catastrophic sintering caused by chlorine or chloride compounds, excessive temperature alone can cause sintering. The very small active crystallites of copper are thermodynamically favored to coalesce into larger crystals and thus produce a less active catalyst. Although the small crystallites are stabilized by the associated ZnO and also alumina, this protection is destroyed at elevated temperatures, either over a period of time or quickly, depending on the size of the temperature rise and the particular catalyst characteristics. Inlet operating temperatures between 175–275°C for the latest formulations have been suggested, but it is wise to always operate at the lowest possible temperature, since sintering is a phenomenon related to both time and temperature. The lowest temperature, however, should be no lower than 20°C above the dewpoint to avoid condensation in the pores and subsequent catalyst damage. As temperature is increased during an operating cycle to overcome deactivation, the growth of crystals increase and catalyst life is shortened. In the lower regions of temperature, the sintering rate is very low but will increase as temperature is raised. Ultimately, the deactivation rate becomes significant, and catalyst life can rapidly be shortened. It is reported, for example, that an LTS catalyst withstood serious sintering when subjected to 300°C for 2 hr.[9] But much longer operation at this or a higher temperature would certainly accelerate sintering and catalyst activity decline.

Process Units

Depending on the desired product gas, shift conversion (high-temperature or both high- and low-temperature) is applied along with other process steps to produce a variety of synthesis gases as summarized in Table 19.1. The specific applications will be noted in more detail as various synthesis products are discussed. The focus in this section is the shift converters and their operation.

Shift conversion is an exothermic process that is conveniently implemented in fixed-bed, adiabatic reactors in most plants. The gas is fed downflow through a bed of catalyst containing layers of ceramic support balls or extrudates at the inlet to serve in creating even flow distribution over the entire cross section and for separating solid contaminants that may have entered the feed.

Ceramic support balls or extrudates are also used to support the bed and provide a means for maintaining even flow in the bed bottom. Alternatively, a metal grid with holes sized somewhat smaller than this catalyst can be used. Multiple beds provide the opportunity to optimize the temperature in each portion of the HTS bed. Such a procedure may be valuable in reaching the desired CO level. The effect of changing temperature of the steam can produce varied results, depending on the equilibrium approach. If the reaction is at equilibrium, raising the temperature causes the reverse reaction to occur, as does a lower steam rate.

Because of the larger temperature rise in the HTS, it can be convenient to provide several beds in the HTS reactor with intermediate cooling. A temperature rise of only 20°C is possible in the LTS reactor, and no such arrangement need be considered for it.

Operating conditions for the HTS and LTS reactors are typically 350–400°C and 200–250°C, respectively. Since there is no net change in moles, pressure has no effect on the equilibrium. Excessive pressure drop, however, can damage the catalyst, but modern shift catalysts are rugged and have a relatively high crush strength.

Catalyst suppliers offer thorough instructions for start-up, catalyst reduction, operation, and shutdown for the particular catalysts purchased. Instructions for catalyst reduction and start-up are particularly critical, since excessive temperatures must be avoided. Reducing gas of H_2 mixed with N_2 or with natural gas is usually recommended for LTS catalysts. By contrast, HTS catalyst is reduced after careful heating by wet process gas from the reformers. Various procedures for heating the bed and then introducing the process gas are practiced. The higher the bed temperature reached, the more critical temperature control becomes, particularly if the process gas is introduced after the bed reaches operating temperature. Sudden rapid rise in temperature can damage the catalyst.

Intermediate cooling and energy recovery is used between the two types of catalyst beds. Further processing of the outlet shift product is described for the various synthesis processes discussed in the following sections.

Process Kinetics

HTS Catalyst

The high-temperature shift catalyst has been studied for many years, and a number of kinetic rate equations have been proposed. Several reviews summarize most of the work in this area.[2,10] It appears that a power-low equation can provide a very good agreement with observed data.[2]

$$r = k(CO)^a (H_2O)^b (CO_2)^c (H_2)^d (1 - \alpha)$$

where   k = rate constant
   $\alpha = (CO_2)(H_2)/K(CO)(H_2O)$
   K = equilibrium constant
   ( ) = designated component in concentration units (A similar equation in partial pressure can also be written.)

The equation form has proven useful. At typical operating pressure, the hydrogen exponent is 0, that on CO_2 varies from −0.3 to 0.56, and that on CO is 1.[2,5] There is merit in fitting the equation based on data obtained on the actual catalyst size to be used. In this manner, transfer into the catalyst pore is taken into account by a lower observed activation energy compared to small catalyst particles, which present little or no significant diffusional gradient in the pellet.

LTS Catalyst

The low-temperature shift catalysts are newer than the HTS catalysts, and there has not been as much effort spent in developing and testing rate forms. Rate equations have been reviewed[2] involving various forms, but a power-law equation of the same type as presented above for HTS catalysts has been used successfully when restricted to a given catalyst size.[10] The order on hydrogen continues to be observed as zero.

One should remember that, for large-volume catalysts such as shift catalysts, the catalyst manufacturer will usually have well documented rate equations and reaction models for a particular catalyst product.

19.4 NAPHTHA STEAM REFORMING

The value of naphtha as a feedstock for other petrochemicals and the emergence of major new discoveries of natural gas off shore and in newer land-based findings has made naphtha steam reforming less important. Areas devoid of cheap natural gas but with a source of naphtha cuts from petroleum will continue to find naphtha steam reforming attractive. Aliphatic naphtha is preferred, since aromatics can cause a great deal of undesired side reactions.

Chemistry

The chemistry of naphtha steam reforming is similar to methane steam reforming with the exception of the initial reaction sequence involving the higher hydrocarbons. To the stoichiometric equations given for methane reforming must be added the first equation below, the reaction of the higher hydrocarbons.

$$C_nH_m + nH_2O \rightarrow nCO + \left(\frac{m}{2} + n\right)H_2 \tag{1}$$

This reaction is irreversible.

$$CH_4 + H_2O \Leftrightarrow CO + 3H_2 \tag{2}$$

Temp., K	ΔH, kcal	K_p
800	53.21	0.031
1000	53.87	26.12
1123	54.33	530.26

$$CO + H_2O \Leftrightarrow CO_2 + H_2 \tag{3}$$

Temp., K	ΔH, kcal	K_p
700	−9.05	9.0
1000	−8.31	1.4

A casual look at these equations leaves one wondering where the methane comes from. In terms of the stoichiometry and the equilibrium values, Reaction 2 is thermodynamically favored to proceed from right to left at the inlet region of the tubes. Thus, the CO and H_2 formed from Reaction 1 react to produce methane. As the reaction mix is heated in the reformer furnace (>1100 K), the reforming reaction of CH_4 to CO and H_2 is favored. As discussed in the section on methane reforming, Reaction 3 (shift reaction) tends to be at equilibrium.

Mechanism

Much has been written on the possible mechanism of steam reforming, and a summary is given beginning on p. 408 for methane reforming. The higher hydrocarbons present in naphtha and even light multicarbon hydrocarbons (ethane, etc.) are more readily converted than methane, since the bonds are more easily broken. A reactivity sequence has been determined as follows for a variety of hydrocarbons.[1]

cyclohexane > trimethylbutane > n-butane and n-decane > n-heptane > ethane > benzene > methane

One can assume that the same description of the steps in methane reforming shown on p. 406 apply. Multiple CH_x fragments from the higher hydrocarbons react with steam to produce CO and H_2. At high temperatures (reforming conditions), Reaction 2 goes to the right and forms CO and H_2 from methane and water. At low temperatures (less than 550°C), the equilibrium reverses to produce methane (methanation) exothermally. Reaction 3 becomes more favorable in the direction of CO_2 and H_2 formation. Lower temperature operation is the basis of the production of synthetic natural gas and of prereforming.

Catalyst Type and Suppliers

Since higher hydrocarbons are susceptible to catalytic cracking under primary reformer conditions, it is important to select a catalyst with little or no acidic sites. Thermal cracking can also begin to occur near the inlet portion of the tubes. A highly active catalyst near the inlet portion of the tube will aid in converting much of the heavier hydrocarbons to lower-molecular-weight unsaturates at lower temperatures before the region of higher temperatures is reached and thermal cracking, along with catalytic cracking, can occur. This catalyst should be based on a low-acid support to deter catalytic cracking of the unsaturated intermediates and subsequent carbon or coke formation.[7] Below this top layer of catalyst, a nonalkalyzed catalyst beginning somewhere between 2–4 m from the inlet is recommended, since the reactions in this region are similar to those in methane reforming.[2]

The amount of alkali included in the inlet catalyst is kept to a minimum (1–1.5 wt% K_2O) since, in sufficient amounts, it reduces the steam reforming reaction rate.[3] Since the K_2O is slightly volatile at reforming conditions, over time it can deposit on nickel sites, reducing activity. When deposited on an acidic support, potassium is held more firmly.[3] If heavy naphtha is used as feed, the second part of the bed is supplied with a smaller alkali loading (0.3–0.5 wt%) to take care of possible breakthrough of heavy hydrocarbons.

The alkali of choice is a potassium compound such as KNO_3, which is added in an aqueous solution along with other components to the carrier. It ends up as K_2O upon decomposition of the nitrate.[4]

Various explanations on the role of potassium have been forthcoming.[5,6]

- On catalyst with acidic supports, the potassium neutralizes the acidic sites and suppresses the cracking of naphtha.
- The promoting role of potassium involves the increase in steam adsorption which, in turn, aids in preventing carbon formation.

Catalyst Suppliers and Licensors

Suppliers
  Same as for methane reforming (p. 411)

Licensors
  Kellogg Brown & Root, Haldor Topsoe

Catalyst Deactivation

See section on methane reforming.

Process Units

The primary reformer description for methane steam reforming applies to that used for naphtha reforming. Of course, the design based on the heat fluxes required and convective energy recovery will vary in detail.

Early in the operation of naphtha steam reforming, British Gas introduced *prereforming,* which consisted of an adiabatic fixed-bed reactor prior to the primary reformer. This reactor operated in a temperature range of 450–500°C. At these conditions in an adiabatic reactor, the temperature declines (~20°C) initially because of endothermic cracking reactions, which are then followed by methanation reactions and water-gas shift that produces rapid increases (~30°C) and then leveling off of temperature due to the two exothermic reactions.[5]

The effect of the prereformer is to convert a major portion of the naphtha components to methane, hydrogen, and CO_2. In addition, it serves as a guard chamber for the primary reformer and downstream units and reduces the heat load of the primary reformer.[8,9] Because of the lower operating temperature, it is a strong adsorber of sulfur compounds that might occur because of some upset in the upstream sulfur removal section.

Although natural gas is the main feedstock these days, prereforming is used in many such units to remove small amounts of heavier hydrocarbons and save on energy costs. The temperature contour is similar to that for naphtha, but the drop in temperature is greater because of the endothermicity of the overall Reaction 1 for the lighter hydrocarbons. The rise in temperature is not very large, because of the small amount of higher hydrocarbons present and the low amount of exothermic heat produced by the reaction of CO and H_2 to methane and carbon dioxide.

Other Processes for Production of Synthesis Gases

Gases from partial oxidation, coal gasification, and coke-oven product have been used in the past, but partial oxidation continues to be used in various locations where heavy hydrocarbon stocks are available at very low cost compared to natural gas. It is a noncatalytic process that can use any hydrocarbon feedstock. Operating costs are higher, but no NO or SO_2 emissions occur.[7] More than 100 partial oxidation plants are in operation.

New Routes

A consortium of companies (Amoco, British Petroleum, Praxair, Statoil, Sasol, and Phillips Petroleum) has been developing a process for producing synthesis gas from methane based on ceramic membranes.[10] The membranes separate oxygen from air and partially oxidize methane in a single step. Such a process should reduce energy and equipment costs. The search for a less energy intensive process than steam reforming has been sought for many decades with little success. Perhaps this development will be the exception.

19.5   METHANOL SYNTHESIS (CARBON MONOXIDE + HYDROGEN → METHANOL)

The sole source of methanol from 1830 to 1923 was the dry distillation of wood (thus the name *wood alcohol*). In 1923, BASF began the first commercial production of methanol at its Leuna Works from synthesis gas, carbon oxides, and hydrogen.[1] It is not surprising that BASF was the first, for during the development of the Haber–Bosch process in the earlier 1900s, researchers noted the presence of ammonia and other oxygenated compounds. Synthesis gas in those days was produced from coal and contained sulfur compounds that were strong poisons for the reaction to methanol. The successful process emerged from the development of a sulfur-resistant catalyst, zinc oxide–chromium oxide. High pressures and temperatures were required (250–350 bar and 320–450°C). The high-pressure process continued as the only viable approach until the 1960s, when ICI introduced a much improved catalyst for use with then readily attainable sulfur-free synthesis gas made mostly from natural gas. Operation was at 50–100 bar and 200–300°C.[1,2] This process is now dominant.

The major use of methanol continues to be the production of formaldehyde, followed by the fuel additive MTBE (methyl tert-butyl ether).[2] Acetic acid is produced from methanol via carbonylation and is the third largest single use. Dimethyl terephthalate, methyl amines, methyl halides, dimethyl ether, and methyl methacrylate all require methanol as a reactant in the manufacturing process. Methanol has many other uses such as a solvent and antifreeze.

Every time the price of oil or natural gas goes up a little, doomsayers call for methanol use as an alternative fuel. In fact, it is now possible to produce gasoline from methanol. The Mobil process using the ZSM-5 zeolite catalyst is operating in one plant in New Zealand. The process, however, is not competitive with hydrocarbon sources such as crude oil, which, instead of increasing in price, remains relatively stable in supply and value. The continuing discovery of new natural gas sources holds the price on natural gas so that, for the present, it is not rational to produce methanol as a fuel or fuel blend when natural gas, a clean-burning fuel, can be used directly.

As more natural gas is discovered in remote areas of the world, there is merit in considering conversion of the gas to methanol so that relatively safe transport can occur. The methanol can then be converted to valuable synthesis gas at the place of delivery.

Chemistry

The modern catalyst is a highly selective copper oxide-zinc oxide-alumina catalyst. Typical operating conditions are 50–100 bar and 200–270°C (473–543 K). The following three equations are used to describe the overall stoichiometric reactions.

$$CO + 2H_2 \Leftrightarrow CH_3OH \tag{1}$$

Temp., K	ΔH, kcal
400	−22.63
500	−23.40
600	−24.01

$$CO_2 + 3H_2 \Leftrightarrow CH_3OH + H_2O \tag{2}$$

Temp., K	ΔH, kcal
400	−12.92
500	−13.85
600	−14.72

$$CO_2 + H_2 \Leftrightarrow CO + H_2O \tag{3}$$

Temp., K	ΔH, kcal
400	9.71
500	9.52
600	9.28

Since Reaction 2 is the summation of Reactions 1 and 3, or the difference between Reaction 1 and the reverse of Reaction 3 (the water-gas shift), equilibrium calculations need consider only Reactions

1 and 3, the independent equations. This statement does not suggest that Reaction 2 does not occur. See "Mechanism" for such issues.

Note that Reaction 3 is endothermic and does not go from left to right except at higher temperatures.

Thermodynamics

Reaction 3 is the reverse shift reaction and is endothermic, but the net production of methanol is exothermic, and lower temperatures and higher pressures are favorable in terms of equilibrium. Fortunately, high-activity catalysts permit operation at lower temperatures and energy-saving lower pressures (50 to 100 bar).

Equilibrium calculations for methanol synthesis have been thoroughly analyzed, with special emphasis on the correct evaluation of fugacity coefficients.[4] Nearly ideal gas behavior has been noted for CO, CO_2, and H_2, because the maximum reaction temperature range of 200–400°C is well above the critical temperature. By contrast, methanol and water vapor are very much nonideal.

The equilibrium expressions for Reaction 1 and 3 are as follows:

$$K_1 = \left[\frac{P_M}{P_{CO}P_H^2}\right]\left[\frac{\phi_M}{\phi_{CO}\phi_H^2}\right] = \left[\frac{n_M n_T^2}{n_{CO}n_H P_T^2}\right]\left[\frac{\phi_M}{\phi_{CO}\phi_H^2}\right]$$

$$K_3 = \left[\frac{P_{CO}P_w}{P_{CO_2}P_H}\right]\left[\frac{\phi_{CO}\phi_w}{\phi_{CO_2}\phi_H}\right] = \left[\frac{n_{CO}n_w}{n_{CO_2}n_H}\right]\left[\frac{\phi_{CO}\phi_w}{\phi_{CO_2}\phi_H}\right]$$

where   n = moles of indicated component
          P = partial pressure of indicated component
          P_T = total pressure
          φ = fugacity coefficient of indicated component subscripts M, CO, H, CO_2, W, T = methanol, CO, H_2, CO_2, H_2O, total moles

Typical published values of K_1 and K_3 are as follows:[4,5,6]

$$K_1 = 9.740 \times 10^{-5} \exp\left[21.225 + \frac{9143.6}{T} - 7.492\ln T + 4.076 \times 10^{-3}T - 7.161 \times 10^{-8}T^2\right]$$

$$K_3 = \exp\left[13.148 - \frac{5639.5}{T} - 1.077\ln T - 5.44 \times 10^{-4}T + 1.125 \times 10^{-7}T^2 + \frac{49170}{T^2}\right]$$

The best values of fugacity coefficients are obtained by using the Peng–Robinson or the Soave–Redlich–Kwong equation of state based on due attention to nonidealities of the various components in the reaction mixture rather than use of pure component fugacities.[4]

Table 19.4 summarizes such calculations of CO and CO_2 conversions and equilibrium methanol mole fractions or volume fractions for a defined mixture at various pressure and temperature conditions. The calculations are based on once-through operation (no recycle). Actual operation takes advantage of the ease of removing methanol from the reactor effluent by water cooling and then recycling the unreacted carbon oxides and inerts. In such an operating mode, much lower conversions and final compositions are obtained because of the diluting effect of the recycle and, of course, the necessity of a practical reactor operating at a viable approach to equilibrium. Typical commercial percentages of methanol in the converter effluent vary from 5 to 8%,[3] depending on reactor design and operation, including the quantity of recycle in the total feed to the converter.

TABLE 19.4
Equilibrium Conversions and Equilibrium Methanol Content of Once-Through Synthesis Gas (No Recycle)[*]

Temp., °C	CO Conversion, %		CO$_2$ Conversion, %		Methanol in Exit (mole%)	
	50 Bar	100 Bar	50 Bar	100 Bar	50 Bar	100 Bar
200	96.0	99.0	25.9	58.5	24.6	32.1
250	71.2	90.6	12.8	29.3	15.3	23.4
300	23.5	57.3	12.8	18.7	5.0	12.6
350	−2.3	14.7	18.5	20.2	1.2	4.2
400	−12.3	−6.9	18.9	21.0	0.3	1.1

[*]Selected values from a more complete compilation using the SRK equation of state for the mixture at equilibrium. The negative sign on CO conversion indicates that the reverse water-gas-reaction is occurring, producing CO rather than consuming it.

Note: Actual operation with recycle will produce lower methanol composition in the converter effluent (typically 5 to 8%) because equilibrium is not normally reached, and the fresh feed is mixed with recycle. Portion of original table reprinted with permission from Chang, W. H., Rousseau, R. W., and Kilpatrick, P. K., *Ind. Eng. Chem. Process. Des. Dev.* 25, 477 (1986). Copyright American Chemical Society.

Mechanism

Over many years, much work has been done on attempts to develop a valid mechanism for the synthesis of methanol from kinetic data. This approach, however, has not been productive. Inevitably, the question arises whether methanol is made from CO or CO$_2$, and the answer is not forthcoming from kinetics, since the shift reaction provides an interchange between CO for CO$_2$. Kinetic data cannot lead to the identification of the carbon source for the methanol produced.

Some excellent studies of the reaction system based on isotopic labeled carbon and various surface science techniques under industrial-type operating conditions have been reported.[7] The interested reader will want to follow the experimental details and reasoning that have led to the following mechanistic scheme:[7]

- Copper metal is the active catalyst component created by the reduction process that is performed to activate fresh catalyst.
- Despite the earlier intuitive concept that methanol is produced by reaction of carbon monoxide with hydrogen, detailed experimental evidence proves that, under typical industrial conditions, methanol is produced by hydrogenation of carbon dioxide. Hydrogenation of carbon monoxide is negligible.
- The following mechanism based on the reported detailed study is proposed.[7]

$$CO_2 \Leftrightarrow CO_{2\,ads}$$

$$H_2 \Leftrightarrow 2H_{ads}$$

$$CO_{2ads} + H_{ads} \Leftrightarrow HCOO_{ads}$$

$$HCOO_{ads} + 3H_{ads} \Leftrightarrow CH_3OH + O_{ads}$$

$$CO + O_{ads} \Leftrightarrow CO_2$$

$$H_2 + O_{ads} \Leftrightarrow H_2O$$

- At the steady state, the copper surface is partially covered by adsorbed oxygen. The remainder of the surface is metallic copper. Adsorbed oxygen facilitates carbon dioxide adsorption. The metallic copper surface is the site for hydrogen dissociative adsorption.
- Adsorbed carbon dioxide reacts with adsorbed hydrogen to create a surface format that then reacts with adsorbed hydrogen to yield methanol and adsorbed oxygen.
- Adsorbed oxygen reacts with CO and H_2 to produce CO_2 and water, respectively. In so doing, additional oxygen formed by the methanol step is removed by these water-gas shift steps.
- When the copper surface is clean with no adsorbed oxygen, as in the case for a CO_2-free mixture, carbon monoxide will be hydrogenated directly to methanol by an entirely different mechanism.
- When CO_2 is present, the copper surface develops a partial coverage of adsorbed oxygen, and the CO present reacts rapidly with adsorbed oxygen to produce carbon dioxide as shown in the above mechanistic scheme.

Catalyst Type and Suppliers

Typical Analysis

	wt%
CuO	55
ZnO	25
Al_2O_3	8

Size: 6 × 4 mm or 6 × 3 mm
Form: tablets

Suppliers

Haldor Topsoe, Synetix, United Catalysts

Licensors

Acid-Amine Technologies, Haldor Topsoe, Kellogg Brown & Root, Kvaerner Process Tech., Linde AG, Lurgi, Synetix

Catalyst Characteristics

As discussed under "Mechanism," copper metal is the active catalyst. It provides both types of surfaces, copper metal for hydrogen adsorption and copper with adsorbed oxygen.[7]
Zinc oxide serves several important functions that enhance the stability and life of the catalysts.[7]

- It is credited with an important role in the proprietary manufacturing procedure that creates a high-surface area of copper.
- Along with alumina, it prevents copper agglomeration.
- ZnO reacts readily with copper poisons such as sulfur and chlorine compounds.
- ZnO, being a basic oxide, neutralizes the alumina acidity preventing thereby acid catalyzed dehydration of methanol to dimethyl ether, which is catalyzed by acidic alumina sites.

$$2CH_3OH \rightarrow (CH_3)_2O + H_2O$$

Alumina provides structural integrity and, along with ZnO, separates copper crystallites so as to hinder sintering.

Catalyst Selectivity

Selectivity to methanol from carbon oxides exceeds 99%, which is most remarkable considering that formation of methanol is not greatly favored thermodynamically compared to many other compounds that could be formed (as is the case for synthesis gas reaction via the Fischer–Tropsch process).[7] Part of this unusual selectivity is due to copper itself, which readily splits the first oxygen from carbon dioxide but cannot split the CO bonds in carbon monoxide and methanol.[7]

Catalyst manufacturers exercise great care in producing the methanol synthesis catalyst to avoid impurities such as free alumina, iron, and nickel, which catalyze hydrocarbon formation and alkalies such as sodium carbonate that catalyzes the formation of higher alcohols.[7] Iron and nickel can react with CO to form volatile carbonyls that readily penetrate the catalyst pores, causing deactivation and some loss of selectivity.

Catalyst Deactivation

Catalyst life is reported to be 3–6 years. Preservation of copper surface is essential. Sintering of copper crystallites into large crystals reduces the active surface area and is the main cause of catalyst deactivation. Unfortunately, the usual operating temperature, especially as temperatures are raised as catalyst ages (250–300°C), is in the range where copper crystallites aggregate.[7] Proper catalyst preparation hinders this deactivating phenomenon by the separation of crystallites by ZnO and Al_2O_3 components of the catalysts. Ultimately, sintering will gradually occur and, since it is irreversible, the catalyst must ultimately be replaced. Catalyst poisoning by sulfur or chloride compounds can be catastrophic for catalyst activity but is normally prevented by upstream removal in the preparation of the synthesis gas. By-products such as waxy hydrocarbon can block access to pores and to active copper surfaces. Great care is exercised in catalyst manufacture to avoid impurities that can catalyze side reactions.

Catalyst Recovery

Dumped catalyst, if not badly contaminated, can be used as a source of ingredients for new catalyst manufacture. The catalyst is oxidized to avoid flammable conditions and then dissolved in nitric or sulfuric acid to produce useful salts.[8]

Process Unit

A typical methanol plant consists of a steam reformer preceded by feed poison removal as described in the section on steam reforming. Because of the high sensitivity of the copper component of methanol catalysts, great care must be exercised in designing sulfur and chloride removal systems. Often, a Co-Mo hydrodesulfurization reactor located before the ZnO bed is used as added insurance against possible higher sulfur containing natural gas that cannot be effectively handled by the ZnO bed alone. The H_2S formed in the hydrodesulfurization reactor is removed by the ZnO bed to final level of 0.1 ppm.[3]

Primary Reformer*

Operating conditions for a primary reformer in a methanol plant are in the range of 10–20 atm, depending on methanol synthesis reactor design and other downstream design requirements. Temperature is in the range of 850–890°C, and the steam-to-carbon ratio 2.5–3.[2] Lower values save energy, but some concern for values below 2.7 has been expressed[3] since, at lower steam rates, the carbon-forming Boudouard reaction occurs. Of course, careful operation at lower temperature and higher space velocities can overcome such a problem.

At the general conditions expressed above, the molar ratio of hydrogen-to-carbon oxides is 3.0 in the reformer effluent. Methanol synthesis requires only two moles of hydrogen. This fact can

---

* See p. 414 for more details on steam reforming.

be seen most clearly from the mechanism or from the stoichiometric equation: $CO + 2H_2 \rightarrow CH_3OH$. Either the excess hydrogen must be removed as purge from the synthesis loop and used as fuel, or it is separated and used for an adjacent ammonia plant. Alternatively, by-product CO_2 from sources such as an adjacent ammonia plant may be added to the reformer feed or product to favor the reverse shift reaction or the direct reaction of CO_2 to methanol, as shown in the mechanistic scheme when added to the methanol converter feed.

$$CO_2 + H_2 \rightarrow CO + H_2O$$

This procedure causes additional methanol production but adds additional water to the synthesis gas and thus to the crude methanol. The loss of hydrogen from the purge gas, if hydrogen is removed by the procedure just described, requires additional natural gas for the reformer furnace.[3] In general, CO_2 addition is a useful alternative if a nearby inexpensive source is available.

Other Reforming Procedures

Designs with other reforming configurations for methanol plants are in operation.

Secondary Reforming   An arrangement similar to that in an ammonia plant is provided, except pure oxygen is used, since nitrogen from air is not required. This procedure is often called *combined reforming*. The primary reformer is designed to operate at higher pressures and lower temperatures.[2,3] It is smaller, has a lower duty and, because of higher pressure operation, reduces synthesis gas compression costs.[3] The partial combustion of the primary reformer effluent consumes excess hydrogen producing close to a 2:1 mole ratio of H_2 to carbon oxides.

The secondary reformer (autothermal reformer, see p. 442) requires experienced design professionals, since proven methods for safe operation must be carefully planned. Since partial combustion of the feed provides the energy for the autothermal process, the combined heat duty is 45% less than for a primary reformer operating alone. Make-up gas power consumption is reduced about 50%. Such savings, however, can be offset by the high cost of air separation. Much depends on the air separation plant and its uses. Several instances of on-site air separation plants operated by major air separation companies have been attractive if other air products are required at the same location.

Gas-Heated Reforming   Gas-heated or heat-exchanger reforming is offered by Synetix, Kellogg Brown & Root, and Haldor Topsoe. In these designs, the fired primary reformer is replaced by a refractory-lined vessel containing tubes filled with catalyst, or bayonets with the annular space filled with catalyst. Heat is supplied by autothermal secondary reformer exit gas. Both combined reforming and gas-heated reforming require higher capital costs but reduce energy consumption.[3] Use of these alternatives is likely to be more common in revamps of existing facilities for the purpose of increasing production.

Heat Recovery

The major areas for energy savings are in gas compression and heat recovery. Effluent from reforming is somewhat above the desired inlet temperature to the methanol converter (~600°C above methanol unit). A waste-heat boiler is used to produce high-pressure steam for steam turbine drives after it is superheated in the combustion gas passing out the furnace stack. Valuable heat remaining is typically recovered sequentially as high-pressure boiler feedwater, reboiler duties, and LP boiler feedwater.[3]

Compression[2,3]

Final water cooling of synthesis gas occurs prior to the make-up gas compressor. A knockout drum removes water and other condensates prior to the entry of the make-up gas to the centrifugal compressor with interstage cooling. Pressure is raised to 50–100 bar. The recovered condensate is fed to a steam stripper, which removes rather pure boiler feedwater from the bottoms.

Synthesis gas from the knockout drum is passed to a multistage compressor with interstage cooling. Compressed make-up gas is combined with recycle and sent to the methanol converter. Product from the converter is cooled by interchange with the incoming make-up gas and then further cooled by a water cooling exchanger.

Conversion and Purification[2,3]

The cooled converter effluent contains condensed methanol and water that is readily removed in a separator vessel, and the crude methanol is passed to a topping column for removal of unreacted gases, which are recycled to the converter after combining with make-up gas via a separate recycle compressor or a stage of the main compressor. A final distillation receives the bottoms from topping column and separates methanol from water with methanol being removed in the upper section and purge gases removed as overhead.

Converter Types

Every known type of reactor designed for exothermic, equilibrium-limited reactions has been used successfully.

Quench Converter

The quench converter was the first converter used by ICI, the developer of the modern low-pressure methanol process. There are still large numbers of these in service. A quench converter consists of a single bed with intermediate quench inlets of the lozenge type. To distribute the quench more effectively, a number of parallel lozenges are arranged on each header. Headers with lozenges are located in three or four positions, producing, in effect, four or five beds.

This design has the advantage of ease of loading and unloading. Quench gas is a portion of the feed gas that has not been preheated. It dilutes the reaction mix at each quenched position and therefore bypasses a portion of the catalyst. This situation is compensated for by using more catalyst than other designs. Up to six beds have been used, with maximum methanol content in the effluent of no more than 7%.

Adiabatic Reactors with Intermediate Cooling

Kellogg Brown & Root and Haldor Topsoe offer a design, amenable to very large plants, consisting of three to four large spherical reactors with heat exchangers in between each. The large size of the spheres provides essentially straight sides for the catalyst bed. Well designed inlet distributors are essential for preventing bypassing. The catalyst charge can rest on inert packing in the bottom curved portion of the sphere. Unloading and loading of catalyst is not hindered by internals. Typical methanol content in the effluent is reported to be 5%.[3] The capital cost is higher than for the quench converter, but the catalyst loading is less.

"Isothermal" Converter

The term *isothermal reactor* is frequently used to describe multitubular reactors with cooling medium exterior to the tubes, which are filled with catalyst. Such reactors typically exhibit a rapid heat-up period over a rather short distance where rates are high because of the distance from equilibrium. In this range, heat produced exceeds heat removal. A hot spot (high temperature) is reached, after which the temperature declines to a rather steady value during which a large and remaining portion of tube is host to a lower rate as equilibrium is approached. In this region, the temperature does remain fairly constant, gradually declining. Higher temperature in the early portion of the tubes increases the rate but, as equilibrium is approached in the lower portion, an increase in temperature will reduce the rate and the methanol produced.

The Lurgi converter is the standard methanol reactor of this type. The cooling medium is boiler feed-water, which circulates between the converter shell and a steam drum that separates water from saturated steam and allows water to circulate back to the converter. Temperature of the cooling water is controlled by the back pressure on the steam drum. Methanol concentrations in this converter effluent are reported to be 6–8%.[3]

Linde, AG offers an *isothermal* reactor with the catalyst in the shell and the cooling water in the tubes. A large amount of tube area is contained in the shell by use of a helical wound tube bundle.

Tube-Cooled Converter

In this ICI (now Synetix) design, tubes connected to a header carry synthesis gas upward through the bed to transfer heat from the bed to the synthesis gas in the tubes. At the top of the tubes, the now heated gas turns 180° and proceeds through the bed. This design has the advantage of ease of loading and unloading catalyst.

Converter Material of Construction

The temperatures and pressures are low enough so that a Mo-steel is a satisfactory choice that will avoid hydrogen embrittlement.

Process Kinetics

Much activity in developing rate equations for methanol synthesis took place prior to the time that CO_2 was shown to play a major role in the synthesis mechanism. In many of these equations, CO_2 does not appear as a variable, despite the fact that early work on the low-pressure process showed that CO_2 increased the rate of methanol product when added to a CO based feed.[7]

Reviews of kinetic studies on methanol synthesis have appeared periodically (e.g., Refs. 9 and 10). Two examples demonstrate the variety of rate forms that have been proposed. The first of these is based on three reactions.[11]

$$CO + 2H_2 \Leftrightarrow CH_3OH \tag{1}$$

$$CO_2 + H_2 \Leftrightarrow CO + H_2O \tag{2}$$

$$CO_2 + 3H_2 \Leftrightarrow CH_3OH + H_2O \tag{3}$$

Based on stoichiometry, Reaction 3 is the sum of Reactions 1 and 2. The rate forms suggested for a commercial Cu/Zn/Al catalyst are

$$r_1 = k_1 K_{CO}[f_{CO}f_{H_2}^{3/2} - f_{CH_3OH}/f_{H_2}^{0.5}K_{p_1}]/B$$

$$r_2 = k_2 K_{CO_2}[f_{CO_2}f_{H_2} - f_{H_2}f_{CO}/K_{p2}]/B$$

$$r_3 = k_3 K_{CO_2}[f_{CO_2}f_{H_2}^{2/3} - f_{CH_3OH}/f_{CH_3OH}/f_{H_2}K_{p3}]/B$$

where
$$B = (1 + K_{CO}f_{CO} + K_{CO_2}f_{CO_2})[f_{H_2}^{0.5} + (K_{H_2O}/K_{H_2O})f_{H_2O}]$$

$k_1 k_2$, k_3 = Rate constants for indicated numbered reactions

K = chemisorption constants for indicated components

K_p = partial pressure based chemical equilibrium constant for numbered reactions

f = fugacity of indicated component

In commercial practice, it is often necessary to raise temperature within limits or use a very active catalyst to maximize production so that a diffusion and mass-transfer controlled regime exists. Studies of this sort have shown that, under such conditions, the exponents on all but H_2 are zero for a power-law equation.[12,13]

$$r = \eta k P_{H_2}^a P_{CO}^b P_{H_2O}^c P_{CH_3OH}^d P_{CO_2}^e$$

where η = effectiveness factor

In such a case, the rate form with the reverse reaction considered is

$$r = \eta k (P_{H_2} - P_{H_2}^*)$$

where $P_{H_2}^*$ = equilibrium partial pressure

For low values of the effectiveness factor,

$$\eta = \frac{1}{\phi}$$

where ϕ = Thiele Modulus

or

$$r = \frac{1}{\phi} k (P_{H_2} - P_{H_2}^*)$$

There is reason to assume that this simple form has been used in commercial practice. Catalyst suppliers and licensors will be helpful in supplying useful kinetics for current catalysts.

19.6 PURE CARBON MONOXIDE FROM SYNTHESIS GAS AND ITS USES

Pure carbon monoxide is widely used in the synthesis of fine chemicals via carbonylation, including aldehydes, ketones, carboxylic acids, anhydrides, esters, amides, imides carbamates, carbonates, ureas, and isocynanates.[1] These carbonylation products involve homogeneous catalysts such as transition-metal carbonyls, including complexes of cobalt, iron, palladium, and rhodium.

Large-scale carbonylation industrial processes using pure carbon monoxide include the production of phosgene, acetic acid, an ethylene route to acrylic acid, and alternate routes to toluene-diisocyanate, 4,4′-diphenyl-methyl isocyanate, formic acid, methyl formate, and propionic acid. Each of these involves homogeneous catalysis except phosgene. The reader is referred to a major reference on carbonylation reactions that considers the many homogeneous catalytic systems.[1] Heterogeneous catalysis in the production of phosgene and pivalic acid are discussed in Chapter 5.

19.6.1 Production of Carbon Monoxide by Steam Reforming

The primary source of carbon monoxide for industrial use is steam reforming. Ultimately, hydrogen produced in the reforming reaction must be removed, but the task may be simplified somewhat if steam reforming operation can be adjusted so as to increase CO and decrease H_2

production. The primary tool in this situation is the recycle of CO_2 so that the reverse shift reaction will occur.

$$CO_2 + H_2 \rightarrow CO + H_2O$$

Additional CO_2 from other nearby sources can be added so that the final ratio of H_2/CO is about 1.3 or less. Normally, values below 1.3 can be attained only if some procedure for preventing carbon formation is implemented. One such procedure uses increased steam in the feed.

$$C + H_2O \Leftrightarrow CO + H_2$$

There are proprietary processes that avoid this solution by, in one case (Haldor Topsoe), using partial poisoning of the reforming catalyst with a sulfur compound that reduces the activity toward carbon formation and allows as low as a 0.6 ratio.[2] Another process (Caloric, GmbH) employs staged catalyst and reports a 0.4 ratio.[2]

In any event, purification of the reformer effluent must be undertaken to produce a truly pure carbon monoxide product. After removing particulates, tars, or heavy hydrocarbons by scrubbing and acid-gas removed by monethanolamine or hot potassium (or other acid-gas processes), the raw CO stream is fed to a final purification section.[3]

19.7 PURE HYDROGEN FROM SYNTHESIS GASES AND ITS USES

Hydrogen is the most abundant element in the universe, and hydrogen is also one of the most used elements in chemical manufacturing. From fine chemicals to major commodity chemicals, catalytic hydrogenation plays a crucial role in many hundreds of syntheses of valuable products (see Chapter 12, "Hydrogenation").

A considerable amount of hydrogen is available as off-gas or purge gas from various processes. Ammonia and methanol synthesis, hydrocracking, hydrotreating, and toluene hydrodealkylation are consumers of hydrogen, but the necessary purge of a portion of the recycle stream yields a stream fairly rich in hydrogen. Catalytic reforming, the production of aromatics from naphtha, has been a major producer of hydrogen in refineries. In recent years, as benzene has been phased out as a component of gasoline, the hydrogen from these units has diminished.

Many refineries and petrochemical plants often require additional sources of hydrogen, and the most frequently used process for such purposes is steam reforming. Partial oxidation is an alternative noncatalytic route that is used particularly when natural gas is expensive and when heavy hydrocarbons are relatively cheap. The process is advantageous, since it can use almost any hydrocarbon feed and does not produce NO or SO_x.[1] The Texaco and Shell processes dominate the field, and more than 100 units are operating worldwide.[1] At the high temperatures employed (1250–1500°C), oxidation to CO and cracking to methane occur along with the methane reforming reaction and water gas reaction and the shift reaction without the need of a catalyst. The water gas and shift reactions and the initial partial oxidation of the hydrocarbon all produce hydrogen.[2]

Steam Reforming

Because of the high ratio of hydrogen to carbon in methane and the wide availability of natural gas, steam reforming of methane is the preferred route to hydrogen. The general characteristics and details of steam reforming are discussed beginning on p. 405.

Process Description

There are two processes for pure hydrogen production, each beginning with desulfurization followed by a standard methane steam reformer operating with a high steam-to-carbon ratio (4–5 mole/mole), and then a high-temperature shift reactor for converting carbon monoxide to carbon dioxide. The standard process then follows with low-temperature shift, methanation, and CO_2 removal by absorption (see p. 416). Final product hydrogen is usually no higher than 98.2% hydrogen with the remainder being mainly methane.[1,3]

Cryogenic, membrane, and pressure-swing adsorption have been used for final purification of the hydrogen stream. Cryogenic recovery is not usually considered for recovery of hydrogen alone, but for hydrogen, carbon monoxide, and light hydrocarbon mixtures. Membrane systems capable of producing a product purity of 95% are most efficient and cost effective for small flow rates, e.g., less than 5 mm SCFD.[1]

Pressure-swing adsorption, which was introduced in the 1960s, has become the preferred purification process, because it can produce the highest hydrogen purity product (up to 99.95%) and requires a lower capital cost than the so-called standard process. With pressure-swing adsorption, the low-temperature shift and methanation reactions are replaced by selective adsorption on molecular sieves. The affinity for hydrogen adsorption is very low on the sieves, but larger molecules, such as CO, CO_2, CH_4 and H_2O as well as nitrogen and light hydrocarbons that may be present, are adsorbed. The total recovery of hydrogen from the raw hydrogen stream depends on the number of stages and can vary from 70–90%.[3]

19.8 AMMONIA SYNTHESIS (NITROGEN + HYDROGEN → AMMONIA)

In 1798, Malthus announced his population principle that world population increases in a geometric progression while that of food production increases in an arithmetic progression. And indeed, clearly, that was definitely occurring, but Malthus did not consider what the ingenuity of chemists and later chemical engineers might create. As summarized in Table 19.5, a number of important discoveries led to today's energy-efficient and worldwide use of ammonia as a fertilizer. The fascinating history has been well documented, including advancements in the science of thermodynamics, which provided the necessary information for a precise definition of feasible operating conditions for this seemingly impossible process.[1-3]

The major use of ammonia remains as a fertilizer, either directly applied as a liquid ammonia or via compounds such as ammonium nitrate, ammonium phosphate, and ammonium sulfate. A significant amount is used to produce urea, which is used in mixed fertilizers and also in the production of urea-formaldehyde and melamine-formaldehyde resins. Nitric acid is produced solely by the direct oxidation of ammonia. Ammonium nitrate is also used as a component in explosives, as is nitroglycerin, a component of dynamite.

Early in the development of the synthetic ammonia process using nitrogen from the air and hydrogen from water gas, many began to describe the process as *bread from air*, which it still is.

19.8.1 PROCESS SEQUENCE (NATURAL GAS)

An integrated plant involves the sequence of operations summarized in Table 19.6. Auxiliary equipment is replete with exchangers and waste-heat boilers that produce high-pressure steam as a power source to drive compressors and pumps. In addition, heat recovery at lower temperatures provides for efficient use of lower pressure steam in heating process streams.

Ammonia synthesis plants based on natural gas feed dominate the ammonia-plant operations of most ammonia plants. They exhibit the lowest energy and capital costs. The difference in design for naphtha feed is primarily related to the steam reformer, which is discussed beginning on p. 426 as "Steam Reforming of Naphtha and Higher Hydrocarbons."

TABLE 19.5
Highlights in Ammonia History

Year	Event
1840	Justis von Leibig demonstrated that nitrogen was one of the three essential elements for plant growth.
1898	Sir William Crookes warns of ultimate exhaustion in 50 years of naturally occurring nitrates, mainly in Chile, which were being used not only for fertilizers but also explosives.
1905	Commercial development of arc process for fixing nitrogen as nitrous and nitric acid (Birkeland process). Located in Norway, where low energy costs made feasible this low-yield (2%) process.
1910	Frank and Caro process for producing calcium cyanamide from calcium carbide and nitrogen (CaC_2 + $N_2 \rightarrow CaCN_2$ + C). Used one-fourth the energy of arc process, and the calcium cyanamide could be used directly as fertilizer or hydrolyzed to ammonia.
1913	First commercial plant (30 tons/day) for the direct synthesis of ammonia from the elements begins operation in Oppau, Germany. Thus ended an over ten-year effort beginning with attempts to determine ideal operating conditions using initially primitive thermodynamic data and early higher-pressure small-scale experiments by Fritz Haber and Walther Nernst. Ultimately, the highest conceivable pressure and the lowest possible temperature became the proven goal. Development of the process was undertaken by BASF under the leadership of Carl Bosch with the cooperation of Haber. Hydrogen from water gas and nitrogen from liquid air were used. Hydrogen was separated from the water-gas by refrigeration later replaced by the water-gas shift reaction, now simply called the shift reaction.
	Such new problems of high-pressure vessel design (200 atmospheres) hydrogen embrittlement, active catalyst (2500 catalysts tested with iron-alumina-potash the best), and corrosion were methodically and successfully solved.
1916	Larger reactors upped capacity to 200 tons/day.
1920	World War I treaties made German patents available to other countries without charge, and the Haber process plants appeared in most industrial countries.
1920–1950	Growth in plant size.
1953	First use of methanation in ammonia plants, Mississippi Chemical Corporation.
1950–1960	Development of more energy-efficient plant designs and the introduction of the integrated plant by M.W. Kellogg Company (now Kellogg Brown & Root) that integrated heat recovery and employed single streams and a large single reactor rather than multiple reactors in parallel. Also use of methane as a source of synthesis gas developed by BASF and ICI reduced capital costs.
1962	ICI used light naphtha feed as alternate to natural gas (methane).
1963	Introduction by M.W. Kellogg Co. and Clark Brothers of large single centrifugal compressor to boost pressure from synthesis gas unit 25 or 35 bar to 150 bar initially and later 220 bar allowed large single-train plants to be built of 1000–1500 metric tons per day.
1968	First plant constructed to deliver liquid ammonia from Texas Panhandle by pipeline. (Hill Chemical Company). The destination was the wheat belt of Texas, Oklahoma, Kansas, and Iowa.
1970s	Large numbers of single-train plants built worldwide of high efficiency to overcome effects of increased energy costs. Improved catalysts available.

19.8.2 NATURAL GAS PURIFICATION

Natural gas is received at the battery limits from a pipeline under pressure of up to 70 bar (1000 psig) in areas of high petrochemical plant concentration. The pressure is then lowered to the level commensurate with the reformer operation. Detailed discussion of the major poisons for reforming units are discussed on p. 413. The following is concerned with poison removal.

Sulfur Removal

Natural gas has the advantage of being free of or having only modest quantities of reformer catalyst poisons: sulfur compounds (mainly H_2S, COS, CS_2 and RSH), chlorides (usually from off-shore

TABLE 19.6
Major Steps in Ammonia Synthesis

Operation	Purpose
Desulfurization of feed (also removal of other catalyst poisons)	Protect downstream catalysts
Primary reforming	Produce H_2 and CO
Secondary reforming	Complete conversion of significant portion of remaining CH_4 to H_2 and CO and supply required nitrogen via combustion air
Shift conversion (HTS and LTS)	Convert CO (an NH_3 catalyst poison) to CO_2
Carbon dioxide removal	Remove inert CO_2 and use it for other purposes
Methanation	Remove last remaining CO and CO_2 to avoid poisoning of NH_3 synthesis catalyst
Compression of N_2-H_2 mixture	Bring pressure up to ideal synthesis condition
Ammonia synthesis	React H_2 and N_2 to ammonia

reservoirs), and mercury (only in some natural gases). In most cases, only ZnO adsorbent beds are required. Although often called a catalyst, the ZnO reacts with the sulfur impurities and ultimately must be discarded or the Zn content recovered by a supplier.[4]

$$ZnO + H_2S \rightarrow ZnS + CO_2 \text{ or } H_2O$$
$$\uparrow$$
$$CS_2$$
$$\uparrow$$
$$COS$$
$$\uparrow$$
$$RSH$$

A second bed is available so that the exhausted bed may be dumped and reloaded. Better utilization of two beds can be attained by installing the two units in series, valved so that either bed can be used as the first bed. In this manner, the initial first bed may be dumped while the second bed remains in operation and then reconnected with new ZnO adsorbent, becoming thereby the second bed in the series.[5]

Although the equilibrium constant declines with temperature increase, it remains high. Higher-temperature operation (up to 400°C) improves the rate of pore diffusion. High porosity ZnO will produce similar net reaction rates at lower temperatures (~100°C). However, at least 350°C is needed to decompose merceptans to H_2S and hydrocarbons.[6]

Adsorbent Characteristics (Typical)

Size and form: 3/16-in. extrudates, 1/8-in. pellets
ZnO content: 80–90% typical
Effluent purity: 0.1 ppm sulfur
Sulfur capacity: 25–30 wt%
Life: usually designed for one year

Suppliers

United Catalysts, Halder Topsoe, Synetix

Chloride Removal

If small in amount, the ZnO will remove chlorides satisfactorily. For larger amounts, a chloride adsorbent is required.

Installation

Usually installed prior to the ZnO bed. In some cases, it is simply a guard bed resting on top of the ZnO bed.

Typical Characteristics

1/8- to 3/16-in. (3 to 4.5-mm) spheres alkalized alumina

Suppliers

United Catalyst, Haldor Topsoe, Syntix

Mercury Removal

Although usually not a problem, some natural gases may contain amounts of mercury such that removal is necessary. Usually, a bed of activated carbon impregnated with sulfur has been successfully employed.[6]

19.8.3 PRIMARY REFORMER

Feed gas, after passing through the sulfur and other poison removal beds, is combined with process steam and preheated in the convection section of the primary reformer. It then flows to the radiant section. A typical steam-to-carbon mole ratio is 3.5, and operating conditions are in the range of 780–830°C and 30–40 bar. The primary reformer and catalyst are described in detail beginning on p. 414.

19.8.4 SECONDARY REFORMER

The outlet temperature of the primary reformer is often around 800–830°C, which corresponds to a methane content (methane leak) of 7.5–8 vol%.[7] The secondary reformer operates as an adiabatic reactor and completes the reforming of the methane operating at high inlet bed temperatures (1250°C). The temperature declines to around 1000°C, corresponding to a low methane gas content of 0.2 to 0.5 vol% dry.[7,25] Depending on primary reformer and secondary reforming operations, a content of CH_4 up to 1% is possible.[8]

The endothermic heat is supplied by the combustion of a portion of the primary reformer effluent with an amount of air necessary to provide the nitrogen in the final synthesis gas in a ratio of three H_2 to one N_2. This dual goal requires careful thermal balance between the primary and secondary reformer. The total heat added by the air-gas combustion is limited by the 3:1 hydrogen-to-nitrogen goal. The gas passes from the combustion zone to the catalytic reforming zone, where reforming and water-gas shift reactions occur, and equilibrium of the reforming and shift reactions take place.

Chemistry (Secondary Reforming)

The following reactions occur in the combustion zone of the secondary reformer, at temperatures of 1200–1250°C, and in the catalytic zone, where the temperature declines because of the endothermic reactions that occur on the catalyst.

	ΔH, kcal	Temp., K	K_p
Combustion Zone			
$CH_4 + 1/2O_2 \rightarrow CO + 2H_2$	−5.31	1273	3.013×10^{11}
$H_2 + 1/2O_2 \rightarrow H_2O$	−59.24	1273	1.153×10^{10}
$CO + 1/2 \rightarrow CO_2$	−67.33	1273	1.378×10^{10}

	ΔH, kcal	Temp., K	K_p
Catalytic Zone			
$CH_4 + H_2O \Leftrightarrow CO + 3H_2$	54.36	1273	9301.0
$CO_2 + H_2 \Leftrightarrow CO + H_2O$	7.67	1273	1.7

For design purposes, the maximum flame temperature is primarily due to H_2 and CO combustion to the extent of the available air. Methane has a much lower flame temperature and is present in relatively low amounts in the feed. In fact, just using hydrogen alone is a quick and acceptable procedure. Although hydrogen reacts more rapidly than CO and CH_4 with air, some combustion of these other two components must occur. But most of the methane content is reacted with steam in the catalytic section.

Stoichiometric equations, beginning with primary reforming, have been presented based on the irreversibility of the combustion and water-gas reactions, the total consumption of oxygen in the air added, the reaction of methane confined totally to the water-gas reaction, and the fraction of CO_2 converted by the water-gas shift reaction.[9] These *empirical* forms can be useful in developing heat balances. Calculations involving an equilibrium approach of 25°C at the outlet of the secondary-reformer bed have been recommended.[3]

Mechanism (Secondary Reforming)

The catalytic mechanisms of the reforming and shift reactions are similar to that proposed for the same reactions occurring in the primary reformer since the catalyst chemical nature is essentially the same (see p. 408).

Catalyst Type and Suppliers (Secondary Reforming)

The catalyst used for secondary reforming is similar to that employed for primary reforming, except it has a lower nickel content (4–8 wt%) and is designed to withstand the high temperatures. Rings and multihole pellets are the usual shapes. Either a refractory alumina or magnesium aluminate carrier is used. These carriers, as a final step in manufacture of the catalyst, are subjected to calcining temperature above that of maximum operating temperature so that the pellet will be rugged with higher crush strength and high-temperature resistance.

Suppliers

BASF, Haldor Topsoe, Synetix, United Catalysts

Form and size

Rings (19 × 9 × 19 mm, 23 × 11 × 25 mm), cylinder (20 × 18 mm with 4-mm holes)

Licensors

Haldor Topsoe, Kellogg Brown & Root

Catalyst Deactivation (Secondary Reforming)

Although subject to the same poisons as described for primary reforming, most of these poisons, such as sulfur compounds, will have been removed upstream. Furthermore, high-temperature operation significantly lowers the potential for adsorption of poisons. Poisons such as sulfur or chloride compounds, which may be introduced in the compressed air, however, will pass through the secondary reformer but then be adsorbed on the low-temperature shift catalyst. Secondary reformer catalyst life has been reported to have been as high as 10 years.

Process Units (Secondary Reforming)

The secondary-reformer reactor differs from a typical adiabatic reactor in that it consists of a reduced diameter top section for homogeneous combustion of a portion of the feed, followed by a bed of catalyst in the larger portion of the reactor.

The typical inlet temperature for secondary reforming is in the range of 800–830°C. The gases leaving the combustion zone are at 1200–1250°C. The inlet gases from the primary reformer can contain 7–12% methane.[7,25] As the major portion of methane is converted to CO and H_2 in the catalytic bed, the temperature declines to about 1000°C at the outlet. At this condition, and with proper quantities of air being fed to the secondary reformer, the H_2/N_2 ratio on a dry basis, less CO and CO_2 content plus the H_2 to be produced in the shift conversion, will be 3:1, and the methane quantity will be low (0.2–0.5%). These conditions satisfy the following major purposes of the secondary reformer, which are only served by higher operating temperatures not possible in the primary reformer.

- Production of a gaseous product that, when purified by subsequent steps, will have the desired H_2-to-N_2 ratio of three.
- Elimination of as much methane as possible. Although it does no harm in the synthesis reactor, methane constitutes an inert that must be vented to a recovery unit to prevent inert buildup and lower equilibrium compositions of ammonia.

Because of the high temperatures, the carbon-steel shell is lined with three layers of refractory brick so that the steel-wall temperature will be within an operating range that is below conditions that could cause hydrogen embrittlement if leakage through the refractory takes place. The refractory should be essentially free of silica, which can vaporize at the operating temperature and deposit in downstream exchangers.

The combustion section consists of an inner tube, which introduces filtered compressed air that has been preheated in the convection section of the primary reformer. The primary reformer effluent is introduced in a baffled annular space between the air tube and the combustion-section wall.[33] The burner tip contains many small air jets that aid in the mixing process. This arrangement assures turbulent mixing as the air contacts the reformer gases.

Alumina balls are placed in several layers on top of the catalyst bed to prevent impinging hot gases from causing catalyst movement and attrition. Also, the inert balls can protect the catalyst from prolonged high-temperature damage.

In some plants, a water-cooled jacket has been used to prevent high-temperature hot spots on the vessel wall. This technique, although effective, makes it difficult to determine if a significant hot spot exits. On installations without cooling jackets, the hot spots are more easily detected and can be remediated by directing steam or a water spray on the hot spot region.[33]

As the hot gases pass through the catalyst bed, the reaction declines because of the endother-micity of the catalyzed reactions. Temperature is the major operating variable. It must be adjusted in accordance with the methane content of the primary reformer product entering the secondary reformer. If the CH_4 content increases, the secondary reformer temperature must be increased by increasing air preheat.[33] Thorough mixing of air and feed gases as well as good distribution in the catalyst bed are essential to prevent runaway reactions in portions of the bed that can damage internals such as thermocouples.

Process Kinetics (Secondary Reforming)

The kinetic forms suggested for primary reforming (see p. 416) should be useful for secondary reforming. Many plant calculations are simply based on equilibrium approach assumptions related to plant performance.

19.8.5 Shift Conversion

The very high temperature effluent from the secondary reformer passes through a waste-heat boiler to recover the valuable high-temperature energy in the form of high-pressure steam. The mixture then enters the high-temperature shift converter, followed by the low-temperature shift converter as described in detail beginning on p. 418. The purpose is to remove carbon monoxide to low levels, since it is a poison for ammonia catalysts.

Carbon Dioxide Recovery

The effluent from the shift reactor train is passed to a waste-heat boiler, followed by further cooling to condense steam and thence to a separator to remove condensate, which can be stripped for recycle boiler feedwater recovery. The gases from the separator are passed to a carbon dioxide absorber using one of a number of amines or amine derivatives or a hot potassium carbonate. The adsorbed CO_2 is subsequently stripped and is a valuable product. Ammonia plants are a major source of commercial carbon dioxide.

The overhead from the absorber, now free of CO_2 but containing carbon monoxide, then passes to the methanator after being preheated to the required inlet temperatures.

19.8.6 Methanator (Carbon Oxides → Methane)

The methanator removes essentially all the carbon monoxide and carbon dioxide remaining in the synthesis gas. This, along with subsequent removal of H_2O, is essential, since each of these compounds deactivate the ammonia catalyst.

Chemistry (Methanation)

The following reactions occur over a nickel catalyst on a refractory carrier.

	Temp., K	ΔH, kcal	Kp
$CO + 3H_2 \rightarrow CH_4 + H_2O$	500	−51.26	1.16×10^{10}
	600	−52.07	2.00×10^6
	700	−52.70	3.77×10^3
	800	−53.21	3.48×10
	900	−53.59	0.775
$CO_2 + 4H_2 \rightarrow CH_4 + 2H_2O$			
	500	−41.73	8.730×10^7
	600	−42.72	2.360×10^4
	700	−43.15	4.207×10^2
	800	−44.42	8.091
	900	−45.04	0.353
Side reaction			
$2CO \rightarrow C + CO_2$	600	−42.53	5.322×10^5

Various operating inlet temperatures are reported.[6,8,10] A typical range would be 250–325°C (523–690 K). Outlet temperature depends on the quantity of CO and CO_2 in the feed. For each mole percent of CO and CO_2, the temperature rise varies depending on the total feed composition from 50–74°C for CO and 40–60°C for carbon dioxide.[9,12] The presence of one mole percent of oxygen due to an upstream upset can cause three times as much increase as do carbon oxides.[9]

Carbon formation is usually not a problem. Although it is favored thermodynamically at temperatures below 800 K (526°C), it is a very slow reaction at the lower temperatures relative to the CH_4-producing reactions, provided that the H_2-to-CO ratio is above a calculable minimum (see Figure 19.7). Increasing pressure reduces the minimum, since pressure drives the methane-producing reactions to the right.[10] Modern commercial operations do not experience carbon formation. In fact, at the increasing temperature that occurs as the reaction proceeds along the length of the bed, carbon formation is not favored thermodynamically.[35]

Mechanism (Methanation)

A number of studies on methanation have been reported, both for low CO concentrations as in ammonia plants and for high concentrations as in synthetic natural gas processes. Reviews related to the former have been published.[10,11,34] There are some general agreements based on surface studies.

1. Carbon monoxide reacts to form CH_4 more rapidly than CO_2 but, as CO is consumed, CO_2 begins to react as the gases pass through the catalyst bed.
2. Both carbon monoxide and carbon dioxide are chemisorbed. Hydrogen is dissociatively adsorbed.
3. Focusing on carbon monoxide, the following steps are assumed at equilibrium:

$$CO(g) \Leftrightarrow CO(ads)$$

$$CO(ads) \Leftrightarrow C(ads) + O(ads)$$

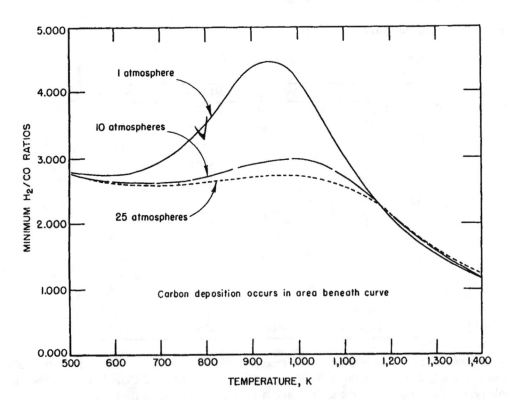

FIGURE 19.7   Carbon Deposition Boundaries. Reprinted from Grayson, M., Demeter, J. J., Schlesinger, M. D., Johnson, G. E., Jonakin, J., and Myers, J. W., U.S. Bureau of Mines: Report No. 5137, July, 1955.

$$H_2(g) \Leftrightarrow 2H(ads)$$

$$2H(ads) + O(ads) \Leftrightarrow H_2O(ads)$$

And the rate-determining step involves the reaction of adsorbed carbon and hydrogen atoms.

$$C(ads) + yH(ads) \rightarrow CH_y(ads)$$

This last step is presumed to occur in a series of steps that end in the production of methane.

The role played by CO_2 has not been well defined. The shift reaction could be reasonably assumed to be at equilibrium. With the large excess of hydrogen in the gas, the endothermic reverse reaction could occur when most of the CO is consumed, despite the fact that the equilibrium constant is low for this reaction at methanation conditions.

The tendency for reduced nickel to react with carbon monoxide to form poisonous nickel carbonyl is negligible at the usual methanator operating conditions for this exothermic reaction.[10]

$$Ni + 4CO \rightarrow Ni(CO)_4$$

$$K_p @ 500\ K = 9.48 \times 10^{-6}$$

$$800\ K = 7.45 \times 10^{-12}$$

At low temperatures (below 150–200°C), the equilibrium constant is in the range of 10^{-2}, which is still low, but the catalyst remains active. Small amounts of nickel carbonyl can thus form, and such results must be avoided, since even small amounts are poisonous. To prevent formation of nickel carbonyl, contact of the catalyst with feed gas containing carbon monoxide should be avoided during the heating period at start-up and the cool down period at shutdown. Suppliers recommend the use of nitrogen, hydrogen, ammonia synthesis gas, or methane, all of which should be free of carbon oxides for heating the bed during start-up. For shutdown, process gas can be used to cool down until reaching 175–200°C, at which time pure hydrogen or nitrogen should be used.

Catalyst Type and Suppliers (Methanation)

Catalyst Description

Nickel-on-alumina or other refractory carrier. Nickel content 20–34 wt%. The higher nickel content is preferred for operations in which the upstream CO_2 removal system is expected to at times of upset introduce poisons in the gas product and thus feed to the methanator. The carrier has high surface area so as to provide maximum active nickel area.

Form and Size

Spheres: 3–6 mm (1/8–1/4 in.)
Tablets: 6 × 6. 5 × 5 mm (1/4 × 1/4 in., 3/16 × 3/16 in.)
Extrusions: 3 mm, 5 mm (1/8 in., 3/16 in.)
Rings: 6 mm (1/4 in.)
Surface areas: 100–250 m²/g

Suppliers

Haldor Topsoe, Synetix, United Catalysts

Licensors

Haldor Topsoe, Kellogg Brown & Root, Lee consulting, Linde, AG, Synetix, Uhde GmbH

Catalyst Deactivation (Methanation)

Poisoning

This nickel catalyst, like all the others, is subject to poisoning by sulfur and arsenic compounds, but removal of these and others has occurred in upstream units. However, there is the problem of carryover from the CO_2 removal system preceding the methanator during an operational upset.[12] Fortunately, carryover from the various frequently used amine, methanol, and glycol units does not poison the catalyst. Potassium-carbonate unit carryover will cause core blockage and is also a poison. Systems using sulfalane (tetrahydrothiophene) can also cause poisoning due to decomposition of the sulfalane carryover into simpler sulfur compounds that poison the nickel catalyst.

Sintering

The reactor vessel is usually designed for a temperature of 450°C, and normal operation is considerably below this temperature.[12] Upsets upstream that cause a significant increase in temperature will produce runaway reactions that can damage both the carrier and the nickel, causing reduction in active area and thus major losses in activity. Short excursions, however, have been experienced up to 765°C without significant damage if the temperature peak is not long lived. However, since the vessel design temperature is 450°C, quick action is necessary to reduce the temperature surge using cold nitrogen, hydrogen, process gas, or ammonia gas.

Catalyst Regeneration

Catalyst life of from 5 to 10 years has been reported, and no attempt is made to reactivate the spent catalyst. Recovery of nickel by a supplier or jobber may be economically feasible.

Process Units (Methanator)

The methanator reactor is a typical adiabatic reactor that requires careful loading and feed distribution so that bypassing is reduced to a minimum. The feed gas usually contains 0.2–0.5 vol% carbon monoxide and 0.01–0.2 vol% CO_2, and the main cause of failure to attain essentially total removal of carbon oxides (less than 10 ppm) is a poorly packed bed that permits significant bypassing.

Temperature is the main operating variable. Pressure is set by the pressure of the reformer less the ΔP across the previous units up to the inlet of the methanator. In the usual inlet temperature (250–325°C) and outlet temperature ranges, the equilibrium constants are high, and the reactions are irreversible. Thus, kinetics control the reaction. If the inlet carbon monoxide is low, the rise in temperature will be low. In such a situation, the inlet temperature must be raised to reach the desired low level of CO and CO_2 in the methanator product. Conversely, higher quantities of carbon oxides due to aging catalysts in the shift converters require a lower methanator operating temperature to avoid excessive temperature rise. The usual maximum continuous operating temperature is 450°C, which is related to the vessel design temperature. The catalyst can actually be operated at higher temperatures (500°C) for extended periods.

A major concern associated with methanator operation is temperature runaway such as that caused by a process upset in the upstream shift converters or in the CO_2 absorber, which creates a large quantity of carbon oxides in the feed to the methanator. Rapid warning and corrective action is required. Stopping feed to the methanator and rapid blowdown to atmospheric pressure, followed immediately by a cold nitrogen purge, protects not only the catalyst but also the vessel, since the vessel can better tolerate high temperatures at low pressure.[13]

The methanation process has been very successful and is used in all modern ammonia plants. It does a good job in reducing carbon oxides.

Dehydration and Compression

Cooling of the methanator product is followed by a knockout drum to remove condensed water, and then the gas passes to the synthesis gas compressor. The reduced volume issues from the first stage and passes through a drier bed loaded with molecular sieves. The effluent from the drier with a combined concentration of less than 1 ppm of water and carbon oxides then enters the second stage, which also receives ammonia converter recycle.[6] The combined compressor discharge flows directly to the ammonia converter after exchanging heat with the ammonia product stream.

Process Kinetics (Methanation)

Various rate expressions for CO methanation have been reviewed.[10] A simple power-law equation has been found to correlate most of the experimental data except when H_2 and/or CH_4 are present in excess.[14]

$$r = k P_{CO} P_H^{0.5}$$

The entire range is apparently covered by the following form:[10,14]

$$r = \frac{k P_{CO} P_H}{1 + K_1 P_H + K_2 P_M}$$

where subscripts CO, H, and M refer, respectively, to CO, H_2, and CH_4.

Carbon dioxide does not react until most of the carbon oxide has reacted. Its mechanism and kinetics are not well developed. Adding to these uncertainties is the fact that, at commercial conditions, the reactions are diffusion controlled. An earlier report provided empirical equations for both CO and CO_2 that may still be in use.[11]

$$SV = \frac{k_{CO} P^{0.5}}{\log_{10}(CO_{in}/CO_{out})} = \frac{C k_{CO} P^{0.5}}{\log_{10}(CO_{2\,in}/CO_{2\,out})}$$

where   SV = the space velocity in reciprocal hours
         P = the pressure in psi absolute
        k_{CO} = the reaction rate constant for CO
         C = the constant = 0.5

19.8.7 AMMONIA CONVERTER (NITROGEN + HYDROGEN → AMMONIA)

Chemistry (Ammonia Synthesis)

Modern ammonia plants operate in the range of 350–550°C and 100–300 bar. The catalyst consists of primarily Fe_3O_4 (magnetite) in the form of granules with several promoters, usually Al_2O_3, K_2O, CaO, and others such as MgO.[12] The single stoichiometric reaction can be written as follows, with a basis of one mole of ammonia.

$$0.5 N_2 + 1.5 H_2 \Leftrightarrow NH_3$$

Temp., K	ΔH, kcal	K_p
600	−12.23	3.98×10^{-2}
700	−12.53	9.036×10^{-3}
800	−12.77	2.905×10^{-3}

The general equilibrium constant is defined in terms of activities, which for ammonia synthesis is as follows:

$$K = \frac{a_A}{a_N a_H} = \frac{Y_A P}{(Y_N P)^{0.5}(Y_H P)^{1.5}}\frac{v_A}{v_N^{0.5} v_H^{1.5}} = K_p K_v$$

where   a = activity
Y = mole fraction
P = pressure in atm
v = fugacity coefficient

and subscripts A, H, and N refer, respectively, to ammonia hydrogen, and nitrogen.

K_p is obtained from recorded data or equations based on the Gibbs free energy, which is a function of temperature only. K_v is the fugacity coefficient term, which is a function of pressure and composition of the reaction system.

Equilibrium composition is a straightforward calculation for modest pressures where ideal gas laws apply. In such cases, K_v is equal to 1. For ammonia synthesis, K_v deviates from unity. It is then necessary to initially use the value of K_p only and calculate an equilibrium composition for the ideal-gas case. Using these compositions and equations of state, fugacity coefficients can be calculated and the actual mole fractions calculated from the above equation with the RHS being $K_p K_v$.

Figures 19.8 and 19.9 present the results of such calculations for, respectively, a pure H_2 and N_2 mixture and a typical feed with argon and methane inerts. Inerts lower the effective partial pressure of the reactants and thus lower the equilibrium conversion and the rate.

The figures also show quite dramatically the effect of pressure on equilibrium for this reaction of declining moles. Over the years, development of improved catalyst, reactors that featured better temperature control, and techniques for minimizing inerts made it possible to operate in more favorable temperature ranges and at lower pressures, which not only improved operations but saved energy.

Mechanism (Ammonia Synthesis)

Because of the crucial importance of assuring adequate crop production for the growing world population, it should not be surprising that ammonia synthesis has received more attention than any other commodity. Many reviews have been published over the years. The interested reader will find several of the more recent reviews the most valuable.[15,16]

The surface science techniques developed in more recent years have led to a better understanding of the catalyst surface sites, the role of the various promoters, and the probable sequence of reaction steps. One such list of steps has proved successful in predicting commercial reactor performance.[15,17] This mechanism was established based on kinetics and observations of adsorption on single crystals of Fe(III).

$$N_2(g) + * = N_2\text{-}* \tag{1}$$

$$N_2\text{-}* + * = 2N\text{-}* \tag{2}$$

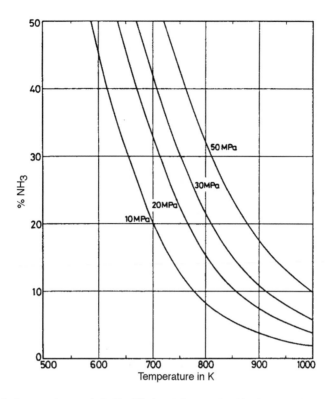

FIGURE 19.8 Mole Percent Ammonia in Equilibrium Mixture. Feed basis: 75% H_2, 25% N_2, and 0% inerts. Reprinted by permission: Christiansen, L. J. in *Ammonia: Catalysis and Manufacture,* A. Nielsen, ed., p. 7, Springer-Verlag, New York, 1995.

$$N\text{-}* + H\text{-}* = NH\text{-}* + * \tag{3}$$

$$NH\text{-}* + H\text{-}* = HN_2\text{-}* + * \tag{4}$$

$$NH_2\text{-}* + H\text{-}* = NH_3\text{-}* + * \tag{5}$$

$$NH_3\text{-}* = NH_3(g) + * \tag{6}$$

$$H_2(g) + 2* = 2H\text{-}* \tag{7}$$

The asterisk indicates an active site.

Step 2, adsorptive dissociation of N_2 on the catalyst surface, was determined to be the rate-limiting step. All other steps are at equilibrium. These conclusions were reached by a number of investigators and confirmed in a most remarkable procedure involving quantum mechanical calculations and studies of adsorption on single crystals of Fe III of N_2, H_2, and NH_3 under ultra-high vacuum.

All parameters in the model were determined without reference to data obtained from ammonia synthesis experiments.[12] Yet, the calculated parameters, when applied to predicting NH_3 mole fraction in the outlet of an experimental reactor, gave good agreement between experimental and calculated NH_3 output in the range of 1–300 atm.[15,17] Such confirmation is compelling, particularly since actual synthesis reactors do not operate under high vacuum, and the commercial catalyst consists of polycrystalline Fe III, not single crystals.

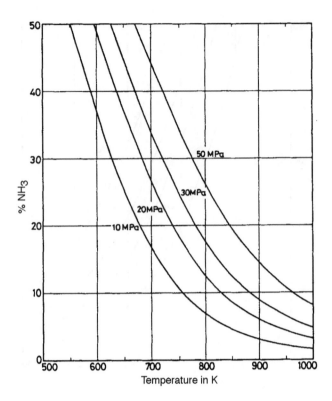

FIGURE 19.9   Mole Percent Ammonia in Equilibrium Mixture. Feed basis: 67.5% H_2, 22.5% N_2, 3% Ar, and 7% CH_4. Reprinted by permission: Christiansen, L. J. in *Ammonia: Catalysis and Manufacture,* A. Nielsen, ed., p. 8, Springer-Verlag, New York, 1995.

Catalyst Type and Suppliers (Ammonia Synthesis)

The ammonia synthesis catalyst is not the typical catalyst with active components dispersed on a carrier. A brief description of its manufacture will be helpful in understanding its particular characteristics.

Catalyst Manufacture and Composition[18]

Haber used natural magnetite, which happened to have impurities of K and Ca oxides and Al_2O_3 present in the range similar to that used today. High-quality magnetite ore is used today. The ore, Fe_3O_4, is crushed and mixed with the promoters (e.g., 1% K as carbonate, 1–3% calcium carbonate, 2–5% Al_2O_3, and in some cases other additives such as chromium oxide (0.5–3%). The mixture is melted in a fusion furnace and cooled slowly, and the resulting ingot is crushed and separated into various sizes.

Catalyst Sizes and Form

In general, the catalyst consists of irregular pieces in the following mesh size ranges: 1.5–3, 3–6, 6–10, 8–12, and 12–21 mm. Selection depends on the reactor design. Short beds can allow small pressure drops and thus use smaller particles, which have the advantage of higher activity. Designs with deeper beds require larger particles to avoid excessive ΔP and higher energy costs. A typical larger size is 6–10 mm mesh.

   Spheroidal shapes are also produced (5 × 10 mm) that have higher low-temperature activity. Polished granules are also produced that eliminate sharp edges that can break off during use and cause unwelcome increases in pressure drop and perhaps uneven flow distribution.

Suppliers

Haldor Topsoe, Engelhard, Synetix, United Catalysts

Process Licensors (Active)

Haldor Topsoe, Kellogg Brown & Root, Lee Consulting, Linde AG, Synetix, Uhde GmbH

Catalyst Surface Characteristics (Single-Crystal Studies)

In recent years, various details of reactant chemisorption have been postulated as shown in Table 19.7. Much of this work has been on single crystals of iron or an iron crystal with promoter present.[15] These observations, along with others related to O_2 and H_2O in the feed, have been described.[15]

TABLE 19.7
Chemisorption on Iron Crystal Surfaces

	Structure	Thermodynamics	Effect of Promoters
Hydrogen	Hydrogen is adsorbed as atoms.	Decreases with increase in temperature.	Chemisorption not affected by presence of potassium on iron single crystal.
	Weakly adsorbed as H*.		Chemisorption increases in the presence of Al_2O_3 when promoted by potassium. Strength of adsorption also increased.
Nitrogen	Nitrogen is adsorbed "side-on" and is more weakly adsorbed than a poison such as CO, but the dissociation to N* is easier.	Coverage is less than physical adsorption.	Potassium increases the stability of N_2* on iron. Adsorption occurs on iron adjacent to the potassium and not on top of a potassium site. N* increases stability of potassium. Potassium increases the strength and the rate of chemisorption on iron. Aluminum decreases chemisorption per unit area.
Ammonia	Species formed on iron upon adsorption of NH_3 are: NH_3* at 120–300 K, NH* at 350 K, and N* at 541 K.	Enthalpy of chemisorption is 20% higher on the 111 face than the 110 face.	

The purpose of such tedious studies on single crystals and, at times, polycrystalline iron is not to enable the development of more accurate kinetic equations. Instead, such studies seek more detailed understanding of the molecular processes with the goal of developing improved catalysts. Over the years, improved catalysts have been placed on the market, all distant cousins of Haber's catalyst.

Catalyst Activation and Structure of Active Catalyst

Activation of commercial ammonia catalyst is accomplished using heated synthesis feed gas in the range of 370–390°C and, ultimately, 400–425°C. As the catalyst begins to become active, ammonia production begins and reaches full production after about one week. Shorter times are possible by using a prereduced or partially prereduced catalyst charge.[22] Catalyst suppliers provide complete details on the optimum procedures.

The next question that comes to mind is, "what does the reductive activation do to the structure of the catalyst?" Briefly, a weight loss of about 25% occurs due to oxygen removal as water, which causes an increase in the crystal density of the iron phase. The overall particle dimensions, however, are not changed.[19] These events result in a catalyst particle with 50% porosity. The iron crystallites, which are about 20–30 mm and have an internal area of $10m^2/g$, are imbedded in a matrix of promoters (with some potassium at the surface) that maintain separation between iron crystals.[19]

A massive amount of excellent work on the structural details has been reported based on such techniques as Mossbauer spectroscopy, x-ray diffraction EXAFS, SEM, and TEM. These studies were designed to define the fundamental differences between the activated catalyst and pure iron powder.[23] Of course, one already knows that the catalyst precursor itself differs because of the addition of promoters.

The sharp differences between pure magnetite and the catalyst precursor at 27°C (300 K) have been described.[23] The precursor is not pure magnetite but contains some metastable wustite (FeO) and textural promoter oxides at the grain boundaries. Upon slow heating, the FeO disproportionates into elemental iron and magnetite. If hydrogen is present, the iron formed acts as a nucleus for the reduction of the other product, magnetite. This series of events creates some porosity in the particle and facilitates the reduction of the bulk magnetite at the higher activation temperature.

The structural features of an activated ammonia synthesis catalyst have been postulated based on a single grain of the original magnetite catalyst precursor.

- The formerly dense grain has become an agglomerate of small iron particles.
- This agglomerate consists of a stack of platelets of single-crystalline iron with Fe(III) as the basal plane.
- The particles are separated, which creates a pore system that is stabilized by spacers of structural promoter (ternary oxides).
- Sintering is prevented by distribution of structural promoter oxides on the surfaces of the platelets. This action is essential in maintaining the metastable state of the platelets, which is the catalytically active state.

Promoters (Ammonia Synthesis)

Aluminum, potassium, and calcium are the main promoters used in the commercial iron synthesis catalyst. In addition to the discovery and implementation of recipes for the promoted catalyst, there has been a great deal of useful investigation on the distribution of promoters in the catalyst structure and the effects they have on various functions of the catalyst.[24] For easy reference, the conclusions that have been reached are summarized in Table 19.8. See also Table 19.7, which is focused on chemisorption and includes the effect of promoters on chemisorption of the reactants.

Catalyst Deactivation (Ammonia Synthesis)

Poisoning[15,19,20]

The most common permanent poisons are sulfur compounds, but chances of sulfur compounds entering the ammonia converter are low for natural gas plants, since many prior steps remove sulfur compounds, and natural gas feed usually has a very low sulfur content. Sulfur containing lubricating oil must be avoided. If sulfur compounds do break through, they are initially chemisorbed in the first bed. Ultimately, sufficient reaction or coverage by sulfur compounds of the iron surface produces an inactive bed. It is reported that 0.2 mg S/m^2 of catalyst surface will completely deactivate the catalyst.[19]

Sulfur deposition on the catalyst surface as high as 6300 ppm completely deactivates the catalyst. Hydrogen sulfide of 20–30 ppm in the feed has been reported to cause significant poisoning over a 24-hour period. All sulfur poisoning is irreversible.

Other permanent poisons include compounds of phosphorus and arsenic. These are rarely detected in natural gas plants. Phosphorus and arsenic both form stable compounds with iron and

TABLE 19.8
Location and Effect of Promoters

Promoter	Location and Form on Reduced Catalyst	Effect
Al_2O_3	Present as the oxide as a thin layer on the reduced exposed surface. Also as pillars of segregated Al_2O_3. Optimum amount variously reported as 2–5% Al_2O_3.	Stabilizes active planes of iron such as Fe(III) and Fe(211) and prevents them from converting to more thermodynamically stable but less active surfaces. Pillars or spacers of Al_2O_3 stabilize pore system.
CaO	Remains as the oxide but becomes segregated in spaces between iron crystallites. Optimum amount variously reported as 2–3% CaO.	Increases activity, surface area, and resistance to impurities.
K_2O	Since K_2O decomposes at ~620 K, it is generally thought that oxygen and potassium form an adlayer on the iron surface.	Increases the rate limiting step of nitrogen dissociative adsorption. Also thought to lower the concentration of adsorbed NH_3 on the active iron, thereby increasing the number of sites available for nitrogen dissociation and the rate of NH_3 formation. Potassium has been shown to decrease the adsorption energy on iron.

Highlights from Refs. 15, 25, and 26.

destroy the activity of the iron with which they react. Chlorine compounds react readily with potassium, producing a KCl that will migrate out of the catalyst and enter the gas stream because of the higher vapor pressure of the chlorides. In so doing, the important function of potassium is irreversibly destroyed over time. Chloride compounds are at times present in lubricating oil additives.

Temporary Poisons

Oxygen compounds (namely, CO, CO_2, H_2O, and O_2) act as significant temporary poisons. The ammonia catalyst is capable of catalyzing the methanation reaction.

$$3H_2 + CO \rightarrow CH_4 + H_2O$$

$$4H_2 + CO_2 \rightarrow CH_4 + 2H_2O$$

$$H_2 + 0.5O_2 \rightarrow H_2O$$

As the reactants move through the beds, the amount of water vapor increases.

Usually, oxygen compounds can form during methanator upsets or from make-up gas issuing from the secondary separator. Oxygen compounds apparently block their nearest neighbor unit cell. The ultimate effect is an increase in the activation energy of the rate limiting step of the synthesis gas reactor to NH_3,[21] thus lowering the rate of production. But the situation is temporary and is corrected when an essentially oxygen-free feed is restored.

In general, operators endeavor to keep the oxygen equivalent in the feed to 0–10 ppm (ppm oxygen equivalent = 2x ppm CO_2 + ppm H_2O + ppm CO + 2 x ppm O_2).

Sintering

Sintering, which is a common problem with many catalysts caused by excessive operating temperatures, is a more complex issue with ammonia synthesis catalysts. Either thermal or chemical stress can change the distribution of promoter oxides such that the activation energy is lowered for the conversion of the required metastable state to the stable isotropic cubic form of iron particles.[23] If that happens to the catalyst, activity plummets, and the damage is irreversible.

From a practical viewpoint, one needs to know what is the most likely cause of sintering in an operating plant. It seems rather clear that sintering can happen when substantial oxygen poisoning reduces the catalyst activity and, of course, forms water. The combined effect of an increased temperature, required to maintain production, and water vapor can lead to a more rapid sintering.[19] Water alone, produced in catalyst reduction at low space velocities that allowed higher water content, can cause significant deactivation.[19]

Unless prolonged operational upsets occur, sintering is usually not a problem.

Catalyst Life

Lives of 10 years or more are normal. Usually, the catalyst is changed when the converter vessel requires inspection.[19]

Process Units (Ammonia Synthesis)

The ammonia synthesis reaction is an extreme example of an exothermic, strongly equilibrium-limited reaction. As such, it requires efficient means for cooling the reaction mixture as the conversion progresses as well as operation at elevated pressure to improve equilibrium conversion because of the decrease in moles (2 to 1) in the synthesis reaction. Furthermore, high reaction rates in the low-conversion region can cause mass-transfer control that is overcome by high mass velocities. But higher mass velocities create higher pressure drop and increase energy consumption, which can be reduced by using larger catalyst particles. Finally, larger particles reduce the catalyst effectiveness factor and thus the apparent activity. Clearly, there are various options in reactor design and catalyst selection that can be used to optimize reactor planning and operation. Catalyst manufacturers and licensors with years of experience have designed their catalysts and commercial processes to optimize the various functions of the reaction system. One major result has been a number of innovative reactor and overall process designs that have been most successful. In fact, many of these reactor design characteristics have been employed for other exothermic, high-pressure, equilibrium-limited reactions in both the petroleum and chemical industries.

Reactor System Overview

Reactor effluent from any equilibrium-limited reactor contains significant amounts of unreacted feed as well as product. Separation of product can be difficult or simple, depending, in the case of a gaseous effluent, on the volatilities of the components. In the case of an ammonia reactor product, ammonia separation requires subambient temperatures. In a frequently used configuration, the reactor effluent first passes through a high-level heat recovery exchanger that produces high-pressure steam. The effluent then interchanges with the feed to the reactor and is subsequently cooled by a cooling water exchanger and then, finally, a evaporating liquid-ammonia chiller where ammonia condenses. The off gas from the condensation contains the unreacted H_2 and N_2 plus inerts (CH_4 and Ar). These inerts will build up in the system, and a portion of the off-gas must be purged continuously to maintain a low inert concentration in the feed to the reactor.

The recycle is conducted to the interstage of the synthesis-gas (N_2-H_2) compressor where it joins fresh synthesis gas from the first stage that has been dried by a bed of molecular sieves. This so-called loop is considered advantageous for the following reasons:[6]

- The molecular-sieve drier removes oxygenated impurities from the synthesis gas as well as other impurities.
- The recycle gas, along with the purge stream, is the overhead from the ammonia product condenser. Hence, the recycle gas has been thoroughly contacted with liquid ammonia and impurities absorbed, thereby ensuring long catalyst life.
- Refrigeration duty is reduced, because the converter effluent gas is not saturated by the make-up gas.
- The net result is a lower ammonia concentration of ammonia in the converter feed, which produces higher initial reaction rates in the converter.

Other loop designs are also used and have been described in detail.[25] Another major type recovers the ammonia product by condensation after make-up gas recycle compression. This design also produces a stream devoid of impurities due to the washing effect of the liquid ammonia but requires higher energy consumption, since the entire ammonia product effluent is compressed.

Synthesis Converter (Reactor)

Operating Conditions   Hundreds of plants are in operation worldwide. Some are older units, many have been revamped, and many are modern high-tonnage, low-energy plants. Operating conditions covering such varied plant types and designs suggest a range of 360–550°C and 100–300 bar. Operating conditions vary, depending on the capacity, catalyst, and energy minimization features in the plant's design. Most plants have been custom designed for the economic conditions appropriate for operation at a particular location. As conditions changed, many units have undergone revamping by replacing internal converter components so as to increase capacity using the original pressure vessel.

Designers are intuitively guided by the effect that operating conditions have on conversion and thus total ammonia production, as summarized in Table 19.9. Design firms and catalyst suppliers have design programs that provide numerical values for changes in operating variables and ammonia production as the feed progresses through the catalyst beds. But general knowledge on the effect of such changes is always valuable in understanding the process and in checking the reasonableness of computer generated results.

Converter Temperature Control   The rate of an exothermic reaction that is reversible increases with temperature at a given composition, reaches a maximum, and then declines rapidly as equilibrium is approached. This situation is depicted in Figure 19.10, where constant rate lines, 0 to 9, are shown with rates increasing in that order. The zero line represents equilibrium ammonia composition with a rate of zero. Select, for example, 14% NH_3 and follow the horizontal line from 700–1000°F. The rate increases as the horizontal line passes through 3, 4, and 5, which is close to the peak rate. Moving on, the lines passes through progressively lower rates 4, 3, 2, and 1 and finally reaches zero rate at close to 1000°F.

It should become apparent that the ideal reactor is one that follows the maxima of the constant rate lines.[26] This observation is essentially the basis for the various attempts to accomplish such a goal by cooling the reactants either directly or indirectly as the reaction proceeds along the reaction path. This cooling can be accomplished directly by a quench stream of feed gas or indirectly by heat exchangers.

Because it is necessary to consider only one reaction, a graphical procedure is easily implemented to quickly overview various strategies involving multiple beds with direct contact quench or intermediate indirect cooling. For the quench case a heat balance on the reaction mixture and quench combination and a mole balance on ammonia yields the following at the exit of a stage.[26,36]

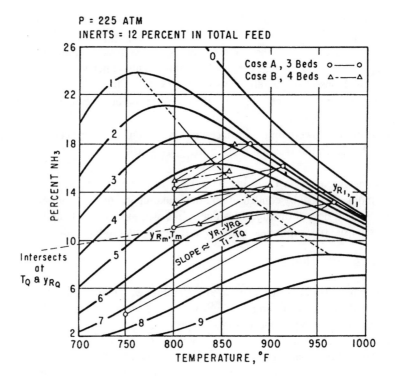

FIGURE 19.10 Percent Ammonia vs. Temperature Rate Plot Ammonia Synthesis. The numbered rate lines increase in value from 0 though 9. The sloping straight lines represent the adiabatic operating lines beginning with the first state at 750°F followed by the direct-contact quench line. If indirect quench is used, the cooling line would be horizontal. Reprinted by permission of John Wiley & Sons, Inc., Rase, H.F., in "Chemical Reactor Design for Process Plants," Vol. 2, p. 69, Wiley, New York. Copyright © 1977.

$$\frac{Y_{RM} - Y_{RQ}}{T_M - T_Q} = \frac{Y_{R1} - Y_{RQ}}{T_1 - T_Q} = \text{slope of quench lines}$$

where T_1 = exit temperature from stage 1

T_M = mixture temperature

T_Q = quench temperature

Y_{R1}, Y_{RM}, Y_{RQ} = mole fraction of R (ammonia) in product from stage 1, in mixture, and in quench, respectively

The same slope may be assumed for each successive quench line. The slope of the reaction line is based on a simple adiabatic temperature calculation relating ΔT to the change in moles of ammonia. For the indirect cooling design, the outlet from a stage is cooled by drawing a horizontal line (constant NH_3 composition) to the intersection with a desired interstage temperature. Clearly, indirect cooling has the advantage of not diluting the reaction mixture and lowering the ammonia composition in the reacting gases. In so doing, higher conversion per bed results. An effluent of higher ammonia content is produced then that from a quench reactor with the same catalyst loading. This more concentrated outlet reduces the recycle quantity lowering energy costs. It also raises the ammonia dewpoint such that a portion of the ammonia can be condensed by cooling water with attendant energy savings.[27]

TABLE 19.9
Effect of Operating Variables on Ammonia Synthesis Convertors ($1/2\ N_2 + 3/2\ H_2 \rightarrow NH_3$)

Operating Conditions	Effect on Increasing	Explanation
Inlet temperature	Increases rate. Lowers adiabatic equilibrium NH_3 concentration.	Typical effect for exothermic equilibrium limited reaction since equilibrium constant (K_p) declines with temperature increase.
Pressure	Increases reaction rate and increases equilibrium conversion and thus production.	Decline in moles (two-to-one) for each mole of NH_3 produced increases possible equilibrium conversion and thus ammonia production.
Space velocity	Decreases conversion as reflected in lower NH_3 concentration in converter effluent, but can increase production.	From a contact point of view, less contact time produces lower conversion, but if total feed flow is high enough, the total production of NH_3 could increase.
Catalyst particle size	Smaller particles produce higher conversion.	Smaller particles are less inhibited by diffusional gradients inside the pellet (high effectiveness factor).
Effect of inerts	Inerts decrease conversion to ammonia, and they cause higher recycle rates and effect separation equilibrium.	Causes lower partial pressure of reactants, which reduces the reaction rate and the equilibrium ammonia composition at equilibrium. An inert increase in the effluent requires an increase in separator pressure or a lower condensing temperature.

Based on Refs. 25 and 26.

Converter Types (Ammonia Synthesis)

A remarkable variety of designs have successfully been applied over many years. In more recent decades, a major focus has been on reduction of energy consumption, improved yield, and increased production. Resulting designs featured low pressure drop improved temperature control, ease of catalyst loading and dumping, and maximum energy recovery. Excellent reviews have been written on the wide variety of designs over three decades of ammonia production history.[4,7,25,27,31] The focus in this section will be on the more recent designs, along with some reference to earlier designs that continue to operate in many locations and are often subjects for revamp.

One Common Feature

Many of the various designs have one feature in common. To reduce the vessel thickness, and also avoid hydrogen embrittlement, a stainless steel inner vessel that contains the catalyst is enclosed by an outer carbon steel vessel. An annular space between the two vessels is used to pass low-temperature synthesis gas to the reactor inlet. In this manner, the inner vessel, which sees feed high in hydrogen at high temperature and pressure, is constructed of stainless steel, which is resistant to attack by hydrogen. Because the ΔP between the annulus and the inner vessel is relatively small, the inner-vessel thickness will be modest. The outer vessel is kept cool by the feed gas flow that it can be safely designed using carbon steel with a thickness based on the lower temperature of the wall. Recently, reactors without an internal shell have been developed (p. 461).

Hydrogen at high pressure and temperatures diffuses rapidly into carbon steels. It reacts with the carbon, producing methane that builds up pressure in the steel, ultimately causing cracks that severely weaken the vessel. Austinitic stainless steels are not affected. Steels containing molybdenum or cobalt-molybdenum are not attacked under many conditions, because these elements stabilize the carbide.[26]

Converter Descriptions[6,7,25,27,28,31]

Licensors of ammonia plants not only offer special converter designs but also various innovations in the entire process of natural gas or naphtha to ammonia. No attempt will be made to examine and compare the various proprietary elements of the entire process. Instead, attention will be focused on the most active converter types used beginning in 1960 and through the 1990s.

Major Trends[6,7,25,31]

- Introduction of the single-train plant by M. W. Kellogg (now Kellogg Brown & Root) with large centrifugal compressors for synthesis gas, which initially were limited to 150 bar (1963) and required reciprocating compressors for the final stage. Located at American Oil in Texas City.
- Second Kellogg single-train plant had total centrifugal compression to 317 bar (1965). Located at Monsanto, Luling, Louisiana. Kellogg developed the large centrifugal compressors for synthesis gas with Clark Brothers Compressors.
- Improved reliability and higher capacities of centrifugal compressors over reciprocating compressors propelled a massive growth of single-train plants of Kellogg design and joined by many other licensors. Improvement in compressor design led to designs of ever increasing capacity, reaching 2000 tons per day using a single compressor.
- In the 1970s, the efforts toward energy savings became paramount. Reducing power costs and improving energy recovery became essential. Higher synthesis gas compression increases power cost, but compensating affects can occur, such as higher temperature ammonia recovery or elimination of refrigeration for ammonia recovery at pressures above 300 bar. No general rule applies, since each process must be analyzed relative to energy costs. In general, low-capacity plants (400–600 tons per day) use lower pressures, because it increases actual volume flow and provides higher compression efficiency. By contrast, higher pressures are used in large-capacity plants (1500 tons per day or more) to reduce the size of piping and that of such major vessels as converters. Typical modern units are optimized in the range of ~140–180 bar.
- Continuing trends include improved energy efficiency and increased production from existing plants and new designs.
- Improved computer control.

Quench Converters

The hundreds of single-train plants built beginning in the 1960s used a quench-type reactor. The synthesis gas enters the annular space between the high-pressure shell and the catalyst containing inner vessel. It then passes through an exchanger heated by the product gas, which rises up from the bottom of the reactor section in a central pipe. The heated gas then passes through a series of catalyst beds, each of which is followed by a quench from a distributor located after each bed.

The internals consist of a central riser pipe, a quench feed line and distributor, a screen support for each bed, and a catalyst dropout tube. As catalyst is released in such a tube from the bottom bed, the tube from the previous bed is uncovered. Catalyst then flows from the previous beds sequentially. At the bottom of the inner reactor basket, a dropout line extends through the outer shell so catalyst can be collected.

Crossflow Kellogg Quench Converter

To make servicing of the reactor simpler and to reduce pressure drop, Kellogg developed a horizontal reactor with longer beds that were entered along the length rather than axially. This crossflow design enabled the use of smaller catalyst of higher effectiveness factor and thus, in effect, high activity.

By this innovation, NH$_3$ production could be increased significantly. The horizontal design included a track for sliding the entire inside catalyst bed module out of the converter shell for easy access.

The initial horizontal-reactor design included quench lines between beds, and the unit was typically 90 ft long or more.

ICI Quench Converter

ICI (now Synetix) designed a quench reactor with lozenge-type quench distributors that enabled use of one continuous easily removable catalyst bed. This design is particularly interesting, because the concept was later employed in controlling highly exothermic reactions in petroleum refineries.

Indirect Cooling Converters

The modern trend in the 1990s continued to favor indirect, heat-transfer converters and also radial or crossflow designs. In so doing, higher concentration of ammonia in the converter effluent is accomplished by avoiding the diluting effect of quench converters. Additionally, flow paths that are short reduce pressure drop and enable the use of smaller catalyst particles (1.5–3 mm), thereby increasing ammonia production per unit volume by dramatically increasing exterior catalyst surface area and effectiveness factor. Practically speaking, if smaller catalyst is installed in the same converter volume, higher concentrations of NH$_3$ will result in the reactor effluent. This result makes it possible to reduce power required for recycle gas circulation and refrigeration and for make-up gas compression.[32]

Kellogg Crossflow, Indirect Cooling Converter

This design is similar to the horizontal-quench converter previously described except that, in place of quenches between stages, a compact exchanger allows cooling of the effluent from bed no. 1 (see Appendix). Also, a heat-exchanger bypass or partial bypass line allows for cold synthesis gas to be mixed with heated gas entering bed no. 1 for accurate temperature control. Unlike quench, no dilution of bed effluents occurs. Again, as in the earlier quench reactor, crossflow provides low bed depths in the direction of flow. Catalyst of low particle size can be used without excessive pressure drop and higher ammonia production per unit bed volume results. The rectangular cross section of the beds assures even flow distribution through the beds when entry distribution and catalyst loading practices are properly executed. As with the earlier horizontal quench reactor, the readily removable cartridge allows for ease in loading and unloading.

Haldor–Topsoe Radial Indirect-Cooling Converter

This reactor uses radial flow through the catalyst (see Appendix). The synthesis gas enters at the bottom of the reactor, proceeds up the annular space next to the pressure shell, and then moves down through the tube side of the interbed heat exchanger. Cold synthesis gas is introduced from a central tube into the outlet of the tube side for temperature control. Then the properly heated gas stream passes to the outside circumference of the first radial bed, through the bed radially, and thence through the shell side of the interbed heat exchanger where cooling occurs. Finally, the cooled gas from the exchanger passes through the second bed, from the outside circumference to the inside and then exits the reactor or passes through a third bed prior to exiting.

The hot product exiting the reactor can be used to produce valuable high-pressure steam. An alternative design employs an outlet heat exchanger in the bottom of the converter and different routing of the converter feed so that it initially passes through the outlet exchanger. Thus, the feed is heated by the product.

Kellogg Radial Converter

In recent years, M. W. Kellogg Technology Co. (now Kellogg Brown & Root) introduced a new process for ammonia synthesis (KAAP System) based on the use of their proprietary ruthenium-on-graphite catalyst, said to be 20 times more active than the traditional iron-based (magnetite) catalyst. High activity is most advantageous in the final stages of a converter so that higher equilibrium concentrations of ammonia can be realized at lower temperatures. The converter design for a new plant contains four beds, the initial bed (the largest) being loaded with standard iron-based (magnetite) catalyst and the final three beds with the high-activity catalyst. The internals are designed to prevent very hot gases from contacting the reactor wall so that an inner shell of high alloy steel is not required (see Appendix).

Revamped plants continue to use the existing converter with the standard iron-based catalyst but add downstream a two-bed converter downstream with high-activity catalyst. This reactor is illustrated in the Appendix, which shows the radial beds and the cooling unit that exchanges heat from the exit of bed no. 1 with the feed gas.

Brown and Root–Braun Indirect-Cooling Converter

The Braun converter consists of two separate adiabatic axial flow reactors with exterior heat exchangers after each converter. The synthesis gas passes up the converter via the space between the catalyst vessel and the pressure vessel after being preheated in a heat exchanger exterior to the converter. The heating medium is the outlet product from the first reactor. A portion of the feed can bypass the exchanger for temperature control. Cooled product passes from the intermediate exchanger to the second and larger converter where further reaction occurs. Product from the second converter is used to generate high-pressure in a closely mounted waste-heat boiler.

Operating temperatures are higher than other converter designs, which provides a self-regulating character due to a closer approach to equilibrium. Such a design benefits from modern computer control (distributed control systems) with algorithms that permit continuous control of reaction temperature so as to maximize ammonia production. The converter inlet temperature is 400°C, which is higher than other designs and requires special vessel design considerations, including selection of proper alloys. The maximum outlet temperature is reported to be 530°C.

Revamp Converters and the Future

Many ammonia plants have been in operation for over 30 years. Fortunately, the original converter pressure vessels are often in good shape, so companies can take advantage of new designs that greatly increase plant capacity. Many quench converters have been revamped by replacing the interior components with arrangements that permit radial or crossflow and smaller catalyst.

After over 83 years of synthesis of ammonia via atmospheric nitrogen and hydrogen from hydrocarbons, the industry, including operations, design, and catalyst manufacture, is alive with new successful ideas that, when implemented, increase ammonia production and lower energy costs.

Process Kinetics (Ammonia Synthesis)

Despite the excellent research involving surface science as well as kinetics as an adjunct to confirming mechanistic proposals, the most frequently used rate equation for converter simulation and design is the original Temkin–Pyahev equation.[30] Mechanistic insights based on the aforementioned research, however, have been most helpful in developing new and improved catalysts.

The original Temkin–Pyahev equation[29,30] was written in terms of partial pressures but, for industrial use, fugacities (f) are more appropriate and produce better results.

$$r_A = 2k \left[K^2 \left(\frac{f_N^2 f_H^2}{f_A^2} \right)^a - \frac{f_A^2}{f_H^3} \right]^{1-\alpha}$$

where   k = rate constant for nitrogen disassociation, the controlling step

          K = equilibrium constant

          f = fugacity; activity = fugacity since the standard fugacity is one for a gas; therefore, fugacity of component $f_j = y_j n_j P$

A, N, H = ammonia, nitrogen, and hydrogen, respectively

        r_A = rate of NH_3 formation in convenient units assigned to k such that r_A, that for example, is in moles of A produced/(time) (volume of catalyst).

        P = pressure in atmospheres

A typical value of a of 0.5 has been found satisfactory in many cases, which then yields the following:

$$r_A = 2k\left(K^2 \frac{f_N f_H^{1.5}}{f_A} - \frac{f_A}{f_H^{1.5}} \right)$$

K is based on the stoichiometry of

$$0.5N_2 + 1.5H_2 \rightarrow NH_3$$

Other values of α are possible depending on the process conditions.[31]

Note that at low values of ammonia in the reaction mixture the rate calculation becomes infinite, but ammonia converters are fed with synthesis gas containing recycled ammonia and unreacted hydrogen and nitrogen. Under these conditions, the equation serves very well.

Licensors have design programs and also plant simulations for their particular process that can be helpful to a new operation. Brief reviews of such programs with copious references are valuable.[25,29] Obviously, such issues as effectiveness factors and transport processes between the reacting gases and the catalyst must be considered.

REFERENCES, SECTIONS 19.1 AND 19.2 (SYNTHESIS GAS AND METHANE REFORMING)

1. Tamarn, K., in *Catalytic Ammonia Synthesis*, J. R. Jennings, ed., p. 1, Plenum Press, New York, 1991.
2. Fullweller, W. H., in *Roger's Manual of Industrial Chemistry*, C. L. Furnas, ed., 6th edition, p. 578 and 736, D. Van Nostrand, New York, 1942.
3. Gunardso, H., *Industrial Gases in Petrochemical Processing*, Marcel Dekker, New York, 1998.
4. Rostrup-Nielsen, J. R., in *Catalysis: Science and Technology*, Vol. 5, p. 1, Springer-Verlag, New York, 1984.
5. Bridger, G. W., in *Catalysis: Specialists Periodical Reports*, Vol. 3, p. 39, The Chemical Society of London, 1980.
6. Rostrup-Nielsen, J. R. and Hojlund-Nielsen, P. E., in *Deactivation and Poisoning of Catalysts*, p. 259, Marcel Dekker, New York, 1985.
7. Bridger, G. W. and Wyrwas, W., *Chem. Process. Eng.*, 48, 101 (1967).
8. Nielsen, B. and Villadsen, J., *Appl. Catal.*, 11, 123 (1984).
9. Manufacturer's literature.
10. Renner, H. J. and Marschner, F., in *Ullmann's Encyclopedia of Industrial Chemistry*, 5th edition, Vol. A12, p. 186, VCH, New York, 1989.
11. Johansen, T., Raghuraman, K. S. and Hackett, L. A., *Hydrocarbon Processing*, p. 120, Aug. 1992.
12. Van Hook, J. P., *Catal. Rev.-Sci. Eng.*, 21 (1), 1, (1980).
13. Ross, J. R. H., in *Surface and Defect Properties of Solids: Specialist Periodical Reports*, Vol. 4, p. 34, The Chemical Society of London, 1975.

14. Stiles, A. B. and Kock, T. A., *Catalyst Manufacture,* 2nd edition, Marcel Dekker, New York, 1995.
15. Ross, J. R. H. in *Catalysis: Specialist Periodical Reports*, Vol. 7, p. 1, Royal Society of Chemistry, London, 1985.
16. Alzamora, L. E., Ross, J. R. H., Kruissink, E. C. and Van Reijen, J. R. H., *J. Chem. Soc., Faraday Trans. II,* 77, 665 (1981).
17. Munster, P. and Grabke, J. J., *J. Catal.,* 72, 279 (1981).
18. Allen, D. W., Gerhart, E. R. and Likins, M. R., *Ind. Eng. Chem. Process Des. Dev.,* 14 (2), 256 (1975).
19. Grenoble, D. C. and Estadt, W. M., *J. Catalysis,* 67, 90 (1981).
20. Schneider, R. V. III and LeBlanc, J., *Hydrocarbon Processing,* p. 51, March, 1992.
21. Giacobbe, F. G., Iaguaniello, G., Lolacona, O. and Liguori, G., *Hydrocarbon Processing,* p. 69, March, 1992.
22. Christensen, T. S. and Primdahl, I. I., *Hydrocarbon Processing,* p. 39, March, 1994.
23. Numaguchi, T. and Kikuchi, K., *Chem. Eng. Sci.,* 43, 2295 (1988).
24. Xu, J. and Froment, G. F., *AIChE Journal,* 35, 88 (1988).
25. Akers, W.W. and Camp, D.P., *AIChE Journal,* 1, 471 (l955).

REFERENCES, SECTION 19.3 (SHIFT CONVERSION)

1. Kalkova, N. V. and Temkin, M. I., *Zh. F. Khim.*, 23, 695 (1949).
2. Newsome, D. S., *Catal. Rev. Sci. Eng.,* 21, (2) 275 (1980).
3. Oki, S., Happel, J., Hnatow, M. and Kaneko, Y., *Proc. 5th International Cong. on Catal.,* Vol. 1, p. 173 (1973).
4. Stiles, A. B. and Koch, T. A., *Catalyst Manufacture,* 2nd edition, Marcel Dekker, New York, 1995.
5. Bohlbro, H., *An Investigation on the Kinetics of the Conversion of Carbon Monoxide with Water Vapor over Iron Oxide Catalysts,* 2nd edition, Gjellerup, Copenhagen, 1969.
6. Rostrup-Nielsen, J. R. and Nielsen, P. E. H. in *Deactivation and Poisoning of Catalysts,* p. 259, Marcel Dekker, New York, 1985.
7. Carstensen, J. H., Hansen, J. B. and Pedersen, P. S., *Ammonia Plant Safety,* 30, 139 (1980).
8. Borgars, D. J. and Campbell, J. S., in *Ammonia, Part II,* p. 25, Marcel Dekker, New York, 1974.
9. Campbell, J. S., *I & EC Process Des. Develop.,* 9, (4), 588 (1970).
10. Bohlbro, H. and Jorgensen, M. H., *Chem. Eng. World,* 5, 46 (1970).

REFERENCES, SECTION 19.4 (NAPHTHA REFORMING)

1. Rostrup-Nielsen, J. R., *Steam Reforming Catalysts*, Teknisk Forlag A. S., Copenhagen, 1975.
2. Manufacturer's literature (Haldor Topsoe).
3. Bridger, G. W., *Catalysis: Specialist Periodical Reports,* Vol. 3, p. 39, The Chemical Society, London, 1980.
4. Stiles, A. B. and Koch, T. A., *Catalyst Manufacture,* 2nd edition, Marcel Dekker, New York, 1995.
5. Renner, H. J. and Marschners, F., in *Ullmann's Encyclopedia of Industrial Chemistry,* 5th edition, Vol. A12, p. 196, VCH, New York, 1989.
6. Rostrup-Nielsen, J. R. in *Catalysis: Science and Technology,* J. R. Anderson and M. Boudart, eds., Vol. 5, p. 3, Springer-Verlag, New York, 1984.
7. Gunardson, H., *Industrial Gases in Petrochemical Processing*, Marcel Dekker, New York, 1998.
8. Clark, D. N. and Henson, W. G. S., *Ammonia Plant Safety,* 28, 99 (1988).
9. Verdnijn, W. D., *Ammonia Plant Safety,* 33, 165 (1993).
10. *Chemical & Engineering News,* p. 10, May 11, 1998.

REFERENCES, SECTION 19.5 (METHANOL SYNTHESIS)

1. Fiedler, E., Grossmann, G., Karsebohm, B., Weiss, G. and Claus Witte in *Ullmann's Encyclopedia of Industrial Chemistry,* 5th edition,Vol. A16, p. 465, VCH, Weinheim, Germany, 1990.

2. English, A., Rovner, J. and Davies, S. in *Kirk-Othmer Encyclopedia of Chemical Technology*, 4th edition, Vol. 16, p. 537, Wiley, New York, 1995.
3. LeBlanc, J. R., Schneider, III, R. V. and Strait, R. B. in *Methanol Production and Use,* W. H. Cheng and H. H. Kung, eds., Marcel Dekker, New York, 1994.
4. Chang, T., Rousseau and Kilpatrick, P. K., *Ind. Eng. Process Des. Dev.,* 25, 477 (1986).
5. Bissett, L., *Chem. Eng.,* 84 (21), 155 (1977).
6. Cherednichenko, V. M., Dissertation, Karpova, Physico Chemical Institute, Moscow, USSR, 1953.
7. Chinchen, G. C., Mansfield, K. and Spencer, M. S., *CHEMITECH,* p. 642, Nov. (1990).
8. Stiles, A. B. and Koch, T. A., Catalyst Manufacture, 2nd edition, Marcel Dekker, New York, 1995.
9. Bart, J. C. J. and Sneedon, R. P. A., *Catalysis Today,* 2, 1 (1987).
10. Chinchen, G. C., Denny, P. J., Jennings, J. R., Spencer, M. S. and Wnagh, K. L., *Appl. Catal.,* 36, 1 (1988).
11. Graaf, G. H., Stamhinis, E. J. and Beenackers, *Chem. Eng. Sci.,* 43, (3) 185 (1988).
12. Lee, S., *Methanol Synthesis Technology,* CRC Press, Boca Raton, Florida, 1990.
13. Dybkjaer, I., Hoilund, P. E. and Hansen, B. J., *Synthesis of Methanol on Copper-Based Catalysts*, paper at 74th AIChE Meeting, New Orleans, Nov. 8–12, 1981.

REFERENCES, SECTION 19.6 (PURE CARBON MONOXIDE)

1. Culguhoun, H. M., Thompson, D. J. and Twigg, M. V., *Carbonylation*, Plenum Press, New York, 1991.
2. Gunnardson, H., *Industrial Gases in Petrochemical Processing,* Marcel Dekker, New York, 1998.
3. Pierantozzi, R. in *Kirk-Othmer Encyclopedia of Chemical Technology*, 4th edition, Vol. 5, p. 97, Wiley, New York, 1993.

REFERENCES, SECTION 19.7 (PURE HYDROGEN)

1. Gunardson, H., *Industrial Gases in Petrochemical Processing*, Marcel Dekker, New York, 1998.
2. Häussinger, P., Lohmüller, R. and Watson, M. in *Ullmann's Encyclopedia of Industrial Chemistry,* 5th edition, Vol. A12, p. 317, VCH, New York, 1989.
3. Czuppon, T. A., Kriez, S. A. and Newsome, D. S., in *Kirk-Othmer Encyclopedia of Chemical Technology*, 4th edition, Vol. 13, p. 844, Wiley, New York, 1995.

REFERENCES, SECTION 19.8 (AMMONIA SYNTHESIS)

1. Tamaru, K. in *Catalytic Ammonia Synthesis: Fundamentals and Practice,* J. R. Jennings, ed., p. 1, Plenum Press, New York, 1991.
2. Topham, S. A. in *Catalytic Science and Technology,* J. R. Anderson and M. Boudart, eds., Vol. 7, p. 1 (1985).
3. Heinemann, H., in *Catalytic Science and Technology,* J. R. Anderson and M. Boudart, eds., Vol. 1, p. 1 (1981).
4. Stiles, A. B. and Koch, T. A., *Catalyst Manufacture,* 2nd edition, p. 153, Marcel Dekker, New York, 1995.
5. Kohl, A. and Nielsen, R., *Gas Purification*, 5th edition, Gulf Publishing Co., Houston, 1997.
6. Czuppon, T. A., Knez, S. A. and Rowner, J. M. in *Kirk-Othmer Encyclopedia of Chemical Technology,* 4th edition, Vol. 2, p. 638, Wiley, New York, 1992.
7. Bakemier, H., Huberich, T., Krabetz, R. and Liebe, M. S., in *Ullmann's Encyclopedia of Industrial Chemistry,* 5th edition, Vol. A2, p. 177, VCH, Weinheim, Germany, 1985.
8. Hojlund Nielseon, P. E. in *Ammonia: Catalysis and Manufacture*, A. E. Nielsen, ed., p. 191, Springer-Verlag, Berlin, 1995.
9. Nobles, E. J. in, *Ammonia, Part I,* A. V. Slack and C. R. James, eds., p. 275, Marcel Dekker, New York, 1973.

10. Mills, G. A. and Steffgen, F. W., *Catal. Rev.* 8, (2), 159, 1973.

11. Phillips, J. R., in *Ammonia Part II,* A. V. Slack and C. R. James, eds., p. 311, Marcel Dekker, New York, 1973.

12. Rostrup-Nielsen, J. R. and Nielsen, E. H., in *Deactivation and Poisoning of Catalysts,* J. Oudar and H. Wise, eds., p. 259, Marcel Dekker, New York, 1985.

13. Campbell, J. S., Craven, P. and Young, P. W., in *Catalyst Handbook*, Imperial Chemical Industries, p. 97, Springer Verlag, New York, 1970.

14. Lee, A. L., Fellkirehner, H. L. and Tajbl, G. D., *Preprints* Div. Fuel Chem., Amer. Chem. Soc., 14 (4), Part 1, p. 126, Paper 31, Sept., 1970.

15. Stoltze, P. in *Ammonia: Catalysis and Manufacture*, A. E. Nielsen, ed., p. 21, Springer-Verlag, Berlin, 1995.

16. Ertl, G. in *Catalytic Ammonia Synthesis: Fundamentals and Practice,* J. R. Jennings, ed., p. 109, Plenum Press, New York, 1991.

17. Stoltze, P. and Norskov, J. K., *J. Catal.* 110, 1 (1988).

18. Stiles, A. B. and Koch, T. A., *Catalyst Manufacture,* 2nd edition, A167, Marcel Dekker, New York, 1995.

19. Hojlund-Nielsen, P. E. in *Catalytic Ammonia Synthesis: Fundamentals and Practices,* J. R. Jennings, ed., p. 285, Plenum Press, New York, 1991.

20. Hojlund-Nielsen, P. E. in *Ammonia Catalysis and Manufacture,* A. E. Nielsen, ed., p. 191, Springer Verlag, Berlin, 1995.

21. Stöltze, P. and Norskov, J., *J. Vac. Sci. Technol.,* A5, 4, 581 (1987).

22. Manufacturer's Literature.

23. Schlogl, R. in *Catalytic Ammonia Synthesis: Fundamentals and Practice,* J. R. Jennings, ed., p. 19, Plenum Press, New York, 1991.

24. Strongin, D. R. and Somorjai, G. A. in *Catalytic Ammonia Synthesis: Fundamentals and Practice,* J. R. Jennings, ed., p. 133, Plenum Press, New York, 1991.

25. Dybkjaer, I. in *Ammonia Catalysis and Manufacture*, A. E. Nielsen, ed., p. 202, Springer Verlag, Berlin, 1995.

26. Rase, H. F., *Chemical Reactor Design for Process Plants*, Vol. 1 and 2, Wiley, New York, 1977.

27. Hooper, C. W. in *Catalytic Ammonia Synthesis: Fundamentals and Practice,* J. R. Jennings, ed., p. 133, Plenum Press, New York, 1991.

28. Strelaoff, S., *Technology and Manufacture of Ammonia,* Wiley, New York, 1981.

29. Gramatica, G. and Pernicone, N. in *Catalytic Ammonia Synthesis: Fundamentals and Practice,* J. R. Jennings, ed., p. 211, Plenum Press, New York, 1991.

30. Hansen, J. D. in *Ammonia Catalysis and Manufacture,* A. E. Nielson, ed., p. 149, Springer Verlag, Berlin, 1995.

31. lack, A. V. in *Ammonia Part III*, A. V. Slack and C. R. James, eds., p. 93, Marcel Dekker, New York, 1974.

32. Dybkjaer, I. and Gam, E. A., *Ammonia Plant Safety* 25. 15 (1984).

33. Slack, A. V. in *Ammonia: Part I,* A. V. Slack and C. R. James, eds., p. 57, Marcel Dekker, New York, 1973.

34. Strelzoff, S., *Technology and Manufacture of Ammonia,* p. 267, Wiley, New York, 1981.

35. Pearce, B. B., Twigg, M. V. and Woodward, C. in *Catalyst Handbook,* 2nd edition, M. W. Twigg, ed., p. 7.2.1, Manson Pub., London, 1996.

36. Brötz, W., *Fundamentals of Chemical Reactor Engineering,* Addison-Wesley, Reading, Mass., 1965.

Appendix
Reactor Types Illustrated

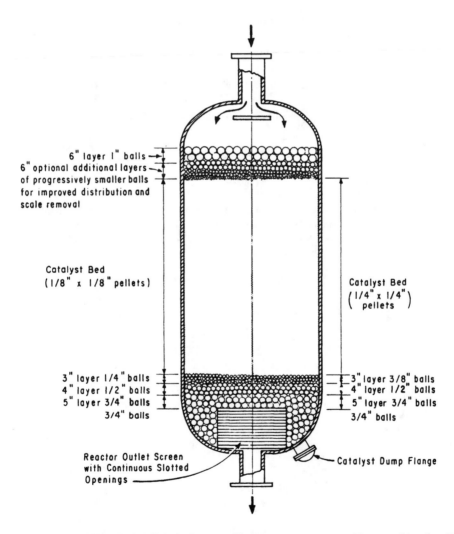

6" layer 1" balls
6" optional additional layers
of progressively smaller balls
for improved distribution and
scale removal

Catalyst Bed
(1/8" x 1/8" pellets)

Catalyst Bed
$\left(\begin{array}{c}1/4" \times 1/4"\\ \text{pellets}\end{array}\right)$

3" layer 1/4" balls
4" layer 1/2" balls
5" layer 3/4" balls
3/4" balls

3" layer 3/8" balls
4" layer 1/2" balls
5" layer 3/4" balls
3/4" balls

Reactor Outlet Screen
with Continuous Slotted
Openings

Catalyst Dump Flange

FIGURE A.1 Typical Single-Bed Adiabatic Reactor. Illustrates arrangement of inert packing for different catalyst sizes designed to ensure even distribution along with inlet and outlet distributing devices. Reprinted by permission of John Wiley & Sons, Inc.: Rase, H. F., *Chemical Reactor Design for Process Plants,* Vol. 1, p. 515, Wiley, New York, 1977. Copyright © 1977.

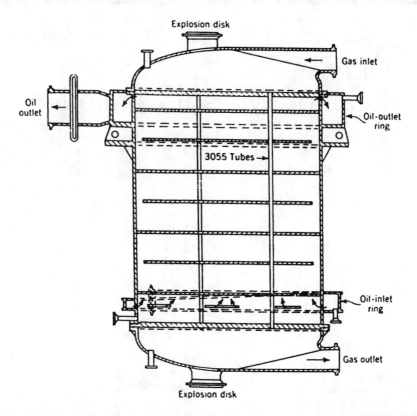

FIGURE A.2 Multitubular Heat-Exchange Reactor. This illustration is of a relatively small oil-cooled reactor for ethylene oxidation to ethylene oxides. Reprinted by permission of John Wiley & Sons, Inc.: Deyer, J. P., George, K. F., Hoffman, W. C., and Soo, H. in *Kirk–Othmer Encyclopedia of Chemical Technology,* 4th ed., Vol. 9, p. 927, Wiley, New York, 1984, copyright © 1984.

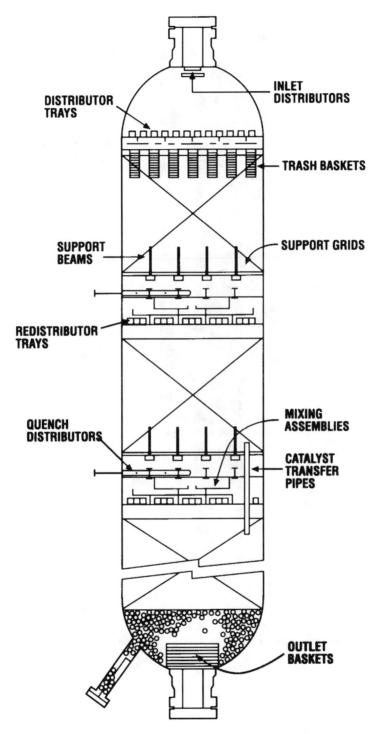

FIGURE A.3 Illustration of a Variety of Reactor Vessel Internals. Items shown are designed and fabricated to distribute and collect fluid flow and to capture particles in the inlet that could cause bed plugging. Reprinted by permission from "US Filter/Johnson Screens" brochure, courtesy of US Filter/Johnson Screens Div., St. Paul, MN.

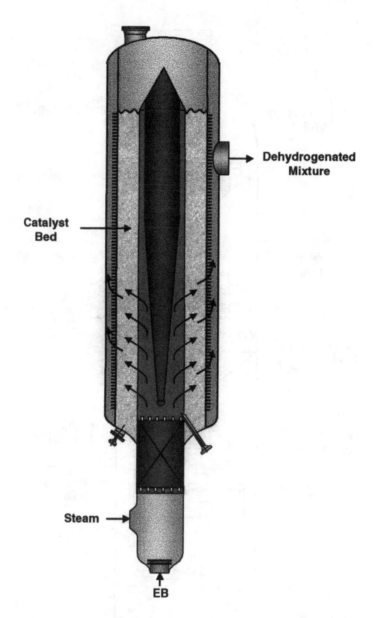

FIGURE A.4 Lummus Ethylbenzene Dehydrogenation Radial Reactor. Central tapered component serves to promote optimum flow distribution and minimize contact time in the noncatalytic region where undesired thermal reactions can occur. Reprinted courtesy of ABB Lummus Global, Bloomfield, NJ.

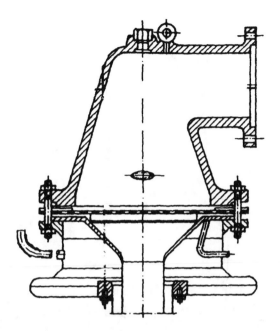

FIGURE A.5 Shallow-Bed Adiabatic Reactor Gauze Packs. These gauze packs made of precious metal constitute the catalyst bed. Ratio of reactor diameter to bed depth is very large. Reprinted from Engelhard-CLAL "Nitroprocess Technologies" brochure, courtesy of Engelhard-CLAL, Carteret, NJ.

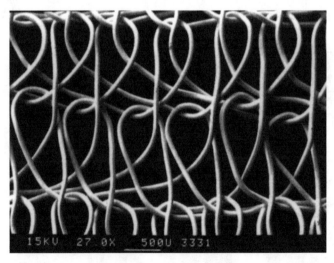

FIGURE A.6 Magnified View of a Portion of Warp-Knitted Catalyst Gauze. Knitted gauze is resistant to tearing and provides a high ratio of available surface per unit weight. Increased void volume assures lower DP, decreased metal loss, and reduced retention of contaminates. Reprinted from "Engelhard-CLAL Nitroprocess Technologies" brochure, courtesy of Engelhard-CLAL, Carteret, NJ.

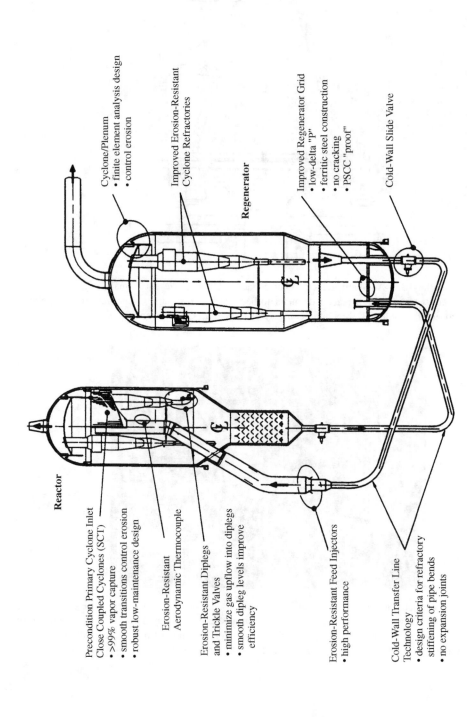

Reactor

Precondition Primary Cyclone Inlet
Close Coupled Cyclones (SCT)
• >99% vapor capture
• smooth transitions control erosion
• robust low-maintenance design

Erosion-Resistant
Aerodynamic Thermocouple

Erosion-Resistant Diplegs
and Trickle Valves
• minimize gas upflow into diplegs
• smooth dipleg levels improve
  efficiency

Erosion-Resistant Feed Injectors
• high performance

Cold-Wall Transfer Line
Technology
• design criteria for refractory
  stiffening of pipe bends
• no expansion joints

Cyclone/Plenum
• finite element analysis design
• control erosion

Improved Erosion-Resistant
Cyclone Refractories

Regenerator

Improved Regenerator Grid
• low-delta "p"
• ferritic steel construction
• no cracking
• PSCC "proof"

Cold-Wall Slide Valve

FIGURE A.7   Exxon Flexicracking FCCU Diagram. Various mechanical hardware improvements are noted along with illustration of major internals. Reprinted by permission: Zaczepinski, S., Shaw, D. F., and Walter, R. E., Exxon Research and Engineering Company, Florham Park, NJ. Paper AM-96-24, NPRA Annual Meeting. March 1996, San Antonio, TX.

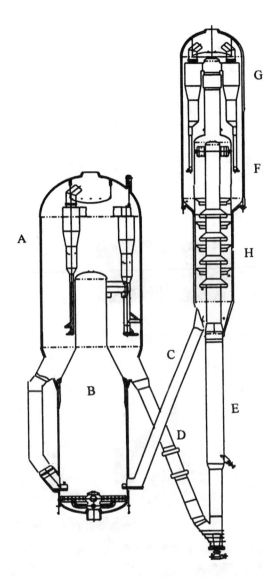

FIGURE A.8 UOP Fluid Catalytic Cracking Unit. A = regenerator, B combustor, C = spent catalyst standpipe, D = regenerated catalyst standpipe, E = riser cracker, F = riser discharge and catalyst separator, G = cyclone separators, H = spent catalyst stripper. Courtesy of UOP LLC, Des Plaines, IL. Identifiers added by HFR.

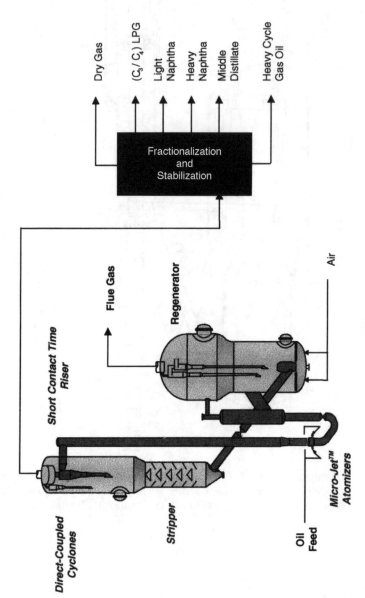

FIGURE A.9 ABB Lummus Global Fluid Catalytic Cracking Reactor and Regenerator. Reprinted by permission: ABB Lummus Global, Inc., Bloomfield, NJ.

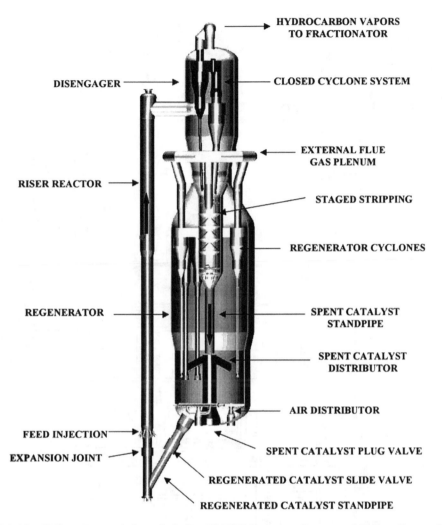

FIGURE A.10 Kellogg Brown & Root Orthoflow™ FCC Converter. Courtesy of Kellogg Brown & Root, Houston, TX.

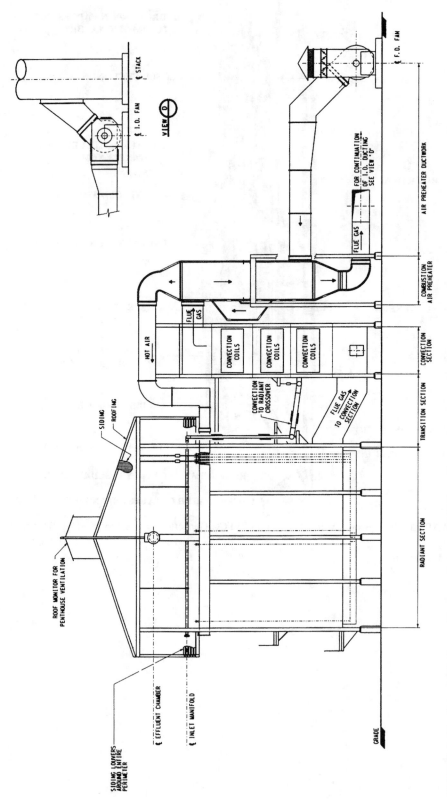

FIGURE A.11 Primary Reformer Furnace. Courtesy of Kellogg Brown & Root, Houston, TX.

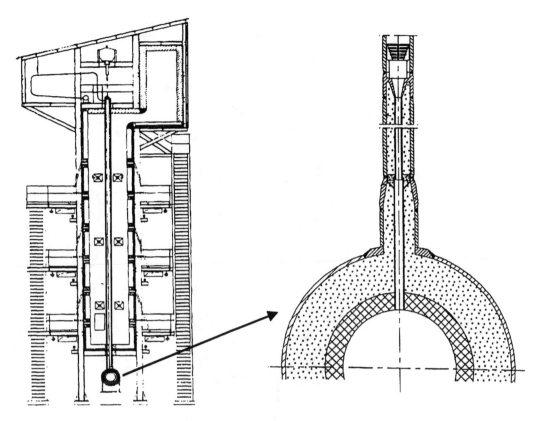

FIGURE A.12 Diagram of Topsoe Reformer Layout. Courtesy of Haldor Topsoe, Houston, TX.

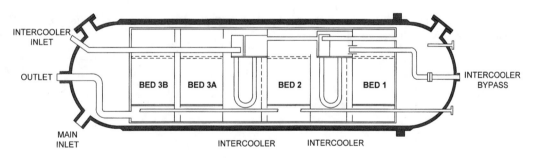

FIGURE A.13 Three-Bed Horizontal Intercooled Ammonia Converter. Courtesy of Kellogg Brown & Root, Houston, TX.

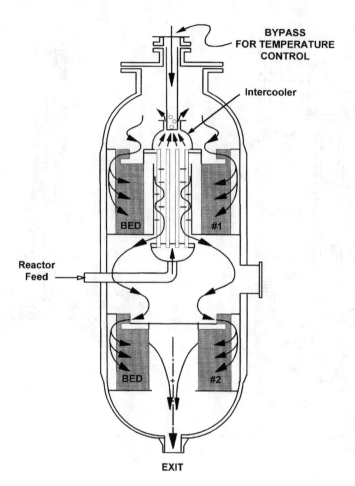

FIGURE A.14 Radial Two-Bed KAAP Reactor for Retrofit Applications. Courtesy of Kellogg Brown & Root, Houston, TX.

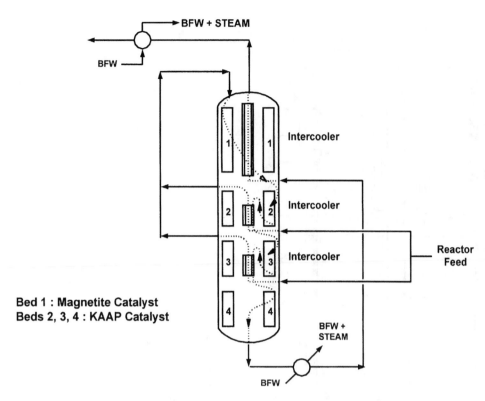

FIGURE A.15 Schematic of Three-Bed KAAP Reactor for Grassroots Applications. Courtesy of Kellogg Brown & Root, Houston, TX.

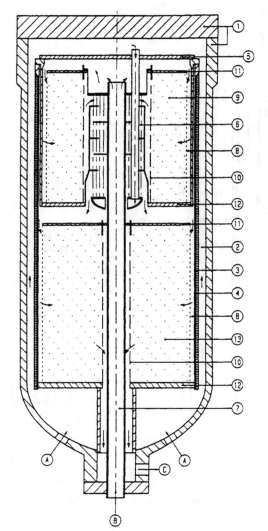

A. main gas inlet
B. cold by-pass inlet
C. gas outlet
1. pressure shell
2. outer annulus
3. outer basket shell
4. basket insulation
5. basket cover
6. interbed heat exchanger
7. cold by-pass pipe
8. screen panels
9. first catalyst bed
10. center screen
11. cover plate
12. catalyst support
13. second catalyst bed

FIGURE A.16   Haldor Topsoe Two-Bed Radial Ammonia Converter (S200). Fresh feed passes up the annulus between the basket shell and the pressure shell and down a pipe leading to the bottom tube sheet and thence up the tubes. Cold by-pass enters through a central pipe and mixes with the heated feed exiting the tubes. This mixture then enters the catalyst basket, thence to the shell side of the heat exchanger and on to the second bed. Reprinted by permission: Haldor Topsoe, Houston, TX, and Lyngby Denmark, Dybkjaer, Ib., and Jarvan, J. E., "Advances in Ammonia Converter Design and Catalyst Loading," paper at Nitrogen '97 Conference, Feb. 1997, Geneva, Switzerland.

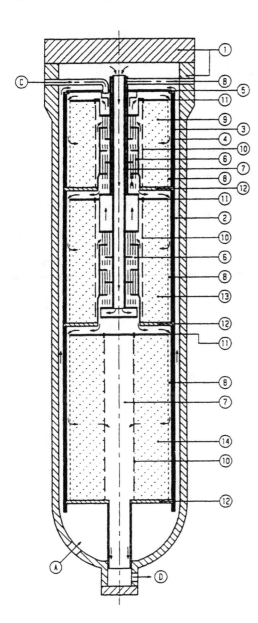

A. main gas inlet
B. inlet for gas to upper IHE tube side
C. cold by-pass inlet
D. gas outlet
1. pressure shell
2. outer annulus
3. outer basket shell
4. basket insulation
5. basket cover
6. interbed heat exchanger (IHE)
7. transfer pipe
8. screen panels
9. first catalyst bed
10. center screen
11. cover plate
12. catalyst support
13. second catalyst bed
14. third catalyst bed

FIGURE A.17   Haldor Topsoe Three-Bed Radial Ammonia Converter (S300). Flow pattern is similar to S200 except central pipe serves as conduit for feed to exchanger tube side and cold by-pass enters from top. Three-bed converters are often used to increase total conversion with the additional bed of catalyst. Reprinted by permission: Haldor Topsoe, Houston, TX, and Lyngby Denmark, Dybkjaer, Ib., and Jarvan, J. E., "Advances in Ammonia Converter Design and Catalyst Loading," paper at Nitrogen '97 Conference, Feb. 1997, Geneva, Switzerland.

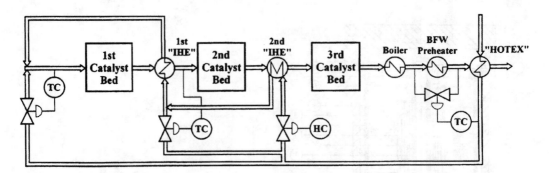

FIGURE A.18   Control Scheme for Haldor Topsoe S300 Ammonia Converter. Inlet temperature to bed 1: controlled by the amount of cold gas by-passing exchanger 1. Inlet temperature to bed 2: controlled by controlling the inlet temperature of the cooling gas to exchanger 1. Inlet temperature to bed 3 is controlled by controlling the temperature at the converter inlet via a by-pass around the boiler feedwater preheater. Reprinted by permission: Haldor Topsoe, Houston, TX, and Lyngby Denmark, Dybkjaer, Ib., and Jarvan, J. E., "Advances in Ammonia Converter Design and Catalyst Loading," paper at Nitrogen '97 Conference, Feb. 1997, Geneva, Switzerland.

Index of Products*

* Note: Numbers in parentheses are chapter numbers.

Index of Reactants*

---

* Note: Numbers in parentheses are chapter numbers.